1 MONTH OF
FREE
READING

at

www.ForgottenBooks.com

By purchasing this book you are eligible for one month membership to ForgottenBooks.com, giving you unlimited access to our entire collection of over 1,000,000 titles via our web site and mobile apps.

To claim your free month visit:

www.forgottenbooks.com/free568007

ISBN 978-0-483-13179-8
PIBN 10568007

SCIENCE PROGRESS

EDITORIAL

In 1894 appeared the first number of *Science Progress*, a review of current scientific investigation, which, conducted by Sir Henry Burdett, K.C.B., and under the editorship of Prof. J. Bretland Farmer, passed through seven volumes and earned a recognised place for itself amongst the science journals of the day. During the years which have elapsed since 1898, the date of the last volume, regret has from time to time been expressed that there was no British science journal devoted to those comprehensive summaries of recent work, to those general discussions of topics of interest to workers in all branches of science, which were the predominant features of the old publication and for which it was rightly valued.

Last year regret gave place to hope, and with the valuable support of the majority of the members of the Advisory Committee of the former journal and others, whose names appear elsewhere, efforts were made to found once again a quarterly science review. These efforts took definite form with Mr. John Murray undertaking the publication of the journal, and we now present to our readers the first part of *Science Progress in the Twentieth Century*, and hope that it will be found worthy to maintain in the new century the traditions its predecessor established in the past.

Specialisation and the multiplication of scientific and technical journals render it increasingly difficult, even for those actually engaged in scientific work, to keep abreast of the advance of knowledge and the trend of thought in more than that portion of their own subject to which their attention is especially directed. The difficulty is much greater for that larger public which, although not avowedly scientific, is interested in science and desirous of obtaining reliable information in not too technical language regarding the results of recent research, the problems which are awaiting or are in course of solution, and the inter-relations between pure science and practice.

It will be the main endeavour in the new journal, as in

the old, to present summaries, as far as possible of a non-technical character, of important recent work in any branch of science, to show the progress achieved, and if possible to indicate something of the line along which further advance is to be made towards the desired end. The chemist, to take an example, will describe for the botanist recent advances in chemistry, the botanist will do the same service for the chemist, often, it is hoped, to the advantage and assistance of both, especially when, as is an essential feature of *Science Progress*, the summaries are prepared by those actually engaged in the work and capable of marshalling the facts in their true perspective.

Past experience has demonstrated the great utility of such summaries to teachers and students. Some are prevented by lack of opportunities from obtaining access to original memoirs and papers, whilst others who are more favourably situated are often at a loss for a guide amongst, perhaps, the numerous and scattered contributions on a given subject. To all, a résumé in which the isolated facts are gathered together, marshalled in their proper sequence, and their general bearings discussed in a somewhat broader manner than is sometimes possible in a more technical journal, should prove of value.

It is especially to be hoped that *Science Progress* will prove useful to scientific workers and others in the more distant parts of the Empire, where, through want of access to much current literature, more and more reliance has to be placed on summarised information. On the other hand, there is in Great Britain often but scanty information regarding the scientific problems confronting our kinsmen beyond the seas, the progress they are making, and the conditions under which their work is carried on; and if this journal should be the means of knitting more closely together and mutually aiding fellow-workers throughout the Empire, it will more than justify its existence.

Just as it is to be hoped that *Science Progress in the Twentieth Century* will provide a common meeting-ground for scientists in different lands, so we trust that those engaged in different branches of science may here come together to discuss subjects of common interest. Nor will the more " practical " aspects of science be lost sight of, but, as the contents of this number sufficiently indicate, efforts will be made to show the relationships of so-called " pure " science to agriculture, mining, health, commerce, education.

N. H. ALCOCK.
W. G. FREEMAN.

A SCIENCE OF COMMERCE
AND SOME PROLEGOMENA

By W. J. ASHLEY, M.A.

Professor of Commerce in the University of Birmingham

"COMMERCIAL education" is in the air; and of late years the idea has made its way in England from the stage of the secondary school to the stage of the university. The enthusiasm of the advocates of "commercial education" in schools has not been without its good results. The insistence upon "commercial" geography, for instance, or "commercial" arithmetic, has contributed in no small measure to the more sensible teaching of arithmetic and geography, even when the adjective is not prefixed; but certainly the movement has had its drawbacks—chief among them the tendency to promote an over-early specialisation. Moreover, until recently it practically aimed at nothing beyond the acquisition of clerkly accomplishments; and it has doubtless stimulated the already excessive pressure into the black-coated occupations of the desk. It is not wonderful, therefore, that the defects of the propaganda positively prejudiced for some time the cause of commercial education of a higher type. But the half-articulate feeling among business people that even in university training, in spite of all the recent reforms, there were still grave deficiencies so far as the needs of their own class were concerned, was too deep and widespread to be frightened out of existence, even by the mistakes of the early advocates of commercial education. Business men had come to realise that for their own sons, who were ultimately to take their places, there was as yet no training devised even remotely comparable with that provided for the future "professional man." The theory that university studies were sufficient for the purpose if they furnished "a *general* mental training," because the powers thus developed could afterwards be applied in any particular direction that might be necessary—supported, as perhaps it was, by a few examples in each generation—had broken down in practice. For the sort of general culture thus secured only too often ended by making business life distasteful

and deterring young men from entering upon it. How far this
may be the fault of business, and how far of the university, we
need not now consider ; but naturally it has not been a pleasant
experience to the manufacturer who expected his son to return
from the university and set to work in earnest.   And to
experiences of this kind, joined to other causes of an historical .
nature, is due the fact that hitherto the great body of substantial English manufacturers and merchants has kept aloof
from the older universities.   A few of the wealthier among
them have sent their sons chiefly for social reasons ; and every
year a few clever boys have been picked out of their native
milieu by means of scholarships and ambitious schoolmasters,
and their ability lost to the commercial world.   These exceptions, however, have hardly affected the class as a whole.   And
yet a good many business men were beginning to realise in an
obscure sort of way that the old practice of putting a boy into
the office at sixteen or seventeen was also no longer satisfactory.
They felt that a boy's time could be more advantageously employed than in the work of a junior clerk, if there was no cogent
pecuniary reason for putting him at once to the desk.

It was the statesmanlike imagination of Mr. Chamberlain
which gave a name to this vaguely felt want when, in the
charter of the new University of Birmingham, he provided for
the future establishment of a " Faculty of Commerce," side by
side with the older faculties of Arts, Science, and Medicine.
Since then the new northern universities of Manchester and
Leeds have also established faculties of commerce ; and a
department so named will soon be a part of the ordinary
machinery of all the modern seats of learning.   The ancient
University of Cambridge, though it clings to its old nomenclature and calls its new degree course the " Economics Tripos,"
means the same thing by it.

So far so good.   But, granted a serious intention on the
part of university authorities to provide a training which shall
tend to fit, rather than unfit, men for business life, in what is
that training to consist?   This is a question which has so far, in
my opinion, obtained nothing like the consideration it requires.

Let us, then, begin by observing that Higher Commercial
Education, or education of a university type, cannot be related
to commercial education in the schools (of whatever grade) in
what is probably the usual sense of " higher " as related to

"lower": in the sense, for instance, in which "higher" mathematics is related to elementary. It cannot mean simply a continuation or carrying-further of the same studies. A boy whose parents can afford to delay his entrance into business life to the age of twenty or twenty-one, after three years at a university, is obviously expected to occupy a different sort of place from the boy who enters the office at seventeen. He is expected ultimately to become an *officer* in the industrial and commercial army. To put it quite frankly, the boys for whom commercial education of a university type primarily caters are the sons and kinsfolk of people who can give them better openings into business life than most clerks can possibly enjoy. The only students without any such backing that, in justice to the fellows themselves, a university dare attract, are boys of more than usual ability. For such boys, indeed, their professors, if they keep in touch with manufacturing and commercial circles, will usually have no great difficulty in finding suitable openings. And though men with a college training, if they prove to have no aptitude for affairs, may remain all their lives in very subordinate positions unless they have family wealth at their back ; and, on the other hand, boys educated simply to be efficient clerks frequently rise to positions of power : yet the training of the one class must differ in its whole tone and spirit from that suitable for the other. Training of the highest kind must aim before everything at guiding and strengthening the powers of *judgment.* And evidently this is possible in a different sense with young men of from seventeen or eighteen to twenty or twenty-one, from what it is with boys of from fourteen to seventeen.

Some subjects will at once occur to most people as suitable constituents of a university commercial course. Foremost among them will probably be modern foreign languages. Some academic authorities will demur. The value of modern languages, they say, is "instrumental" only. They are tools which will be found useful, just as arithmetic will be ; but a faculty of commerce ought to assume, so they assert, in the one case as in the other, that students will acquire these tools for themselves, and need not exact an acquaintance with them for its degrees. But the parallel with arithmetic is very incomplete. We can assume that boys will bring with them to the university sufficient arithmetical knowledge ; we can enforce the requirement by an entrance examination. But it is not yet practical politics to insist on "a working knowledge" of even one modern language at matriculation. In spite of the years boys spend in "doing" modern languages at school, not one boy in three can translate readily an ordinary piece of French prose when he comes up to the university. The number who know anything of German is far smaller; and yet German would be much more useful to most business men than any other language, for several reasons. Hence the wise policy would seem to be to give students an opportunity at college, as part of their commercial course, to learn the language or languages they are likely to find of service.

One or two general principles may be laid down with regard to the modern language teaching. First, the specifically "commercial" knowledge—by which is meant chiefly the technical terms and the forms of correspondence—should be superimposed on a sound knowledge of the rudiments of the language. And secondly, though the student, when he comes to take his degree, should have reached the point of being able both to converse and to correspond in the foreign tongue, this must not be regarded as the sole purpose of his study. A main purpose, and one that should be kept steadily in mind during the course, should be to enable a business man easily to keep abreast of the foreign literature of his occupation— economic, financial, or technical. It should be as usual for a business man who has dealings with Germany to read the *Industrie-zeitung* or the *Wirthschaftszeitung* as the *Iron and Coal Trades Review* or the *Statist*, and no more difficult.

The place of languages in a commercial curriculum cannot be uniformly determined for every student in advance. There will be many who are not likely to have much to do with foreign markets; and for them something may properly be substituted for part of the language requirement. Most of them will probably be entering into *manufacturing* life; and for them what is desirable in their curriculum is a strong infusion of the "science" or "sciences" (in the curious English use of that word) which are directly or indirectly applicable to the sort of manufacture they are likely to enter. This proposal meets with even less sympathy in some academic quarters than the foregoing. This is partly due to the contempt the "scientific" teachers themselves entertain for a "mere smattering"; and among economists it meets with the same objection as the introduction of linguistic teaching—that the knowledge is only "instrumental," and that manufacturing technique must be altogether subordinate, from the point of view of commercial success, to business judgment. This is very true; and yet, if the knowledge *is* likely to be useful, and if it can be acquired during the college course without hindering other and more important acquisitions, surely this is the time for it. Otherwise it will commonly not be obtained at all. But if such studies are to be permitted as part of the work for a commercial degree, it will be necessary (1) to allow them so considerable a place,—say as much as a third of his time, if a student so chooses—that a really adequate amount of knowledge can be secured within the three years; and (2) to allow the largest possible freedom of choice in accordance with the student's previous preparation and future needs. And throughout it must be borne in mind that the purpose is not to train the technical expert—the professional engineer or chemist. Students who look forward to careers of that kind will naturally enrol themselves in the scientific and technological departments of the university from the outset. The "science options" are intended for men who are expected ultimately to take part in the *commercial* conduct of businesses, but who wish for some general understanding of the processes that go on in the workshops, some notion of the tendencies of technical progress, some ability to use "experts" without succumbing to them.

We may pass hurriedly over Commercial Law, merely remarking that the brief course, which is all there will be time for, should come towards the end of the period of study, and that it will not aim at anything so foolish as to make a man "his own lawyer."

So far we have remained on the circumference of the subject; we have not yet penetrated to the centre, which is the training of the judgment to deal with the actual problems of commercial life. We approach more closely to the essential purpose we have in mind when we come to the subject of Accounting. "Accounting" was a new term when it was brought into England four or five years ago, though by this time it seems to be finding general acceptance in 'the more progressive academic circles. It may conveniently be used to designate that command of accounts which every business man should possess, as distinguished from the professional information and expertness included in "accountancy." And it is quite clear that, in the sense here meant, accounting deserves a large place in any scheme of higher commercial education. It begins with book-keeping in the ordinary sense of the term; but it does not stop there. As soon as the student has been well drilled in the necessary rudiments, accounting proceeds to become a critical study of financial stability and prosperity as revealed (or concealed) by balance sheets; and starting from a mere arrangement of figures by single and double entry, it insensibly makes its way to those questions of expediency and efficiency which are suggested by such words as "depreciation," "reserves," and the like. Properly taught, it is a subject of high educational value; for it makes its appeal in the last resort not to arithmetical dexterity, but to a sound judgment of a business situation. Moreover, accounting, adequately taught, will handle in a thorough manner the whole difficult question of "costing." And though, to use the term already employed more than once, even cost accounts are only "instrumental"—mere tools in the hands of their users—yet they are tools which possess the quality of themselves stimulating reflection. To determine the wise policy to adopt with regard to selling price at a particular juncture is a different thing, it is true, from the mere knowledge of what the thing cost to make. But there is nothing so likely as a knowledge of what a thing costs under varying circumstances, and of the proportion to be assigned to fixed charges, to promote a wise decision as to the price to be asked for it.

The really constitutive and most characteristic part of a commercial curriculum at the university must, however, after all, be found in Economics. Yet, without desiring to provoke controversy, I am bound to express the opinion that economics, as that subject has generally been taught in this country, will hardly satisfy the needs of the new academic situation. Whatever may be its value as an instrument of sociological investigation, political economy, as represented by the usual textbooks, is defective both in its character and in its scope for the purposes of business education. In its character; because of its tendency—with "marginal utility" and "consumers' rent," and the like—to become a branch of psychology; in its scope, because it gives a quite inadequate amount of attention to the concrete facts of industrial and commercial life. In commercial teaching the abstract political economy hitherto current in England should certainly find a place—but reduced to its narrowest limits and in its simplest terms; not as a great matter in itself, but rather as one of the means of mental discipline, and as furnishing suggestive points of view for the further examination of economic conditions. That place having been given to it, the main lines of work appropriate to a commercial faculty will be found in two directions. The first and most obvious is the descriptive survey of the actual forms of economic activity. Our faculties of commerce must aim at giving an exposition of the really large facts of all the great industries of England and its rivals, as well as of typical smaller trades, and of the marked tendencies in their historical development. I am well aware of the criticism such an assertion will provoke. It will be said that, even when courses of instruction of this kind have been created, they will be "merely narrative," merely "informational." That there is a real danger here, I readily recognise. But, to begin with, information is among the things our future business men most require. They ought to know far more than they do of what is going on in the world; at present many of them are so limited in their outlook that it would be an undeserved compliment to call them even "insular." And in the next place, it should be the distinguishing note of a "descriptive economics" worthy of a university that it is so bent on selecting the larger features of the phenomena and relating them to one another as to suggest all the time the idea of causation. After all, the ultimate purpose of our economics is to know the economic world. The prevailing method hitherto in England has been to pursue certain abstract lines of argument as to cause and effect, and then occasionally to look out into the noise and turmoil of real life and find there bits of concrete illustration. The method I urge—not as the only desirable one, but as the one peculiarly appropriate to commercial training—is the exact opposite: it is that of simple observation of actual life, with recourse, whenever it seems useful, to abstract explanation.

But there is another direction in which, in my judgment, the current economics require to be supplemented for the purposes we have in view; and with this we come to the very heart of the matter. What is absolutely requisite and quite feasible, though certainly difficult, is the creation of a "science of commerce," in the sense of a systematic consideration of the problems of business policy. What is wanted is "*private* economics" for the business man, as distinguished from "political" or "social" economy. Is it our aim, then, to "teach men to make money"? Yes, in the sense in which it is the aim of a law school or medical school to teach men to make money as competent lawyers or physicians. The aim of a faculty of commerce—let us be quite clear about it—is in the first place frankly utilitarian : it is to turn out competent men of business. In each case no training is of much use unless a man has some natural aptitude ; and there are such great differences between "business" and "the professions" that the analogy is certainly not complete. But it is valid for some distance.

The name to be given to this systematic consideration of business problems is unimportant. And the kind of problems handled will vary from place to place. The field of business is too wide to allow of any simple schematic treatment applicable to all types and scales of enterprise. A provincial university faculty will naturally have regard, first of all, to the activities of its own district. In most districts it will probably be found best to make the position of the *manufacturer* the central theme. Following a manufacturing business with the eye of imagination (guided by recorded experience) through its career, a series of problems of policy will be seen emerging —financial, commercial, administrative. It will not be the function of the academic teacher to lay down *a priori* rules as to how the questions are to be decided as they arise. His function is rather to collect examples, as a naturalist collects specimens, of the way in which they have actually been dealt with, successfully or unsuccessfully, in real instances. These he will classify and arrange ; he will bring to bear upon them all the knowledge of industrial history, all the power of abstract analysis, his previous studies may have given him : until at last there rises into view a series of wide generalisations deserving the name of "principles." The materials lie all around us in the reported proceedings of companies ; and American economists are already showing us how to utilise them.

It should not need to be remarked that the final result will not be a set of recipes for success ; and it will not be so presented. The chief function of such a course is simply to make men think systematically; and it is only so far as it succeeds in being systematic that it will deserve the term "science," if we care to give it the name. Business men, in the midst of the pressure of affairs, are too apt—if I judge of them aright—to regard a grave question of policy which comes and confronts them as if it had presented itself for the first time; whereas, in three cases out of four, conditions substantially similar have arisen again and again. So that the function of the economist is not to dictate to the business world, but to set forth in all its essentials, and thereby to explain, the experience of the business world to itself. And what is within the power of a faculty of commerce is to turn out men who will realise that a commercial career is going to be intellectually interesting, just because it will furnish them with opportunities for the exercise of judgment.

Is all this woefully "banausic," as Aristotle might say; a mere "bread-and-butter" business? I think not. We in England have too long aimed at culture, and hoped that utility would appear as a by-product. The result has been that the great body of the English middle-class has left the culture severely alone. Let us now, for a change, not be ashamed to aim at utility, and let us trust that culture will appear as a by-product. It will, if the avowedly utilitarian subjects are taught sensibly.

Moreover, I am convinced that the introduction into the universities of these new and very practical disciplines will revivify economic studies, and contribute both to knowledge and to social progress. To knowledge—passionless, unselfish knowledge—because English political economy has in some measure lost its attractiveness for men of ability, because it has drifted out of touch with actuality. A mental picture of the Steel Industry may be a lower thing than a Doctrine of Value; but unless the Doctrine of Value is felt to have a bearing on things like the Steel Industry, men will find it difficult to continue to take an interest in it. And it will conduce no less to social progress. It may seem an odd thing to say at the very time a Labour Party has made its appearance in the House of Commons, but I believe it to be true, nevertheless, that the path to social reform will lie in future as much through the administrative expediencies of business as through humanitarian sentiment.

CHLOROFORM A POISON

By B. J. COLLINGWOOD, M.D. (CANTAB.)

Demonstrator of Physiology, St. Mary's Hospital Medical School

CHLOROFORM has been rightly regarded as the most dangerous of anæsthetics, for it undoubtedly has been accompanied by a greater number of fatalities than any other narcotic. Nevertheless, the light which has recently been thrown on the causes of its danger has done much to render it possible to reduce its risks to a minimum. The growth of knowledge has been at once followed by increased safety in administration. It is of the very nature of anæsthetics that they should be poisons, for their name implies the power to abrogate one of the functions of the central nervous system—namely, the function of sensation. They are thus in their very essence toxic to the nervous system; it is only in a secondary sense that they are poisons of other systems, such as the respiratory or vascular. It is because of its extremely potent influence on these latter two systems that chloroform is the most poisonous of these drugs.

To administer a poison in unknown doses is to court disaster, with every chance of attaining it; and to judge of the amount administered solely by its effects must ultimately lead to an unjustifiable study of the phenomenon of death. The skilled anæsthetist who is acquainted with the earliest signs of an overdose of chloroform may legitimately use a rough method of administration; but the drug is frequently given in a similar manner by those who make no claim to be regarded as specialists in anæsthetics. It is therefore no matter for wonder that the number of deaths from chloroform is at a high level. To drop chloroform on a Skinner's mask may be the most convenient and the safest method for the specialist, but one can only protest against such a procedure when employed by unskilled hands.

Much has recently been said in favour of some mechanical device for supplying graduated percentages of chloroform and air; and it appears to the writer that an apparatus of such a nature, if in truth reliable as to its accuracy, would do more than anything else to diminish the dangers of overdosage. Yet it must always be remembered that no apparatus, however accurate, can ever do away with the necessity of constantly observing the condition of the patient. All that an apparatus can accomplish is to remove *one* of the dangers of chloroform anæsthesia. If the alternatives were either to watch the patient or to use an apparatus, there can be no question but that the former would be by far the safer course. Fortunately no such alternatives exist, and it is possible to combine the use of an apparatus with the most careful observation of the patient. Nevertheless, from lack of such a warning as this it might well happen that the use of an apparatus might become a source of added danger rather than of increased safety.

One may adopt the view that the same percentage of chloroform in the lymph produces the same effect on all patients, thus reducing idiosyncrasy to conditions which alter the rate of absorption. Yet such an opinion, although receiving considerable support from experiments on isolated nerves, is of little practical value, for an apparatus can only tell us the percentage of chloroform in the air inspired—no information is given as to the percentage in the lymph. The varying relationship between these two percentages thus becomes a matter of great importance to the anæsthetist, and it is for this reason that a study of the absorption of chloroform is of considerable value as throwing light on all questions concerned with its administration. It is one of the objects of this paper to attempt to investigate this subject in detail.

During the actual administration of chloroform the percentage in the lymph can only be gauged by the condition of the patient. How, then, it may be asked, does an apparatus afford any assistance to the anæsthetist? The answer to this question is contained in the statement that if the percentage of chloroform inhaled be known and kept at a value that experience has shown to be legitimate, the condition of the patient can scarcely become serious without ample signs of warning, whereas if the percentage of chloroform inhaled be unknown, the patient's condition may suddenly become one of grave danger from overdosage of the drug. In other words, the use of a graduated apparatus renders sudden changes of a dangerous character in the patient almost impossible. It is not the slow, but the sudden changes that the anæsthetist has learnt to dread in chloroform anæsthesia.

A properly constructed apparatus should remove this dread from his mind.

The writer would suggest that three factors are concerned in the safe administration of chloroform, namely :

    (1) A knowledge of the physics of the absorption of chloroform.

    (2) A knowledge of the percentage of chloroform inhaled.

    (3) A careful observation of the patient's condition.

He proposes to deal with these three factors in the above order:

I. The Absorption of Chloroform

Unless we are to assign a selective action to the pulmonary epithelium, an action similar in character to that which Haldane and other workers have shown probably to exist in regard to oxygen, we must conclude that chloroform passes through the epithelium of the alveoli, and is dissolved in the blood according to the physical laws which govern the diffusion of vapours and their solution in liquids. The recent researches of Moore and Roaf (*Proceedings of the Royal Society*, 1904, vol. lxiii. p. 382) have, however, demonstrated that the blood proteids form loose compounds with chloroform, and hence that chloroform exhibits a greater solubility in blood than in water. We thus establish a similarity between the relation of oxygen and chloroform to blood, oxygen being associated with the hæmoglobin and chloroform with the proteids. Further, Byles, Harcourt, and Horsley have found that blood with corpuscles has a greater retaining power for chloroform than simple plasma. Such observations indicate that chloroform exists in the blood not only in solution in the plasma, but also in a condition of combination with the proteids and some constituent or constituents of the corpuscles. Hence the solution of chloroform in blood does not solely depend on the physical laws governing the solution of vapours in liquids, but, as in the case of oxygen, other factors come into play. But of one thing we can be certain—namely, that slight increases in alveolar concentration of chloroform produce appreciable changes in the pressure of chloroform in the blood, as shown by the condition of the patient. In this respect chloroform differs widely from oxygen, for large differences in oxygen alveolar concentration can occur without producing any alteration in the individual (Haldane, *Journal of Physiology*, vol. xxxii. p. 225). The fact, then, that chloroform exists in the blood

partly in loose chemical combination does not radically affect
the problem of its absorption, for we can make the definite
statement that any increase in alveolar concentration produces
an appreciable increase in the amount in the blood.

During the absorption of chloroform by the organism the
vapour passes from the alveoli to the blood, and thence to the
lymph. It follows that the vapour pressure of the drug must be
higher in the alveolar air than in the blood, and higher in the
blood than in the lymph. We note, in addition, that if the chloroform vapour pressure in the inspired air be kept constant, this
pressure must be higher in the expired air, which again must
be higher than in the alveolar air. There is thus a descending
scale of vapour pressure in the inspired air, the expired air, the
alveolar air, the blood, and the lymph. So long, then, as chloroform is being absorbed it is with such a condition as this that
we are dealing. This scale can be rendered diagrammatically
as follows :

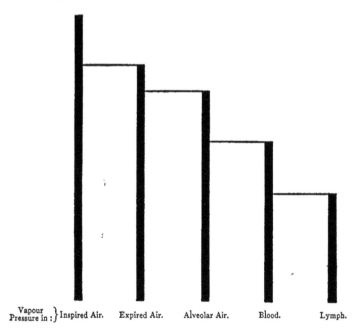

Vapour } Inspired Air.   Expired Air.   Alveolar Air.   Blood.   Lymph.
Pressure in :

The steepness of this scale will be greatest at the beginning
of anæsthesia, and will gradually diminish as anæsthesia
advances. It is only the first step in the scale that we are

able to estimate directly—that is, the difference between the chloroform vapour pressure in the inspired and in the expired air. But still we can deduct that when this difference is great the whole scale will be steep, and that the vapour pressure in the lymph will be considerably less than that in the inspired air. Early in anæsthesia we should expect, accordingly, a marked difference in the chloroform content between the inspired and the expired air; and this is precisely what one finds by experiment to be the case (A. D. Waller and B. J. Collingwood, " Estimation of Inspired and Expired Chloroform," *Proceedings of Physiological Society*, February 25, 1905).

There is another point which is worthy of notice—namely, the ratio between the first and second step in this scale. Now, so far as the writer is aware, no attempt has been made to estimate the chloroform vapour pressure in the alveolar air during anæsthesia; thus one can only arrive at its value at the present time by deductive methods. Oxygen fortunately affords a most instructive parallel in which the alveolar tension has been estimated by Haldane and Priestley. The figures given by these observers show that the difference between alveolar and expired air oxygen pressure is much less than that between inspired and expired air oxygen pressure. Such a result is only to be expected when one considers that the " dead space" in breathing is estimated by these observers to be only 30 per cent. of the volume of tidal air. To take an example :

Let the tidal air　　　　　 = 600 c.c.
Then the "dead space" will = 200　 „
Let the inspired air contain 2 per cent. of chloroform.
Let the expired air contain 1 per cent. of chloroform.

Then, as the dead space will contain 4 c.c. of chloroform, the alveolar air must contain 2 c.c. of chloroform (6 − 4), and the percentage of chloroform in the alveolar air must be 0·5 per cent. We thus arrive at the following figures :

Inspired air.	Expired air.	Alveolar air.
2 per cent.	1 per cent.	0·5 per cent.

Substituting $x + y$ for percentage in inspired air, and x for percentage in expired, we find the following result :

Inspired air.	Expired air.	Alveolar air.
$x + y$ per cent.	x per cent.	$x - \dfrac{y}{2}$ per cent.

Thus the difference in chloroform percentage between alveolar and expired air is only half the difference between the expired and the inspired air. This ratio is represented in the diagram given above.

It has already been suggested that the steepness of the first step in the scale gives an indication of the vapour pressure of chloroform in the lymph. This fact is well illustrated by the two following experiments, the first of which is taken from a table in the paper already quoted by Dr. Waller and the present writer on the estimation of chloroform in inspired and expired air, the second of which was performed by the author at a later date:

	Percentage of $CHCl_3$ inspired.	Percentage of $CHCl_3$ expired.
Dog	3·3 per cent.	1·5 per cent.
Cat	1·3 „	1·2 „

Firstly, we can deduct from these figures that the descending scale of vapour pressures was much steeper in the dog than in the cat. Secondly, we can deduct that the chloroform percentage in the alveolar air of the dog was approximately 0·6 per cent., and in the cat approximately 1·15 per cent. We therefore conclude that the chloroform vapour pressure in the lymph of the dog was considerably lower than that in the lymph of the cat. Our conclusion is in accordance with the facts of the case; for the dog was in a condition of light anæsthesia, and the cat in a condition of deep anæsthesia.

We will now consider what factors influence the first step in the scale—namely, the difference between the percentage of chloroform in the inspired and the expired air. There are two main factors with which we must deal:

(1) The rate of absorption of chloroform by the blood.

(2) The frequency and the depth of the respirations.

To discuss the first of these two factors. The rate of absorption of chloroform is again under two influences: (*a*) the difference of vapour pressure of the anæsthetic in the blood and in the alveolar air; and (*b*) the rate of blood stream through the lungs.

Early in anæsthesia, when the vapour pressure in the blood is low, the rate of absorption will be rapid, and, as a consequence, if the percentage in the inspired air be kept at a constant value, the difference of chloroform concentration between the inspired and expired air will be greater than in the later stages of anæsthesia. In other words, if the percentage of chloroform in the inspired air be kept at the same level, the concentration of the anæsthetic in the expired air will be constantly rising, and, therefore, also rising in the alveolar air, in the blood, and in the lymph. The conclusion is obvious, that *a safe percentage to administer early in anæsthesia is not necessarily safe in the later stages; and further, that the percentage of anæsthetic in the inspired air can be slowly diminished without diminishing the concentration in the alveolar air, for as the alveolar air loses less and less anæsthetic to the blood as anæsthesia advances, the alveolar air requires less and less vapour to be added to it to maintain its concentration.*

One cannot, therefore, answer the question, "What do you consider a safe percentage of chloroform to administer?" without first asking, "To what period of anæsthesia do you refer?" Even on obtaining an answer to this, one is unable to give a dogmatic reply, until one is aware of the amount of lung ventilation that the patient may possess. This matter will be considered anon. At the present stage of our investigation we can declare that *what is a safe percentage of chloroform to administer early in anæsthesia may be a fatal percentage in the later stages of narcosis.* The Special Chloroform Committee of the British Medical Association in 1904 reported that 2 per cent. of chloroform was sufficient to induce anæsthesia in the human subject, and 1 per cent. of chloroform to maintain it. This may be taken as a fairly accurate statement in reference to the majority of cases that require the administration of an anæsthetic. But as it will be shortly pointed out that the question of lung ventilation plays an important part in the rate of absorption, no such statement can be of anything approaching universal applicability. It must always be remembered that a vapour pressure of chloroform in the *lymph* corresponding with a 2 per cent. mixture with air is *certainly a fatal concentration.* Consequently the administration of 2 per cent. of chloroform must, if continued, ultimately prove fatal, the length of time intervening being proportional to the rate of absorption, which rate will vary with variations in lung ventilation.

In this connection Waller has conclusively shown that in the case of cats a 2 per cent. vapour of chloroform is safe to administer during the induction of anæsthesia. Yet from experiments the writer has performed on the same animals he can state that such a percentage may be fatal after the lapse of about two hours.

The writer has conducted a series of experiments again on the same animals to determine the percentage of chloroform required to abolish the corneal reflex after varying periods of anæsthesia. In these experiments an apparatus designed by the writer for the administration of known percentages of chloroform was used (*Transactions of the Royal Medical and Chirurgical Society*, 1905). The method adopted consisted in diminishing the concentration of chloroform administered so long as the corneal reflex did not reappear. The following are the approximate values obtained:

Percentages of Chloroform in the inspired air required to abolish the corneal reflex in cats.

After ½ hour	.	.	.	.	.	.	.	1·0 per cent.
,, 1 ,,	.	.	.	.	.	.	.	0·8 ,,
,, 2 hours	.	.	.	.	.	.	.	0·7 ,,
,, 3 ,,	.	.	.	.	.	.	.	0·6 ,,
,, 4 ,,	.	.	.	.	.	.	.	0·5 ,,
,, 5 ,,	.	.	.	.	.	.	.	0·4 ,,

These figures show how marked is the diminution of percentage required to abolish this reflex as anæsthesia advances.[1]

We must now turn to the second factor which influences the rate of absorption of chloroform by the blood—namely, *the rate of blood stream through the lungs*. It is clear that the slower the blood stream the less vapour of chloroform will be absorbed, for there will be a deficiency of supply of fresh blood with a low chloroform vapour pressure. But although less will be absorbed, yet the blood will contain a greater amount of chloroform, since the blood will have a longer exposure to the vapour in the alveolar air. This point appears to the writer to be of extreme importance, for it shows that the beginning of heart failure will cause the blood which

[1] We must not, however, deduct from these figures that a concentration in the lymph corresponding to a 0·4 per cent. vapour is sufficient to abolish the corneal reflex after five hours purely by virtue of its anæsthetic properties, for the recent researches of Sutherland Simpson and Herring have shown that the diminution of temperature following prolonged administration of anæsthetics is in itself narcotic (*Journal of Physiology*, vol. xxxii. p. 305).

reaches the left auricle to contain an increased concentration of anæsthetic, and this will occur without any rise of percentage of anæsthetic in alveolar air, and, accordingly, cannot be guarded against by maintaining the chloroform in the inspired air at a constant level. It thus seems certain that weakening of the heart is the first link in a vicious circle, for the weaker the heart the stronger the concentration of chloroform which it will receive. Many observers have shown that chloroform has a directly toxic influence on the heart; amongst them may be mentioned Embley, Tunnicliffe, Ronenheim, Schafer, and Scharlieb. Accordingly there can be little doubt that the matter discussed above is one well worthy of attention.

We will now consider the second factor which influences the differences between the percentage of chloroform in the inspired and in the expired air—namely, *the frequency and the depth of respirations.* It is sufficiently obvious that increased frequency of respirations will raise the percentage of chloroform in the expired air, and so in the alveolar air, more rapidly than diminished frequency. As to the depth of respiration, the deeper the breathing the more rapidly will the concentration in the expired air, and so in the alveolar air, rise. In both the above cases it is premised that the percentage of vapour in the inspired air is kept constant. From such considerations it has been argued that it is actually an advantage to give a higher concentration of vapour when the breathing is shallow; in addition, methods of administration which have a tendency to deliver a higher percentage when the breathing becomes more shallow have been defended as possessing an error in the right direction. The following table throws some light on this question. The data on which it is constructed are given below:

(1) Residual + supplemental air = 2,000 c.c.

(2) Percentage of chloroform in alveolar air = 1·3 per cent.

(3) No absorption of anæsthetic taking place. This is merely to simplify the issue, for absorption would not alter the main result, but would only diminish the rate of increase of alveolar concentration.

(4) To simplify the calculation complete mixture is supposed to take place between the inspired and the alveolar air. Corrections for "dead space" are accordingly not made.

	Respirations of 250 c.c. of a 2 per cent. vapour.	Respirations of 100 c.c. of a 4 per cent. vapour.
	Alveolar concentration.	Alveolar concentration.
After 1st respiration .	1·38 per cent.	1·43 per cent.
„ 2nd „ .	1·45 „	1·55 „
„ 3rd „ .	1·51 „	1·66 „
„ 4th „ .	1·56 „	1·77 „
„ 5th „ .	1·61 „	1·88 „
„ 6th „ .	1·65 „	1·98 „
„ 7th „ .	1·69 „	2·08 „
„ 8th „ .	1·72 „	2·17 „

The calculation was made as follows : The total amount of vapour in the residual and supplemental air was calculated, and to this amount was added the amount of vapour in the inspired air. Having then estimated the total ₜvolume of inspired, residual, and supplemental air, the percentage of vapour in them was calculated from the above data. The following equation expresses the method :

$$\left(\frac{x+y}{a+b}\right) \times 100 = \text{percentage of chloroform in alveolar air after one respiration.}$$

Where x = amount of anæsthetic in the inspired air in c.c.
Where y = amount of anæsthetic in supplemental + residual air in c.c.
Where a = inspired air in c.c.
Where b = supplemental + residual air in c.c.

It will be noted that in the above table in the case of the 100 c.c. respirations, although actually less vapour is being inspired than in the 250 c.c. respirations, nevertheless the concentration rises more rapidly in the first than in the second case. In the first case only 4 c.c. of chloroform are being inhaled at each respiration, whilst in the second case 5 c.c. are being taken in. The reason for the more rapid rise of alveolar concentration in the first case is to be sought in the fact that although less is being taken, still less to a greater proportion is given out at each respiration. Taking the 250 c.c. respirations, and the first of the series given, 5 c.c. is taken in, and 3·45 c.c. given out, the amount retained being accordingly 1·55 c.c. Taking the 100 c.c. respirations, and the first of the series given, 4 c.c. is taken in, and 1·43 c.c. given out, the amount retained being 2·57 c.c. If it were desired to exactly compensate for the diminished depth of the respirations, the 100 c.c. respirations would have to contain not 4 per cent. but only 3 per cent. of chloroform approximately.

When it is remembered that as a general rule the more shallow the respirations the more rapid they become, the rise of concentration in alveolar air in the 100 c.c. respirations would be relatively still more rapid than the table indicates. For instance, if eight of the shallow respirations corresponded in point of time to six of the deep, then a percentage of 1·65 in alveolar air in the case of the deep respirations would occur after the same period of time as a percentage of 2·17 in alveolar air in the case of the shallow respirations.

Considerations such as these abundantly prove the need of extreme caution in administering a higher concentration of chloroform when the breathing becomes more shallow. More especially is caution necessary since it so frequently happens that the respirations become more shallow because the patient *has already received an overdose of chloroform.* It seems to the writer that the only legitimate alteration to make on the respirations becoming shallower is *a reduction and not an increase of concentration administered.*

So far we have regarded the supplemental and residual air as constant in amount. Any variation in their bulk, however, will alter the rate of increase of alveolar concentration. The following table illustrates this fact. The data on which it is constructed are given below:

(1) Percentage of anæsthetic in alveolar air . 1·3
(2) Percentage of anæsthetic in inspired air . 2·0
(3) Depth of respirations . . . . . 250 c.c.

	Supplemental + Residual Air. 1,000 c.c.	Supplemental + Residual Air. 2,000 c.c.	Supplemental + Residual Air. 3,000 c.c.
	Alveolar per cent.	Alveolar per cent.	Alveolar per cent.
After 1st respiration . .	1·44	1·38	1·35
„ 2nd „ . .	1·55	1·45	1·40
„ 3rd „ . .	1·64	1·51	1·45

It is thus apparent that the smaller the proportion between the tidal air and the sum of the supplemental and the residual air, the slower is the rise in the alveolar concentration. Pulmonary emphysema supplies an excellent example of an increase in supplemental and residual air. One would, therefore, expect that an emphysematous patient would be difficult to anæsthetise, and this seems to be the general experience. The sufferer from emphysema should experience the same difficulty in getting chloroform as he does in getting oxygen.

Something has already been said as to the result of alteration of rate of respirations on the absorption of chloroform. The following experiment by the writer may throw some light on this matter. A cat, two kittens, a guinea-pig, a rabbit, a rat and a mouse were placed in a closed glass box, into which a 2 per cent. vapour of chloroform was pumped. The animals were carefully watched, and the time of death in each case noted. Below is given a table showing the results obtained :

Animal.	Death after	Weight.
Mouse	30 minutes	17 grams.
Rat	40 ,,	200 ,,
Kitten, No. 1	120 ,,	200 ,,
,, No. 2 . . .	123 ,,	200 ,,
Rabbit	125 ,,	1500 ,,
Guinea Pig	125 ,,	500 ,,
Cat	240 ,,	2330 ,,

It will be observed that although the animals were inhaling exactly the same concentration of chloroform, the periods which elapsed before their deaths varied very widely. At first sight, then, it would seem that a genuine idiosyncrasy—that is, an idiosyncrasy depending on a special susceptibility to the chloroform which reaches the tissues by the lymph—exists in some members of the animal kingdom. But a more careful scrutiny of this experiment does not support such a conclusion. Although the animals all inspired the same concentration of chloroform, the rate of their respirations showed very marked differences. It has already been pointed out that the rate of chloroform absorption varies directly as the rate of respirations ; one would, accordingly, expect that the fastest-breathing animal would die first : this is exactly what occurred. The animal which died first was the mouse, and this animal was breathing much more rapidly than any of the others. Unfortunately the rate of respirations of the various animals was not recorded numerically, since it was the experiment itself that suggested that the relative rapidity of respiration was an important consideration.

The following rates of normal respiration are taken from

a table given by Paul Bert (*Leçons sur la Physiol. Comp. de la Réspiration*, p. 393, Paris, 1870) :

Rat.	.	.	.	210 respirations per minute.
Rabbit .	.	.	.	55 ,, ,, ,,
Cat.	.	.	.	24 ,, ,, ,,

These numbers quite bear out the contention that the faster-breathing animals are the first to die.

The rapidity of respiration is dependent on the rate of heat loss, and consequently one would expect that small animals which have a large surface for heat loss in comparison with their bulk would breathe the faster, unless the larger proportion of surface be compensated by the possession of thick fur, such, for instance, as clothes the guinea-pig. Thus the writer would suggest that animal idiosyncrasy is largely dependent on the rate of heat loss, since that governs the rate of respirations. The very steps that the rapid heat loser must take to procure an extra amount of oxygen to maintain the temperature will also procure for him an extra amount of chloroform in his blood. The young are more rapid heat losers than the fully developed : the kitten, therefore, is more quickly killed than the cat. The writer, applying these considerations, found that a concentration of 1 per cent. of chloroform was quite sufficient to induce anæsthesia in a young child, whilst 0·7 per cent. was sufficient to maintain narcosis. These figures are only about half the magnitude of those commonly accepted in regard to adults.

It might be well here to give a list of the factors which influence the rate of absorption of chloroform, and accordingly its concentration in the lymph.

(1) Rate of respirations.
(2) Depth of respirations.
(3) Relation of depth of respirations to bulk of supplemental and residual air.
(4) Rate of blood stream through the lungs.

When we consider this list we see how careful we should be not to declare any case to be one of idiosyncrasy until we have assured ourselves that it is not one of the above factors which is in reality the cause. The writer believes that *the vast majority of cases of so-called idiosyncrasy are as a matter of fact to be explained by variations in one of the factors which influence the rate of absorption.*

II. The Percentage of Chloroform inhaled

It has already been pointed out that a knowledge of the concentration of the chloroform administered affords a great safeguard to the anæsthetist. Such a knowledge can only be obtained by the use of a thoroughly reliable apparatus. The essentials, in the writer's opinion, of such an apparatus are given below.

(1) The delivery of the anæsthetic quite independently of the patient's respirations, so that no variation of concentration will occur with variations in depth or in frequency of respirations.

(2) The nominal concentrations should *under no possible circumstances* vary more than *very* slightly from the actual.

(3) The delivery of a large enough stream of the mixture to prevent the deepest respirations of the patient becoming mixed with room air. (If such an event occurs, deep respirations will contain a lower concentration than shallow respirations.)

(4) An easy method of altering the concentrations.

(5) Simplicity of mechanism and ease of portability, so far, but *only* so far, as these may be consistent with accuracy.

It will be seen that the writer does not favour any method which depends on the patient sucking air over chloroform. He bases his objection on the fact that such a method tends to deliver a stronger concentration of chloroform when from any reason the breathing becomes weaker. He is, however, well aware of the extreme ingenuity that has been brought into play to overcome this difficulty; but he regards the difficulty as being a needless one, since in methods depending on the *pumping* of a mixture of chloroform and air into a mask the difficulty is non-existent.

A list of the apparatus of modern times for the delivery of known percentages of chloroform might be usefully added here.

(1) *The Dubois Pump.*

The percentage administered is in no way dependent on patient's respirations.

(2) *The Vernon Harcourt Apparatus.*

This is an apparatus in which the patient sucks air over a surface of chloroform (*Proc. of Royal Soc.*, vol. lxx. p. 504, 1903).

(3) *Dr. Levy's Apparatus.*

A suction apparatus with a most ingenious device to overcome the objection mentioned above (*Trans. Royal Medical and Chirurgical Soc.*, 1905).

(4) *Dr. Waller's Wick Vaporiser.*

The percentage administered is in no way dependent on patient's respirations (*Journal of Physiology*, vol. xxxi. p. 6 of Proceedings).

(5) *The Author's Apparatus.*

The percentage administered is in no way dependent on patient's respirations (*Trans. Royal Medical and Chirurgical Soc.*, 1905).

III. Observation of the Patient's Condition

There can be no doubt that this is by far and away the most important duty of the anæsthetist. Whether we adopt the "Scotch view" that the respirations alone should engage our attention, or whether we follow the "London view" that the pulse should be our guide, we cannot doubt for a moment the duty of watchfulness. These pages are not the right place to discuss the clinical aspects of the question. One can, however, bring forward the opinion that where a reliable apparatus is used the respirations should in the vast majority of cases afford a safe guide as to the patient's condition. *Unless the recent researches on chloroform anæsthesia are entirely misleading, the administration of mixtures of low concentration should render cases of sudden heart failure the rarest of accidents.*

If this should prove to be the case, as one has every right to expect that it will, those who have been engaged in these investigations will indeed reap an ample reward.

PHYSICAL GEOGRAPHY AS AN EDUCATIONAL SUBJECT

By J. E. MARR, Sc.D., F.R.S.

University Lecturer in Geology, Cambridge

GEOLOGY has long been regarded by those who are concerned with education as the Issachar of the sciences, "couching down between two burdens "—to wit, the weight of the mineralogical knowledge on the one side and the biological knowledge on the other necessary for its right comprehension. It is many years since Ruskin wailed that the energies of the Geological Society were diverted to palæontology, and since then they have been largely devoted to the mineralogical branch of the study. Little wonder, then, that school-teachers have looked askance at the subject, especially since it is customary for the writers of even the most elementary text-books to attempt to cover the whole range of the science, forgetting that a study of rocks and fossils by boys who have not mastered the elements of mineralogy and biology is conducive to slipshod methods. It is true that a number of facts may be readily learned, but the significance of these is not fully grasped—ay, and worse than this, the students are often led to suppose that they have been grasped, when the contrary is the case.

An attempt on the part of youth to study all branches of the science is positively harmful, and those who wish to see geology take its proper place in the educational curriculum should discountenance any such endeavour. But though this detailed study is not to be commended, there is an ever-increasing number of those who are interested in education who believe that the principles of the science can be taught in a manner which is, from the educational point of view, thoroughly beneficial.

The alternative title of Sir Charles Lyell's great work *The Principles of Geology* is " The Modern Changes of the Earth and its Inhabitants considered as illustrative of Geology." In that work petrology and palæontology find no place, and yet to it we largely owe the present position of geology among the sciences.

Now the study of the modern changes of the earth and its inhabitants is physical geography pure and simple; and if physical geography be properly taught, it introduces us rightly to the study of geology in general.

Its value as an educational instrument was clearly recognised by Huxley, when in 1877 he published his *Physiography*, as any one who will read the preface to that work will see. At that time the term "physical geography" was under a cloud, owing to the character of many of the educational works which were devoted to it; and Huxley borrowed his title from another science because he "wished to draw a clear line of demarcation, both as to matter and method, between it and what is commonly understood by 'physical geography.'"

Huxley's method is well known: it was to give his readers "in very broad but, I hope, accurate outlines, a view of the 'place in nature' of a particular district of England, the basin of the Thames," and he expressed his opinion that any intelligent teacher would "have no difficulty in making use of the river and river basin of the district in which his own school is situated for the same purpose."

Unfortunately a large number of teachers have been unable to apply this method, owing to the issue by the Science and Art Department of a syllabus of physiography which necessitates a treatment of the subject very different from that which Huxley advocated. His treatment was simple and scientific: that of the trainer of candidates for the examinations held by the Science and Art Department complex, and too often, alas! of the nature of "cram." Who can give boys and girls a truly scientific explanation of the theory of tides? How much better are they for such information as they can obtain concerning terrestrial magnetism? And what matters it to them whether the shape of the earth approaches more closely to that of the orange or of the pear, and if it be oblately spheroidal or geoidal?

The latter-day physiography is, no doubt, like Latin repetition or a list of the reigns of the kings and queens of England with dates, excellent for cultivating the memory, but for developing the reasoning powers it leaves much to be desired. Its teachers have adopted the methods and included the subjects which Huxley wished to avoid; but, in the meantime, text-books on physical geography have appeared which, while avoiding the use of illustrations drawn from a limited area like the Thames basin, treat the subject to a large extent from his standpoint. " Physiography " and " physical geography" have, indeed, changed places since the 'seventies ; and it is the claims of the latter, as regarded by present-day writers, that I wish to press upon those who are concerned with the teaching of science.

These claims have been put forward prominently of recent years. The Royal Geographical Society, largely owing to the energy of its late President, Sir Clements Markham, has been actively engaged in advocating the proper teaching of geography, physical, political, historical, and commercial, and has been largely instrumental in obtaining the establishment of Boards of Geographical Education at the older universities.

In 1893 a meeting of public-school masters was held at Oxford, and at that meeting the " Geographical Association " was founded, and now flourishes. Its secretary is Mr. A. J. Herbertson, the Reader in Geography at Oxford. Its aim is " to impress the teaching of geography by spreading the know-ledge of all such methods as call out the pupils' intelligence and reasoning powers, and make geography a real educational discipline, instead of merely loading the memory with names and isolated facts." Every term the association issues a magazine, *The Geographical Teacher*, for the discussion of methods of teach-ing geography and the diffusion of information useful to teachers, which is sent free to all members of the association, and in many other ways gives assistance to geographical teachers. The annual subscription is 5s. (and may be compounded by the payment of £3 10s.), and intending members should address themselves to the Hon. Treasurer, Mr. J. S. Masterman, St. Margaret's, Dorking, and if engaged in teaching should state in what capacity. The association now numbers over 500 members.

But notwithstanding what has been done by this and other bodies in establishing the claims of geography, and especially of the physical side of the study, to take its right place in the schools, progress, it must be admitted, is but slow. Nor is this at all surprising when we consider the difficulties to be encountered. It is clear that the schoolboy cannot be expected to get any real knowledge of the whole range of sciences, physical and biological, but must confine his attention to one or two subjects; and there is general agreement as to the subjects which are of the greatest importance. They are chemistry and physics, of which some knowledge is requisite for the proper study of any other science, including those which are concerned with biology. It is right, therefore, that any boy beginning the study of science should be taught the elements of chemistry and physics; and fortunately such teaching is simple, and does not require any particular originality on the part of the teacher, for the line of study is by now well established, good text-books exist, and laboratories are easily equipped in accordance with known plans. We cannot wonder, therefore, that masters in schools as a whole prefer to confine their pupils' attention to these studies, which entail comparatively little trouble on the part of the teacher, and to avoid such subjects as botany, zoology, and physical geography, where the teaching depends to a considerable extent upon the nature of the surrounding country, and where accordingly it is necessary to supplement the bookwork by outdoor work, which is different for various localities, necessitating a certain amount of originality of observation and of thought on the part of the teacher himself.

It may be supposed that the above remarks ignore the present "boom" in "nature study"; but this study seems to be pursued largely in the elementary schools only, and to be neglected in the higher schools, though true nature study is one which can be pursued throughout a lifetime, for all natural science is nature study, though all "nature study" is not natural science.

Agreeing as we do that an elementary knowledge of chemistry and physics is essential to the proper study of other sciences, should schoolboys be discouraged from beginning the study of these sciences, even if they have a bent towards one of them? It would seem from the action of some of our educational authorities as though they desired to keep the natural sciences, other than chemistry and physics, out of the school curriculum, and I feel most strongly that this is harmful.

Specialisation is the feature of the day, but it is carried too far, and, above all, begun too early; for it seems to be forgotten that generalisation is also important, and that the highest scientific mind is one capable of generalisation as well as of specialisation—in proof of which assertion mention of the name of Charles Darwin is in itself sufficient.

A fairly good method of gauging the trend of modern scientific teaching in the higher schools is afforded by a consideration of the qualifications of candidates who present themselves for the Entrance Scholarship Examinations in the older universities. In former days it was common for a candidate to present himself for examination in four subjects, and quite usual for him to take at least three ; whereas now the majority simply take chemistry and physics, and occasionally only one of these. This is surely a sign of the times. Is it a good sign ?

Many of the brilliant discoveries of recent date are undoubtedly due to the work of the chemist and physicist ; but it has not always been so. The discoveries of geologists in the past—as, for instance, in settling the antiquity of man—and of biologists in throwing light on the origin of species, have profoundly affected the thoughts of man ; and what has happened will no doubt occur again.

Also it must be remembered that research is not the only goal of scientific teaching. The practical applications of science are in many cases in the hands of those who are not engaged in research work. Each of the natural sciences has its practical bearing, and discouragement of its study by those who are responsible for the education of the people of a nation will have its effect in checking the progress of that nation.

In considering the value of a subject in connection with education, we are not, however, concerned so much with the possibilities of research, or of the importance of that subject on account of its applications, as with its actual use as a means of education. We have therefore to consider what value physical geography possesses in this aspect, and whether its value is in any way different from that of the sciences which are usually taught in schools—namely, chemistry and physics.

As a means of quickening the memory, the subject is neither better nor worse than any other in which a large number of facts must be acquired by oral instruction, reading, and observation, and so co-ordinated as to give us that "exact, regular, arranged knowledge" which is science.

In cultivating the powers of observation, again, physical geography shares much in common with the experimental sciences. Inasmuch, however, as it depends little upon experiment, but chiefly upon study of natural physical features, the kind of observation required is somewhat different from that demanded by the sciences of chemistry and physics, and each kind supplements the other, a combination of the two being desirable for the proper quickening of the observing power.

The great importance of developing the faculty for observing the phenomena which are within the domain of the physical geographer is becoming more and more recognised, but upon this head something will be said in a later part of this article.

Let us now pass to the consideration of the value of our subject as a means of development of the reasoning powers. We find here a marked difference between the influence of physical geography and that exacted by the study of chemistry and physics.

The sciences are sometimes divided into two groups according as they are "certain" or "uncertain"—the former being concerned with matters which are capable of mathematical demonstration, while the natural sciences fall into the latter group. But the uncertainty of the sciences of this group varies in degree, and we might speak of the less uncertain and the more uncertain studies. In the former we should place chemistry and physics, and in the latter physical geography. The uncertainty of the latter constitutes a danger as far as education is concerned, but also gives it a very real value. The danger to be avoided is that of loose reasoning, and it is one which concerns the teacher rather than the taught. The teacher should utilise this very danger as a means of instruction, pointing out that an explanation may contain the truth, and yet not "the whole truth and nothing but the truth." Take the case of the formation of those coral islands which are known as atolls. The theory put forward many years ago by Darwin to explain these structures on account of its simplicity and beauty gained general acceptance, and was often explained in elementary text-books without the reservation which its illustrious author himself made. It was subsequently attacked, but that it contains part of the truth is clear. Probably not the whole truth, however, for, as Huxley once remarked (as recorded by Prof. Judd): "I am convinced, from all that is being done now, that we shall not find any simple easy explanation of all coral reefs; that the study of coral reefs is one of the very greatest complexity; that the conditions under which they were formed would have varied greatly in different cases; and that one theory of their origin will probably not be found to suit all cases."

The loose reasoning in this case is to be found in those books which attempted to explain Darwin's theory in a few brief paragraphs.

The theories of the physical geographer have, in fact, been established or partly established as the result of cumulative evidence, and they are often finally established by "the method of successive approximations," thereby differing from many of the theories of the more certain sciences. From the point of view of education, it is this weighing of the evidence which is of so great value to the student, for the process is one which has often to be applied to the affairs of everyday life, and the power of right judgment is of inestimable benefit to its possessor. I may illustrate this by an anecdote recorded by the late Frank Buckland in his *Curiosities of Natural History*, which refers to the sister science of geology. His father, Dean Buckland, was once lecturing at Oxford on the remains of hyænas in the Kirkdale caverns of Yorkshire, and among his audience was " one of the most learned judges of the land." Buckland, "after having, with his usual forcible and telling eloquence, put his case to prove not only the former existence of hyænas in England, but even that they were rapacious, ravenous, and murderous cannibals . . . turned round to the learned lawyer and said, 'And now, what do you think of that, my lord?' 'Such facts,' replied the judge, 'brought as evidence against a *man*, would be quite sufficient to convict and even hang him.'"

It is, however, as an aid to the appreciation of the beauties of nature that the study of physical geography differs most markedly from that of other sciences which are usually taught in schools. The artistic temperament may appear to have little to do with the spirit of scientific inquiry; but one usually finds that the lover of natural beauty has an insight into the meaning of the objects which call forth his admiration, and at all periods of human history the lovers of nature seem to have had a desire to explain what they saw, though the craving for explanation in early days often gave rise to speculations far removed from the truth.

The appreciation of natural beauty has undoubtedly spread greatly in our country within recent times, and the desire to know something of the changes which brought about the present scenic features is equally marked. Wordsworth's well-known sneer at the geologist was unmerited when he penned it, and few would approve it at the present day. The author of *Modern Painters* has written chapter upon chapter with reference to the importance of our subject to the lover of scenery, for, as he observes, "the real majesty of the appearance of the thing to us depends upon the degree in which we ourselves possess the power of understanding it."

Many books have been written to satisfy the wish to know something of the causes which produced the earth's features. How many of our intelligent countrymen have read Sir Archibald Geikie's *Scenery and Geology of Scotland*, Lord Avebury's *Scenery of England* and his *Scenery of Switzerland*! Many of these have no doubt wished that they had been taught somewhat of the subject in their school-days.

There is yet one other difference between the study of physical geography and that of physics and chemistry, in that the laboratory of the former is no confined room, but the open country.

In these days, when some deplore the extreme devotion of our countrymen to athletic sports, all are agreed as to the value of outdoor exercise. To the student of our subject this exercise is obtained in the pursuit of his science, which often leads him into the most health-giving places on upland moor and mountain side, or by the cliffs which overlook the ocean's marge.

When the study is further pursued, another element which gives zest to sport is also encountered—that of danger. That this is present is sadly manifest, owing to the geologists who have lost their lives in pursuit of their calling ; but it is slight, and its presence calls into play other important qualities— namely, the care which should lead people to avoid such danger as is avoidable, and the presence of mind which enables them to cope with it when it cannot be avoided.

I have laid stress upon the importance of physical geography in that it is largely an open-air study, for I look back to very happy days spent in attempting to elucidate the causes of various physical features in many a fair part of the country, and recall the envious tone in which students of other sciences have remarked, "What a lucky man you are, being able to do your work in places like these!"

It was before remarked that the whole of geology is not

of a character suitable for teaching in schools, and the same must be said of physical geography. It will be well, therefore, if we briefly consider what parts of the subject are eminently adapted for that purpose.

The relationship of the earth to celestial bodies must be touched upon during some part of a course, though it is well to do so briefly, and not at the beginning of the study. Apart from this, the physical geographer is specially concerned with climatology, oceanography, geomorphology (especially that part which deals with the study of land forms), and the distribution of organisms. The last named is not suitable for the elementary student of physical geography, as its proper cultivation necessitates some knowledge of biology. Such parts of it as can be satisfactorily acquired by the young are best treated as lessons in zoology and botany.

Oceanography in turn is open to the objection that little can be learned of it by the general body of schoolboys as the result of actual observation. Such parts as are essential to the right study of climatology and geomorphology can be taught in their proper places when those branches are under consideration. It is, then, to climatology and parts of geomorphology that we turn as specially suitable for conveying to the young the principles of the science.

Let us begin with climatology. Those parts which may be studied with advantage are concerned with cloud formation and weather forecasting.

How often do we hear the question asked, " What is the weather going to be like to-day ? " How rarely is a satisfactory answer obtainable ! Where the pursuits of the natives of a district necessitate some knowledge of weather changes, one usually consults a native ; elsewhere we turn to the forecast for the day issued by the Meteorological Office. Each of these sources of information may be unsatisfactory. That given in the weather forecast is for wide areas, and may be useless for a limited district; and, on the other hand, that culled from the native is obtained by observation of local changes, without any knowledge of the general distribution of weather conditions over wide areas. If the student will observe the sky signs for himself, and has obtained sufficient knowledge of his subject to understand not only the weather forecasts, but the observations on which they are founded as displayed on charts, he is in a position to foretell with considerable certainty the changes which will take place in the immediate future over a limited area.

There is much available literature on this branch of the subject. For general meteorology we have Mr. R. H. Scott's *Elementary Meteorology*, and the Hon. R. Abercrombie's book, *The Weather*, both in the International Scientific Series; and for information about clouds we can turn to Mr. A. W. Clayden's *Cloud Studies*.

It is when we come to the study of land forms that the student's inability to range at will through all the quarters of nature's laboratory is most strongly felt. In our own country no glaciers are found and no active volcanoes. The work of wind as an agent of sculpture is insignificant when compared with its action in desert regions, and the action of frost is not very marked. Again, we can only study certain features in the British Isles by journeying over wider tracts than can usually be traversed. The dweller in the Fenland may know nothing of mountains and lakes by observation, and those who inhabit the interior may be practically unacquainted with the action of the sea along the coasts. But some of the operations of weathering may be observed in all places, and the action of rivers may be everywhere partially grasped from actual observation; and in this way a sound practical acquaintance with some of the principles of the science may be obtained.

Even where certain agents are no longer in operation in our own country, the effects of their action in the past may be studied, and more is often learned in this way than if the agent itself were visible. The effects of glaciers are seen in many parts of the country, and are in many cases more easily studied than when the glacier itself is in existence, for the glacier covers its works to some extent, as the case of a watch conceals the machinery. Again, in several spots in our islands we find the remains of ancient volcanoes in all stages of dissection, so that we can find out the nature of the interior of a volcano in a manner impossible with one now in activity. By means of such illustrations the intimate connection between physical geography and geology ("the sum of the physical geographies of past times") is brought home to the student.

When studying any branch of physical geography, the use of maps is obvious, and it is most desirable that the teacher should accustom his pupils to this by giving lessons in "map-reading," taking the Government ordnance maps of his own district as a basis, and accompanying the students into the country map in hand. In addition to this, contoured and hill-shaded maps of various types of country should be studied, and the significance of the features explained.

The understanding of maps is a subject of national import-ance, and it is astonishing to find how generally it is neglected in educational establishments, with consequences which at times have proved nothing short of disastrous.

That part of the subject which cannot be directly observed in the student's district must needs be acquired from books and from study in the museum.

We have an admirable presentment of the kind of physical geography which Huxley had in mind when he wrote his *Physiography* in Prof. W. Morris Davis's *Elementary Physical Geography*, which may well be used as a text-book for schools.

Every school in which natural science is taught should possess its museum as well as its laboratories, and this museum should be exclusively fitted out for teaching purposes. Such museums are becoming a recognised feature in some schools; the writer well remembers the pleasure with which he examined the museum of Winchester College, in which provision is made for the teaching of physical geography among other sciences.

The outlay for furnishing the requisites for teaching physical geography need not be great. The chief instruments for clima-tological study will be in the laboratories. Typical weather charts may readily be collected and displayed, and also photo-graphs of the principal types of clouds. For geomorphological purposes photographic and other illustrations are readily obtainable, and the students should be encouraged to make simple models for themselves. Temporary ones may be formed of plasticene or other material; indeed, much may be done with ordinary unbaked pastry. Collections of materials which have been modified by wind, water, and ice action, and the products of volcanic and other actions, will be gradually brought together; these will enable the teacher to give instruction of real value in those cases where illustrations cannot be obtained in the open country.

It must be remembered that common objects are of the greatest value for teaching purposes. A museum of rarities illustrates phenomena of limited distribution or occasional occurrence; the products of the widely distributed changes which it is the province of the beginner to study are easily acquired, and it is these which find their proper place in the teaching museum.

I have in this article advocated the use of physical geography as an educational instrument. Apart from that, it is surely the duty of every educated person to know something about the earth on which he lives.

ON THE OCCURRENCE OF PRUSSIC ACID AND ITS DERIVATIVES IN PLANTS

BY T. A. HENRY, D.Sc. (LOND.)

Principal Assistant, Scientific and Technical Department, Imperial Institute

THOUGH the poisonous character of bitter almonds appears to have been well known in early times, and they were regularly prescribed by physicians in the Middle Ages, it was not till 1800 that Bohm,[1] a pharmacist of Berlin, detected prussic acid in the volatile oil produced when crushed bitter almonds are macerated in cold water and the liquor obtained, subsequently distilled. Curiously enough, Scheele,[2] who discovered prussic acid in 1782, did not realise that it was poisonous, and the final proof that prussic acid was the toxic constituent in volatile oil of bitter almonds was made by Schrader in 1803.[3]

Since that time the occurrence of free prussic acid, or of derivatives which readily yield this acid,[4] has been noted in a large number of plants belonging to such different natural orders as the Gramineæ, Liliaceæ, Salicaceæ, Ranunculaceæ, Passifloreæ, Bixineæ, Tiliaceæ, Sterculiaceæ, Linaceæ, Sapindaceæ, Rhamnaceæ, Celastrineæ, Euphorbiaceæ, Saxifragaceæ, Rosaceæ, Leguminosæ, Sapotaceæ, Oleaceæ, Asclepiadaceæ, Convolvulaceæ, Rubiaceæ, and Compositæ. In an incomplete list of plants known to contain cyanogenetic compounds, published by Jouck in 1902,[5] over one hundred species are enumerated, and this number has been increased since.

Most of these cases are only of importance at present as pointing to the fact that the occurrence of cyanogenetic compounds in plants is far more common than is generally believed;

[1] *Neues allgemeines Journal der Chemie*, 1803.
[2] *Nova acta Acad. reg. sued.*, 1782.
[3] *Trommsdorf's Journal*, 1803.
[4] Dunstan and Henry have used the term "cyanogenesis" to describe this process, and the adjective "cyanogenetic" to describe the precursors, usually glucosides, of prussic acid in the plant.
[5] Jouck, *Beiträge zur Kenntnis der Blausäure abspaltenden Glycoside*, Strassburg, 1902.

but there are a few which are of historical and general interest, and which may, therefore, be referred to more particularly. In 1731 Madden,[1] in a paper communicated to the Royal Society, drew attention to the toxicity of cherry-laurel water (a flavouring agent prepared by macerating broken cherry-laurel leaves in water), with the result that a number of investigations were undertaken to ascertain the nature of the toxic constituent; and the problem was finally solved by Schrader in 1803,[2] who showed that cherry-laurel water contained prussic acid, which he had recently proved to be poisonous.

Another interesting case is that of Cassava (*Manihot utilissima*), widely cultivated throughout the tropics for the sake of its starchy, edible root, from which the tapioca of commerce is prepared. The occurrence of a volatile poisonous product in cassava root was first recorded by Fermin[3] in 1764, and in 1836 Henry and Boutron-Charlard[4] identified this poisonous constituent as prussic acid; this observation was subsequently confirmed by Francis[5] in the West Indies, who also showed that the production of prussic acid is not confined, as had previously been supposed, to " bitter " cassava, but also takes place in the " sweet " variety, now referred to *M. palmata* or *M. Aipi*. Reference may also be made to Jorissen's[6] discovery of the presence of a cyanogenetic glucoside in embryonic flax plants.

It is remarkable that the production of prussic acid should have been noted in so many plants which are of economic importance. Cassava, flax, and bitter almonds have already been instanced, and in addition to these may be mentioned the seeds of *Taraktogenos Kurzii*,[7] which are the source of the chaulmugra oil used as a remedy for 'leprosy and certain skin diseases; the seeds of *Schleichera trijuga*,[8] from which " macassar oil " is prepared; alder bark,[9] used to some extent in medicine; the roots of *Manihot Glaziovii*, the plant yielding the Ceãra rubber of commerce[10]; the seeds, leaves, and flowers

[1] *Phil. Trans.* 1731, **37**, 84.
[2] *Trommsdorf's Journal*, 1803.
[3] *Mem. Acad. Sci. Berlin*, 1764.
[4] *Mem. Acad. Med. Paris*, 1836.
[5] *Analyst*, 1878, **2**, 4.
[6] *Ann. Agron.* 1885, **10**, 468.
[7] Power and Gornall, *Journ. Chem. Soc.* 1904, **85**, 838.
[8] Thümmel, *Archiv. der Pharm.* 1891, **229**, 182.
[9] Gerber, *ibid.* 1828.
[10] Van Romburgh, *Ann. jard. bot. Buitenzorg*, 1899, **16**, 1.

of *Hevea brasiliensis*,[1] from which Para rubber is procured; the seeds, leaves, and flowers of numerous rosaceous plants, grown either for the sake of their flowers or fruit; the seeds of several species of *Vicia*,[2] largely used as feeding-stuffs; the leaves of *Ipomœa dissecta*,[3] used in the preparation of "noyau"; and, lastly, the leaves of the "great millet," *Sorghum vulgare*,[4] and of maize, both widely cultivated for the sake of their edible grain.

In all plants in which "cyanogenesis" has been thoroughly investigated, it has been ascertained that the prussic acid occurs for the most part not in a free state, but combined, usually in the form of a glucoside.

Comparatively little progress has been made till recently in isolating and characterising these glucosides, which appear to be the form in which prussic acid is temporarily stored by plants in which cyanogenesis occurs, and still less work has been done in characterising and classifying the enzymes, which usually occur with the glucosides and possess the property of decomposing them, thus generating prussic acid. Cyanogenesis opens out in these two directions extensive fields of investigation for the chemist. But the subject is probably of even more interest to botanists, since Treub[5] has suggested, as the result of his investigations of *Pangium edule* and *Phaseolus lunatus*, that the formation of prussic acid is probably the first step in the process by which these plants convert the "inorganic" nitrogen of nitrates into the "organic" nitrogen of proteids.

Apart from these purely scientific aspects cyanogenesis is of great, though perhaps rather grim, interest from the agricultural point of view. One of the problems which every agricultural department is sooner or later called upon to deal with is that of plants poisonous to farm animals, and there is quite an extensive literature on this subject scattered through the various agricultural journals.[6] Recently several plants of this type obtained from British colonies and dependencies have been examined at the Imperial Institute, and it has been shown that their toxicity is due to the formation of prussic acid when the plants are eaten

[1] Van Romburgh, *Ann. jard. bot. Buitenzorg*, 1899, **16**, 1.
[2] Ritthausen and Kreusler, *Journ. für prakt. Chemie*, 1870, **2**, 333.
[3] *Med. uit 'slands plantentuin*, 1888, vol. **29**.
[4] Dunstan and Henry, *Phil. Trans.* 1902, A. **199**, 399.
[5] *Ann. jard. bot. Buitenzorg*, vol. 13, and series ii. 1905, **4**, 86.
[6] Maiden, *Plants Poisonous to Stock*.

by animals, the acid being formed in the usual way by the decomposition of a cyanogenetic glucoside by an enzyme, both these substances being simultaneously present in the plants.

Investigations have been somewhat actively prosecuted along these various lines during the last few years, and results of some general interest have been obtained, the most important points of which are dealt with in the succeeding paragraphs of this article.

CYANOGENETIC GLUCOSIDES

AMYGDALIN.—This, the best-known member of the group of cyanogenetic glucosides, was isolated from bitter almond seeds by Robiquet and Boutron-Charlard[1] in 1830, and was subsequently investigated by Liebig and Wöhler,[2] who first explained its real nature.

It can be decomposed by enzymes in two stages, which may be represented by the following equations:

$$(1)\ C_{20}H_{27}O_{11}N + H_2O = C_6H_{12}O_6 + C_{14}H_{17}O_6N$$

Amygdalin.   Dextro-glucose. Mandelic nitrile glucoside.

$$(2)\ C_{14}H_{17}O_6N + H_2O = C_6H_{12}O_6 + HCN + C_6H_5 . CHO$$

Mandelic nitrile glucoside. Dextro-glucose. Prussic acid. Benzaldehyde.

Oil of bitter almonds.

The first stage is brought about by the action of the enzyme *maltase* of yeast,[3] and the second by the action of the enzyme *emulsin*, which occurs in both sweet and bitter almonds; but it should also be stated that emulsin has the property of decomposing amygdalin directly into glucose, prussic acid, and benzaldehyde without the apparent intervention of mandelic nitrile glucoside, and it is to the interaction of these two substances in ground bitter almond seeds that the production of "natural oil of bitter almonds" is due.[4]

It is generally believed that sweet almond seeds contain the enzyme emulsin but not the glucoside amygdalin, although Flückiger and Hanbury[5] state that moist sweet almond seeds invariably yield traces of prussic acid when comminuted. Jorissen[6] has also observed that the embryo of the sweet

[1] *Ann. chim. phys.* 1830 [ii.], **44**, 352.
[2] *Ann. Chem. Pharm.* 1837, **22**, 11.
[3] Fischer, *Ber. der Deutsch. Chem. Ges.* 1894, **27**, 2989 ; 1895, **28**, 1809.
[4] Liebig and Wöhler, *loc. cit.*
[5] *Pharmacographia*, London, 1879, 247.
[6] *Ann. Agron.* 1885, **10**, 468.

almond, immediately after germination, yields considerable quantities of prussic acid (probably produced by the decomposition of amygdalin), and that the quantity of acid yielded by the embryo steadily increases to a maximum. It is interesting in this connection to point out that the seeds furnished by wild *Phaseolus lunatus* yield, like bitter almond seeds, considerable quantities of prussic acid, whilst those produced by the cultivated *P. lunatus* resemble sweet almond seeds in yielding only traces of the acid or none at all.[1] If these can be regarded as strictly parallel cases, it would appear that possibly the sweet almond may have been produced by the cultivation of the bitter almond.

It has been asserted that amygdalin occurs in other plants besides the two varieties of almond,[2] and in particular a so-called "amorphous amygdalin" has been regarded as the source of the benzaldehyde and prussic acid yielded by cherry-laurel leaves when macerated in water;[3] and the same product was stated by Greshoff[4] to be present in *Gymnema latifolium*, *Pygium parviflorum*, and *P. latifolium*, and by others[5] in the seeds of various rosaceous plants, such as *Malus communis*, *Cydonia vulgaris*, *C. japonica*, *Sorbus Aria*, and *S. Aucuparia*. With the exception of *Gymnema latifolium*, in which no emulsin was detected, all these plants were stated to contain both emulsin and amygdalin; but the evidence adduced of the presence of the glucoside is usually merely the production of benzaldehyde and prussic acid from the seeds after maceration in water. The fact that such evidence is insufficient to warrant the conclusion that amygdalin is present in any particular case has been illustrated recently by Bourquelot and Danjou,[6] who have shown that the berries of the common elder (*Sambucus nigra*), which yield benzaldehyde and prussic acid on maceration in water, contain not amygdalin but a new glucoside, *sambunigrin*, which is isomeric with the mandelic nitrile glucoside of Fischer.

DHURRIN.—This substance, which is closely related chemically to the three glucosides referred to in the preceding

---

[1] Dunstan and Henry, *Proc. Roy. Soc.* 1903, **72**, 285.

[2] Greshoff, *Ber. der Deutsch. Chem. Ges.* 1890, **23**, 3548.

[3] Lehmann, *Über das Amygdalin*, 1876.

[4] Greshoff, *loc. cit.*

[5] Lutz, *Repertoire de Pharm.* 1897, 312. Riegel, *Ann. Chem. Pharm.* **48**, 361. Wicke, *ibid.* **79**, 81 ; **81**, 241.

[6] *Compt. rend.* 1905, **141**, 598 ; cp. Guignard, *Compt. rend.* 1905, **141**, 236.

paragraphs, was isolated by Dunstan and Henry[1] from the common sorghum, or "great millet" (*Sorghum vulgare*), a plant widely cultivated in tropical and sub-tropical countries for the sake of its edible grain, which forms a staple food-stuff of the great majority of the natives of India, Egypt, and other countries. The glucoside is secreted in the young green parts of the plant, and the amount present increases up to the time that the plant is about 12 inches high, after which it diminishes, the mature plant and the seed being quite free from it.

Dhurrin is decomposed by hot dilute hydrochloric acid or by emulsin (which also occurs in the plant) in accordance with the following equation :

$$C_{14}H_{17}O_7N + H_2O = C_6H_{12}O_6 + HCN + C_6H_4(OH) \cdot CHO$$

Dhurrin.   Dextro-glucose. Prussic acid. Para-*hydroxybenzaldehyde.*

This reaction indicates that the relation of dhurrin to mandelic nitrile glucoside (or sambunigrin) is that of a *para*-hydroxy derivative. Young sorghum plants have long been known to be occasionally poisonous to cattle, and much speculation has at different times been indulged in as to the nature and origin of this poison, some authorities asserting that it was a secondary effect only exhibited by diseased sorghum, and others that what was called "sorghum poisoning" was really "hoven" (a form of indigestion), produced by immoderate consumption of the young green plant. There can be no doubt now, however, that the toxicity of young sorghum is due to the generation of prussic acid by the interaction of "dhurrin" and emulsin in the plant.

The glucoside has as yet only been isolated from sorghum grown in Egypt ; but Slade[2] and Brünnich[3] have shown independently that green sorghum grown in the United States and in Queensland yields prussic acid, and there can be little doubt therefore that *Sorghum vulgare* always contains the glucoside at least in the earlier stages of growth.

Dhurrin is the only cyanogenetic glucoside so far isolated from a grass ; but several members of the same natural order have been shown to yield prussic acid, notably "Guinea grass" (*Panicum maximum*), "Para grass" (*P. muticum*),[4] and ordinary maize (*Zea mays*), in which, as in sorghum, the amount of prussic acid obtainable increases up to a certain stage, and

[1] *Phil. Trans.* 1902, A. **199**, 399.   [3] *J. Chem. Soc.* 1903, **83**, 788.
[2] *J. Amer. Chem. Soc.* 1903, **25**, 55.   [4] *Ibid.*

then diminishes to zero. A number of the grasses have been found not to contain any cyanogenetic compounds, notably *Cynodon Dactylon*, *Paspalum distichum*, and sugar-cane tops grown in Australia. Samples of young oats, wheat, and barley, grown in Essex, have also been examined at the Imperial Institute and found not to yield any prussic acid.

PHASEOLUNATIN.—This glucoside was isolated by Dunstan and Henry[1] from the seeds of uncultivated plants of *Phaseolus lunatus* grown in Mauritius, where the plant is used as a green manure. It is widely cultivated in tropical and subtropical countries, and yields beans which may vary in colour from pale pink with purple spots (Rangoon beans) to pale cream or white (Lima beans). The seeds from cultivated plants are lighter in colour and less shrivelled than those produced by the wild plant, and yield mere traces of prussic acid or none at all, whereas the wild seed may yield as much as 0·13 per cent. by weight of the acid.[2]

Phaseolunatin is decomposed by hot dilute acids or by an emulsin-like enzyme, which also occurs in the seeds, in accordance with the following equation :

$$C_{10}H_{17}O_6N + H_2O = C_6H_{12}O_6 + HCN + (CH_3)_2CO$$

Phaseolunatin.   *Dextro-glucose.* *Prussic acid.* *Acetone.*

Phaseolunatin is therefore a glucose ether of acetonecyanhydrin.

It is of interest to recall that in 1888 van Romburgh,[3] who has been associated with Treub at Buitenzorg in the study of cyanogenesis, recorded that in many plants in which cyanogenesis occurs acetone is produced simultaneously with prussic acid. It is probable that phaseolunatin or a similar acetone glucoside is present in these plants. Among the species mentioned by van Romburgh are *Manihot utilissima* (cassava), *M. Glaziovii* (Ceãra rubber plant), *Hevea brasiliensis* (Para rubber plant), and *Hevea Spruceana*. More recently Jouck[4] has shown that the glucoside isolated by Jorissen and Hairs[5] from embryonic flax-plants when decomposed by acids yields acetone and prussic acid. Van Itallie[6] observed that the same products are yielded by the leaves of *Thalictrum aquilegifolium*, and suggested that phaseolunatin may be present in this plant.

GYNOCARDIN.—This glucoside was isolated by Power and

[1] *Proc. Roy. Soc.* 1903, **72**, 285.
[2] *Ibid.*
[3] *Ann. jard. bot. Buit.* 1899, **16**, 1.
[4] Jouck, *loc. cit.*
[5] *Bull. Acad. roy. Belg.* 1891 [iii.], **21**, 529.
[6] *Journ. pharm. chim.* 1905, **22**, 337.

Gornall from the seeds of *Gynocardia odorata*, an Indian plant, which belongs to the natural order Bixineæ and was long supposed to be the source of chaulmugra oil: the production of prussic acid both by the leaves and seed of this plant had been previously observed by Greshoff and by Desprez.[1] The glucoside was subsequently investigated by Power and Lees,[2] who showed that it differed from all the known cyanogenetic glucosides in being decomposed by acids only with great difficulty. Gynocardin is, however, readily decomposed by a specific enzyme, *gynocardase*, occurring in the same seeds, yielding one molecule each of prussic acid, *dextro*-glucose, and a very unstable third product whose nature has not yet been ascertained.

Lotusin.—This substance was isolated by Dunstan and Henry[3] from the annual leguminous plant *Lotus arabicus*, which grows along the valley of the Nile, where it is known by the vernacular name "khuther." The Arabs have long known that "khuther" is poisonous in the early stages of its growth, and that it becomes innocuous and, indeed, a useful and nutritive fodder as it matures. The plant was a great source of trouble to the Anglo-Egyptian armies during the first and second Sudanese wars, owing to the fact that many of the transport animals were poisoned by eating it.

This led eventually to its investigation, and to the discovery that it contained a glucoside, lotusin, which, under the influence of an enzyme, *lotase*, also present in the plant, underwent decomposition yielding prussic acid as one product. The same decomposition occurs when lotusin is treated with hot dilute hydrochloric acid, and the reaction which takes place may be represented by the following equation:

$$C_{28}H_{31}O_{16}N + 2H_2O = C_{15}H_{10}O_6 + HCN + 2C_6H_{12}O_6$$

Lotusin.   *Lotoflavin. Prussic acid. Dextro-glucose.*

The first of these decomposition products—viz. lotoflavin—is of considerable interest, since it is a yellow dye belonging to the same group as quercetin, the dyeing principle contained in the well-known dyewood obtained from *Quercus tinctoria*.

Further investigation of the plant confirmed the accuracy of the Arab belief that "khuther" is only poisonous in the early

[1] Jouck, *loc. cit.*
[2] Power and Lees, *Journ. Chem. Soc.* 1905, **87**, 349.
[3] *Phil. Trans.* 1901, B. **194**, 515.

stages of its growth, for it was found that lotusin is only present in the young plant, and that as this matures the glucoside gradually diminishes in amount and finally disappears.

It is worth noting that a similar glucoside appears to occur in the allied plant *Lotus australis*, a native of Australia, which has acquired some notoriety there as being poisonous to cattle.

The cyanogenetic glucosides so far known fall naturally into two classes—(*a*) those in which the cyanogen group is associated with the non-sugar residue, and (*b*) those in which this group is combined with the sugar residue. The essential difference in the constitution of these two types may be represented graphically thus:

Group A.

AMYGDALIN $C_6H_5CH\begin{smallmatrix} O\!\!-\!\!-\!\!-\!C_{12}H_{21}O_{10} \\ \\ CN \end{smallmatrix}$

SAMBUNIGRIN $C_6H_5CH\begin{smallmatrix} O\!\!-\!\!-\!\!-\!C_6H_{11}O_5 \\ \\ CN \end{smallmatrix}$

    (Benzaldehydecyanhydrin residue.)   *(Sugar residue.)*

DHURRIN $C_6H_4(OH)CH\begin{smallmatrix} O\!\!-\!\!-\!\!-\!C_6H_{11}O_5 \\ \\ CN \end{smallmatrix}$

(Para-hydroxybenzaldehydecyanhydrin residue.)   *(Sugar residue.)*

PHASEOLUNATIN $\begin{smallmatrix} CH_3 \\ \\ CH_3 \end{smallmatrix}C\begin{smallmatrix} O\!\!-\!\!-\!\!-\!C_6H_{11}O_5 \\ \\ CN \end{smallmatrix}$

    (Acetonecyanhydrin residue.)   *(Sugar residue.)*

Group B.

LOTUSIN $C_{15}H_9O_5\!\!-\!\!-\!\!O\begin{smallmatrix} \\ \\ CN \end{smallmatrix}CH.C_{11}H_{21}O_{10}$

(Lotoflavin residue.)   *(Sugar-cyanhydrin residue.)*

The substances enumerated in the preceding paragraphs are all the cyanogenetic glucosides so far definitely known, but there can be little doubt that compounds of this type will eventually be isolated from all the plants in which cyanogenesis is known to occur. Allusion may be made here to two such plants, which are of special interest. In *Pangium edule*[1] Treub asserts that prussic acid occurs in a free or in a very loosely combined state, and he draws attention to the occurrence in the same plant of a sugar possessing reducing properties which may be the substance with which the acid is combined. No real attempt has, however, yet been made to determine whether or not

[1] Treub, *Ann. jard. bot. Buitenzorg*, 1896, vol. 13.

P. edule contains a cyanogenetic glucoside. The seeds of *Taraktogenos Kurzii*,[1] which are now recognised as the source of chaulmugra oil, have been shown by Power and Gornall to contain a cyanogenetic compound, from which prussic acid is readily obtained by the action of acids or the specific enzyme also present in the seeds; according to Jouck the related plant *T. Blumei* has also been found to yield prussic acid.

GLUCOSIDOLYTIC ENZYMES

It is unfortunate that so far no satisfactory method of characterising enzymes has been discovered, and consequently it has become almost customary to assume that where a glucoside occurs in a plant in association with an enzyme which decomposes it, the latter is specific in its action.

This tendency has been rather less marked since Fischer discovered that glucosides were capable of existing in two isomeric forms derived respectively from *a*-glucose and *β*-glucose, and that there appears to be a certain relationship between the configuration of the glucoside and that of the enzyme which is capable of decomposing it. Thus Fischer[2] found that yeast ferment (maltase) was only capable of hydrolysing *a*-glucosides and emulsin of decomposing *β*-glucosides.

The enzymes associated with cyanogenesis in plants have, however, not as yet been systematically studied, except in the case of the emulsin of almonds, which is now known to be a *β*-enzyme—*i.e.* it is capable of decomposing *β*-glucosides. The action of emulsin on amygdalin presents some features of interest. Since this glucoside is decomposed both by yeast enzyme (maltase) and by emulsin, it must be at once an *a*- and a *β*-glucoside— *i.e.* of the two glucose residues present in the molecule, one must exist in the *a* and the other in the *β* form. It is curious, however, that while maltase attacks amygdalin in such a way as to liberate only one molecule of glucose (presumably *a*-glucose), leaving the rest of the molecule (mandelic nitrile glucoside) intact, emulsin attacks amygdalin in such a way as to lead to complete decomposition, producing presumably a molecule each of *a*- and *β*-glucose. This complete decomposition of amygdalin by emulsin can be accounted for if it may be assumed that the enzyme first splits the glucoside into benzaldehydecyan-

---

[1] Power and Gornall, *loc. cit.*   [2] Fischer, *loc. cit.*

hydrin and the $\alpha\beta$-disaccharide, the latter being then resolved by the further action of the enzyme into the two simple glucoses.

This essential difference shown by maltase and emulsin in their methods of attacking amygdalin appears to negative the view sometimes expressed that maltase and emulsin are complementary to each other in activity.

The only other glucosidolytic enzymes calling for attention in the present connection are *lotase*, present in *Lotus arabicus*, and *gynocardase*, occurring in gynocardia seeds. The activities of gynocardase are very similar to, if not identical with, those of emulsin. Lotase differs markedly from emulsin in activity, as might be expected, since it occurs with *lotusin*, in which, as already pointed out, the cyanogen group is associated with the sugar residue.

The Physiological Significance of Cyanogenesis

The earlier investigators who devoted attention to cyanogenesis were usually inclined to regard the production of prussic acid by a plant as a means of protection, and later on this view was succeeded by the belief that prussic acid and its immediate precursors are merely waste products of metabolism, stored in places such as barks, epidermal cells of leaves, etc., where they are harmless to the plant. Recently, however, several investigators have assigned a more important rôle to the acid, and Treub,[1] in particular, has asserted that it is probably the first recognisable product of the assimilation of the nitrogen of nitrates by plants. Treub's conclusions are based principally on the results of his elaborate investigations of the plants *Pangium edule* and *Phaseolus lunatus*. It would take too long to describe these in detail, but the most important points established seem to be that in the green parts of both these plants both free prussic acid and cyanogenetic glucosides occur, the latter serving apparently as temporary reserve products; and that with more active assimilation in the plant, whether brought about by improvement in environment or by increased supplies of suitable nutrition, an increase in the amount, both of free prussic acid and of cyanogenetic glucosides, follows. Some confirmation of Treub's views is also afforded by the results of Soave's[2] experiments with bitter almonds, which show that *free*

[1] Treub, *loc. cit.*   [2] Soave, *Nuov. giorn. bot. ital.* 1899, **6**, 219.

prussic acid begins to appear as soon as the embryo germinates, implying that the amygdalin contained in the seeds is used as a reserve material. Soave deduces from his observations the conclusion that cyanogen compounds are transitional substances furnishing the plant with nitrogenous food. In support of such a view one may urge from the chemical side that the cyanogen group is a highly reactive one, and in particular when suitably combined is readily convertible by simple reactions into amino-derivatives of the type now known to form the nucleus of some of the more simple proteids, and this is probably one of the principal facts which has led Gautier[1] and others to assign a preponderating place to cyanogen and its derivatives in the synthesis of proteids in plants.

An important point upon which practically nothing is known at present is the method by which prussic acid is first produced in plants, though Gautier has suggested that it may be formed by the reduction of nitrates by formaldehyde. In this connection Treub has observed that, although nitrates are usually almost entirely absent from the laminæ of the leaves of *Phaseolus lunatus*, there is always a supply of these salts in the petiole, which seems to act as a storehouse for this material.

[1] *Leçons de chim. biol.* Paris : Masson et Cie. 1897.

THE SOLVENT ACTION OF ROOTS UPON THE SOIL PARTICLES

By A. D. HALL, M.A.

Director of the Rothamsted Experimental Station

THOUGH it has always been recognised that the roots of plants in the main derive their nutriment from substances dissolved in the water within the soil, yet the possible direct solvent action of the roots themselves upon the solid materials has long been a debated question amongst plant physiologists and agricultural chemists. The problem has been attacked from several distinct points of view, and the investigations find applications in one or two rather unexpected directions.

The starting-point in the discussion may be taken to be the classical experiment of Sachs (*Bot. Ztg.* 1860, 117), repeated nowadays in every botanical laboratory, in which a slab of polished marble is placed vertically in the soil of a pot carrying a growing plant. When it is removed after a few weeks' growth of the plant it will be found to be etched wherever the network of fine roots has been in contact with the smooth surface. Sachs further found that such etched figures were produced upon plates of dolomite, magnesium carbonate, and calcium phosphate, but not upon silicates; while a gypsum plate showed a reversed effect of raised lines beneath the roots, which seemed to have protected the original surface from the solvent action of the water in the soil. Since it is well known that the expressed sap of the roots of most plants contains a certain amount of free organic acid, these etching effects have often been put down to the exudation of the acids in the sap through the thin cell walls of the roots. Sachs, however, regarded the carbon dioxide, which is always being excreted by the plant's roots, as the origin of the corrosion. He pointed out that the substances attacked were all soluble in water containing carbon dioxide, which alone would be able to effect the phenomena observed. Again, the soft cell membrane of the root hairs, which is in such intimate contact with the soil particles, must contain a relatively strong solution of the carbon dioxide that is passing outwards and would possess a correspondingly active solvent action. Sachs's experiments were extended by Czapek (*Prings. Jahr. f. wiss. Bot.* 1896, **29,** 321), who contrived an excellent method of making plates for the study of these corrosion figures by floating a mixture of equal weights of plaster of Paris and the substance to be examined upon a smooth surface of plate glass. In this way could be obtained delicately polished surfaces containing various carbonates and phosphates of known solubility. On exposing them to the action of the roots of a growing plant it was found that while the phosphates of calcium, magnesium, and iron were attacked, aluminium phosphate remained untouched. All the inorganic acids, and also oxalic, malic, citric, lactic, and tartaric acids, will attack precipitated aluminium phosphate, just as they will the phosphates of calcium and the other bases specified. Excluding carbon dioxide, there remain among the possible plant acids in the sap only two which have but little solvent action upon aluminium phosphate—acetic and propionic acids, and these give an intense blue colour with Congo red. But when a plaster plate stained with Congo red was exposed to the attack of the roots, there was no blueing to mark the line of contact. Czapek thus confirmed Sachs's original conclusion that the carbon dioxide secretion of the root is the seat of its solvent action, and will sufficiently account for its etching effect.

Some experiments of Schloesing (*Ann. Sci. Agron.* 1899, 316), in which he showed that plants could perfect their growth in the extremely dilute solutions of phosphoric acid that can be extracted from soils or result from the action upon them of slightly carbonated water, provided the solution 'were continually renewed, also indicate the sufficiency of the carbon dioxide secreted from roots to effect the necessary attack upon the soil without invoking the intervention of any more permanent acid.

Kossowitsch (*Ann. Sci. Agron.* 1903, 220) showed that plants could not succeed in a nutritive solution made up with no phosphoric acid, but which was first allowed to flow gently through a pot containing sand and certain ground mineral phosphates, although the same plants grew well in a similar pot when the roots were in contact with the ground phosphate. But in these experiments the solvent power of the carbon dioxide and of any permanent acid excreted from the root are equally eliminated; they only prove that the aqueous

THE SOLVENT ACTION OF ROOTS

solution of the phosphates in question contains insufficient phosphoric acid for the needs of the plant.

Another class of experiments are often quoted in support of the theory of the excretion of a permanent acid. If the roots of a young seedling be pressed between blue litmus paper, a permanent red coloration ensues, and that this is not due to breakages of the cells and simple extrusion of the sap may be seen from the fact that water in which seedlings have been germinated also possesses an acid reaction. Czapek (*loc. cit.*) satisfied himself that the acid reaction found in these cases was due to potassium hydrogen phosphate. It has again been stated, though there is some doubt about the correctness of the data, that young seedlings put to grow in a solution of ammonium chloride induce a permanently acid reaction—a fact which, if true, points much more to the preferential selection of the ammonia than to the excretion of any acid from the root. It is, indeed, not always possible to observe the alleged blueing of litmus when the roots of young seedlings are brought in contact with litmus paper, and Czapek admits that the drops of liquid, always to be seen standing on the root-hairs of seedlings developing in a damp atmosphere, possess no acid reaction. In any case, however, these phenomena exhibited by seedlings have but little bearing on the question at issue; the life of a seedling goes on by the breaking down of elaborated reserve materials and their transport to the growing points where reconstruction takes place—a wholly dissimilar process to the inflow of simple nutrients and their removal from free solution which marks the action of a normal root. Nor is it credible that any growing plant, once past the seedling stage, will part with so valuable a material as potassium phosphate.

Another method of attacking the problem has been to sum up the actions of the plant upon the soil by taking out a balance-sheet between the acids and bases contained within the completely developed plant. In a normal soil the plant has to obtain its nutrition from neutral salts, except so far as their solution is rendered acid ;by the presence of carbon dioxide; if, then, the plant excretes any fixed inorganic acid, the plant itself must be left with an excess of base. Now, though the ashes of a plant are alkaline in reaction through the presence of an excess of base, yet the ash no longer contains the nitrogen, all of which entered the plant in the form of nitrate and should therefore be reckoned among the acids. On summing the constituents with nitrogen included, and setting off equivalents of the acids against the bases, it is almost always found that the acids are in excess in the plant. This means that during nutrition in a neutral solution of calcium nitrate, such as naturally prevails in the soil, the plant withdraws an excess of nitric acid over lime. Presuming the solution to be ionised, the nitric acid ions continue to pass into the plant by osmosis to a greater extent than the calcium ions, because the former are more rapidly withdrawn from solution and utilised by the protoplasm. But if an excess of acid enters the plant, an excess of base must be left behind, and the net action of the plant on the medium in which it is growing will be the production of an alkaline rather than an acid reaction. There might still, however, be a secretion of an organic acid, equivalent to or even in excess of the bases unabsorbed.

This, however, does not prove to be the case on experiment; plants grown in water cultures of neutral or even slightly acid reaction at starting induce a gradually increasing alkalinity, eventually to the destruction of growth, unless the solution be changed or the alkalinity neutralised. An example (*Proc. Roy. Soc.* 1905, B 77, 1) may be given of wheat grown from seed to seed in a water culture the liquid of which was never changed, though its reaction became in the end faintly alkaline, instead of the distinct acid with which it started.

DISTRIBUTION OF ACIDS AND BASES AFTER GROWTH OF WHEAT IN WATER CULTURE. 93·7 GRAINS OF DRY MATTER PRODUCED.

| | Quantities found, reduced to Equivalents of Hydrogen. | | |
	Culture medium.	Plant.	Total.
Potash	0·0107	0·0780	0·0887
Magnesia	0·0220	0·0081	0·0301
Lime	0·0468	0·0514	0·0982
	0·0795	0·1375	0·2170
Nitrogen	0·0096	0·0967	0·1063
Phosphoric Acid . .	0·0121	0·0176	0·0297
Sulphuric Acid . . .	0·0161	0·0100	0·0261
Chlorine . . .	0·0229	0·0379	0·0608
	0·0607	0·1622	0·2229
Bases in excess . .	+0·0188	−0·0247	−0·0059 [1]

[1] Represents nitrogen in the seed, dust, etc.

Here the culture medium, which represents the soil and its water, has gained alkalinity, nor did it show any indication of an organic acid excreted to neutralise the excess of base. A large bulk of the culture liquid on evaporation and heating gave no signs of charring; the residue also possessed the same reaction before and after ignition. The fact that water cultures will become alkaline has been noticed from time to time; some books assert that they become acid, but this is only likely to happen when ammonium salts are used as a source of nitrogen and the plant takes in the ammonia as such, without previous nitrification.

The analyses, of which many exist, of our chief farm crops show that the normal action of plants upon the soil is to leave behind a basic residue from the neutral salts on which they feed, and that, so far from being excretors of acid, their action upon the soil is precisely the contrary. They must leave behind in the soil after their growth quantities of base equivalent to 100—300 lb. of calcium carbonate per acre; and this affords an explanation of several facts in the field hitherto difficult to understand. For example, many soils possess but traces of base (calcium carbonate, etc.) available for the neutralisation of the acids produced during nitrification, a process which is always going on in nature, and is indeed the normal preliminary to the supply of nitrogen to the plant. Despite this constant draft upon the small amount of base in the soil, these soils maintain their neutral character when in arable cultivation, and show no signs of becoming sour and infertile. There must be some recuperative process at work, and this we may now attribute to the growth of the crop, which annually takes from the soil such an excess of acid as will leave behind an amount of base of the same order of magnitude as that consumed in the nitrification process.

Again, it has been noticed that the use as manures of neutral salts, like sodium nitrate and potassium sulphate, induces certain injurious changes in the texture of heavy soils; the clay lands become more sticky when wet, and dry into harder and more intractable clods; the water running from the land drains is more turbid and carries a comparative excess of suspended matter (*Trans. Chem. Soc.* 1904, **85**, 964). Effects of this kind are due to the "deflocculation" of the clay in the soil—*i.e.* its resolution into the ultimate particles of extreme fineness of which it is composed; they are produced by the action upon the clay of very small quantities of alkalis or alkaline carbonates, such as would be left in the soil by the withdrawal of an excess of the acid constituents of the neutral salts used as manure. Similar effects have been obtained by Krüger (*Landw. Jahrb.* 1905, **34,** 761) by pot experiments in the laboratory.

The attack of the acid root sap upon the mineral particles of the soil formed the theoretical basis of the method of determining the "available" plant food in soils, which was proposed by Dr. B. Dyer in 1894 (*Trans. Chem. Soc.* 1894, **65,** 115), and has subsequently been very widely employed. Dyer determined the acidity of the sap of a large number of roots, and found that it might be approximately represented by a 1 per cent. solution of citric acid. He therefore proceeded to attack the soil with this solvent, assuming that it would bring into solution just those compounds of phosphoric acid and potash which the plant could obtain by a similar mode of attack. Hence the materials determined in this way could be regarded as "available" for the plant, and would afford a more accurate means of judging whether a soil required manuring with phosphoric acid or potash than an estimation of the gross amount of these substances, however combined, in the soil. Dyer's method has undoubtedly proved of practical utility, and though for various reasons other acids and other concentrations have been proposed, they have not been generally accepted nor yielded more useful results. With the abandonment, however, of the view of the direct solvent action of the acid root sap, the theoretical basis of Dyer's method falls to the ground, and it must be judged empirically by its success in interpreting the soil conditions established by crop experiments. More recently (*Trans. Chem. Soc.* 1906, **89,** 205) it has been shown that the citric acid solution will continue to extract constituents like phosphoric acid from the soil indefinitely if the solvent be removed and repeatedly renewed. This solvent does, however, differentiate between the various phosphates of calcium, aluminium, iron, etc., present in the soil, removing some of them entirely in the earlier extractions and establishing with others solutions of approximately constant composition.

From several points of view, then, it will be seen that it

is not necessary to assume the existence of an excretion from the roots of the plant of a permanent acid, organic or inorganic, to attack the solid mineral particles of the soil and to bring them into solution for the nutrition of the plant. The growing portions of a plant root are always giving off carbon dioxide, and carbon dioxide, especially in the concentrated solution which must be momentarily formed in the cell wall of the root hairs, has an appreciable solvent effect upon the majority of the minerals composing the soil. This carbon dioxide alone is capable of giving rise to such solutions as are required for the nutrition of the plant. As the direct evidence is also adverse to the idea of an excretion of acid, the principle of not seeking remote causes would lead us to attribute to carbon dioxide, and to carbon dioxide only, the long-recognised solvent power of the plant upon the soil.

SOME NOTABLE INSTANCES OF THE DISTRIBUTION OF INJURIOUS INSECTS BY ARTIFICIAL MEANS

By FRED. V. THEOBALD, M.A.

Vice-Principal of the South-Eastern Agricultural College; President of the Association of Economic Biologists of Britain; Foreign Member of the Association of Economic Entomologists, U.S.A.; etc.

ONE of the most noticeable phenomena amongst the insects injurious to fruit, farm and garden crops, stores, man and his domesticated animals, is the almost world-wide distribution of certain of these pests. That this distribution is not a natural one we may safely infer from what we know of the general range of insects. Some species spread over a very wide area naturally, such as the Cotton Boll Worm (*Heliothis obsoleta*, Fabricius), which is found in Europe, Africa, and America; the Army Worm (*Leucania unipuncta*, Haworth), which is common to both hemispheres; and the Migratory Butterfly (*Danais archippus*). Such instances are comparatively few in number. In spite of these, we can say definitely that there is no insect really cosmopolitan by nature. Just as with birds and other animals, so with insects—each species has a definite area of distribution. Many may increase this area by natural means, as we see has taken place with the Mexican Boll Weevil (*Anthonomus grandis*, Boheman), and the Colorado Beetle (*Doryphora decemlineata*, Say.), but only within certain limits. Hence when we find some insects that attack plants, man, or animals almost world-wide in distribution, we may be sure that their range is due to some artificial cause or causes. On studying this subject we can at once see how easy it is for certain pests to be carried over the face of the earth by man's agency.

This distribution takes place by means of boats and trains and all other ways of intercourse. The more rapid these means of communication become, the more likely we are to see a concomitant increase and spread of many injurious insects, unless checked by stringent regulations. This dispersal has

58

taken place mostly from north and south towards the Equator. We find that many temperate-climate insects will live and flourish in sub-tropical and tropical climates, but the reverse only applies within certain narrow limits according to each species.

It is extremely unlikely that many tropical pests would live and flourish in the warmer climates of Europe, although we have an instance of such in the Yellow-fever Mosquito (*Stegomyia fasciata*, Fabricius). On the other hand, sub-tropical species may do so, and even penetrate into still more temperate regions—as, for instance, the San José Scale (*Aspidiotus perniciosus*, Comstock), which is found spreading as far north as Canada. The Yellow-fever Mosquito (*Stegomyia fasciata*, Fabricius), however, does not seem to occur farther north and south of the Equator than 48°. It has evidently spread outwards from the Central American States. We also see that the Brown Spotted-Mosquito (*Theobaldinella spathipalpis*, Rondani) has spread from Europe to the Sudan, and also far into the Cape.

Of wider distribution still are some insects which attack stored goods. We find the Corn and Rice Weevils (*Calandra granaria* and *Calandra oryzæ*, Linnæus) now in almost all countries from the Equator to Norway and New Zealand, because they are so easily carried in grain.

The manner in which an introduced insect may behave in a new country cannot be foretold. It may increase very rapidly, such as did in California, the Cottony Cushion Scale (*Icerya purchasi*, Comstock), which came from Australia. On the other hand, it may die out sooner or later, as happened with the Tasmanian Lady-birds (*Leis conformis*), which I introduced into this country. The Cushion Scale found a comfortable home, and there being none of its natural enemies to prey upon it, the Scale increased at an enormous rate. The Tasmanian Lady-birds, although they survived two seasons, found the climatic conditions unsuited to them, and consequently died out.

Many insects from the tropics and sub-tropics may be imported to temperate climates, such as our own, and will flourish under glass, where they find congenial heat and moisture. Scale insects, or *Coccidæ*, are particularly prone to do so. Most of the palm- and hot-house Scale insects we have in Britain are foreign importations. Only recently a Mealy Bug

(*Dactylopius nipæ*, Maskell) new to Britain has been sent to me on palms from Blackheath. It was originally described from Demerara, and has also occurred at Bournemouth in Hampshire.

The introduction of injurious insects can only be stopped by legislation. We find that most of the countries of the world have regulations protecting them from foreign importations, such as "An Act to prevent the introduction and provide for

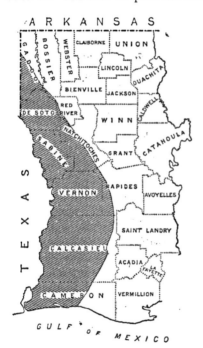

FIG. 1.—Area in Louisiana infested by the Cotton Boll Weevil in December 1904 (after Newell).

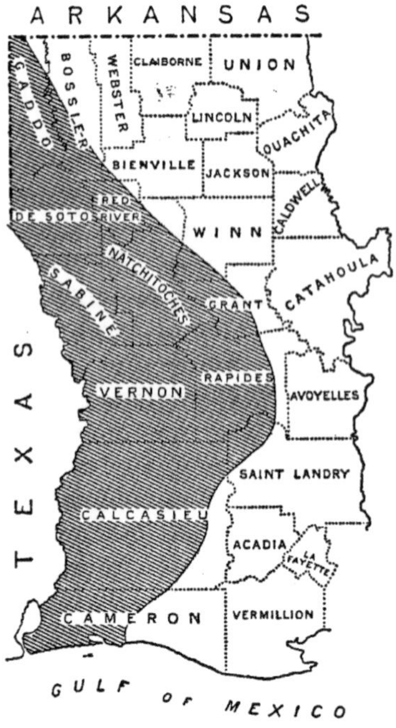

FIG. 2.—Area in Louisiana infested by the Cotton Boll Weevil in November 1905 (after Newell).

the eradication of disease affecting orchards and gardens" in New Zealand.

We may also get greatly increased local distribution by natural methods. As an instance of rapid natural distribution, we may mention the Cotton Boll Weevil. The beetles migrate in definite directions, and at certain times of the year. How great this distribution may be we can best judge by a glance at the figures showing the range of this serious cotton pest in Louisiana in 1904 and 1905 (figs. 1 and 2).

First, we will consider insects that attack fruit-trees in orchards and gardens, which may easily be carried from one country to another over sea, or from one state to another over land. The following are some of the most important in this category :[1]

The Codling Moth (*Carpocapsa pomonella*)*; the Bud Moth (*Tmetocera ocellana*)†; the Pith Moth (*Laverna atra*); Pistol and Cigar Case-Bearers (*Coleophoridæ* spp.); Bagworms (*Thyridopteryx*, etc.); Lackey Moths (*Clissiocampidæ* spp.); the Peach Borer (*Ægeria exitiosa*, Say.); and Currant Borer (*Ægeria tipuliformis*, Linn.); the Shot-borer Beetles (*Xyleborus dispar*, etc.); Bark Beetles (*Scolytus rugulosus* and *S. destructor*, etc.)†; Currant Sawflies (*Nematus* spp.); the Pear Slugworm (*Eriocampa limacina*). Various fruit flies, such as the Apple Fruit Fly (*Trypeta pomonella*)†; the Mediterranean Fruit Fly (*Ceratitis capitata*); the Australian Fruit Flies (*Dacus tyroni*, etc.); and the Indian Fruit Fly (*Dacus ferrugineus*). Psyllidæ, such as the Apple Sucker (*Psylla mali*) of Europe and the Pear Sucker (*Psylla pyri*) of America; the Woolly Aphis (*Schizoneura lanigera*)†; the Currant-root Louse (*Schizoneura fodiens*); the Phylloxera or Vine Louse (*Phylloxera vastatrix*). Various aphides of all kinds, such as the Cherry Aphis (*Myzus cerasi*); Apple Aphides (*Aphis pomi*, *A. fitchii*, and *A. sorbi*). Of scale insects or Coccidæ, the most important are the Mussel Scale (*Mytilaspis pomorum*)*; the San José Scale (*Aspidiotus perniciosus*); the Japanese Cherry Scale (*Diaspis amygdali*)†; the Peach Scale (*Aspidiotus persicæ*); the Cottony and Egyptian Cushion Scales (*Icerya purchasi* and *I. ægyptica*); and very many Citrus Scales, such as *Aspidiotus aurantiæ* and *Mytilaspis citricola*.

Besides true insects we get the ova of Red Spiders (*Bryobia pretiosa*); various gall-mites or Eriophyidæ, such as the Currant-bud Mite (*Eriophyes ribis*) and the Pear-leaf Blister Mite (*Eriophyes pyri*).

These various pests may be carried in three different ways— namely (1) in and on fruits and seeds; (2) on living plants; and (3) in the cases and packages in which fruits and plants are sent. It will at once be said that the first and last are of little importance, as such cases of fruits are taken to large towns and markets.

---

[1] Those marked with an asterisk come in abundance to Britain ; those with a dagger may do so now and again.

But they *are* very important, for these cases are distributed from the central markets to stores in small towns and villages. The stores, especially of the latter, are frequently close to gardens and even large orchards. Thus the pests that come over with the fruit may easily reach trees and bushes near by, either by flying or crawling, or by being carried by birds, other insects, or by man. Where the pests in the fruit are active insects, they often crawl to the baskets, that home-grown fruit is sent to the market in, for shelter. In this way many insects are carried back to our plantations, and many insect enemies have thus been spread over the earth. Nursery stock has also played, and does still, an important part in distributing diseases.

The most important insect spread by means of fruit is the Codling Moth (*Carpocapsa pomonella*). Originally this apple pest seems to have come from Europe. Its life-history is too well known to recapitulate fully here. Suffice it to point out that the larvæ, or maggots, occur in the fruit at all stages. When they are quite young they do not show any very marked symptoms of attack; but a careful examination will always reveal the presence of a small quantity of brown "frass" around the "eye," a sure sign of the maggots' presence. The larvæ always leave the fruit when full grown to pupate in convenient shelter, which is normally found under the bark of apple trees. They leave the fruit just the same when it has been picked, and then seek shelter in store-rooms and in the cases and barrels in which the apples are dispatched. Failing this, they will leave the barrels in large numbers when opened and seek shelter elsewhere, such as in market baskets near by. In the latter way they are conveyed to the country. Quite exceptionally the maggots may be carried with nursery stock, they having spun their cocoons in some fork of the branches. That the Codling Moth is sent in numbers from country to country any one can verify for himself by examining apples imported into this country when the barrels are opened in the markets. It must not be imagined that *all* apples come thus infested. I have not detected any in Tasmanian apples, but have very frequently in American and Canadian. Perhaps most occur in the shipments of Portuguese and Madeira fruits. The result of this means of transit has been that we now have the Codling Moth not only all over Europe, but in America, Canada, Madeira, Teneriffe, Cape Colony, Australia, and New Zealand.

The amount of damage done by it must be millions of pounds a year in the collective countries.

Fortunately, many of our colonies are alive to the importance of this subject. The Codling Moth Act of Tasmania, passed in 1884 and amended in 1891, has been so far successful that the colony is now almost free of this pest. All possible steps are taken to prevent its importation.

Natal has also legislated concerning the importation of this and other fruit pests. An Act (Law 15) was passed in 1881 " To regulate the introduction in this country of plants or cuttings which by reason of disease or otherwise might be injurious to the interests thereof." The benefit of this legislation we see in the case of the Codling Moth. Natal is one of the few countries where apples are grown in which the Codling Moth does not occur. It has been said to do so, but I am informed that these statements are erroneous. Mr. Fuller, the Government Entomologist, has shown that the diseased apples have been attacked by Fruit Flies.

Why is this? Simply because, under the Act referred to, a proclamation (79, 1897) was made prohibiting the introduction of all plants, portions of plants, cuttings, and anything taken off or from apple trees in the island of Madeira. Later, all diseased apples were prohibited coming from any region. Under these powers we find that in September 1903 seventy-five cases of Portuguese apples were destroyed; in September 1904 thirteen baskets of apples from Madeira were immediately reshipped on account of this pest. As a result, the colony is kept free from one of the most insidious apple and pear enemies the grower has to put up with.

The same is done in New Zealand, where 2,257 cases of apples were destroyed in 1901. In 1904 we find that the authorities in the Hawaiian Islands, acting under legal powers, destroyed all infested apples sent there.

It is almost useless to deal with this insect in a country unless protection is given from invasion from without.

The other important insects distributed in fruit are the numerous species of Fruit Flies (*Trypetidæ*). Different species attack a great variety of fruits, such as apples, pears, peaches, citrus fruits, guavas, bananas, etc. There are three genera that are destructive as Fruit Flies—namely, *Dacus*, *Ceratitis*, and *Trypeta*. The former has clear wings with one or two dark lines, and has been wrongly referred (so I am informed by Mr. Austen, Dipterologist at the British Museum, by Mr. Froggatt and others in Australasia) to the genus *Tephritis*; *Ceratitis* has ornamented wings with a dense fine network of veins at the base; whilst the last named has mottled wings with normal venation. These Fruit Flies lay their eggs in both sound and rotting fruits; the maggots live in the pulp, and can at once be told from Codling Maggots by being apodal. Numbers are frequently found in a single struck fruit.

The great importance of these pests cannot be overestimated, as we have no remedy for them. All that can be done to guard against Fruit Flies is to protect the trees by means of muslin tents, an operation too costly to be carried out in most places.

That these insects have been and are still imported with the fruit we know quite well. The Mediterranean Fruit Fly (*Ceratitis capitata*) has thus been distributed into Australia and Cape Colony and elsewhere, and has become a serious pest in its new homes. The various Australian Fruit Flies (*Dacus tryoni*, Froggatt) and others are found in numbers in exported fruit. As a result, New Zealand prohibits all fruit infested with Fruit Flies from being landed. We find in the official records that 3,700 cases of fruit were destroyed in 1901 on account of the presence of these enemies.

Similarly, St. Helena has passed an Ordinance (1904) prohibiting the importation of all fruits from South Africa, Mauritius, Cape Verde, and Malta, and for the extermination of the Peach Fly (*Ceratitis capitata*).

Neither *Ceratitis* nor *Dacus* are likely to flourish here, if imported; but they may do so in many warm countries where they are not yet known as pests, and might even exist in hothouses here.

We find these insects also in Mexico and many other countries. During 1904 many cases of fruits were destroyed at the Hawaiian Islands on account of the presence of Mexican Fruit Flies. Recently they have been found attacking melons in the Sudan (*Dacus* sp. ?).

One species may easily be imported into Britain—namely, the Apple Fruit Fly of America (*Trypeta pomonella*, Walsh). I feel confident some apples sent me from the Isle of Thanet were attacked by this pest; but, unfortunately, the dipterous larvæ did not hatch out. Should it appear in this country, it would prove another serious enemy in our already much-smitten orchards, and *we* have no law to prevent any such incursion.

Scale Insects, or *Coccidæ*, are also largely distributed on fruit, the majority on citrus fruits, but many on other kinds. We find amongst these the Red Scale (*Aspidiotus aurantiæ*), the San José Scale (*Aspidiotus perniciosus*), and the Mussel Scale (*Mytilaspis pomorum*).

The Mussel Scale occurs most abundantly on apple and pear, but also on many other plants, both cultivated and wild, such as thorns, in this country. By means of imported stock and fruits this insect has become almost world-wide in distribution. It is now found in Cape Colony, Natal, Egypt, all over America, Australia, and New Zealand, as well as in Europe, which is probably its original home. This Coccid is found on trunks, boughs, leaves, and fruit. It is not so often seen on the fruit in this country, but it comes over from abroad in large numbers in this way on apples, pears, etc. The great difficulty of keeping the Mussel Scale in check is well known, as it lives on wild as well as cultivated plants. Moreover, it is difficult to kill, except with strong paraffin emulsion; hence it is very important to keep it from entering a country or district where it does not occur. This can only be done by absolutely prohibiting any plants infested with it. Fumigation with hydrocyanic acid gas, I have found, has no effect upon the egg stage. When trees are lifted for removal or export, the scale is usually found in this condition. I do not think it is yet found in Central, Eastern, or Western Africa, so these areas should guard against its introduction in no uncertain manner.

The danger of introducing a scale insect is best seen in the case of the White Cushion Scale (*Icerya purchasi*, Comstock). This insect was imported into America from Australia. Its rapid increase, with such disastrous results to fruit-growers, is now a matter of history; and also its subsequent check by bringing over its natural enemy from Australia—the Lady-bird (*Vedalia cardinalis*). We now have this Icerya working in other places, notably in Egypt, in company with an allied species (*Icerya ægyptica*), and also in Portugal.

The San José Scale (*Aspidiotus perniciosus*, Comstock) is perhaps most feared of all Coccidæ. Its original home cannot be said to be definitely known. Most probably it came from Northern China, where it has been recognised so long that there is no record of its origin. It did not call for much comment there, because it was held in check by its natural enemies. The introduction of one of these enemies (a lady-bird beetle) into America has not had the same good results as the Australian Vedalia, partly because the beetles themselves were attacked by a parasite.

The actual introduction of the San José Scale into America seems to have been made from Japan, on some Japanese plums sent to California with the hope they would prove Cuculio-proof. From California it was sent with nursery stock to Pennsylvania in 1890 and 1891, and also to Virginia and a few other Eastern States. From that time it has gradually spread over America until it now does millions of dollars' worth of damage every year. So important is this pest that not only most of the American States, but most of the European countries, have laws safeguarding them against its introduction—even Turkey. Britain stands alone in not fearing the advent of this pest! There is no reason why it should not flourish in the West of England just as it does in Canada. It occurs also in Australia. Not only may it be spread by means of nursery stock, but also on fruit, especially apples and pears.[1] It also infests countless hardy plants, trees, shrubs, and vines.

Thus in three ways we see how insects have been spread from one country to another by artificial means—namely, on stems of plants, in and on fruit, and in packages. There is yet a fourth way—namely, on the roots.

The Woolly Aphis (*Schizoneura lanigera*);' or, as it is wrongly called, the American Blight, not only lives on the trunk and twigs of apple and pear, but also on the roots, where it produces galls of a similar form to those above ground. This woolly aphis lives during winter in two ways—(1) as active insects hidden in crevices in the bark, both above and below ground, and (2) as eggs, which are few in number and which are placed close to the base of the stem. It is extremely difficult to see a few of these insects or ova during the winter, when they are hidden away. As no steps were taken formerly to check insect introduction into our colonies and America, it is not surprising to find this Schizoneura wherever apples are grown. It abounds

---

[1] This is doubted by many of the chief authorities in America.

in the United States and Canada, it is very common in Australia and in New Zealand, and we find it in Natal and in Cape Colony, as well as all over Europe, from north to south.

The Phylloxera is another root-form which has been spread artificially over the face of the earth, carrying ruin with it throughout the vineyards. The Vine Louse also seems to be an European insect, and has been spread to America, Australia, the Cape, etc., with vine plants and cuttings. With this vine pest we get, as in the Woolly Aphis, a subterranean and an aerial race, and it is probably on the roots that it has been distributed so widely. Every country save our own has some Act or Ordinance forbidding the entry of vines, cuttings, etc., from foreign countries, or else limiting the introduction in certain drastic ways.

Another European Phylloxera (*Phylloxera corticalis*) which attacks oaks in Europe has recently been found in South Africa doing considerable damage. It has doubtless been introduced with seedling oaks.

Many Aphides have been spread on fruit trees and plants. One that has the widest distribution is the Black Fly of the Cherry (*Myzus cerasi*), which is now found in America, Australia, New Zealand, and South Africa, as well as all over Europe. The small black eggs of this aphis are not at all easy to detect on young cherry trees. Thus they are easily passed unnoticed with nursery stock into new areas. Living aphides may also be imported from some distance. A consignment of strawberries from England were examined (as all imported stock is) on arrival at Durban, and all the plants were found to have many living aphides upon them. Had there been no fumigating regulations in vogue, a serious pest would have been introduced into the colony which previously did not exist. The Apple Aphides (*Aphis mali, A. sorbi,* and *A. fitchii*) common to Europe and America doubtless were imported from one country to another with nursery stock many years ago.

The earth around the roots of plants must also bear its quota of insect enemies, for we find the Pear Midge (*Diplosis pyrivora*) in a few localities in America. It cannot have entered in any other way. We know that the larval midges fall from the fruitlets when they are mature and pupate in the soil. These puparia are very small, and are easily overlooked. Mixed up with particles of earth that stick to the roots, they may easily be carried over land and sea for any distance. It is surprising that this pest has not made its way to other parts than America before now.

In a similar way we can account for the wide distribution of the Pear and Cherry Sawfly, or Slugworm (*Eriocampa limacina*), which exists in America and Cape Colony. The Slugworms pass the winter in small earthen cocoons in the soil. They are often very abundant on nursery stock, and some cocoons are sure to be lifted with the stock and may remain attached to the roots with small clots of earth. Taken to a fresh country, they will hatch out in due course, unless the roots are previously treated.

As there is no doubt about the dispersal of such pests on rootage, it is very important that all imported stock should have the roots well cleansed before being planted. This is, I fancy, a point generally overlooked, but one of great importance in the artificial distribution of insect enemies.

Just as with hexapods, so with Eriophyid mites. These minute acari, formerly known as *Phytoptidæ*, live either in buds alone, as we see happens with the Big Bud Mite of the Currant (*Eriophyes ribis*, Nalepa), or in galls formed on leafage and blossom, as in the Pear-leaf Blister Mite (*E. pyri*, Sch.) and the Cotton Mite (*E. gossyperi*) respectively. In the pear pest the winter is passed in the buds, and the same applies to those that gall the leaves of the plum. It is extremely difficult to detect the presence in winter buds of the leaf-living forms, as the invaded buds do not swell. On the other hand, we all know the well-marked, swollen appearance of the "Big Bud" attack in black currants. It is not surprising, therefore, to find that the pear phytopt has a wide distribution. The Pear-leaf Blister Mite (*Eriophyes pyri*) is now found in America, Canada, and in Cape Colony. There is no doubt that it has been imported into all three countries with nursery stock, grafts, etc.; but we have no evidence showing when this took place. In some of its new environments this mite seems much more destructive than it is in its original home.

The Big Bud Mite of the Currant (*Eriophyes ribis*) so far does not occur outside Europe, but its distribution there has increased very rapidly. No natural causes can account for it. On the other hand, we can do so by the distribution of infested stock. That this has taken place in a most persistent manner for some years we know. Some growers at one time even maintained that the "Big Buds" were signs of increased vitality. The importation of such diseased plants into any new district has always been followed by the disease spreading out around the infected plantation by natural means. The minute acari are carried about by bees and birds. They are also spread artificially around an infested plantation by means of the baskets in which fruit is picked, by the clothing of the pickers, and even with the mud on people's boots.

The Colorado Beetle (*Doryphora decemlineata*, Say.) has been distributed artificially to this country, and there is not the least doubt that if it had not been rigorously stamped out by the authorities of the Board of Agriculture, acting under the Colorado Beetle Order of 1877, it would have spread to a disastrous extent. First it was found breeding luxuriously in Tilbury Dockyard in 1901, and continued until 1902, in spite of most drastic measures. In the latter year it was effectually stamped out.

The spread of this beetle by natural means has been remarkable. Originally it lived on wild solanaceæ in the Rocky Mountains; but as settlers pushed forward with their patches of potatoes, the beetles left the wild plants for the cultivated, and so spread farther and farther afield until now it is found flying in the neighbourhood of New York, and has even been seen in the city. How it was conveyed to England we do not know, but probably some specimens flew on to a ship and left it again on arrival at Tilbury.

The Hessian Fly (*Cecidomyia destructor*, Say.) is another instance showing the great harm caused by the introduction by artificial means of an injurious insect into a new country. There is plenty of evidence to show that this corn pest is European in origin. It attacks not only cultivated graminaceæ, but also wild kinds. We hear, for instance, of it devastating "twitch" or "couch" grass in Siberia. We know it occurs on the same and on other grasses in Britain. One heard so much of the Hessian Fly in the papers in 1891 that one was brought to believe that it had been imported into this country from America. In reality the insects were imported into America with straw by the hired Hessian troops, and first took up their abode at Long Island, and have gradually spread over the greater part of the wheat-growing areas of North America.

Thus again we find great damage done by an introduced insect. In England it never will be a very serious pest, because the chief damage it does is to autumn-sown corn, which in America is sufficiently up to allow of attack by the second brood, but with us a second brood would die off before the seed had germinated. This pest may be disseminated in two ways—one on the straw, as we see happened in regard to America, and another with seed corn, for some of the puparia may be found in it, although most come away in the " tailings."

Here again we find that somehow the natural enemies were not imported with the pest into America. For with us and in Russia the puparia are frequently found parasitised by chalcid flies to such an extent that one breeds out often more parasites than flies in this country ; hence it is kept in check with us.

The Wheat Midge (*Diplosis tritici*) of Europe, also found in North America, doubtless owes its origin there to similar factors.

Of the pests of animals we may still more expect to find a wide distribution due to artificial causes.

The Sheep Scab Mite (*Psoroptes communis*, var. *ovis*, Fustenberg) has undoubtedly been spread to all sheep-farming countries with the imported stock. Thorough quarantining would have prevented this, such as the regulations which are in force in most countries now concerning the introduction of animals. We can see similar results in regard to the insect enemies of stock. The Sheep Nasal Fly (*Œstrus ovis*), which lives in its maggot state in the nasal cavities and sinus of the sheep, has been spread with the sheep, and we now find it in America and Australia. The same has happened with the Ked (*Melophagus ovinus*).

Of all human injurious insects the Mosquitoes, or *Culicidæ*, stand foremost in this subject of artificial distribution. This is not so much so on account of the great importance they are to man as annoying agents and as the means of conveying such diseases as yellow fever, malaria, filariasis, and dengue fever, as on account of the wide knowledge we have of their distribution, and the means by which they have been and still are spread over the globe. It is only natural to expect that those with a wide distribution are intimately connected with man ; such is what we find to be the case. It is those which we may class as domesticated mosquitoes which we find have a wide range. The three most noticeable are the Yellow Fever Carrier (*Stegomyia fasciata*, Fabricius), the Household Brown Mosquito (*Culex fatigans*, Wiedemann), and the Brown Spotted-Mosquito (*Theobaldinella spathipalpis*, Rondani).

The Yellow Fever Carrier (*Stegomyia fasciata*) is commonly called the "Banded" or Tiger Mosquito. It occurs in the following countries : *Europe*—in Spain, Portugal (south), Southern Italy, Greece, the Mediterranean Islands; *Asia*—in India, the Malay ports, China and Japan; *Africa*—in Natal, Transvaal and Orange River Colonies, in Rhodesia, Uganda, on both east and west coasts, in Egypt and up the Nile past Khartoum, in Algeria and Tunis; *North America*, in the central and southern States, and, as we proceed south, still more abundantly

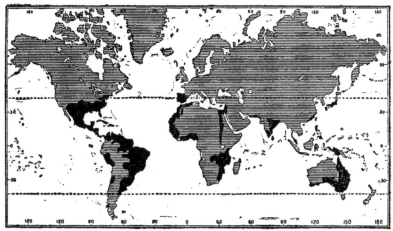

Fig. 3.—The Distribution of the Yellow Fever Mosquito (*Stegomyia fasciata* Fab.).

in *Central* and *South America* and the West Indies, which regions seem to be its original home. It is abundant in *Australia*, and passes up the East Indies to join the others at Malaya. Most oceanic islands also have this insect in their fauna. There is little doubt that the Central American States and West Indies are the home of *Stegomyia fasciata*, and that it has spread out from this area by artificial means. This is easily explained by the fact that we frequently find the Tiger Mosquito on board ship, and often in considerable numbers. In the old days they probably bred in the ship's tanks, just as we find them doing now on the Nile steamers. Coming to port, a certain number may fly to land, and so set up a new colony. In this respect

we may note that in some countries—Malaya, for instance—
they so far only occur in the ports and along the littoral.
Their further advance inland is more gradual. This takes place
mainly along the river courses by boats and along the rail tracts.
In both river steamers and trains we often find this mosquito
in numbers. Not only by artificial means do we get a most
annoying insect spread, but there is the concomitant danger of
yellow fever as long as we have the insect, that carries it, present.
Luckily this mosquito is found breeding almost exclusively in
and around houses and dwelling-places, and so can easily be
destroyed.

In a similar way the Brown Household Mosquito (*Culex
fatigans*, Wiedemann) seems to have been distributed. Skuse
tells us that its advance inland in Australia has followed the
opening of the railways. We know, as with the former insect,
that it is often a fellow-passenger on board ship. Both these
insects have nevertheless their finality of distribution, and we
find that they will not live if they reach farther than somewhere
near 48° north and south of the Equator (under normal conditions). They have both spread outwards from the warmer
regions.

On the other hand, by similar artificial means the Brown
Spotted-Mosquito (*Theobaldinella spathipalpis*, Rondani) of
Southern Europe has spread in the reverse way, and we now
find it at Khartoum and at the Cape. We know of no records
until the last few years of it at the Cape, its advent being
probably due to the large number of transports running there
during the recent war. This species is found in abundance at
Teneriffe and other islands on the way to Africa.

The instances quoted here are merely a few which show from
recent observations how insect enemies may and have been
distributed from country to country, and how they may or may
not increase to such an extent that they even out-rival the
damage they do in their native lands. To repeat once more,
"One can never prophesy how an introduced insect may act
in its new home." It is therefore essential to the well-being of
mankind that this insect dispersal by artificial means should
be dealt with universally, in regard to those pests which attack
farm and garden produce, stores, stock, and man, to save further
loss and danger.

THE BLOOD-PLATELETS

By G. A. BUCKMASTER, M.A., D.M. (Oxon.)

University College, London

In thinking over the problems in physiology of which so many await solution, we seem almost compelled to assume that the amount of positive, unquestionable knowledge that exists on any given subject stands in inverse ratio to the quantity of contentious literature devoted to it. In this paper it is our intention to present the views of those who have made the blood-platelets a subject of particular study, and, by considering the exact way in which the observations have been carried out, to examine whether all the statements, together with the inferences which various observers of undoubted ability have drawn from the histological appearances of shed human blood, are beyond criticism. Since a purely literary study of any scientific question is of doubtful value, we may take this opportunity of stating that the majority of the observations and experiments about to be described have been repeated and extended by us during the past three years, and it will be sufficient for the present to review some of the work which has been done on the nature and origin of the blood-platelets, and to omit any consideration of the part which these bodies play in the process of coagulation of the blood.

According to E. Schwalbe,[1] the platelets were originally discovered in 1846 by Fr. Arnold. A detailed description of these bodies as they appeared in human blood was first given in 1865 by Max Schultze, who spoke of them as " Körnchenbildungen," noticed that they resisted the action of weak acids, were destroyed by weak alkalis, and occurred in exceptionally large numbers in the blood of an anæmic individual. Between these observations and those made in 1878 and 1882 by Hayem and Bizzozero, which formed the starting-point for all the various controversial papers that have appeared up to the present time,

---

[1] *Untersuchungen zur Blutgerinnung Braunschweig*, 1901.

the blood-platelets were seen and described by Riess, Kölliker, Birch-Hirschfeld, and Osler. To this period also belongs the work of Zimmermann and the unconfirmed observations of Lostorfer. The "elementary particles" described in 1860 by Zimmermann[1] were seen in the supernatant liquid which collected above a sediment obtained by allowing blood to flow from a cut vessel into a solution of neutral salt. As far as it is possible to form an opinion from his description of the aspect and size of the particles, it is probable that many, if not all, of these were undoubted blood-platelets.

Osler's paper was published in 1874, but in the preceding year a short account by Osler and Schäfer, entitled " Uber einige im Blute vorhandene Bacterien-bildende Massen," had appeared in the *Centralblatt für die medicinischen Wissenschaften*. It is this paper which is generally quoted by recent Continental observers. More than thirty years have elapsed since Osler's observations were made in the Physiological Laboratory of University College, and the unfortunate title to his paper, "An Account of Certain Organisms occurring in the Liquor Sanguinis," is probably responsible for its neglect. The bodies he described so fully and accurately are undoubtedly blood-platelets, and some of his figures appear to-day in the text-books. When we remember how uncertain and inconclusive was our knowledge of bacteria, the ignorance as to whether these occurred in the normal fluids and organs of the body, and call to mind that Koch's work on *Wundinfectionskrankheiten* did not appear until 1878, it is easy to understand that Osler considered the motile transformations of the platelets were minute organisms within the blood and blood-vessels. In the blood of rodents, and particularly well in that of the rat, when the blood was mixed with serum and examined continuously for one to four or five hours, the formed bodies or platelets were noticed to alter. They developed processes, and the "developed forms" resembled spermatozoon-like forms, some of which possessed two or three tails and moved freely among the corpuscles. Osler's figures leave no doubt but that these transformations are exactly similar to what we have observed to occur when human blood mixed with oxalates is examined for the same length of time. They are particularly well seen in human blood, and also in that of other mammals, while the admixture of oxalates with the blood

[1] "Zur Blutkörperchenfrage," *Virch. Archiv*, xviii. 1860.

of *Sauropsida* and *Ichthyopsida* in our hands does not yield similar bodies. The platelets figured by Osler within the vessels are often reproduced as illustrations, but since they were only occasionally observed in the excised venules in pieces of subcutaneous tissue removed from the back of rats and spread out on a slide, it is open to question whether this is convincing evidence that they exist normally in blood.

It is by no means certain that different authors who have contributed to our knowledge of the platelets are really describing the same things. The varied terminology which has been employed also leads to some confusion. The name " Blut-plättchen" we owe to Bizzozero; the French School have accepted Hayem's term, or hæmatoblast; Eisen speaks of these bodies as plasmocytes; Dekhuyzen and Kopsch as thrombo-cytes; while the use of such names as endoglobular bodies (Hirschfeld), true and false platelets, Arnold's bodies, Woold-ridge's bodies, Deetjen's bodies, all of which, together with that of microcytes (Pappenheim), have been employed, is almost useless without a specific description. The essential discovery made by Bizzozero was not that the blood contained what he described as a third morphological constituent, but that the process of coagulation, as the experiments of Zahn had demon-strated, was due to the formation of white thrombi, which were dependent on the presence of blood-platelets. The coagulation of shed blood *in vitro* and the local formation of a thrombus are not processes which can fairly be compared together. The knowledge which is obtained from experiments on extra-vascular clotting can certainly not be applied without some reservation to explain the formation of clots when a vessel is transfixed by a needle, or the wall of a vessel is injured by a ligature, or by the application of some damaging reagent such as crystals of sodium chloride or silver nitrate.

From many experiments it is beyond dispute that masses of platelets can be caused to collect and form a thrombns. A conglutination or aggregation of these bodies occurs at the damaged region. When the rate of flow is artificially reduced, the figures of Eberth and Schimmelbusch[1] show that although leucocytes can be seen at the edges of the vessel, no platelets are to be observed, these only becoming displayed when the rate of flow is quite slow and just precedes stasis in

[1] *Die Thrombose nach Versuchen und Leichenbefunden*, Stuttgard, 1888.

the process of inflammation. The experiments of Wlassow[1] led him to the idea that the more fragile of the erythrocytes in circulating blood shed out a nucleo-proteid which forms platelets, while the remnants remained as microcytes. From the fact that a vessel may be plugged by a mass of platelets, Dekhuyzen[2] regards these bodies or thrombocytes as specific agents which exist in blood and protect the organism against accidental hæmorrhage. According to this observer, in all vertebrates hitherto examined with the exception of mammals, a cell or thrombocyte homologous with a platelet of mammalian blood is found which possesses the following features. It is a non-amœboid spindle-shaped body with an oval nucleus; the protoplasm is finely granular and furnished with fine radiating processes, which coalesce with those of adjacent thrombocytes. They are familiar to every one as the "Spindeln" of frog's blood, but are also found in worms, echinoderms, crustacea, and molluscs. Conclusions drawn from purely histological studies are always to be received with caution, and we must suspend our judgment before entirely accepting Dekhuyzen's views. It is certain that the thrombocytes of amphibian blood are not the homologues of mammalian blood-platelets (E. Neumann, J. Arnold, Eisen, Löwit, Engel, Maximow, Carl Marquis, Pappenheim, E. Schwalbe), if any bodies do exist which have this relationship it is the plasmocytes which G. Eisen[3] has described in the blood of *Necturus* and *Amphiuma*. Delezenne's discovery, confirmed by every one, that blood free from tissue-lymph remains unclotted, is absolutely opposed to the idea that any bodies of the nature of thrombocytes can be present in normal uninjured blood.

The view of Hayem and his school that hæmatoblasts contain hæmoglobin and are the antecedents of red blood corpuscles is still found in recent French text-books. All the evidence adduced as to the existence, origin, and destiny of hæmatoblasts is entirely histological, which is a most uncertain and unreliable foundation for the building up of hypotheses. The methods of histology are those of the morphologist and anatomist; the

[1] "Untersuchungen über die histologischen Vorgänge bei der Gerinnung und Thrombose mit besonderer Berücksichtigung der Entstehung der Blutplättchen," Ziegler's *Beiträge*, xv. 1894.

[2] "Über die Thrombocyten (Blutplättchen)," *Anatomisches Anzeiger*, xix. 1901.

[3] "On the Blood-Plates of the Human Blood, with Notes on the Erythrocytes of Necturus and Amphiuma," *Journal of Morphology*, Boston, xv. 1899.

only recognised methods in physiology are those of the chemist and physicist. Views which rely on purely histological evidence often have but a short-lived existence, and though it would be both invidious and profitless work to point out how frequently the inferences derived from histological observations have been shown to be misleading, the history of physiology and pathology abounds with such examples.

Descriptions of the platelets as they occur in mammalian blood are not in complete agreement. As to their size, aspect, peculiar green tint, and amount of granulation in a hyaline plasm, most observers are in accord. Among disputed points we may mention the presence or absence of hæmoglobin, the possession of a nucleus, their alteration in shape and power of locomotion ; and with reference to all these differences, some of which are of cardinal importance, there is absolutely no agreement. For example, Hayem affirms that the platelets contain hæmoglobin, a statement unconfirmed by Bizzozero. The nucleus described by Deetjen and Dekhuyzen, and shown in the illustration given by Kopsch,[1] is undescribed by Bizzozero, and, according to Kemp, is only a central collection of granules. Schneider[2] denies that the platelets are locomotive, a statement with which we entirely agree. The number of platelets in human blood is exceedingly variable ; the average number per cubic millimetre is given in the following table :

Date.	Blood.	Number per c.mm.	Author.
1882	Dog	200,000	Bizzozero
1884	Human	200,000—300,000	Afanassiew
1887	Human	253,000	Hayem
1887	Human	500,000	Prus
1887	Human	180,000—250,000	Fusari
1891	Human	200,000—250,000	Muir
1896	Human	180,000—256,000	Van Emden
1897	Man	635,300	Brodie and Russell
1901	Man	835,000	Kemp and Calhoun
1901	Woman	862,000	Kemp and Calhoun
1902	Man	457,000 (Paris)	Kemp
1902	Man	1,206,900 (72 hours hours later on the Görner Grat)	Kemp

In mammalian blood the platelets vary in size and aspect

[1] " Die Thrombocyten des Menschenblutes," *Anat. Anzeiger*, xix. 1901.
[2] " Beiträge dur Frage der Blutplättchengenese," *Virch. Archiv*, clxxiv. 1903.

according to the method which is employed for their demonstration. In human blood mixed with Hayem's fluid they are
for the most part biconvex granular bodies about 3μ to $3\cdot5\mu$ in
length. They may be isolated, or clumped together into groups.
Some remain stationary; others float freely in the slow currents
which are always in progress in any wet preparation. In our
opinion, they are entirely destitute of hæmoglobin, and persist
after the erythrocytes are carefully hæmolysed. Their behaviour
to stains also varies according to the nature of the fluid with
which blood has been mixed. With eosin they tint faintly; but
pyronin, iodine-green, dahlia, Spiller's purple, methylene-blue,
or polychrome methylene-blue (Goldhorn) stains these bodies
most effectively. The reaction of Mylius for alkali in blood
shows that the erythrocytes appear as unstained discs, while the
plasma acquires a faint, and the platelets an intense, pink colour.
The platelets also show some degree of iodophilia, and consequently have been described as containing glycogen. This
reaction, however, is, as Hüppert has shown, an inconclusive
test.

In 1883 Wooldridge[1] discovered that a molecular precipitate
settled from iced peptone plasma. This is now known to occur
when blood-plasma is mixed with oxalates, citrates, or fluorides,
or even cooled (Bürker). This morphological separation, beyond
any doubt, forms the bulk, if not all, of the bodies described as
platelets; and though this contention may not be universally
accepted, we may state that the admixture of blood with oxalates
is one accepted method for the demonstration of platelets
(Langley, W. Stirling, and Druebin). Another is the addition
of ·6 per cent. peptone (Afanassiew, Böhm and Davidoff), while
A. E. Wright points out the ease with which these bodies can
be obtained when blood is mixed with citrates; and Bürker[2]
simply allows a drop of blood to remain uncoagulated on a disc
of paraffin, when the summit of the drop is found to consist of
plasma with a quantity of platelets; but from this experiment
we are certainly not justified in assuming that blood is not
rapidly dying and altering in nature, even though coagulation
is delayed. Whether the accumulation of platelets in an excised
vein (Osler), or those which collect at the damaged wall of a

---

[1] "Über einen neuen Stoff des Blut-Plasmas," Du Bois Reymond's *Archiv f.
Physiologie*, 1884, and *Proc. of the Royal Soc.* p. 70, 1885.
[2] "Blutplättchen und Blutgerinnung," Pfluger's *Archiv f. Physiologie*, cii. 1904.

blood-vessel, and are the foundation for a white thrombus, are comparable to Wooldridge's bodies, must remain at present uncertain; but as far as aspect and staining properties are concerned, this question might well be answered in the affirmative. Comparatively few observers have attempted an answer to the question whether the "Blütplattchen" described by various histologists are or are not identical with the granular material which forms in plasma; but we may call to mind that Wooldridge considered this molecular precipitate was a mass of platelets, and believed these coalesced into rosette-shaped bodies, which, except in colour, closely resembled red blood-corpuscles.

The claim made by Bizzozero that platelets form definite morphological elements in living unmixed blood has met with considerable hostile criticism. M. Löwit and Wlassow in particular denied that they existed in normal blood. Both these observers maintained that shed blood protected against the contact of glass by the use of liquid paraffin or castor oil (Löwit) or a mixture of paraffin and vaseline (Wlassow) showed none, or exceeding few, platelets. According to Löwit these bodies are separations from the plasma, and in Wlassow's view this procedure hinders their formation from the erythrocytes. In our observations these experiments, when carried out exactly as they were described, yielded specimens free from any platelets, and we have noticed the same fact when human blood was examined as a thin stretched film on a loop of platinum wire in a moist chamber at 36° C. Wlassow believes that these bodies originate from erythrocytes, a view which assumes that many of the red corpuscles are somewhat fragile bodies; but this can scarcely be the case, for, as Karl Boden[1] has shown, the erythrocytes are the most resistant of all the morphological constituents of shed blood. He collected blood in sterile tubes, and found that the granular leucocytes, lymphocytes, and red corpuscles constantly disintegrated within a definite time—even after a hundred days many red corpuscles remained undamaged—and he considered that these were the most resistant to destruction of all the cells of the blood.

The evidence as to the pre-existence of the platelets in the normal living blood of mammals rests on comparatively few

[1] "Die morphologischen und tinctoriellen Veränderungen nekrobiotischer Blutzellen," *Virch. Archiv*, clxxiii. 1903.

observations. In shed blood Bizzozero[1] believes that it is easy
in one and the same preparation to distinguish " Blutplättchen "
from Wooldridge's bodies, for both these, he states, occur in
peptone blood. The former "are swollen and granular, somewhat pale, collect into heaps, and on treatment with water or
dilute acetic acid can be differentiated 'into two substances ;
a hyaline, rounded, swollen portion and a granular half-moon-
shaped body applied to the periphery of the hyaline part, while
Wooldridge's bodies are homogeneous, refract light strongly, are
arranged in small rosettes, and by the action of dilute acetic
acid become more apparent and glistening ; with water they
vanish from the specimen." In the capillaries of the living
mesentery of rodents spread out in warm ·6 per cent. sodium
chloride at 37° C., and also in an arteriole when the rapid flow
is checked by pressing a glass rod on the main vessel as it
leaves the abdomen, and therefore in blood which comes "direct
from the heart," Bizzozero states blood-plates can be seen.
Weigert's criticism of this latter experiment may be given in
his own words : " If it is beyond contention that under physiological conditions blood-platelets occur in circulating blood, so
is it probable that the requisite manipulations produce so many
vascular lesions that a disintegration of the white cells of the
blood is induced or augmented, and when these products within
the circulation occur, so may any arterial branch contain these
platelets if, as is probable, they are disintegration products of
the leucocytes." [2]   J. Arnold and others who have repeated these
experiments consider that the greater the care which is exercised
in preparing the mesentery the greater is the difficulty of
recognising any platelets within the vessels. To meet the
objection that injury to the mesenteric capillaries induces the
formation of platelets, the rhythmically pulsating vessels in
the patagium of two species of bats (*Vespertilio murinus* and
Plecorus auritus) were examined by Bizzozero and Laker,[3] by
both of whom platelets were recognised among the red
corpuscles. Löwit[4] states that this experiment yielded him a

---

[1] " Über die Blutplättchen," *Festschirft*, Rudol Virchow, Bd. i. 1891.

[2] "Die neuesten Arbeiten über Blutgerinnung," *Fortschulte de Medicin*, i.
1883.

[3] " Die Blutscheiben sind constante Formelemente des normal circulirenden
Säugerthierblutes," *Virch. Archiv*, 116, 1889.

[4] " Ueber die Präexistenz der Blutplättchen und die Zahl der weissen Blutkörperchen im normalem Blute des Menschen," *Virch. Archiv*, 117, 1889.

negative result, and when we consider that the skin of the bat is thickly pigmented, and two layers of this together with the vessels have to be observed with transmitted light, it is evident that the experiment is one most difficult of repetition. N. H. Alcock, who has carefully examined several bats for the purpose of verifying this particular statement, was quite unable to observe platelets within the vessels, even under the most favourable conditions of illumination, and with the use of the highest apo-chromatic objectives which could be used. The proof given by Bizzozero and Laker is, however, direct and positive, and without making any attempt to diminish the value of their work it is most desirable that this particular observation should be repeated. Any stretching of the patagium must be avoided, so that the vessels shall remain uninjured. The platelets are stated by Bizzozero to be seen immediately this structure is examined.

Into the further history of this somewhat acrimonious controversy it would be profitless to enter. Just as it is beyond any question possible to obtain films of mammalian blood which do not show a single platelet, so Löwit states that when the mesentery is spread out and bathed with castor oil at first " besides red and white corpuscles no other morphological elements are to be seen." As the observation is continued platelets may appear, but never to the amount noticed when physiological saline containing ·6 per cent. NaCl is employed instead of oil. Possibly in the light of our present knowledge the assumption that normal saline is in any sense an indifferent fluid must remain uncertain, even if the percentage of salt renders the fluid isotonic with that of blood plasma.

In shed blood platelets may occur but sparingly, indeed may be entirely absent. To explain this latter fact it is taught that these bodies largely disappear when blood leaves the vessels. To preserve them, enable them to be counted, stained, and examined, a variety of methods have been devised. These fixing, preservative fluids are often added in considerable quantity. For example, "a drop of blood allowed to fall into 5 c.c. of a mixture containing ·5 per cent. osmic acid and ·75 per cent. NaCl " is an accepted method for their demonstration. The fluids employed by Hayem, Marcano, Determann, Acquisto, Afanassiew, Brodie and Russell, the last of whom tried a large number of different media ranging from 33 per cent. caustic potash to mixtures of alcohol, glycerine, and water, cannot be regarded as otherwise than damaging to blood. The idea of all, and the contention of most of these observers is that pre-existing bodies which otherwise easily disintegrate are preserved, but this view that the platelets *rapidly* disappear is not in accord with the facts.

Any one who repeats the simple procedure suggested by Bürker, in which a drop of blood is allowed to fall on a disc of paraffin and any evaporation checked by placing the preparation in a moist chamber, can see that the blood remains, if not unaltered, at any rate unclotted for half an hour or longer, during which time the corpuscles settle and leave a cap of plasma loaded with platelets at the summit of the drop.

Moreover, it is easy to see that bodies which all observers would recognise as the blood-platelets figured by Hayem, Schiefferdecker, and Bizzozero actually form when, under the microscope, a drop of fresh human blood comes in contact with any one of the fluids which are considered to fix them. We are satisfied that a variable length of time, ranging from about five seconds to a minute according to the fluid used, elapses between the contact of the fluids and the appearance of the platelets at the contact edges, and the only inference which appears reasonable is that these bodies arise as the result of damage to the blood by the sublimate, iodine, formol, or peptone, which may be in the so-called fixing fluids. Moreover, the number is at first small, but subsequently increases as the fluids become more intimately mixed. This observation can be carried out in a variety of ways, and with 1 per cent. potassium oxalate the phenomenon is particularly well seen. The experiment uniformly fails with the blood of birds, fish, or frogs; for *Amphibia* this fact has been established by Löwit and Druebin.[1] Neither of these observers, nor Mosen,[2] could obtain any evidence of blood-platelets in the lymph.

That amphibian blood yields no platelets similar to those of mammalian blood, though Eisen considers that the plasma-cytes shed out of hæmic cells of *Necturus* are similar, may be explained by the great differences which exist between the proteid quotients of mammalian and amphibian blood-plasma;

[1] "Uber Blutplättchen des Säugerthieres und Blutkörperchen des Frosches," *Du Bois Archiv f. Physiologie*, Suppl. 1893.

[2] "Die Herstellung wägbarer Mengen von Blutplättchen," *Du Bois Archiv f. Physiologie*, 1893 (full literature).

for in the observations we have described there is no fragmentation of either the white or red corpuscles, and we are compelled
to consider the platelets as related to some change in the plasma.
The single or multiple thromboses which may be met with in
pathological conditions may not improbably be due to some
toxic material in the blood, which, acting somewhat in the
way described in the above experiments, may induce profound
changes in the plasma with a resulting accumulation of platelets
—an addition to those which may already exist.

If we assume that blood-platelets pre-existent in blood disappear so rapidly that their absence in a blood-film can be
explained away by such an hypothesis, certain experiments of
Bizzozero which are seldom quoted are opposed to such a view.
These observations were made to ascertain if it was possible
to entirely remove the blood-platelets from the circulation, and
then to see with what rapidity these were regenerated in the
blood. In the process of defibrinating blood Bizzozero distinguishes two periods; during the first "lagen sich an den
Schlagstäbschen eine dichte Blutplättchenschicht ab"; while
during the second, laminæ of fibrin are spread *upon* this aggluti-
nated granular mass of platelets. The blood-platelets, therefore,
do not rapidly disappear in shed blood, but can be removed
from blood *before* fibrin collects upon them. By bleeding dogs
from the carotid to the extent of almost half the total mass of
blood, robbing this of platelets and fibrin, returning the strained
and warmed fluid to the animal by the external jugular vein,
and repeating this procedure about nine times in two hours,
Bizzozero effected an almost entire disappearance of the platelets.[1] Five days later the normal number of 200,000 per c.mm.
was not only restored, but actually exceeded by 165,000. From
the fact that the animal, except for a transient hæmoglobinuria,
remained in health, it is a logical inference that the blood-
platelets, unlike the red and white corpuscles, are in no way
essential for existence.

It appears to us that this experiment, which is one of
undoubted interest, demonstrates in the first place that the
platelets possess considerable resistance to damage; secondly,
that the blood returned to the animal is robbed of much of its

---

[1] The blood was examined and the platelets counted after dilution with several
times its bulk of 14 per cent. MgSO,, or with a mixture of 1 part of 1 per cent.
osmic acid and 3 parts of 1 per cent. NaCl.

proteid; thirdly, that it is quite impossible to either affirm or deny that the platelets existed either in the living or shed blood; and lastly, that if the blood contains cells or "Blutplättchen," removable by whipping with twigs, the regeneration of these cells occurs with a rapidity to which there is no parallel among the other cells of the body. If they are pre-existent structures, no one has hitherto suggested any site for their formation other than the blood-stream itself. Since they do not contain hæmoglobin, all the leucocytes, if estimated at 10,000 per c.mm., would have to fragment, and under these circumstances the cytoplasm of each one would yield thirty platelets; further, if we conceive that this actually does occur, any specimen of blood would in consequence become free from white corpuscles. Other inferences can be drawn from these experiments of Bizzozero, and we would suggest that the blood-plasma becomes deficient in proteids, and especially in nucleo-proteid, inasmuch as, at the end of the experiment, the blood was found to have lost its power of spontaneously clotting. The amount removed by whipping did not reduce the proteid contents of the blood below 6·5 per cent., for this is the limit to which the proteids of the serum can fall (Roscher and E. Grawitz). It is certain that the blood-plasma regains its normal composition rapidly after hæmorrhage, or an intravenous injection of dextrose or proteoses, and this, possibly, is what occurred in the experiment we have described. It is very probable that plasma abnormally poor in proteids reacts like serum on admixture with fixing fluids containing osmic acid or 14 per cent. $MgSO_4$. Just as serum shows no platelets, so may a plasma which is poor in proteids. With the restoration of the blood-proteids, on the admixture of fixing fluids, the platelets will again be present in the same or slightly greater amount. Mechanical damage to the blood-plasma by whipping may therefore not only produce platelets, but so rob the blood of its proteids that the fluid in which the corpuscles float approximates in its composition to that of blood-serum. Under these circumstances the platelets cannot be demonstrated in a specimen of shed blood. As restoration occurs progressively, so will the number of platelets which can be demonstrated in shed blood gradually augment until by the fifth day the normal number will be found. Bizzozero's figures actually show that such a progressive increase does occur.

The term "Blutplättchenfrage" is well employed by German

writers, for the whole subject is beset with difficulties. Renewed interest in the subject originated with Deetjen's discovery in 1901 of a new method for studying the platelets; but while many observers agree that these exist in living blood, a great difference of opinion prevails as to the origin of these bodies both within and outside the body. Not only is there a difference of opinion as to the existence of platelets in living blood, but most irreconcilable views have been advanced to explain the origin of these puzzling bodies.

First View.—The hæmatoblasts of Hayem, or platelets of Bizzozero, or thrombocytes of Dekhuyzen, are independent bodies or cells which exist in normal living blood. They are of equal morphological importance to the red and white corpuscles. In mammalian blood they possess certain peculiarities which distinguish them from the nucleated, spindle-shaped, homologous bodies in amphibian blood.

Second View.—They are not pre-existent structures in normal blood, but are artefacts. They may, and frequently do, occur in disease, either as a consequence of disease or damage to the blood or the vessels which contain it. The blood-platelets are separations from the blood-plasma, being of the nature of a molecular deposit of one or more of the proteids of the plasma. From the fact that they possess a relatively large amount of phosphorus they may be composed of nucleo-proteid.

Third View.—The platelets are pre-existent bodies in circulating blood, and their number varies both in physiological and pathological conditions. They are in no sense equal in morphological importance to the red and white blood-corpuscles. They are either extrusions from or fragment of the leucocytes or erythrocytes—in the latter case they will possess hæmoglobin, and may contain an inner body or nucleoid, which is a remnant of the nucleus which the erythrocyte contained in an early stage of its development. The Blutplättchen, therefore, may be found in any one of four different types:

1. Platelets with hæmoglobin and an inner-body.
2. Platelets with hæmoglobin and no inner-body.
3. Platelets without either hæmoglobin or an inner-body.
4. Platelets without hæmoglobin and with an inner-body.

According to E. Schwalbe,[1] the author of this grouping of the

---

[1] *Virchow's Archiv*, clviii. 1899, clxviii. 1902 ; *Wiener med. Rundschau*, No. 10, 1903 ; *Ergebnisse der allg. Pathologie*, xiii. 1902 (full literature).

platelets, all the forms can be and actually are derived from the erythrocytes, a view which is largely founded upon the appearances of the detached fragments of red cells, which are very similar to the platelets described by Deetjen. Such fragments occur in quantities when dogs are treated with small doses of toluylenediamine.

Fourth View.—From a lengthened series of observations on human blood, together with a consideration of the literature, we incline to the view that the platelets cannot be regarded as genuine morphological constituents of the blood; but under pathological conditions large numbers are to be found in the circulation. They are destitute of hæmoglobin. Shed blood at first shows few or no platelets. The admixture of fixing fluids damages the blood, and produces platelets which had no pre-existence. These platelets vary in size and shape, and the centre may react to stains in such a way as to contrast with the peripheral parts of the body. Human leucocytes of normal blood are resistant cells, and the same is true of the erythrocytes. It is much easier to hæmolyse the latter than to fragment them, for in order to effect this, even pressure is inefficient, but reagents such as chromic acid or Wlassow's fluid can give rise to such fragments (Wlassow-Sacerdotti phenomenon).[1] Since no actual damage of the blood-cells can be observed to take place when, for example, Hayem's or Afanassiew's fluid or oxalates is added to blood, we are compelled to believe that the blood-plasma is the chief, if not the exclusive, source of the platelets. The essential cause of the appearance of these bodies is either damage to the plasma or removal of blood from the body. A temperature below that of the body is favourable to their appearance. The peculiar shapes assumed by the platelets, which are such that they appear to be genuine amœboid cells, cannot be regarded as a criterion of their independent character, for absolutely amorphous bodies can be observed to become transformed to Osler's platelets, or to those figured by Deetjen, Puchberger, and others.

A partial compromise between these various views is that the blood contains two kinds of platelets, which have been termed true and false; the former would correspond to the Blutplättchen, and, either as cells or cell-fragments, rank as actual constituents of the blood; while the latter would comprise the separations

[1] Sacerdotti, " Erythrocyten und Blutplättchen," *Anat. Anzeiger*, xvii. 1900.

from the plasma, which Wooldridge recognised, or the bodies described by J. Arnold[1] as artificially produced when blood is treated with 10 per cent. potassium iodide. The only possible evidence which can be adduced for the decision of this question, which we admit is of cardinal importance, is the aspect of the so-called true and false platelets, together with their behaviour to various stains. We believe that it is difficult, if not impossible, to make this distinction, and consider that since the platelets can be seen to actually form under the microscope in numbers sufficient to give a count of about 500,000 per c.mm., they arise from admixture of the fixing fluids with the blood-plasma. Human plasma cannot be collected free from platelets, for either the lowering of temperature or the substance which is added to prevent coagulation induces damage and consequent alterations in the fluid, and our own observations lead to the belief that plasma alters more rapidly outside the body than any living cell of the organism.

In 1897 H. Deetjen[2] described a new method for the study of blood-platelets. When human blood comes in contact with a sheet of agar which contains ·7 per cent. of NaCl, the preparation shows innumerable amoeboid platelets either separate or in groups of ten to twenty. In 1901 he published the composition of another medium, which consisted of:

Agar-agar	1-2 grammes.
Distilled water	100 c.c.
NaCl	6 grammes.

This is filtered, and 7 c.c. of a 10 per cent. solution of sodium metaphosphate and 5 c.c. of a 10 per cent. solution of potassium phosphate are added. The mixture must not be boiled after the addition of these solutions.[3]

A drop of blood is allowed to spread out on a thin sheet of the medium, and the preparation is covered. Under the microscope large numbers of nucleated flattened platelets, each of which may possess several fine processes, can be immediately seen; many of the bodies adhere to the cover-glass, and, in Deetjen's words, convey the impression of a miniature starry

[1] "Zur Morphologie und Biologie der voten Blutkörperchen," *Virch. Archiv*, cxlv. 1896.

[2] "Eine Methode zur Fixirung der Bewegungszuslände von Leukocyten und Blutplättchen," *Münch med. Wochenschrift*, s. 1192, 1897.

[3] "Untersuchungen über die Blutplättchen," *Virch. Archiv*, clxiv. 1901.

heaven. He describes them as nucleated amœboid bodies, which protrude processes and are actually locomotive, though this property is impeded by their adhesion to the cover-glass. The normal shape is a small sphere, which flattens out under the cover-glass. Deetjen's platelets can be fixed with osmic acid vapour, and will then be found to stain well with basic dyes. For the further demonstration of their structure he employed eosin and hæmatoxylin.

All these statements can be and have been abundantly verified. Puchberger,[1] who used a weak solution of brilliant kresyl-blue, which is allowed to dry on a slide, has examined the platelets in various diseases. A drop of blood is simply allowed to spread out on the stained surface. He states that Deetjen's plates are admirably stained by this procedure, and finds that they are completely absent in pernicious anæmia, in purpura hæmorrhagica and certain other cases of purpura. His figures also show that in myelogenous leukæmia the platelets may be so hypertrophied and swollen as to nearly equal the size of an erythrocyte.

When repeating Deetjen's work we have noticed that should the agar become altered in its salt-content by any bacterial growth, the demonstration of the platelets is greatly impaired, or may even be impossible. With the medium when used without the agar, it is equally easy to observe the platelets ; and these can be seen to actually form when a drop of this fluid and one of mammalian blood (cat, dog, or rabbit) is allowed to come in contact under a cover-glass. The experiment, as might have been anticipated, entirely fails if serum, defibrinated blood, or oxalated plasma free from platelets is used instead of blood. The experiment is negative with the blood of frogs or toads, and no bodies which in any way resemble those described by Deetjen can be seen.

Since Deetjen's observations were made many observers have contributed to the " Blutplättchenfrage," and there is a tendency to return to the older views of Engel, Arnold, and Wlassow, and attribute the origin of these bodies to a process of fragmentation of the red corpuscles. Outside the body the red discs, as Hayem observed many years ago, often execute spurious movements. Nothing similar to these has been

[1] " Bemerkungen zur vitalen Färburg der Blutplättchen des Menschen mit Brilliantkresylblau," *Virch. Archiv*, clxxi. 1903.

described in circulating blood, though the figures of E. Schwalbe show that bodies which closely resemble those described by Deetjen may be detached from the erythrocytes under the influence of certain toxic substances. The purely histological observations of Preisich and Heim,[1] which rest on the somewhat insecure basis of a special form of staining, lead them to believe that platelets which stain somewhat like the granules of the polymorphonuclear leucocytes can be seen either separately or in groups of two or three within the red corpuscles. The platelets may even be the actual extruded nucleus or nucleoid-material of the red corpuscles. A somewhat similar view is taken by Schneider,[2] who considers that while a few arise from leucocytes, the chief number originate from the red discs. On grounds of micro-chemical tests, Petrone[3] holds that the platelets are bodies quite different from either the red or white corpuscles. According to Heim, the assertion of Mondino and Scala[4] that mitosis occurs in the platelets remains unverified. The experiments of Marino,[5] which were carried out in Mechnikoff's laboratory, possess some interest. He has observed that when rabbit's blood is received into absolute alcohol, platelets are entirely absent, but if the experiment is repeated with alcohol containing varying percentages of water they appear. Within limits, the lower the content of alcohol in the mixing fluid the larger the number of platelets.

When blood-films are prepared by any of the accepted methods and stained in such a way that platelets, if present, can be recognised, a specimen may show not a single one of these bodies ; another may show a few, but normal human blood in our observations never contains anything like the proportion of one platelet to ten erythrocytes. In the blood of cases of chlorosis an abnormally large number is generally present, while in pernicious anæmia they appear to be entirely absent. In some cases of post-hæmorrhagic anæmia a blood-film gives the impression that they are more frequent. Until our knowledge becomes more definite and exact, it appears that the figures

[1] " Über die Abstammung der Blutplättchen," *Virch. Archiv*, clxxviii. 1904.
[2] " Beitrag zur Frage der Blutplättchengenese. Eine erweitere Nachprufung der Versuche Sacerdottis," *Virch. Archiv*, clxxiv. 1903.
[3] " Sur le Sang," *Arch. ital. de Biologie*, xxxvi. 1901.
[4] *Arch. ital. de Biologie*, xii. 1889.
[5] " Recherches sur les plaquettes du Sang," *Comptes rendus, Soc. de Biologie*, No. 4, 1905.

which have been published as to the numbers of the platelets
in disease must be received with even greater caution than
that which we are accustomed to give to the enumeration of
white or red corpuscles. From experimental evidence it is
beyond question that under pathological conditions blood-
platelets may exist in circulating blood ; some of these bodies,
such as those which are seen in cases of poisoning by phenyl-
hydrazine or toluylenediamine, are probably derivates from
the red corpuscles since they possess hæmoglobin. Even
in shed blood similar bodies are produced, for the process of
separation can be witnessed. But no observations on shed blood
can decide this question whether " Blutplättchen," whatever their
nature or origin, are normal constituents of the blood. While
not denying the value of the evidence which has been adduced in
favour of their independent existence, we are unable to wholly
accept the view that they are cells of equal morphological
importance to the red and white corpuscles. We can find no
information which gives even a clue to their actual number
in circulating blood ; and since we are familiar with the exceedingly rapid changes which blood-plasma exhibits when it leaves
the vessels, and also realise that no cells which we can examine,
not even nerve cells nor those of muscle, undergo quite so rapid
a change, we are justified in regarding the plasma as peculiarly
susceptible of damage both when within and outside the body.
And when damaged in a variety of ways, the change, whatever
may be the chemical or physical explanation of the process,
becomes one which can be actually observed. From all our
observations on the behaviour of human blood when treated
outside the body with the various fixing or so-called indifferent
fluids, we can draw no other conclusion but that the vast
majority, if not all, of the bodies which have been regarded
as platelets are, as far as histological evidence is of any
value, probably of the nature of plasma-separations. It is
admittedly difficult to prove a negative, but on the question as
to whether the platelets are genuine constituents of normal
blood we must for the present be content to suspend our judgment. The evidence as to pre-existent hæmic cells other than
those of the leucocytes and erythrocytes is, at the present time,
inconclusive.

SOME RECENT PROGRESS IN CHEMICAL AND STRUCTURAL CRYSTALLOGRAPHY

BY A. E. H. TUTTON, M.A., D.Sc. (OXON.), F.R.S.

IN no branch of science has the refining influence of sceptical criticism been more active than in crystallography. Time after time in the history of the subject has a theory, founded on a keen estimation of the probabilities rather than on rigid and then unattainable proofs, been generally rejected, and only now are we beginning to obtain evidence of a truly unassailable character which places the oldest of these theories on a firm basis of fact. In the year 1669, Nicolaus Steno observed that the angles between the corresponding faces of different specimens of rock crystal were equal, and within ten years subsequently Guglielmini established to his own satisfaction that the constancy of the interfacial angles of the crystals of the same substance was a general law of nature. Yet for a hundred years afterwards the idea was rejected as absurd, the great variation in the relative development of faces, which causes such remarkable difference in the exterior appearance of crystals, causing its rejection even by such great minds as those of Boyle (1690) and Werner (1770), in the absence of adequately precise means of measurement. The invention of the contact goniometer, about the year 1780, by Carangeot, assistant to Romé de l'Isle, at length, however, compelled its acceptance as a fact, within the limits of accuracy of that instrument.

In another couple of years the important work of the Abbé Haüy, Professor of the Humanities at the University of Paris, was made known to the French Academy, and published in 1784 in a volume entitled *Essai d'une Théorie sur la Structure des Crystaux*. Haüy may truly be called the father of crystallography, for he showed that all the varieties of crystalline forms could be referred to a few simple types of symmetry; he enunciated the laws of symmetry, proved that all the apparently different forms of the same substance are based on one of these simple fundamental forms, and demonstrated that difference in chemical composition is accompanied by difference of crystalline form. Moreover, he discovered the all-important law of rational indices— that is, the fact that the lengths cut off along the three principal axes of the crystal by the various faces, other than the primary ones parallel to the axes, are relatively expressed by very simple whole numbers, usually only 1, 2, 3, or 4. Haüy further elucidated the simple relation between crystalline form and cleavage, which had been discovered in 1780 by Bergmann and Gahn, and laid the foundation of the idea of the molecular structure of crystals at a time when chemistry was in its infancy.

Many of Haüy's conclusions, however, were exposed for years to grave doubt, and even open scepticism, owing to the remarkable discoveries of Mitscherlich, and their but partial comprehension. In the year 1819, Mitscherlich, a young student of Berlin, pursuing his first research, made the accidental discovery that two different substances, acid ammonium phosphate and acid ammonium arsenate, possessed apparently the same crystalline form. To be on safe ground, he studied crystallography under Gustav Rose, and was subsequently able to carry out the necessary crystal measurements. He then found that two further substances, the two corresponding potassium salts of phosphoric and arsenic acids, were likewise apparently identical in crystalline form. Later, under the guidance of Berzelius at Stockholm, he illustrated his new principle of "isomorphism" by many further examples. Moreover, in 1821, he demonstrated that the same substance, sodium dihydrogen phosphate, crystallises in two distinct forms; and in 1823 followed this up by discovering the much better known case of sulphur, which crystallises from fusion in monoclinic prisms, while the natural form and that deposited from solution is rhombic.

These striking results appeared to demolish at once the theory that any one substance of definite chemical composition is characterised by a specific crystalline form. During the period which followed, new isomorphous series of salts and dimorphous, or even polymorphous, substances were discovered, one after the other, at a rate which was only commensurable with the rapid advance of chemistry. The use of the reflecting goniometer, which had been invented in 1809 by Wollaston, who had thereby placed an infinitely more accurate measuring instrument at the disposal of crystallographers, soon revealed the fact that the members of isomorphous series exhibited slight differences in their angles between analogous faces, which could scarcely be accounted errors of measurement or as being due to slight distortion. Still, no law or regularity was observed among these small differences, and the measurements by different observers had been so various that results of the most conflicting kinds were adduced. Indeed, so late as the year 1891, there was insufficient evidence on which to base an irrefragable decision as to whether the interfacial angles of the crystals of any one substance are a specific property of that substance or not. As generally happens, however, in the case of careful, conscientious work, both Haüy and Mitscherlich were substantially right, and it is one of the happiest results of the past fifteen years' work on the part of the writer of this article that the views of these two master-minds are now reconciled and harmonised.

The accumulations of crystallographic literature consequent on the swift progress of chemistry presented to the student a most perplexing mass of detached data, much of it of merely approximate accuracy, and in which neither law nor order was apparent. The publication by Newlands in 1863 of his law of octaves, however, prepared the way for the development by Mendeléeff in 1869 of the periodic law of the elements, a generalisation which, in the form elaborated by its author in the Faraday Lecture of 1889, is of the most far-reaching importance. That its value is appreciated in this country has recently (November 1905) been emphasised by the conferment on the illustrious Russian chemist of the Copley Medal by the Royal Society. It is not too much to say that this law reduces the hitherto chaotic mass of chemical knowledge to order. It may be remarked, however, that its usual form of enunciation, that "the chemical and physical properties of the elements are periodic functions of their atomic weights," conveys perhaps the idea that there is some particular virtue in the property of atomic weight, whereas it is only used as a very convenient property by which to identify an element; for the point of the generalisation is that all the properties of the elements hang together, and that the periodic progressive arrangement is identical for all—namely, that of an octave of elements in a horizontal row, the series of such octaves or periods when arranged in orderly succession below each other thus producing eight vertical columns. Now it is the members of these vertical columns, the family groups of very similar elements, which give rise to the isomorphous series of Mitscherlich, for they are capable of replacing each other in their compounds without altering the symmetry of the crystals.

It was at this stage, in the year 1891, that the writer commenced an organised attempt to bring some order into the domain of chemical crystallography, by an attack on certain series of isomorphous salts of definitely ascertained chemical relationship. The results of five other crystallographical investigations had previously been published, the last of which, concerning the effect of the addition of the CH_3 group in an organic homologous series, suggested the mode of tackling the greater problems to be solved. The four objects placed in the forefront were: (1) to obtain definite information as to the nature of the relationship between chemical composition and crystalline form; (2) to settle the question whether or not each chemical substance has its own specific crystalline form or (if polymorphous) forms; (3) if the answer to (2) is in the affirmative, and the members of isomorphous series produce crystals which, while similar, are not identical, to ascertain the law which governs their differences; (4) to extend the investigation to the whole of the physical properties of the crystals, in order to discover the relationships of the various salts with respect to those properties, and to determine whether or not they are in line with the structural relationships.

It was decided that the series which offered the best chance of obtaining definite results were those containing the three alkali metals—potassium, rubidium, and cæsium—which not only belong to the first of the vertical groups of family elements in the periodic table, but to the same section (even series) of the group. They are thus related in the most intimate manner possible, and are, moreover, the most electro-positive of the metals. Their atomic weights are separated by considerable intervals, and the atomic weight of rubidium (84·9) is almost exactly the mean between the atomic weights of potassium (38·85) and cæsium (131·9). It was to be expected, therefore, that the crystallographical differences between the members of isomorphous series would be here found at a maximum. It was also decided to include the analogous salts of ammonium in the investigation, as the radicle NH_4 was known to be capable of replacing the alkali metal without apparently altering the crystallographical form, and the result has proved highly interesting. The particular salts chosen were the rhombic normal sulphates and selenates, R_2SO_4 and R_2SeO_4, in which R may be K, Rb, Cs, or NH_4; and the monoclinic double sulphates and selenates which the salts just mentioned form with the sulphates and selenates of magnesium, zinc, iron, nickel, cobalt, manganese, copper, and cadmium, and whose crystals contain also six molecules of water of crystallisation, the molecular generic formula being $R_2M(\genfrac{}{}{0pt}{}{S}{Se}O_4)_2 \cdot 6H_2O$. The inclusion of these double salts has proved a very happy one, for the alkali metal has been found to exert such a preponderating influence in determining the characters of the crystals, that the salts furnish numerous independent confirmations of the crystallographical deformation accompanying the displacement of any one alkali metal by any other or by ammonium.

A very large number of crops of the crystals of each of the forty salts so far investigated have been prepared in order to study every possible variety of habit. They were all of the highest attainable chemical purity, and were grown under conditions involving quite unusual precautions against mechanical or thermal disturbance, in order to avoid distortion. Special care has also been taken from the beginning to exclude crystals showing any traces of the vicinal-face phenomenon recently worked out by Prof. Miers, and which has long been observed by the writer. At least ten perfect specimens were most carefully selected from different crops, and their interfacial angles measured on a goniometer of the highest accuracy, over 20,000 different angles having been measured in all. The specific gravity of the crystals of each salt has been determined both by the pyknometer method and by the liquid immersion method of Retgers, in order to ascertain the volume relationships, internal structural dimensions, and molecular optical constants. The position and dimensions of the optical ellipsoid of the crystals of every salt have been ascertained by stauroscopic measurements, determinations of the refractive index for six wave-lengths of light, and measurements of the optic axial angles for similar spectrum intervals. The optical determinations have, moreover, been repeated for temperatures higher than the ordinary, in order that the results may be independent of temperature; and measurements of coefficients of thermal expansion have also been carried out for the three principal directions within the crystals. For the optical and thermal work no less than 600 section plates, parallel-faced blocks, and 60° prisms have been cut or ground out of the crystals, accurately orientated to 2' of arc as regards the theoretically desirable positions within the crystals.

Improvements in Experimental Methods

The meagreness and inaccuracy of previous optical work on the crystals of artificial chemical preparations was largely due to the limitations imposed by the only available method of preparing plates and prisms—namely, by holding the small crystal between the finger and thumb and grinding the required surfaces on a plate of ground glass. It was necessary, therefore, to devise an instrument for the absolutely accurate preparation of such surfaces before any real progress could be made. After having had several constructed, the final form of the cutting and grinding goniometer is shown in fig. 1. It has proved all that can be desired, after several years' hard work with it, and without it the investigation would have been impossible.

The crystal is attached to an adjustable holder by means of hard optician's wax, or is held in a grip holder, and then suspended from a delicate apparatus, which serves not only for the adjustment of a zone of the crystal's faces to the vertical axis of the instrument, as determined by observation through the telescope of the reflected images of the collimator signal slit, but also, as the movements are graduated, for its setting to any position with respect to the axis. Separate and interchangeable cutting and grinding gear is provided, and also a delicate means of varying the pressure of the crystal on the grinding disc, so that the most fragile crystals can be manipulated without danger of fracture.

Another original piece of apparatus devised for these researches is a spectroscopic monochromatic illuminator, which is represented in fig. 2. This has not only enabled the optical observations to be made for six different spectrum wave-lengths of light in as short a time as was formerly occupied in taking two series with coloured flames, but it has also enabled determinations to be made of the exact wave-lengths for which many interesting special phenomena occur, in which this research has been rich. It is essentially a compact spectroscope, con-

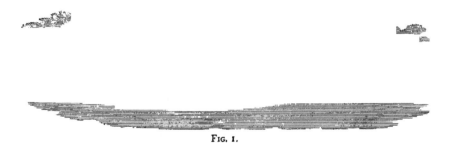

Fig. 1.

structed to transmit as much as possible of the light which streams from the condenser of an electric lantern, and with the observing eyepiece replaced by a second slit, although the eyepiece can be readily attached for calibration purposes. A

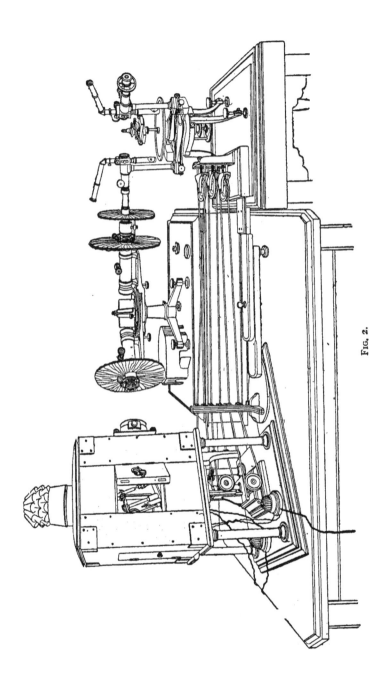

FIG. 2.

broad spectrum is produced by a very large and exceptionally highly dispersive, although perfectly colourless, prism, and is focussed on the back of the second slit, which permits only a selected line of the spectrum to escape, in the same manner as in Sir William Abney's well-known apparatus. By rotation of the prism on a delicately divided circle, which is calibrated for wave-lengths, the spectrum is moved over the exit slit, so as to cause any desired colour to escape, whose wave-length is given by the circle. The line of issuing light is suitably diffused by means of a screen of very finely ground glass, carried as an attachment in front of the exit slit. In the illustration the goniometer-spectrometer is shown in position before this screen, as when refractive indices are being determined.

For the determination of the thermal expansions the interference dilatometer shown in fig. 3 was devised. It is as equally applicable to the measurement of the thermal expansion of any small bodies as to that of crystals, and depends on Fizeau's principle of measuring the displacement of interference bands. The optical apparatus, the interferometer, is indeed common both to the dilatometer and to the elasmometer, the writer's elasticity apparatus. The light from a hydrogen Geissler tube is made to traverse an auto-collimation telescope, which directs it horizontally to a train of prisms carried at the head of a separately supported vertical tube largely constructed of porcelain. The red C-light is here selected and directed down on to two surfaces designed to reflect it in such a manner as to produce interference bands. The latter are visible in the eyepiece of the telescope owing to the reflected light being caused to retrace its steps with just sufficient lateral displacement to enable it to pass through the open half of the auto-collimation aperture, instead of striking the little totally reflecting prism whence it originally proceeded which closes the other half. The crystal block is laid on a small table of platinum-iridium, as shown in fig. 4, and its upper surface, if polishable, forms the lower of the two reflecting surfaces; if unpolishable, a small disc of aluminium is laid over it, whose upper surface furnishes the desired reflection with exactly the right intensity. The upper reflecting surface is the lower surface of a glass plate (slightly wedge-shaped), which rests on a tripod of three platinum-iridium screws

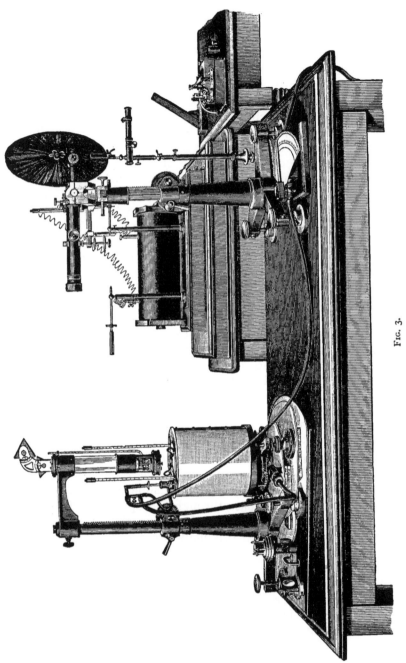

FIG. 3.

passing through the little table of the same metal. These are so arranged that the two reflecting surfaces of glass and either crystal or aluminium are only separated by a thin film of air, and have the necessary minute tilt out of strict parallelism to produce interference bands of the desired width. The lower half of the porcelain tube, together with the chamber suspended below it containing the interference tripod, is immersed in a double air bath which can be heated to the required temperature, during which operation the movement of the bands is observed. For every band which passes the reference mark in the centre of the field the air film at that spot has become altered in thickness, owing to the differential expansion of screws and crystal (and aluminium, if used), by an amount which is equal

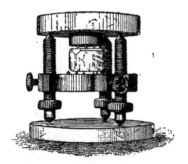

FIG. 4.

to half the wave-length of the red hydrogen light employed. Hence the half wave-length of light is the grosser unit of the scale, and as the one-hundredth of the distance between two bands can be measured with accuracy by the micrometer of the eyepiece, the measurements can be carried to the three-hundred-thousandth of a millimetre, or the eight-millionth part of an inch. With this instrument the thermal expansions of the sulphates of the alkalis have been carried out.

The elasmometer, which has since been devised by the writer for the determination of the elastic constants, is shown in fig. 5, without the interferometer. It is designed to measure the amount of bending suffered by a thin plate of the substance investigated, when pressed up into contact, near its ends, against a pair of platinum-iridium knife-edges, by a known weight

applied under its centre.   It consists of an elaborate apparatus for
the support and adjustment of the plate and knife-edges; a
measuring microscope reading in two rectangular directions by

FIG. 5.

a new method to the thousandth of a millimetre, for measuring
the dimensions of the plate *in situ*; a specially constructed form
of balance, one end of the beam of which carries an agate

point, through which a pressure is applied under the centre of the plate equal to the weight suspended from the other end; and a delicate control apparatus, shown in fig. 6, which only permits the weight to operate with extreme slowness, so that the bands may be readily counted as they pass the reference mark. The elasmometer also carries its own inter-ference tripod in rigid connection with the knife-edges, for the support of the glass disc, and the second reflecting surface is formed by a black glass disc at the top of a vertically sliding

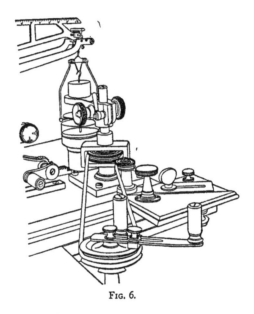

Fig. 6.

aluminium rod, whose pointed lower end rests upon the centre of the substance plate, and consequently moves with the latter. These parts are more clearly shown in fig. 7. This instrument has been fully tested in a research just concluded on the elastic constants of pyrites, and proves as fully satisfactory as the dilatometer, so that it may be employed with confidence in future work on the sulphates.

The Results of the Investigations

As regards the external morphology of the crystals it has been conclusively demonstrated that each salt possesses its own specific interfacial angles, constant for that salt to within 2′ of arc, and which constitute a definite property by which the salt can be identified. The amount of the difference introduced in these angles by replacing one element of the alkali group by another varies in the two series, being smaller in the rhombic simple sulphates and selenates than in the monoclinic double salt series. The maximum difference in the former series, between a potassium and a cæsium salt, is only 41′, whilst in the latter series of lower symmetry it is as much as 2° 21′. These maxima, however, are but rarely attained, and the average change is only a quarter of a degree in the rhombic series, and does not exceed a degree in the monoclinic series.

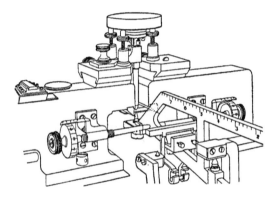

FIG. 7.

With respect to the relation between the amount of the change of angle and the atomic weights of the metals inter-changed, three very interesting facts have been revealed :— (1) The magnitude of the angle between any pair of faces of any rubidium salt is intermediate between the values of the corresponding angles on the analogous potassium and cæsium salts. (2) In both series the average of all the angular changes, and likewise the maximum amount of change, whenever potas-sium is replaced by cæsium is almost exactly twice as much as when potassium is replaced by rubidium ; in other words, both the average and maximum change of angle are directly pro-portional to the change of atomic weight. (3) In the monoclinic series, where the three axes of the crystals are not all fixed at right angles to each other as in the rhombic series, but in which one axis is otherwise inclined, the amount of this inclination changes when one alkali metal is replaced by another to an extent which is directly proportional to the change in atomic weight.

The result of the investigation of the exterior form of the crystals is thus to reinstate permanently, as a law of nature, Haüy's view that difference of chemical composition is accompanied by difference of crystalline form. At the same time, Mitscherlich is shown to be approximately correct, for his isomorphous series are so nearly alike in external morphology that it has required the most careful research to discover and establish the fact that there are real differences, other than fortuitous ones due to malformation. The three above-mentioned quantitative laws, which have been shown to govern the differences between perfect crystals of the various salts, form the harmonising bond of reconciliation between the work of Haüy and Mitscherlich.

 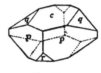

FIG. 8.   FIG. 9.   FIG. 10.

A further fact has been observed with regard to the relative development of the various faces of the crystals, which determines what is known as the "habit." It is most marked in the case of the monoclinic double salts, and figs. 8, 9 and 10, representing respectively typical crystals of potassium, rubidium, and cæsium salts, will at once make it clear. The faces of the principal forms, c, q, and p, show a definite progression, any one of them on the rubidium salt being developed to an intermediate extent compared with its development on the potassium and cæsium salts.

The question of the internal structure of the crystals must now be introduced. Crystallographers had been somewhat thrown off the scent for a time by the idea, prevalent among physicists, that in solid substances we were dealing with edifices composed of units which were aggregates of chemical molecules, and attempts had been made from time to time to determine the number of chemical molecules contained in the physical one.

No definite success, however, attended these efforts; and in a memoir on *The Nature of the Structural Unit*, in 1896, the writer proved that for the series in question the structural unit was simply the chemical molecule, thus clearing the ground of all further complication.

The structural units (chemical molecules) of the various members of isomorphous series are obviously built up into crystals according to the same plan, and the "elements" of the crystals—that is, the relative lengths of the crystallographical axes a, b, and c, cut off by the primary faces, and the three mutual inclinations of the axes a, β, and γ—are relative measures of the sides and also the angles of the elementary parallelopipedon of the structure. The axial ratio $a : b : c$ is usually expressed so that $b = 1$. Hence it only affords the mutual relation of the three axes to each other in that particular substance, and gives no idea of the relations between the dimensions of the elementary parallelopipeda of different members of the series. But if we take the relative volumes of the molecules into consideration, by combining the molecular volume (molecular weight divided by density) of each salt with the crystallographical axial ratios and axial angles, we at once obtain the relative dimensions of the elementary parallelopipeda of the various members of the isomorphous series. These new axial dimensions expressing the true relations in space are termed the topic parameters or axes, and are represented by χ, ψ, ω. They were employed simultaneously by Muthmann in Germany and by the writer, in the year 1894; and the latter was indebted for the idea to Prof. Becke of Prague. They are expressed by simple mathematical formulæ involving only the crystallographical axial values a, b, c, the axial angles a, β, γ, the density of the crystals, and the molecular weight of the salt.

If the whole space of the elementary parallelopipedon were filled with the matter of the molecule, the topic parameters of the members of isomorphous series would indicate the relative sizes of the chemical molecules. But the writer has shown that the space is by no means filled with matter; hence they do not represent the sizes of the molecules, but the relative distances apart of the centres of gravity of contiguous molecules along the directions of the three crystallographic axes. In other words, a point within the molecule, its centre of gravity, is taken to represent it, and the distances apart of these points along the three principal directions within the crystal are represented relatively by the topic parameters. We thus arrive, on purely experimental grounds, at a system of points as representing the crystal structure, and are quite independent of the as yet unknown shape and size of the molecules. This brings us in touch with a remarkable development of the geometrical theory of crystal structure, derived from a study of the possible modes of partitioning space, which demands a brief explanation.

The fundamental crystalline elements of Haüy were supposed to be arranged in a parallel or net-like manner, and in 1833 Frankenheim investigated the possible varieties of such space lattices, and concluded that there were fifteen possible, corresponding with fifteen types of cleavage and of crystal symmetry. Bravais in 1850 reduced them to fourteen, by demonstrating the identity of two of the fifteen, and also showed that the fourteen fell by pairs into only seven truly different systems of symmetry. These seven types are, however, entirely holohedral, and leave hemihedral and tetartohedral crystals unexplained. Camille Jordan, in a memoir in 1869 on "groups of movements," most materially assisted the solution of the problem, from the purely mathematical standpoint, by defining possible types of regular repetition in space of identical parts. He showed that when such identical repetition of parts is exhibited by a rigid system, a definite series or group of correlated movements may be employed, each term being such a movement as, while shifting the system, leaves the appearance the same as before, every point being moved to a position previously occupied by a homologous point.

This principle was applied in 1879 to crystal structures by Sohncke. He showed that it was unnecessary to assume with Bravais that the molecules are all arranged parallel to each other in lattices, but merely that the arrangement about every molecule, as represented by its "point," is the same as about every other. He thus arrived at sixty-five "regular point systems," of which the space lattices of Bravais are special cases, and which include nearly, but not quite, all the thirty-two known types of crystal symmetry. The next and final step is a very remarkable one, for no less than three independent workers, Fedorow in Russia, Schönflies in Germany, and

Barlow in this country, in the years 1890 to 1894, showed that, by the introduction of another factor, the principle of mirror-image symmetry (enantiomorphism), which is characteristic of the crystalline forms not explained by Sohncke's sixty-five point systems, all the thirty-two forms of crystalline symmetry are explained. The geometrical methods adopted are different in the three cases, yet all arrive at the same result—namely, that there are 230 types of structure (groups of movements) possible, all of which fall into one or other of the thirty-two classes of crystal symmetry—which completely explains the whole of those classes. Whether it is the molecules or the atoms composing them which must be considered as the parts forming the crystal structure, they must be arranged according to one of the 230 types of symmetry, of which both Bravais space lattices and Sohncke point systems are special cases.

Sohncke has further developed the theory by considering the atoms instead of the centres of gravity of the molecules, and derived a general theory of crystal structure which was stated by Prof. v. Groth, in his memorable address to Section B of the British Association at the Cambridge meeting in 1904, in the following terms : "A crystal—considered as indefinitely extended—consists of n interpenetrating regular point systems, each of which is formed from similar atoms; each of these point systems is built up from n interpenetrating space lattices, each of the latter being formed from similar atoms occupying parallel positions. All the space lattices of the combined system are geometrically identical, or are characterised by the same elementary parallelopipedon." The combined system conforms to one or other of the 230 possible homogeneous arrangements in space, and is expressed outwardly in the symmetry of one of the thirty-two classes of crystals, while the space lattice determines the crystal system, the angles, and the concordance with the law of rational indices.

The completion of the geometrical theory renders it unnecessary to assume the existence of any molecular forces to keep the structure together, beyond the interatomic forces. This is entirely in accordance with the conclusion of the writer, from direct experiment, in the 1896 memoir on the structural unit, that there is no chemical union between the molecular constituents of the double salts, and that there is merely aggregation in accordance with such a particular type of homogeneous structure as ensures that they are always present in the same proportion. Moreover, the explanation of poly-morphism is at once clear, for the equilibrium of the inter-penetrating atomic point systems will be affected by temperature, and a combined system, possible and stable at one temperature, may become unstable at another considerably higher or lower, when, if other conditions are suitable, another form, possibly belonging to an altogether different system of symmetry, may be produced.

Having now shown that, from both the experimental and the geometrical standpoint, a crystal may be structurally regarded as an organised assemblage of points, which may be formed either by the centres of gravity of the chemical molecules or by the atoms of which the molecule is composed, and that the topic parameters express, in the case of an isomorphous series, the relative separations of these points along the axial direc-tions, we are in a position to discuss the results of the writer's experimental work as regards structure.

In both series of salts, when potassium is replaced by rubidium or cæsium, a deformation of the crystal structure occurs in the form of an extension of all the topic parameters— that is, an increase occurs in the separation of the centres of gravity of the molecules along each of the directions of the crystallographical or topic (not always identical) axes. The increase augments as the atomic weight rises, so that the intermediate position for any rubidium salt is always closer to the potassium salt position than to that for the cæsium salt— that is to say, the function of the atomic weight is of an order higher than the first. Moreover, a similar increase of the distances separating the points (molecular centres) is observed when sulphur is replaced by selenium.

It has been shown by Fedorow to be very important that the correct type of homogeneous structure should be diagnosed before the proper topic parameters can be arrived at. Now the rhombic sulphates and selenates exhibit a marked pseudo-hexagonal character, the primary prism angle being within half a degree of 60°. Hence, while ω represents as usual the vertical topic parameter, and ψ that along one of the horizontal axes of the rhombic system, χ, instead of representing the other horizontal axis at right angles, represents the parameters along a pair of horizontal axes inclined at nearly 60° to each other and to ψ. The above rules governing the relations of the members of the series are, if anything, more elegantly exhibited on this assumption than if the parameters are merely calculated for the directions of the three rectangular axes of the rhombic system, but the facts shown on both suppositions are identical. The monoclinic double salts do not exhibit any pseudo-hexagonal tendencies, and the directions of the topic parameters are here those of the three crystallographic axes.

As regards the very interesting question of the internal structure of the molecule, and of the orientation of the individual points representing the various atoms, which are all averaged into the single point representing the molecule, there is some evidence already available. In the case of the simple sulphates and selenates, when one alkali metal is replaced by another of higher atomic weight, the separation of the molecular points increases to a far greater extent along the vertical axis than along the three pseudo-hexagonal horizontal axes, the smaller increase along the ψ direction being very nearly the same as that along the pair of axes χ. On the other hand, when sulphur is replaced by selenium, while there is also an increase of separation in every direction, and the amounts along the three lateral directions are again nearly equal, that along the vertical axis is less than the latter. Thus, when the metal is changed there is a preponderating change of dimensions of the molecule in the direction of the vertical axis, and when the dominant negative element, sulphur, is replaced by its family analogue, selenium, the change is chiefly along the lateral axes. This would appear to suggest that the two atoms of the alkali metal, or their spheres of influence, are situated along the vertical axis, while the sulphur atom or its sphere of influence occupies the centre. The graphic formula,

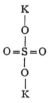

although only representing the molecule in one plane, will at once show how likely this is. The actual arrangement in space is probably that of a structure consisting of three regular point systems, with similar "coincidence movements" (Deckschiebungen), the first being of sulphur or selenium atoms, the second of four times as many oxygen atoms, and the third of twice as many alkali metallic atoms—the three systems interpenetrating in such a manner as to produce equilibrium, and being so mutually arranged that the sulphur or selenium system lies centrally with respect to the other two, and the metallic system in such wise as would affect chiefly the vertical direction. Barlow's type $57a_1$ would appear to fulfil these conditions, and such a supposition is in accordance with a friendly communication recently made to the writer by Mr. Barlow, who hopes shortly to publish his views on the geometrical structure of the alkali sulphates.

That these facts, and others which have been adduced by Prof. v. Groth—such as the case of ammonium iodide NH_4I, in which the replacement of the hydrogen atoms by CH_3 radicles causes equal increases along two parameters, but no increase along the third—are pregnant with meaning as to the positions of the atoms or their spheres of influence in the molecule, admits of no doubt.

One further highly interesting result with regard to topic parameters has been revealed by the investigation of the ammonium salts—namely, that the replacement of potassium by ammonium in either series is only accompanied by about the same amount of separation of the molecular centres as is observed when rubidium is substituted for potassium. Thus two atoms of the alkali metal are replaced by ten atoms of two ammonium radicles NH_4, without more extension of the structure than if merely two rubidium atoms had been introduced instead of two potassium atoms. The amount is far less than when two cæsium atoms are introduced. Hence there has been room in the structure for eight additional hydrogen atoms without pushing the molecules farther apart, and the writer has therefore concluded that the space defined by the topic parameters is not filled with matter, but that relatively large interatomic and intermolecular spaces occur. There appear to be only two other alternatives possible. One is that, owing to the different sizes and shapes of the atoms of the different elements composing the molecule, it is just possible to pack into the crannies eight extra atoms of hydrogen. The other is that room is made for the NH_4 groups by the sulphur-oxygen layers being thrust further apart, but that the resulting totally different structure happens by pure coincidence to be capable of exhibiting similar planes arranged according to the same order of symmetry as the metallic salts. The optical similarity of the rubidium and ammonium salts, presently to be referred to, would appear, however, to negative the idea of mere chance determining the development of similar faces. We know so little yet as to the actual character of the atom in the molecule, and its state of rest or motion, that the writer has always regarded it as a sphere of operations rather than as an immobile entity. Even if the packing is as close as it is held to be by Barlow, it will only probably be so as regards these spheres of operation or motion, within which there may at any given moment be a large amount of free space compared with that occupied by the atomic matter.

Before leaving this subject of the structure of crystals, however, it may be remarked that the experimental evidence of intermolecular and interatomic spaces brings the work of the writer into close touch with the important work of Prof. J. J. Thomson, whose discovery of the complex nature of the atom, and its constitution out of corpuscles of negative electricity (the so-called electrons, immensely smaller than the atom itself), shows that we have now a third kind of inter-space to consider—namely, the interelectronic. Moreover, the discovery of the corpuscular structure of the atom renders it the more probable that atoms may take up definitely orientated positions in the molecule, and consequently enhances the value of the teaching of directional changes in topic parameters as indicative of such positions.

The results of the optical investigations may now be briefly referred to. The optical properties of both series may be represented by an ellipsoid with three unequal rectangular axes. We may employ either the ellipsoid of Fresnel, or that of Fletcher termed the "indicatrix," the two being polar reciprocals and exhibiting the same facts. The indicatrix is preferable, because the relative lengths of its axes are immediately expressed by the refractive indices of the crystal along those directions. In the rhombic system the positions of the axes of the ellipsoid are fixed, being identical with the directions of the crystallographic axes. In the monoclinic series the ellipsoid is free to rotate about the rectangular symmetry axis, and is generally not found to have either of its other two rectangular axes coincident with either the inclined axis or the vertical axis. The first optical result has been to show that this possible rotation of the optical ellipsoid does actually occur in the double salt series, to an extent which varies for the different salts from 4° to 33°, the actual amount being dependent on the atomic weight of the alkali metal present, so that the position of the ellipsoid in any rubidium salt is intermediate between the positions for the corresponding potassium and cæsium salts. Further, in both the rhombic and the monoclinic series the ellipsoid expands in every direction when potassium is replaced by rubidium, cæsium, or ammonium, which is a graphic expression of the fact that the refractive index increases for every direction within the crystal. This inflation is likewise according to the order of the atomic weights of the alkali metals, and in the case of ammonium the amount of extension is almost exactly the same as for rubidium—a result similar to that for the topic parameters, the bearing of which on interatomic spaces has just been discussed. Moreover, the molecular refractions, obtained by combining the molecular volume with the refractive indices in accordance with the formula either of Lorenz or of Gladstone, obey a similar law of dependence on the atomic weight of the alkali metal, and the molecular refraction of any ammonium salt is also almost identical with that of the corresponding rubidium salt.

The double refraction, which is measured by the difference between the maximum and minimum of the three refractive indices of the crystals of any one salt for the same wave-length of light, has been likewise shown to be a function of the atomic weight. and obedience to this law has resulted in some of the most beautiful and interesting optical phenomena in convergent polarised light which have ever been observed. It is responsible for no less than five cases of the rare phenomenon of crossed-axial-plane dispersion of the optic axes. So complicated are the interference figures, particularly with rise of temperature and change of wave-length, that without the law they would be quite unintelligible, but by it they are reduced to perfect order and simplicity.

It only remains now to mention briefly the results of the determinations of thermal expansion. Owing to the presence of water of crystallisation, and its ready efflorescence at only slightly elevated temperatures, the double salts are unsuitable for such determinations, and the rapid deliquescence of the simple selenates also excludes them. But the simple rhombic sulphates have been found suitable, and adequately large blocks were eventually obtained out of perfect crystals. The result was to show that the co-efficients of cubical expansion exhibit a progression corresponding to that of the atomic weights of the three alkali metals. The differences in the expansions of the three salts are only one and a half per cent. of the total expansion, yet this is five times as great as the possible error of the interference dilatometer, so that the results are absolutely conclusive.

Conclusion

The outcome of the writer's investigations has thus been to demonstrate that the exterior geometrical, internal structural, optical and thermal properties of the crystals of each of these series of isomorphous salts exhibit progressive variations which follow the order of progression of the atomic weights of the alkali metals which the different salts contain, and that similar variations attend the replacement of sulphur by selenium in the acid radicle present. Further, the applicability of this rule to such widely different series as the rhombic anhydrous sulphates and selenates, and the monoclinic double sulphates and selenates containing six molecules of water of crystallisation, indicates its probable validity as a general law of nature. This law has been defined in one of the writer's memoirs in the following terms: "The difference in the nature of the elements of the same family group, which is manifested in their regularly varying atomic weights, is also expressed in a similarly regular variation of the characters of the crystals of an isomorphous series of salts of which these elements are the interchangeable constituents."

This law definitely allocates specific angles, structure, and physical properties to every member of isomorphous series, thus removing the only supposed exceptions to the original conception of Haüy, that difference of chemical composition involves difference of crystalline form. At the same time, it explains the real meaning of Mitscherlich's principle of isomorphism, and indicates the limited and exact sense in which it is true.

In addition, the position of ammonium near rubidium in the alkali series has been established, together with the interesting corollary of the existence of intermolecular and interatomic spaces. Lastly, the conception of topic parameters has been developed into a valuable means of elucidating the nature of the internal structure of crystals, and of expressing the dimensions of the point system according to which the molecules are arranged, besides affording some positive evidence as to the arrangement of the atoms composing the molecule.

BIBLIOGRAPHY

(1) W. MUTHMANN, Beiträge zur Volumtheorie der krystallisirten Körper, Zeitschr. f. Krystallographie, 1894, xxii. 497.

(2) A. BRAVAIS, Mémoire sur les systèmes formés par les points distribués régulièrement sur un plan ou dans l'espace, Journ. de l'Ecole Polytech. Paris, 1850, xix. 127.

(3) C. JORDAN, Mémoire sur les Groupes des Mouvements, Annali di matematica pura ed applicata, Milano, 1869, Series 2, ii. 167 and 322.

(4) L. SOHNCKE, Entwickelung einer Theorie der Krystallstruktur, Leipzig, 1879.
—— Erweiterung der Theorie der Krystalle, Zeitschr. f. Kryst. 1888, xiv. 426.
—— Die Structur der hemimorph-hemiëdrischen, bezw. tetartoëdrischen drehenden Krystalle, ib. 1896, xxv. 529.

(5) A. SCHÖNFLIES, Krystallsysteme und Krystallstructur, Leipzig, 1891.

(6) E. VON FEDOROW, Symmetry of Finite Figures, and Symmetry of Regular Systems of Figures (both in Russian), Transactions of the Russian Mineralogical Society, 1885-90 : xxi. 1, xxv. 1. Abstracts in Zeitschr. f. Kryst. 1892, xx. 39 ; 1893, xxi. 679.

(7) W. BARLOW, Ueber die geometrischen Eigenschaften homogener starrer Structuren und ihre Anwendung auf Krystalle, Zeitschr. f. Kryst. 1894, xxiii. 1 ; and 1895, xxv. 86.

(8) —— and H. A. MIERS, The Structure of Crystals, Report of Committee of British Association, Section C, 1901.

(9) H. HILTON, Mathematical Crystallography, and Theory of Groups of Movements, Oxford, 1903.

(10) P. VON GROTH, On Crystal Structure and its Relation to Chemical Constitution, Report of British Association, Section B, Cambridge meeting, 1904, and Chemical News, 1904, xc. 142.
—— Einleitung in die chemische Krystallographie, Leipzig, 1904.

(11) H. A. MIERS, An Enquiry into the Variation of Angles observed in Crystals, especially of Potassium-Alum and Ammonium-Alum, Phil. Trans. Roy. Soc. 1904, A, ccii. 459.

(12) A. E. H. Tutton, Connection between the Atomic Weight of Contained
Metals and the Magnitude of the Angles of Crystals of Isomorphous
Salts, *Journ. Chem. Soc.* 1893, lxiii. 337.

—— Connection between the Atomic Weight of Contained Metals and
the Crystallographical Characters of Isomorphous Salts, *Journ. Chem.
Soc.* 1894, lxv. 628 ; also 1896, lxix. 344 and 495 ; and 1897, lxxi. 846.

—— An Instrument for producing Monochromatic Light of any desired
Wave-Length, *Phil. Trans. Roy. Soc.* 1895, A, clxxxv. 913.

—— An Instrument for Cutting, Grinding, and Polishing Section-Plates
and Prisms of Crystals accurately in the desired Directions, *Proc. Roy.
Soc.* 1895, lvii. 324.   Improvements on the same, *Phil. Trans.*
1899, A, cxcii. 457.

—— The Nature of the Structural Unit, *Journ. Chem. Soc.* 1896, lxix. 507.

—— A Compensated Interference Dilatometer, *Phil. Trans.* 1898, A,
cxci. 313.

—— The Thermal Deformation of the Sulphates of Potassium, Rubidium, and Cæsium, *Phil. Trans.* 1899, A, cxcii. 455.

—— Double Selenates containing Zinc and Magnesium with Potassium,
Rubidium and Cæsium, *Proc. Roy. Soc.* 1900, lxvi. 248, and lxvii. 58 ;
and *Phil. Trans.* 1901, A, cxcvii. 255.

—— Ammonium Sulphate and the Position of Ammonium in the Alkali
Series, *Journ. Chem. Soc.* 1903, lxxxiii. 1049.

—— The Elasmometer, a New Interferential Elasticity Apparatus, *Phil.
Trans.* 1904, A, ccii. 143.

—— Ammonium-Magnesium and Ammonium-Zinc Sulphates and
Selenates, *Journ. Chem. Soc.* 1905, lxxxvii. 1123.

—— Topic Axes, and the Topic Parameters of the Alkali Sulphates and
Selenates, *Journ. Chem. Soc.* 1905, lxxxvii. 1183.

THE GEOLOGICAL PLANS OF SOME AUSTRALIAN MINING-FIELDS

BY J. W. GREGORY, F.R.S.

Professor of Geology, University of Glasgow

		PAGE
1.	THE MOUNT LYELL COPPER-FIELD	118
2.	PYRITE-SMELTING AT MOUNT LYELL	124
3.	THE MOUNT BISCHOFF TIN-MINE	126
4.	THE AUSTRALIAN SADDLE REEFS AND THEIR DISTRIBUTION	128
	(*a*) BENDIGO	128
	(*b*) CASTLEMAINE	130
	(*c*) BROKEN HILL	131
	(*d*) THE SEPARATION OF THE BROKEN HILL ZINC ORES	134
5.	THE INDICATORS OF BALLARAT	135

THE character of the Australian people is well illustrated by their management of their mining-fields. Their mining methods show the daring and originality of the pioneer; the honesty inspired by adventurous comradeship is still conspicuous in their business management; and the progress of the mines is described in a Press which is unsurpassed in integrity in the whole world of mining journalism. Australian mining often lacks the precision, thoroughness, and finish of German metallurgy; its sampling is often slapdash, and its bookkeeping looks slovenly compared with the elaborate specialisation of mining accounts on the Rand. But the highly experienced Australian miners are able to do by instinct, work that, in some fields, must be done by rule; and the trained insight and practical capacity found among every grade of Australian miners have achieved economic successes, which are among the glories of modern mining, and discovered new methods, which have helped the development of the industry in all quarters of the globe. Australian mining has, of course, its record of blunders, and some of them have been disastrous; the chief have been due to the misguidance of false analogies regarding the geological structure of the mining-fields, and to mistaken views as to the plan of distribution of the ore deposits.

1. The Mount Lyell Copper-Field

The Mount Lyell district illustrates Australian mining at its best and at its worst. It is littered with costly failures, owing to a wrong idea as to the geological structure of the field; and its richest high-grade mine has been harassed by a succession of the mistakes that are probably inevitable, when a distant management attempts the development of a new and exceptional mining-field. The mass of ore known as the "Iron Blow" was discovered at Mount Lyell in 1883, and it was believed by eager speculators to be one outcrop of a lode, which would have a length and uniformity, equivalent to its width and purity at the place exposed. Streaks of ore running through the adjacent country rocks strengthened this belief; and so the mining-field was pegged out among over thirty companies, of which only three have ever paid a dividend, and only one can yet be regarded as an established success. The one company has, however, saved the reputation of the field. It has succeeded in the profitable working of an especially difficult low-grade ore by a combination of sound business management, expert scientific skill, and brilliant originality in mining methods. At the invitation of the Mount Lyell Company I had the opportunity to study this field and inspect all the mines then under its control. The results of that study have been stated in a book published by the Australian Institute of Mining Engineers, describing the geology of the field and giving a new explanation of the genesis of its ore.[1]

The ores of the Mount Lyell field consist of a series of isolated masses of sulphides, including copper pyrites, gold, and silver. The most famous of the deposits, the "Parent Mine" of the field, is a vast mass of pyrites, 800 ft. long and 200 ft. broad. Mining work and bores have shown that the mass first tapers and then ends abruptly below. The main mass of ore is pyrites; but on the hanging-wall side is a wedge-shaped block of hematite, known as the "Iron Blow." This mass of iron oxide contained some 15 oz. per ton of silver and 15 dwt. per ton of gold. Mining was begun on the Iron Blow, its ore being crushed in order to obtain the gold by ordinary milling. This

[1] J. W. Gregory, *The Mount Lyell Mining-Field, Tasmania; with some account of the geology of other pyritic ore bodies*, Melbourne, 1905, viii + 172 pp., 16 plates and maps. A Bibliography is given on pp. 164-8.

process failed, owing to the presence of barite in the ore, and both ore masses were acquired by a Melbourne mining company as a copper-mine, the gold and silver being recovered as by-products. The attempt was encouraged by the discovery of a huge bonanza of rich silver ore below the Iron Blow in a shoot, doubtless formed by the concentration of silver, leached from the hematite above.

The geological relations of the ore mass were hidden by the dense forest, nourished by the heavy rainfall on the western coast of Tasmania. But the ore mass was seen to be enclosed in schists, and to occur close beside their contact with a series of quartzites and quartz conglomerates.

The rocks of the field may be classified as follows :

Pleistocene : Alluvium, glacial clays and moraines.

Mesozoic : Diabase of Mount Sedgwick.

Palæozoic : Fenestella shales of the Linda Valley (carboniferous).

    Devonian quartzites and conglomerates, isolated blocks of which form Mount Lyell, Mount Sedgwick, Mount Owen, etc.

    Silurian limestones and shales at Zeehan.

    Silurian or Ordovician. The quartzites and limestone of Queenstown.

Pre-Silurian : The Mount Lyell schists : a series of intensely
(Archean ?) altered quartz porphyrites and acid volcanic tuffs, with intrusive dykes and masses of diabase porphyrite. Some of these rocks have been well foliated and altered into quartz chlorite schists with masses of margarodite, with no traces of the original minerals or rock structures.

The schists form the basis of the whole mining-field, and are exposed beneath the Devonian conglomerates, blocks of which form the summits of the West Coast range. On the west the schists are faulted against the Silurian or Ordovician limestones and quartzites of Queenstown. The schists have been regarded as of various ages, from the Archean to the upper Devonian or post-Devonian. As to their precise age, there is as yet no final evidence. They are clearly pre-Devonian and pre-Silurian, and they are the oldest rocks in the district. In places they overlie the conglomerates; but this superposition is due to a vast overthrust fault, which traverses the whole district, and is one of the most important features in the geology of Western Tasmania.

The ores occur in two conditions—in fahlbands, or long bands of mineralised schist, and as thick, short masses. The fahlbands are confined to the Mount Lyell schists, but some bands of pyritiferous quartzite in the Devonian rocks are of similar origin. The fahlbands are best developed in the neighbourhood of great transverse faults, which cut across the schists. Their ores are pyrite, chalcopyrite, and fahlore, containing some gold and silver. The minerals usually occur in bunches, in bands of crushed schists, and the bunches may be arranged in shoots. The wide distribution of these fahlbands led to the whole of the schist country being taken up by mining companies, but they have never paid to work independently. They have, however, been profitably worked as metal-bearing fluxes; for the extremely basic, pyritic ores of the Mount Lyell Mine require a siliceous flux; the acid ores of the larger fahlbands serve this purpose, and contribute their own copper at the same time. Hence the South Tharsis, Royal Tharsis, and Lyell Tharsis mines have been acquired and successfully worked by the Mount Lyell Company.

The ore masses are economically of far greater importance than the fahlbands. The largest is that of the Mount Lyell Mine, often known as the "Parent Mine," at Gormanstown. It consists of a somewhat boat-shaped seam of pyrites, containing quartz and barite, with chalcopyrite, some gold and silver, and insignificant amounts of galena and blende. The ore mass trends north-west by west, and at the surface is 800 ft. long and 200 ft. wide. It is worked by an open cut, about 300 ft. deep, quarried to the extent of 1,000 tons a day. The ore mass underlies to the west. It tapers below, and is abruptly cut off by a thrust plane. It is twisted, the major axis varying from north-west by west at the surface, to north-west in the fifth level, and to almost due west at the eighth or lowest level. The amount of pyrites above the fifth level was originally 4,200,000 tons, the whole of which, however, would not pay to extract as ore; for on the western or hanging-wall side of the deposit, the average value is only ·64 per cent. of copper, ·06 oz. of gold, and 1·6 oz. of silver to the ton. The ore on the footwall side yields 2˙35 per cent. of copper, 2 oz. of silver, and ˙0725 oz. of gold per ton. The poorer pyrites, however, pays to extract as fuel for the smelting of the rich acid ores of the North Mount Lyell Mine, and for the manufacture of sulphuric acid, used in the preparation of superphosphate, of which the Mount Lyell Company has a factory at Melbourne.

The ore mass is entirely surrounded by schists, but it is close to the junction between the schists and the conglomerates, which form the footwall side of the ore mass, though separated from it by a thin layer of schist. The ore mass was therefore at first regarded as part of a long lode, which was expected to continue across the field, like a gold-quartz fissure lode.

Dr. Peters, however, regarded the schists and conglomerates as all part of one geological series, and explained the ore mass as a lenticular-bedded precipitate, deposited on the floor of a Silurian swamp. This theory necessarily fell, when it became apparent, that the conglomerates and the schists belong to different geological systems.

The occurrence of the chief ores along the junction of the schists and conglomerates then led to their description as contact ores. But though the schists are of igneous origin, their relations to the conglomerates are due to faulting and not to intrusion. Contact ores are those which occur beside igneous contacts.

The clue to the distribution and origin of the ores is given by the fault system of the Mount Lyell field (see fig. 1). The Great Mount Lyell fault runs across the field from north to south; it is a powerful fault, in places reversed, so that the rocks are locally inverted, while slices of the conglomerate have been thrust in to the softer schists. Parallel faults have let blocks of conglomerate down into the schists, as. in the outlier that forms the North Lyell crags. The north and south faults are crossed by about fifteen transverse faults, by some of which the conglomerates that form Mount Lyell have been thrust westward as a great buttress, overhanging the Queen River Valley; the track to the Sedgwick Valley climbs over this buttress almost level with the summit of the Mount Lyell ridge. The transverse faults are associated in places with thrust planes, which have carried masses of the conglomerate over the schists.

The chief ore masses occur in the acute angles between the transverse faults, the great overthrust fault, and the thrust

planes (fig. 2). In such positions the schists have been crushed
and rendered permeable to solutions containing copper and iron
sulphates, with some gold and silver. The solutions no doubt
rose along the Great Mount Lyell fault, and were reduced by

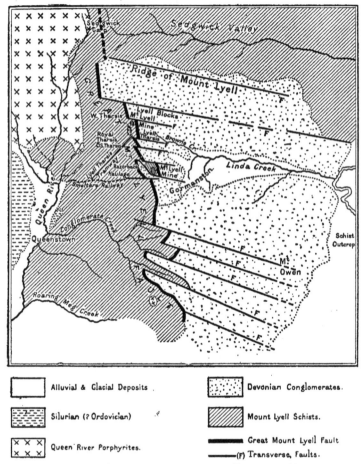

	Alluvial & Glacial Deposits			Devonian Conglomerates.
	Silurian (? Ordovician)			Mount Lyell Schists.
	Queen River Porphyrites.			Great Mount Lyell Fault
				(F) Transverse, Faults.

Fig. 1.—Sketch-map of the Mount Lyell Mining-Field.

reactions with the decomposing chloritic and felsite material in
the crushed schists, masses of which were replaced by sulphides.

The Iron Blow was on the hanging-wall side of the pyritic
mass, and these two ore deposits were clearly connected in
origin. The hematite is not an ordinary gossan, or it would

have been limonite and not anhydrous oxide of iron. It is probably an ore deposit due to the interaction between water percolating through the conglomerate and ascending ferruginous solutions along the fault plane.

The North Lyell Mine has been described by Mr. W. T. Batchelor,[1] the engineer in charge of mining operations at Mount Lyell. The mine is situated in a mass of schists, which has been intensely crushed and contorted, by having been faulted between blocks of the hard conglomerates. The ores are siliceous, and consist of quartz and broken schist charged with chalcopyrite, bornite, and chalcocite. The average of the ore that pays to extract under present conditions has been brought down as low as 6·25 per cent. of copper; but in places there

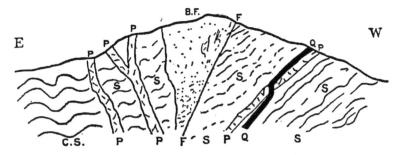

FIG. 2.—Diagrammatic Section through the Mount Bischoff Mine.

c.s. Contorted Silurian. P. Porphyrite. F. Fault.
s. Silurian. Q. Queen Lode. B.F. Brown Face.

are rich veins of almost pure bornite, which are in places 2 ft. in thickness; they follow the strike of the schists. The bulk of the ore is in masses, of which about eight were known in March 1904. One of them, the "Eastern Ore Body,' is a brecciated mass of quartzite and schist, occurring between the main conglomerate mass of Mount Lyell and a faulted conglomerate outlier to the west. The genesis of the ores is doubtless similar to that of the Mount Lyell Mine; but the ore masses are more irregular, owing to the more intense disturbance of the rocks by the more intricate nature of the faulting; and they will probably continue to greater depths.

[1] *Op. cit.* pp. 108-16.

2. Pyrite-Smelting at Mount Lyell

The Mount Lyell mining-field is, metallurgically, remarkable as the first in which pyrite-smelting has been successful on a big scale. The copper ores are now smelted there without the use of any other fuel than their own sulphur and iron—an achievement due to the high skill and patience of the general manager, Mr. R. C. Sticht. The history of the process has been told by Mr. Sticht, with an historical completeness only too rare in mining literature, in his recent presidential address to the Australasian Institute of Mining Engineers.[1] He traces the development of pyrite-smelting from the first beginning of Roharbeit, or pyritic-smelting, by Barthold Köhler, of Freiberg, in 1555. Mr. Sticht insists on the essential differences between pyritic-smelting, a term introduced by Percy in 1880, and the newer process of pyrite-smelting. Pyritic-smelting is a blast-furnace process, in which iron pyrites is added to form a regulus, in which the precious metals of the ores are collected. The process is adapted to ores which do not contain sufficient copper or lead to collect their own gold and silver; and thus it enables ores to be treated by means of valueless pyrites, which are too poor to afford the use of lead and copper. In this process carbonaceous fuel is used as the main source of heat, and the smelting is conducted in a reducing atmosphere.

Pyrite-smelting, on the other hand, is the blast-furnace smelting of sulphide ores by means of the heat generated by the oxidation of their sulphur and iron, and not from the combustion of carbonaceous fuel, and the smelting takes place in an oxidising atmosphere. A block of wood may be added occasionally; but its function in pure pyrite-smelting is mechanical, as it breaks up the crust in the furnace, and it is not added as a fuel.

Partial pyrite-smelting is an intermediate process. It agrees with pyritic-smelting in the use of some carbonaceous fuel; it agrees with pyrite-smelting in the use of an oxidising atmosphere in the furnace, and in using all the heat that can be generated by the oxidation of the sulphides in the ore. Partial pyrite-smelting is, therefore, applicable to ores of which the fuel capacity is insufficient for their own smelting.

[1] R. C. Sticht, "The Development of Pyrite-Smelting: An Outline of its History," *Proc. Austral. Inst. Min. Eng.* vol. ii. 1905, pp. 9-78.

Pyrite-smelting was brought within the range of practical possibility by the invention of bessemerisation. In the very year (1856) in which Bessemer announced his process, Keates proposed the treatment of sulphide ores on the same principle—viz. keeping the metal molten by the heat generated by the combustion of the sulphur and iron in these ores. Later, attempts on the same lines were made at Ducktown, in 1866, by Raht; in Hungary, in 1867, by Rittinger; and by several metallurgists from 1866 to 1868 at Bogoslowsk, in Western Siberia. Mr. Sticht gives most of the credit of the pioneer work to John Hollway, whose paper before the Society of Arts in 1879 he describes as the "pyrite-smelters' gospel, voicing early aspirations." Hollway used a Bessemer converter, and his scheme failed because he tried both to smelt the ores by the burning of the sulphur as fuel, and to recover sulphur as an element at the same time. He therefore used a low-power blast and low temperatures. The attempt to burn sulphur and recover it at the same time, led to the failure of this system when it was tried in Servia in 1881. This failure and the dismissal of the proposal by Hussey Vivian as a wild-cat scheme, "quite inapplicable to copper-smelting," did not prevent the fresh attempts of Bartlett in Maine, and Austin in Montana; but the ores they used were not suitable or the supplies were inadequate; and when Mr. Sticht went to Mount Lyell in 1895 not a single pyrite-smelting plant had been successful. He, however, knowing that according to chemical principles the process was theoretically possible, refused to believe that mere mechanical difficulties would always necessitate the waste of the fuels in these self-fuelled ores. He persuaded the directors to let him put aside the smelting process he had been brought to Tasmania to conduct, and to adopt pyrite-smelting instead. He began with partial pyrite-smelting, with the use of coke and a hot blast; but, one by one, he has overcome the difficulties in the way of the theoretical ideal. Early in 1903 the hot blast was finally discarded; four furnaces do the work previously done by eight, and that without the use of any carbonaceous fuel. The ores of the different mines are so mixed that they, in the main, form their own fluxes and supply the whole of their own fuel; and Mount Lyell accordingly profitably mines and smelts copper ores at the unprecedentedly low cost of 13s. a ton.

3. The Mount Bischoff Tin-Mine [1]

Mount Bischoff is another Tasmanian mine of unusual geological interest. For the opportunity of examining it I am indebted to the courtesy of the manager, Mr. T. Kayser, and to the hospitality and guidance of Mr. H. Herman, the assistant manager.

Mount Bischoff is a hill, 2,600 ft. above sea-level, which rises above the basalt plateau of North-western Tasmania, beside the head-waters of the Waratah River, a tributary of the Arthur River. The adjacent country consists of folded Silurian rocks, pierced by quartz porphyry dykes. The summit of Mount Bischoff is formed by part of one of these dykes, the course of which is like a horseshoe, and surrounds the mine on the west, south, and east. Branches from the dyke radiate into the country outside it, while some of the dykes outside the mine dip inward as if they all met below it.

The tin ores are of three main types: (1) A quartz vein charged with tin forms the Queen lode, on the north-eastern side of the mine; this lode follows along a porphyry dyke, which it overlies near the outcrop; when it was followed downward, it was found to cut across the dyke and then continue beside, but underneath it. (2) Cassiterite also occurs in masses and veins in contact with the porphyry dykes, in pockets and veins in the Silurian slates, and as impregnations in the porphyry. (3) The main wealth of the mine comes from a large mass of iron-stained material, known as the Brown Face, which includes some coarse bedded sand, yielding up to 10 and even 15 per cent. of cassiterite. This Brown Face is bounded to the east by a sloping face of the Silurian slates, the junction being a strong fault plane. To the west it is bounded by altered Silurian rocks, separated from unaltered contorted slates by the western arm of the quartz porphyry dyke.

The special problem in the geology of the Mount Bischoff Mine is in the nature of the Brown Face. Its coarsely bedded, loose material at first inevitably suggests that this brown, iron-stained mass is a sedimentary deposit, filling up a crater-shaped basin within the dykes. But this simple hypothesis unfortunately does not seem adequately to explain the facts. Part

---

[1] Mr. Kayser has given a general account of the geology of the mine in "Mount Bischoff," *Rep. Austral. Assoc. Adv. Sci.* vol. iv. 1892, pp. 342-58.

of the gossan is certainly porphyry in the last stage of decomposition. The grains in the tin-bearing sands are angular, and do not appear water-worn, and they may be derived from rocks containing quartz grains, which have broken down into rotten-stone by the removal of the cement simultaneously with the introduction of the tin. The whole Brown Face seems to me to be a gossan. It was probably formed by the oxidation of a mineralised mass of quartzites and slates, with some intrusive porphyry dykes ; most of this mass had been impregnated with pyrites and the rest of it with cassiterite. The pyrites has been converted into limonite, the bases removed (probably as soluble sulphates), and the settling of the decaying rock mass has in places produced a false bedding. The intensity of the chemical changes that have taken place in the rocks of this mine is shown by the alteration of some of the porphyry into radial clusters of topaz, and the partial alteration of the rock into a topaz greissen.

The economic problem of the mine is, What will happen when all the Brown Face material has been removed? It will be exhausted some day, and the mine will then depend on the veins and masses of cassiterite developed along the course of the quartz porphyry dykes. If the view be correct that the Brown Face is a decomposed gossan, there seems no reason why shoots and veins of rich tin ore should not go down along the course of the quartz porphyry dykes. We may be, at any rate, confident that if the shoots are there, they will be found, and worked with the economic skill, which has given Mount Bischoff such a long career of financial prosperity.

4. THE AUSTRALIAN SADDLE REEFS AND THEIR DISTRIBUTION
(a) Bendigo

The mining-field of Bendigo illustrates the valuable results attained by combination of skilled, economic management, and sound knowledge of the geological structure of a mining-field. Years ago Bendigo was described as exhausted; but the mining industry is still flourishing there. It shows the deepest gold-mining in the world, and low working costs, which are regarded with wonder by some other fields.

Bendigo is a mining town on the northern slope of the Victorian Highlands, near the edge of the Murray Plains, 100 miles north of Melbourne. Its mining dates from 1851. It was at first famous for the wealth of its alluvial gravels, and then for the exceptional richness of its blows of quartz, which stood up in crags, owing to the wearing away of the softer slates. The quartz "blows" seen on the surface were soon crushed, and the miners followed them underground. This work led to the disappointing discovery that the quartz veins, instead of continuing to unknown depths, became thinner, and soon pinched out. The quartz occurred in wedge-shaped bands, scattered with such apparently inexplicable irregularity through the slates, that it was thought they could only be found by

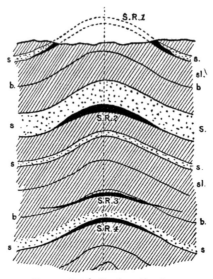

Fig. 3.—Saddle Reefs at Bendigo

chance. Rich though the quartz often was, it would never repay the dead work of haphazard blind-stabbing through the slates. So the field was thought to be done.

The yield, which averaged nearly 500,000 oz. a year up to 1855, and amounted to 527,407 oz. in 1857, fell fairly steadily to under 200,000 oz. from 1877 to 1880, and down to 137,964 oz. in 1890.

But the miners found in time, that the quartz was not so erratic in its distribution as had been thought. The wedges were found to be the lower sides of arches of vein quartz, which were always connected with arch-like foldings in the country rocks. The lodes were therefore called "saddle reefs."

At the end of the 'eighties the field was surveyed by Mr. E. J. Dunn,[1] on behalf of the Victorian Geological Survey. His report showed that the saddle reefs were characteristic of the whole field. The slates and sandstones which form the country rocks have been contorted into many parallel folds; and the quartz reefs have been deposited in the spaces left beneath a hard bed, along the summits of the anticlinals. The reefs are not always exactly along the bedding planes, for during the bending of a band of slates, cracks will be formed obliquely across the bedding, in a flatter curve than that of the main arch (as in fig. 3, s.r.$_3$).

Now at Bendigo the Ordovician rocks are a thick series of

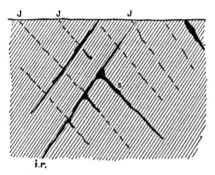

FIG. 4.—False Saddle Reefs formed of an interbedded reef (*i.r*) and a spur (*s*) along a joint plane.   J J. Joint planes.

alternating, graptolitic slates and sandstones, in which the same sequence of deposition frequently recurs, owing to the repetition of the same geographical conditions in the old Ordovician Sea. Hence a line sunk downward along the axial plane of an anticlinal, the vertical line in fig. 3 will meet many positions where interspaces occurred beneath sandstone arches; and in all such positions it is natural to find that saddle reefs have been formed. Shoots of gold occur at intervals along these reefs.

The discovery of the origin of the quartz reefs gave the clue to their distribution. After the exhaustion of one saddle reef, it is only necessary to sink the mine shaft downward to the next similar position, and there another saddle reef is found.

[1] E. J. Dunn, *Reports on the Bendigo Goldfield*, No. 1, 1892 (reprinted 1896), 20 pp. and plans ; No. 2, 1896, 42 pp. with plates and plans.

9

Instead of the reefs being scattered haphazard, they are distributed on a beautifully simple and regular plan. Reef after reef has been discovered at lower levels by sinking along "centre country," as the axial plane of the anticlinals is called, until reefs are now being worked at the depth of 4,250 ft.

The saddle-reef plan has enabled the reefs to be followed northward as well as downward, along the axial lines of the folds; the saddle reefs have been traced for 6 miles north and south, and the most flourishing mines at present are at Eaglehawk, on a formerly neglected extension of the Bendigo goldfield.

The discovery of the plan of the field has enabled the miners to follow the reefs with certainty, instead of playing with them a game of blind man's buff; and the system of local management, under men whose instincts have been trained by long and close study of the ground, enables the mining to be conducted with such economy that, at some of the mines, 2 dwt. ore has been made to yield dividends. *South African Mines*, the Johannesburg mining paper, recently quoted (September 9, 1905 : vol. iii. p. 624) the results at two of the Bendigo mines with expressions of wonder and envy.

(b) Castlemaine

Saddle reefs are also well developed in Victoria, in the goldfield of Castlemaine, of which a valuable account has been published by one of the most rising of the younger Victorian geologists (Mr. W. Baragwanath, jun.).[1] The Castlemaine goldfield presents a striking contrast to the neighbouring field at Bendigo. At Castlemaine the surface gravels were phenomenally rich, whereas the quartz lodes have been poor. The alluvial deposits were richer than that at Bendigo, and the lodes much poorer. Deep quartz-mining at Castlemaine has, on the whole, been disappointing.

So far as I know, no adequate reason for this anomaly has been offered. The explanation seems to be that at Castlemaine the gold-mines are close to the intrusive granitic rocks of the Harcourt Range. There was, therefore, much greater secondary concentration in rocks, which have been removed by denudation,

---

[1] W. Baragwanath, jun., "The Castlemaine Goldfield," *Mem. Geol. Surv. Victoria*, No. 2, 1903, 36 pp., 13 plates and 19 plans and maps.

forming the rich alluvial deposits;[1] and only the roots of the gold shoots remain.

(c) Broken Hill

The success of the saddle-reef mining at Bendigo has led miners elsewhere, to watch eagerly for the occurrence of these delightful simple conditions; and saddle reefs have been reported which have proved to be arched reefs of different origin. It is not uncommon for a quartz reef formed along a bedding plane to give off a branch across the bedding, along a big joint or a fault plane. In such cases the main reef will usually continue above the junction of the two branches, and thus a transverse section shows three rays instead of two. It will be shaped like an inverted Y instead of a saddle (fig. 4). Even if such an arched reef has only two limbs, they are different in origin, the bedding being parallel to one limb and transverse to the other. Such reefs are therefore known as "false saddle reefs." It is all-important to distinguish between true and false saddle reefs, because there is no certainty that false saddle reefs will recur in depth.

A second type of false saddle reefs occurs in contorted metamorphic plutonic rocks. The most important Australian type of these reefs is at the silver, lead, and zinc mines of Broken Hill. Broken Hill is a low ridge, about 2 miles long, of foliated rocks, and forms part of the Barrier Ranges in the western plains of New South Wales. It lies 70 miles north-west of the Darling River at Menindie, and 30 miles from the South Australian border. The hill was crossed by Sturt in his famous inland journey in 1844, but no one suspected its mineral wealth until 1883. It was then part of the Mount Gipps sheep-run; and some of the station hands, their attention having been turned to mining by the discovery of silver in the Thackeringa Ranges, 20 miles to the west, began to work the manganese gossan of Broken Hill as a tin-mine. They failed to get any tin, but found silver, and mining was begun by the Broken Hill Proprietary Silver Mining Company in 1885.

The rocks of the field have been described as altered Silurian

---

[1] Saddle reefs of a similar type to those of Bendigo occur in Nova Scotia, and the interesting reports of Faribault show that the reef distribution is in some respects like that of Bendigo and Castlemaine; but the distribution of the gold in this field is still imperfectly known.

sediments; but they appear to me to be altered igneous rocks, and there seems no adequate reason for separating them from the similar rocks that form the Archean plateau of Australia. The rocks are gneiss, and mica- and hornblende-schists, and in the lode are masses of garnetiferous quartzites. The rocks have been traversed by powerful faults. The lode outcrops as a single vertical lode, which has been worked by an open cut. The lode widens below and forks ; as a rule, the north-western branch is the larger and more persistent. The south-eastern branch is sometimes a short, thick pocket (fig. 5, *b*); at others a spur, which follows the foliation planes in places, and sometimes crosses them; and in some places (fig. 5, *c*) there is no trace of the eastern branch.

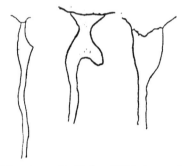

FIG. 5.—Three sections across the Broken Hill Lode near the McBryde's Shaft
in the Proprietary Mine.

A. Shows a thickening on a vertical lode.
B. A saddle-shaped arch of which the eastern limb ends in a pocket.
C. Shows no trace of the second branch of the lode.

The division of the lode occurs at depths of from 200 to 400 ft., where there is a great increase in the width of the lode. In 1892, when the field was examined by Mr. E. F. Pittman, the Secretary of Mines of New South Wales, the most striking feature in the lode was this thickening, from which two branches went downward, along the structural planes of rocks. He made the very useful suggestion that the lode was a saddle reef; and if that hypothesis had been substantiated, similar arches in the lode might have been expected to recur below. Pittman's suggestion was adopted in the valuable report on the field by Jaquet, published in 1894, as one of the Memoirs of the Geological Survey of New South Wales, the finest serial yet published on Australasian geology. But the work that has since been carried on has exposed the structure of the lode to a greater depth and length than in 1894,[1] and we know more about the Archean rocks of Australia than we did then. The later workings show in places the arch-like fold of the reef, which was especially well exhibited at the time of my visit in the South Broken Hill Mine. But the general evidence now available seems to me to show that the lode is not a true saddle reef. In the first place, the rocks are not, in my

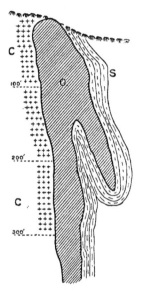

FIG. 6.—A Saddle Branch or False Saddle Reef at the North Lyell Mine.
o. Ore body.   c. Conglomerate and quartzite.   s. Schist.

opinion, altered Palæozoic sediments, but foliated Archean igneous rocks; and there is no reason to expect the same persistence in the crumplings of gneisses, as in the folding of a thick mass of sediments. It seems to me inadvisable to extend such terms as anticlinals, from bedding planes in sediments, to the less regular foliation planes, developed by flow or metamorphism, in igneous rocks.

Further, the structure of the lode seems to me essentially different from that of a true saddle reef; the sections on fig. 5

[1] An excellent account of the field up to 1900 is given by E. F. Pittman, *The Mineral Resources of New South Wales*, 1901, pp. 92–107.

show that the lode is in the main a vertical lode, and it seems to me to have been formed along a powerful fault plane. A branch of the lode passes off to the south-east, but is not always developed (as in fig. 5, c); the branch sometimes follows a folia-tion plane, but elsewhere cuts directly across the foliation planes.

Similar saddle-shaped offshoots are not uncommon in mines; the same branching is shown in the North Lyell Mine (fig. 6), and another occurs at Kalgoorlie, where, in September 1904, I had the opportunity of studying a similar saddle-shaped arch in the telluride lode of the Oroya Brown Hill Mine.

The use of the term saddle reef is, of course, only a matter of definition; and the term could be so defined as to include lodes in igneous rocks and any lodes giving off saddle-shaped flaps or pockets. But the term was proposed for isolated arch-shaped lodes in folded sediments; and its extension to branches off a vertical lode in foliated igneous rocks seems to me inconvenient, as likely to encourage erroneous views as to the distribution of the ores, and to lead to the laying out of mines on unprofitable lines.

(d) The Separation of the Broken Hill Zinc Ores

The Broken Hill mines have a special interest at the present time from the ingenious processes, which have at length con-quered the difficulties, that have hitherto prevented the utilisation of the zinc in the ores. The blende is associated with rhodonite, a mineral which has almost the same specific gravity. Hence it was impossible to separate these minerals by ordinary washing processes, and the zinc ores have been thrown on to the waste heaps. The tailings heap of the Broken Hill Proprietary Mine alone is estimated to contain 2,500,000 tons of ore already crushed and lying on the surface. Efforts were made to separate the blende magnetically, with only partial success. In one of the new processes the ore is treated in an acid solution, whereby bubbles of gas are formed on the blende, and not on the rhodonite. The gas keeps the blende floating on the surface of the bath, whence it is skimmed off as a scum, while the rhodonite falls at once to the bottom of the tank. This process and others are now being discussed in the law courts; but to whomsoever the credit belongs, this method has conquered the difficulties which have hitherto rendered it impossible to utilise the zinc of the Broken Hill ores.

5. THE INDICATORS OF BALLARAT

That Australia was one of the great goldfields of the world was first proved in 1851 by the discovery of the gold-bearing gravels of Golden Point, at Ballarat East. The quartz veins from which these gravels gained their gold are of interest in mining theory, owing to the indicators, which have secured for Ballarat mention in most recent books on ore deposits.

The early miners of Ballarat were faced by the paradox of finding gold in large nuggets and coarse grains in the gravels, and yet the adjacent hills, though the rocks were exceptionally well shown, contained no quartz veins comparable in size to the nuggets shed from them. Hence arose the belief that the nuggets grew by slow accretion in the gravels from percolating gold solutions. This theory has been now almost universally abandoned; but it is of historic interest as indicating the difficulty of believing, that the nuggets could have been derived from the small, irregular quartz veins, that seam the hills of Ballarat East.

The nuggets found in the surface gravels and soils of Ballarat have been found along a line running north and south through the town and suburbs of Ballarat East. The nuggets were no doubt formed in small quartz veins, where they are cut across by narrow seams called indicators. These indicators have been generally regarded as originally layers of sediment especially rich in organic matter. Each indicator was thought to have been deposited as one bed in the succession of clays and sandstones laid down on the floor of a Lower Palæozoic sea. The decomposition of the organic matter was considered to have formed iron pyrites in these particular bands of clay. Later on the rocks have been tilted, until they are nearly or quite vertical, and at the same time the clays have been cleaved into slates and the sandstones altered to quartzites. Solutions percolating through horizontal cracks in these rocks deposited vein quartz; and it was thought that the gold in the solutions precipitated patches or nuggets opposite the seams of iron pyrites.

This theory has been so widely expressed in mining literature that it was a great surprise to me to find, on surveying the Ballarat East mines for the Victorian Mines Department, that it cannot be confirmed. The indicators are not interstratified beds; but as had been maintained by Mr. William Bradford, of Ballarat, they cross the bedding planes irregularly. They are secondary in origin, and, though they are limited to particular bands of slates, they do not occupy uniform positions in these bands.

Microscopic examination of the indicators, moreover, show that they are not layers of pyrites. This view was based on the fact that they appear, in the upper levels of the mines, as thin iron-stained bands; and the iron was thought to be derived from decomposed pyrites. The typical indicators are seams of chlorite; one of them, the Pencil Mark, is a thin band, rendered black by the abundance of fine needles of rutile.

The indicators are seams of secondary minerals, generally chlorites, developed along slip-bands traversing the slates.[1]

[1] Since this article was written a somewhat longer account of the Indicators, with some illustrations of their microscopic structure, has been published in *The Mining Journal*, vol. lxxix. p. 78.

THE CORN SMUTS AND THEIR PROPAGATION

By T. JOHNSON, D.Sc.

Professor of Botany in the Royal College of Science, Dublin

A FEW years ago the reading public was startled by the statement that if the population of the world continued to increase at the present rate there would come a time when a famine would occur owing to a shortage in the supply of corn. This shortage would be due to the fact that as the corn plant cannot be grown without nitrogenous food materials in its mineral food supply, and that as the amount of nitrogenous matter in the soil is limited, the increasing nations, using more and more of the earth's surface for their corn crops, would in time reach the limit of supply.

I do not propose to stop to consider the successful steps now being taken to utilise the atmospheric nitrogen to supplement the soil's supply, but to devote attention to the advance made by botanists in the detection and prevention of the loss of nitrogenous matter already gained. Probably as long as corn has been grown a cause of loss of the nitrogenous food material in the grain of corn, gained by the corn plant's metabolism from the nitrogenous food material in the soil and stored in the grain in a form available for utilisation by man, has been at work. I mean the loss due to the diseases of the cereals we now recognise as caused by certain fungal pests known as "smut" and "rust."

The loss inflicted by these two scourges in the corn-growing countries of the world is enormous, and yet it is only within the last two generations that our knowledge of their nature has become at all accurate. I shall in this paper confine my remarks almost entirely to the smut disease.

My own practical (as distinguished from scholastic) interest in the smut question was first aroused by the sight of the oat crops in the west of Ireland, where, in the islands on the seaboard, the crops were, in many places, in a deplorable state

owing to the prevalence of smut. The appearance of a smutty
oat plant is well known. When the disease is most advanced
the oat grain, first converted into a black powder, is completely
destroyed, and the powder—the smut spores—are dispersed,
leaving of the original flowering panicle merely a skeleton. Such
smutty ears do not tell the whole story of the loss; there are
many heads not present at all, because, in a still earlier stage
of disease, the oat plants have been killed off by the smut or the
oat grains have failed to sprout because of it.

As late as the middle of the eighteenth century the black
powdery substance of the grain was regarded as part of the
corn plant itself—a degeneration of the substance of the grain,
deposited by a change in the juices of the grain, and called
" vitium, morbus, et pestis."

In the middle of last century the brothers Tulasne published
a beautifully illustrated account of their investigations of the
rusts and smuts, and showed how closely allied the two forms
of fungi responsible for these diseases are. They showed
that in each case the resting spore (teleutospore or teleuto-
gonidium of the rust, chlamydospore of the smut) on germination
produces a short filament, jointed or unjointed, and that from
each joint a secondary spore arises. When the filament is
unjointed, then the secondary spores arise from its apex or free
end as a rosette. The thread, jointed or unjointed, arising from
the resting spore has been called the promycelium, and the
secondary spores or conidia arising on this promycelium the
sporidia. It is only within quite recent years that the affinities
of the rusts and smuts to the mushroom, toadstool, and puff-
ball group—the Basidiomycetes—have been emphasised. The
promycelium of Tulasne is now called a hemi-basidium. The
rust and smut (*Ustilago*) jointed promycelium has its counterpart in the jointed basidium of the Proto-Basidiomycetes, and
the unjointed promycelium of the bunt (*Tilletia*) has its counterpart in the basidium group of the ordinary mushroom group,
the Auto-Basidiomycetes.

The rust or smut sporidium is the homologue of the
basidiospore or basidiogonidium of the mushroom, and the
rusts and smuts are now called the *Hemibasidii*. Where close
homologues in the superstructure or after-development are
indicated the foundations should be sure and distinct. The
teleutospores of rust and the chlamydospores of smut are

definite bodies of well-marked origin. The same cannot be said of the cell or "spore" from which the basidium, jointed or unjointed, of the Basidiomycetes arises. As the apex is the only part in the basidium with a free surface in the compact hymenial layer of the Basidiomycetes, I do not think too much stress should be laid on the apical origin of the basidiospores in the comparison with the apical origin of the sporidia in the *Tilletia* promycelium. The affinities of the rusts and smuts are admitted, and I am, I realise, in a heterodox position in suggesting caution before accepting the view of similar affinities with the mushrooms, etc.

The next advance in our knowledge of the smuts after the Tulasne work was made by Kühn, the director of a large estate in Silesia. Kühn attacked them from a practical point of view, and though he protests he writes as a practical man, he shows a thoroughly scientific knowledge of his subject, and in his book, *Die Krankheiten der Kulturgewächse*, published in 1848, he makes many valuable contributions to knowledge. He convinced himself and his neighbours that by treating the seed before sowing with such a fungicide as copper sulphate the loss from smut could be reduced to a minimum.

He gave himself a great deal of trouble in trying to ascertain at what stage in the corn plant's life-history the smut spore attacked it, and was finally rewarded by the discovery that it is the seedling (he thought the seedling only) that is attacked, and that, when the corn plant has passed the seedling stage, the mature cuticularised and hardened tissues of the corn plant are proof against penetration by the smut spore. This fundamental point, of the greatest practical importance, is indicated in a few lines only in the book mentioned. The discoveries of Tulasne brothers and of Kühn gave an apparently satisfactory and complete account of the life-history of the smut fungus, when Brefeld in 1878 turned his attention to the group. He began by testing the germination in water of as many different kinds of smut spores as he could get, and was struck by the fact that while some spores (*e.g.* those of Indian corn smut) do not sprout at all in water, the spores of other forms germinate, giving secondary spores or conidia, which, however, show very feeble power of any further growth. On the one hand, the smut disease was world-wide and very abundantly represented wherever a corn crop was grown; on the other hand, the smut spores apparently responsible for the disease refused to germinate, or germinated poorly in water. This line of thought led to the next important discovery, based on **Brefeld's** experiments, that when smut spores are placed not in pure water, but in a nutritive solution or on a nutritive medium, they germinate readily and abundantly, producing enormous numbers of conidia which bud and rebud in a yeast-like manner, as the illustrations in **Brefeld's** most valuable *Botanische Untersuchungen über Schimmelpilze* (v., xi., xii.) abundantly show. Brefeld saw at once the great practical importance of his discovery, and an explanation of the oft-repeated remark of farmers that farmyard manure had something to do with the spread of the smut disease. Brefeld's nutritive culture medium of the laboratory corresponded to the organic matter of the manure, and the generations of conidia to similar bodies in the manured soil. Further, smut spores retain their vitality or viable power several years. If, in the rotation of crops, a corn crop followed too closely a smutty corn crop, the latter would have left spores in the soil which the manure would feed in preparation for the later corn seedlings.

It will occur to the reader that there may be many other sources of contamination of the soil. Brefeld next satisfied himself that the smut spores do not attack the host plant directly, but that the conidia arising from them do. These make their entrance into a host through its soft surface tissues, by means of branching penetrating hyphæ. This attack, to be effective, must be delivered before the first blade of the corn-plant is more than half an inch long. The story now seemed complete as regards the ætiology of the smut-fungus. Corn smut is caused by a parasitic fungus, the spores of which need, for vigorous development, to live saprophytically in an organic nutritive medium which causes abundant production of conidia. These, on coming in contact with the delicate surface tissue of the host, send through it their branching hyphæ, and so the host becomes diseased, though the fact is not revealed until harvest time.

I do not think we know much more now than in 1750 as to the predisposing causes of smut. Practically, all the explanations offered have been proved not to hold in one case or another. As in the case of rust, valuable work has been carried on with the object of producing races of corn plants which would prove immune to smut. Another line of investigation has been also followed with considerable success. Various fungicides have been utilised. They all depend for their success on their power to kill the adhering smut spores without injuring the grain. I carried out in 1900, for the Congested Districts Board, a number of experiments with various fungicides of proved value.

Hot water, potassium sulphide, "sar" (essentially sodium sulphide), copper sulphate, and formalin were all tried with success. Smutted oats soaked in water at a temperature of 132° F. for five minutes are not injured appreciably in their sprouting power, but the smut spores clinging to the grain are killed. Formalin (0·2 per cent. solution), used for two hours, has a similar action. Though every precaution was taken in these experiments, and the time, trouble, and expense were fully justified by the results, yet the crops were not entirely free from smut. In the west of Ireland this result was easily explicable. In many cases potatoes and oats alternate year after year, and I felt I might be putting treated seed into a smutty bed. In other cases such a solution is not available. What, then, is the explanation here? Is the grain harbouring within itself some stage in the smut fungus which is not killed by the surface action of the various fungicides?

This same persistence of disease beyond the known explanations available has troubled rust investigators also. Ericksson, who has done so much both in the investigation of the rust diseases, and in the attempt to produce races of corn immune to rust attack, has proposed an explanation in his well-known Mycoplasm theory.

According to this theory the rust fungus exists in the corn plant in a dormant state—in a plasmodium-like form—in the protoplasm of the host cell, and takes on the definite mycelial (thread-like) form when the (unknown) conditions become favourable to its further development. If the conditions are not favourable, it remains in its dormant condition; the corn plants grow to maturity and appear to be free from rust. I cannot help comparing the various arguments and explanations by which this theory is supported with the successive explanations offered in the case of fertilisation in flowering plants thirty or forty years ago.

The rust and smut fungi are so closely allied that one would expect a hypothesis such as the mycoplasm theory to be applied to the smut question. That this has not occurred is, in part, due to the fact, I think, that it is only quite recently that one has realised that the failures in the attempt to eradicate smut from grain by treatment with fungicides are not due to the imperfection of the method of treatment, or to the presence of smut spores in the soil in which the treated seed is sown. It is well known that the adaptation of the smut fungus to the corn plant is so complete that there is, in the growing corn plant nursing the smut pest, practically nothing to distinguish it from a growing smut-free plant until the corn plant goes into flower and fruit, when the smut reveals itself by its dark colour and destructive action. Is it possible that just as the growing corn plant while actually harbouring the smut is apparently healthy, so the corn grain may in some cases, owing to the difference in the time of attack of the host plant, be also apparently healthy, but yet in reality have in it the mycelium of the smut fungus? The latest work in the investigation of smut diseases supplies in part an answer to this question. Already in 1895 Brefeld had published a further instalment of the results of his smut investigations which gave the key to the next advance. We have seen that the oat plant is infected in its earliest stage of germination, and that the adaptation of the smut fungus to the host is so complete that it is only when the host forms its flowers the smut reveals itself in these flowers and fruit—*i.e.* at the end of the plant opposite to that at which it enters.

In the Indian corn (*Zea Mais*) the smut is not confined to the flowering organs. The smut tubercles or warts found in the "cobs," and sometimes the size of a man's head, may be accompanied by others found in other parts of the plants, *e.g.* on the adventitious roots at the basal nodes, and even on the styles and stigmas of the flowers. This is not a general outburst of smut galls in a plant due to one attack. Each is a local development produced by a local attack. If adventitious roots are attacked, the disease is local. Germination of the *Zea* smut spores gave Brefeld the key. The spores do not germinate in water; they require a nutritive solution in which to sprout. The conidia in the end form a mycelium on threads of which certain aërial conidia arise. It is these which are the sources of infection: they are carried by the air twenty yards or more, and wherever, in the Indian corn plant, they come into contact with an exposed surface of young tissue, there they are capable of setting up a local smut attack. Strange to say, they do not attack the seedling. They do, however, attack the flower and can cause a purely local stigmatic tubercle. The generalisation that young exposed host tissue was liable to attack from wind-borne conidia, and that in *Zea* the flowers presented such vulnerable tissue, led Brefeld to suspect that probably other cereals are also liable to infection in their flowers. Experiments in this direction were started, but owing to difficulties of various kinds came to nothing until Brefeld was appointed to Breslau, where, with the help of Dr. Falck, an experimental plot, and a botanical laboratory, the two observers during the past five years have carried out their experiments successfully. The oat, wheat, and barley were the three chief subjects of investigation. Two methods of inoculation were tried. In the first the inflorescence, when the flowers were judged to be open and the stigmas ready for pollination, were placed, in a glass cylinder closed below by cotton wool. A number of smutty flowers was, in the meantime, placed inside a rubber bag with suitable tube and opening. By means of this sprayer the smut spores were puffed into the cylinder and allowed time to settle. The operation was not repeated for the same spike or panicle, and as the flowers were not all in an open receptive state the action could not be compared for effectiveness with the possibilities in nature. Here the smutty heads and the blooming healthy ones are simultaneously ripe, and the smut spore can, one would expect, bombard the healthy flowers from time to time as they become receptive, if flower inoculation, as in *Zea*, occurs. The other artificial method tried is more delicate. With the help of a lady's hand (says Brefeld) and a fine camel's hair brush smut spores are placed on the stigma and ovary wall of an oat or other cereal, as little force as necessary being used to open wide enough for the operation, the bracts of the flowers. In this way many individual flowers can be, with certainty, inoculated. Those not treated can be cut off. The method is more reliable but slower than the cylinder one.

Examination by microscope of some of the wheat flowers a few days after inoculation showed that the smut spores had sprouted in the stigmatic secretion and had sent hyphæ into the stigma and through the stigmatic tissue towards the ovary. When the harvest of inoculated plants was gathered, it was found that both the cylinder-infected and the hand-infected specimens gave grains which seemed quite healthy and normal. They were stored and in the following year sown under strict sterilisation conditions. The result was astonishing, in that in the case of some of the pots of grains from hand-infected flowers every plant showed a smutty flower spike,[1] and in the case of the cylinder-infected plants the resulting grains gave on germination, on an average, as many as 26 per cent. of smutted plants. The evidence was conclusive that wheat plants can be infected in their flowers with air-borne smut spores, and that the resulting infection reveals itself not in the same year, but first in the following season in the smutted ears at harvest time. Microscopic examination showed the abundant presence of smut mycelium in the apparently healthy grains, and this mycelium became more evident in the growing point, etc., as the grains sprouted. Experiments under various forms showed, too, that sprouting wheat seedlings, unlike oat ones, are almost or quite free from attack by smut spores. Thus in wheat the chief cause of smut is the infection of the flowers by air-borne spores. Infection of the wheat seedling is an almost, if not quite, negligible cause. Infection experiments on barley similar to those on wheat were carried out. Barley proved less favourable than wheat, as the flower proper is less exposed and there is more difficulty in consequence in making sure that the smut spores actually reach the stigma or ovary wall without injury being done to the flower in the operation. The flowers, too, in the same spike are not so uniformly and simultaneously open. Still the results of infection are in principle the same. In the case of hand infection of individual flowers total infection resulted in one case, and in cylinder infection as many as 20 per cent. of smutted grains were obtained. Infection of barley grain or barley seedlings was unsuccessful. Hence in barley, as in wheat, the chief, if not the only, source of smut infection is the smutty ears in the field. The wind which carries the pollen grains to the stigmas, and thus gives cross and wind pollination, carries also the smut spores from

[1] This is well shown in the illustrations accompanying Brefeld's paper (Bd. xiii. *op. cit.*).

diseased ears to healthy ones, and the ripening grains thus actually infected look in the harvest apparently healthy. They give, however, smutty plants the following season. Both in barley and in wheat the smut mycelium has been found in the apparently healthy grains of the artificially infected flowers. There is thus no necessity here for the introduction of any hypothesis such as the rust mycoplasm theory. Rather the discovery of a definite smut mycelium tends to throw further doubt on the possibility of the existence of the rust mycoplasm.

Another point of interest seems worthy of attention. At one time the powdery smuts of wheat, barley, and oats were regarded as one species, *Ustilago Carbo.* Twenty years ago spore cultivation showed that the spores from these three hosts did not behave alike. The oat smut spores sprouted quite differently from the other two, and accordingly the oat smut was called *Ustilago Avenæ.* The smut spores of barley and wheat are indistinguishable in appearance, their mode of germination is the same, and, in the opinion of Brefeld, they are one and the same species. They are, however, now known as *Ustilago Tritici*, Rostrup (wheat smut), and *Ustilago Hordei*, Bref. (barley smut).

It would be a useful and interesting piece of work to test how far they are species of distinct morphological value, or if, being one morphological species, they are biologic forms or species. I am not aware that any attempts at infection of wheat plants by barley rust or the converse have been made. The matter is of sufficient biological and practical interest to deserve experimental attention.

From what has already been said, it follows that in wheat and barley the danger of infection of seedlings from smut spores in the soil or clinging to the grains is slight or non-existent, and that grain treatment by fungicides is valueless. It is impossible to pluck out of an infected field the diseased plants, so that the farmer should, as far as possible, use for sowing purposes grain coming only from fields known or guaranteed to be smut free. Soaking with fungicides of wheat or barley grains harbouring the smut mycelium does not reach the mycelium, and the death of the adherent smut spores is objectless, as they are harmless.

Oat smut is in its general characters like barley or wheat smut; its spores, too, are, as seen microscopically, similar.

10

Biologically, however, in two essential points the differences are marked. The smut spores of barley and of wheat do not retain their vitality long—beyond some twelve months—*i.e.* probably not into the second year after their formation. The oat smut spores, on the contrary, are capable of sprouting for several years after their production. Further, in a nutritive medium, the oat smut spores sprout, and form the jointed pro-mycelium or hemi-basidium, the segmented members of which give rise to conidia. These conidia go on forming endlessly as long as the food material lasts, budding and forming yeast-like conidia. These conidia finally send out the penetrating hyphæ. In wheat and barley smuts this abundant conidial formation does not occur. Manured soil is then the most suitable medium oat smut spores could have for their almost interminable proliferation. Artificial infection of the healthy oat plant in flower, and of oat grains sprouting, shows that, while in nature infection of the flower from smutted oat panicles by the help of the wind may happen, the main source of infection is from the spores left in the soil or clinging to the sown seed. Hence in cultivation, treatment of oat seed with a fungicide and appropriate rotation of crops are the proper steps to take to ensure the crop against smut.

These interesting and important results of the investigation of the mode of infection of the wind-pollinated grasses led naturally to a re-examination of cases of certain insect-pollinated flowering plants showing smut in their flowers. The *Caryophyllaceæ* or Chickweed order—many with white night-opening flowers—furnish convenient examples, and one of those, *Melandrium album*,[1] was specially selected by Brefeld for experiment. Though the experiments were not carried out on a very large scale, and suffered from various disturbances, the general results show that the sticky smut spores are carried by insects from the diseased anther to the stigma of the female flower, that the spores sprout in the stigmatic juice, and finally infect the ovules. The seeds planted next year give smutty plants. Here too we have infection of flower from smutty flower, and smut spores living saprophytically. The chief difference from barley and wheat is that the smut spores are conveyed with the pollen grains from one flower to another by insects, not, as in these grasses, by the wind. A third

[1] *Lychnis alba*, Mill.

mode of conveyance of pollen from flower to flower is known— by water. Is it possible that smut spores can be similarly carried from diseased to healthy plants? Brefeld has satisfied himself that in the case of *Doassansia* and other smut genera found infecting water-plants like *Alisma* and *Sagittaria*, the conidia reach the submerged young leaves of the host plant by means of water. Imitating the terms used in pollination, one might say that the host plants are anemosporous, entomosporous, and hydrosporous according as the smut spores reach the host by wind, insect, or water respectively.

The relationship of the millet to its smut fungus (*Ustilago Sorghi*) is interesting. Evidence is not yet sufficient to decide whether the millet flowers, like those of barley and wheat, can be infected. The evidence goes to show that in millet (*Sorghum*), *Panicum*, and *Setaria*, infection through the flower is rare or absent. The seedlings, however, like those of oats, are infectible, and the result is not evident until the millet comes into flower. If the millet seedling grows very quickly the smut mycelium which has already "taken" cannot grow fast enough to reach the growing point, and though the fungus is in the joints of the millet stem, the inflorescence itself appears healthy. If, however, as soon as the healthy flowers appear, the plant is cut across at about two-thirds its height, side shoots arise, and these, becoming occupied by the smut mycelium from the nodes, give rise in their flowers to smut. Further, if old seed of millet be sown and, when sprouting, be infected artificially, the slower rate of growth of this old seed allows the smut mycelium to reach the growing point. In such cases the millet flowers on appearing are smutted to such an extent that no single flower escapes. Old seed of barley and wheat, similarly treated, gave no trace of successful infection, *i.e.* retardation of rate of growth of barley and wheat by the use of old seed does not help the smut mycelium to intrude itself into their growing points, because, as we know, the grains when sprouting are not infectible, and there is thus no mycelium in the host to be favoured by its slower rate of growth.

The now well-known discovery of Hellriegel that beans, clover, lupine, and other *Leguminosæ* possess the power, through the presence of microbes, known as *Rhizobiæ*, living parasitically in their roots and forming root nodules, of fixing the free nitrogen of the air and utilising it as food material, has suggested to others that possibly other green hosts with their accompanying parasites might possess similar powers. Perennial shrubs and trees with their mycorhiza suggested themselves for investigations.

In such an inquiry several conditions suggest themselves as desirable. The host should quickly respond to the presence or absence of nitrogenous food, and should harbour the parasite without, for the most part, suffering any disturbances in its normal mode of life. There should also be a tolerable degree of certainty that the host has been actually invaded by the parasite. These conditions are admirably met by artificially smut-infected cereals. Brefeld cultivated during several successive years several kinds of millet, also barley and wheat, using in some cases infected seedlings, in others infected grains. All the plants were annuals, and reached maturity in the one season. In one series, A, the plants were grown in pure sterilised sand soaked with mineral food solution, no nitrogenous compounds being present. In the other series, B, such compounds were given. After three or four weeks the seedlings in series A ceased to grow, and would have died had not nitrogenous food material in the form of calcium nitrate been added. In series B the plants grew normally and compared favourably with the plants from healthy grains grown in the open, until the flowering time, when all the heads in B showed themselves smutty. Brefeld's experiments show that probably ordinary fungi have no power of utilising for their hosts the free nitrogen of the air as the bacteroids (*Rhizobiæ*) do for their leguminose ones.

Summary

1. Smut and rust are diseases due to fungi, the spores of which were examined and caused to sprout by the brothers Tulasne. Their well-illustrated investigations of sixty years ago hold good to-day, and their hypothesis of close affinities between the two groups has been accepted as an established fact.

2. Kühn's observation of the infectibility of oat seedlings holds good, but does not bear the general (and at the same time restricted) application to all corn seedlings botanists were inclined to give it.

3. Brefeld showed that the spores of some species of smut will not germinate at all in water, but that all kinds germinate in an organic food solution and generally produce conidia or secondary spores, sometimes very abundantly. Thus the smut fungus growing parasitically in its host tissue produces one kind of spore—the chlamydospore; and this spore growing saprophytically in an organic food solution produces another kind—a secondary spore, the conidium, or sporidium of older writers. I propose, in order to indicate the existence of these two sources of food supply, to call the smut fungus "heterositic" (Greek *heteros*, one of two, the other; and *sitos*, food). The ordinary rust fungus is heterœcious, needing two different hosts, to complete its life-cycle.

4. The infection of the host in the case of oat plants takes place in the seedling stage—the fungus incubates or leads a concealed life in the host and only shows its presence by its dark-coloured spores in the oat inflorescence.

5. In the case of wheat and barley the seedling is not attacked by the smut spores. Here it is the flowers which give entrance to the pest, which reveals itself first in the crop in the following season. Hecke has infected barley flowers by moistening them with water containing smut spores in suspension.

6. In Indian corn the seedlings are immune, but localised smut tubercles occur, owing to local attack, wherever young tissue is exposed to the invasion of the aërial conidia of the smut of this host.

7. Fungicides and rotation of crops are effective means of protecting oats.

8. Fungicides are useless for wheat and barley grain treatment, because the seedlings are immune to external attack and the internal mycelium is not reached by the fungicide. Safety lies only in securing seed from healthy crops.

9. Fungicides are useful for Indian corn, because by them any adhering spores are killed and the aërial conidia are thus prevented from forming.

NEHEMIAH GREW AND THE STUDY
OF PLANT ANATOMY

By AGNES ROBERTSON, D.Sc.

Fellow of University College, London

THE condition of any branch of science at the present moment can hardly be understood without a consideration of its history and a reference to the work of those inquirers to whom it owes its origin. I think we may fairly say that the pioneers in each subject are to-day held in due honour, but still the fact remains that for the most part they are left unread! This is much to be regretted, since their work is often most suggestive, and not infrequently exhibits a breadth of outlook for which we seek in vain in the writings of modern specialists. And in any case it cannot fail to be of value to workers of the present day to thoroughly understand what aims and expectations were in the minds of the founders of their subject when they first broke the ground. The subject of plant anatomy has made gigantic strides of late years, and it may perhaps be of some interest, in view of the records of recent progress in the subject which will appear in this journal, to turn for a moment to that classic of botanical literature, Nehemiah Grew's *Anatomy of Plants*. This work, which is the first in the English language dealing seriously with the subject, is a large and leisurely folio published in 1682, and dedicated by the author (who was M.D. of Leyden, Fellow of the Royal Society and of the College of Physicians) to " His most sacred Majesty Charles II." In view of the lengthiness and comparative inaccessibility of the volume, which practically comprises the author's collected botanical works, it may be useful to place on record some slight sketch of its contents. " Your Majesty will find," says Nehemiah Grew in his dedication, "that there are Terræ Incognitæ in Philosophy as well as Geography. And for so much, as lies here, it come to pass, I know not how, even in this Inquisitive Age, That I am the first, who have given a Map of the Country. . . . In sum,

Your Majesty will find, that we are come ashore into a new World, whereof we see no end."

Nehemiah Grew has sometimes been accused of borrowing from the Italian, Marcello Malpighi, whose great work on the same subject was laid before the Royal Society in 1674. But it is certain that the first part of Grew's work was in the hands of the Bishop of Chester a year before Malpighi's earliest communication reached England, and there seems, I think, no sufficient reason to doubt the essential independence of the English botanist's work. He explains as follows the reasons which induced him to take up the study of plant anatomy: " The first occasion of directing my Thoughts this way, was in the *Year* 1664, upon reading some, of the many and curious Inventions of Learned Men, in the *Bodies* of *Animals*. For considering, that both of them came at first out of the same *Hand*; and were therefore *Contrivances* of the same *Wisdom*: I thence fully assured my self, that it could not be a vain Design; to seek it in both. And being then newly furnished with a good stock of *seeds*, in order to raise a *Nursery* of *Plants*; I resolved, besides what I first aimed at, to make the utmost use of them for that purpose: that so I might put somewhat upon that side the *Leaf* which the best *Botanicks* had left bare and empty. . . . And although it seemed at first an Objection in my way, That the first projectors seldome bring their business to any good end: yet I also knew, That if Men should stay for an Example in everything; nothing extraordinary would ever be done."

Grew refers to the remarks on plant anatomy which occur in Robert Hooke's *Micrographia*, but justly points out that such observations as he made were merely by the way, and that he never set out definitely to make a complete investigation of the subject. With regard to Grew's methods there is little to be said. He used both naked eye and microscope, and refers to perpendicular, transverse, and oblique sections, " all three being requisite, if not to Observe, yet better to Comprehend, some Things." He held particularly sane views on the education of students: " what is learned," he says, " by their own Observation, will abide much longer in their mind, than what they are only Poynted to, by another."

When we turn to the special part of the book—the anatomical descriptions of the several organs,—we find that in the root he describes the "Lignous Body" as forming a "slender Wyer or Nerve," pithless in the smaller roots, but containing a pith in the thicker parts. "The *Parenchyma* of the *Barque*, is much the same thing, as to its conformation, which the Froth of *Beer* or *Eggs* is, as a fluid, or a piece of fine *Manchet*, as a fixed Body." The vessels, he finds, are not twisted together or branched, but run straight and parallel side by side. The spiral thickening in the vessels he describes as consisting of "Two or More round and true *Fibres*, although standing collaterally together, yet perfectly distinct. Neither are these Single *Fibres* themselves *flat*, like a *Zone*; but of a *round* forme, like a most fine *Thred*." Grew observed the radial arrangement of the wood in the root, and propounds a rather mysterious theory to account for it. "Some of the more Æthereal and Subtile parts of the Aer, as they stream through the *Root*, it should seem, by a certain *Magnetisme*, do gradually dispose the *Aer-Vessels*, where there are any store of them, into Rays." He distinguishes between the original skin of the root and the corky covering of old roots, and appears to have had a fairly clear idea of the mechanism of contractile roots. "The *String-Roots*, . . . which descending themselves directly into the Ground, like so many *Ropes*, lug the *Trunk* after them."

Nehemiah Grew clearly understood the main anatomical distinction between the stem and the root, for he explains that the central position of the ligneous body in the root makes it pliable to an oblique motion, whereas its circumferential position in the stem strengthens it in its upright growth. He also grasped the difference between the origin of stem and root branches. "In the Growth of a *Bud*, and of a *Trunk-Root*, there is this observable difference; That the former, carries along with it, some portion of every *Part* in the *Trunk* or *Stalk*; whereof it is a *Compendium*. The latter, always shoots forth, by making a Rupture in the *Barque*, which it leaves behind, and proceeds only from the inner part of the *Stalk*."

In the case of the stem we find that he recognised the compound nature of each vascular bundle ("fibre"). Each, he says, is "sometimes perforated by 30, 50, 100, or hundreds of *Pores*. Or what I think is the truest notion of them, That each *Fibre*, though it seem to the bare eye to be but *one*, yet is, indeed, a great number of *Fibres* together; and every *Pore*, being not meerly a space betwixt the several parts of the Wood, but the *concave* of a *Fiber*." The phloem (bast) he naturally did not detect as easily as the wood. He generally records "Sap Vessels," as he calls them, occurring both outside the wood and at the edge of the pith. The medullary rays he calls "Insertions of Parenchyma." "These *Insertions* are likewise very conspicuous in Sawing of *Trees* length-ways into Boards, and those plain'd, and wrought into *Leaves* for *Tables, Wainscot, Trenchers,* and the like. In all which, . . . there are many parts which have a greater smoothness than the rest; and are so many *inserted Pieces* of the *Cortical Body*; which being by those of the *Lignous,* frequently intercepted, seem to be discontinuous, although in the *Trunk* they are really extended, in continued Plates, throughout its Breadth." Nehemiah Grew had more than an inkling of the nature of secondary thickening. He tells us that "every year, the *Barque* of a *Tree* is divided into Two Parts, and distributed two *contrary* ways. . . . The outer Part falleth off towards the *skin*; and at length becomes the *Skin* in itself. . . . The inmost portion of the *Barque,* is annually distributed and added to the *Wood*; the *Parenchymatous Part* thereof making a new addition to the *Insertions* within the *Wood*; and the *Lymphœducts* a new addition to the *Lignous pieces* betwixt which the *Insertions* stand. So that a *Ring* of *Lymphœducts* in the *Barque* this year, will be a *Ring* of *Wood* the next; and so another *Ring* of *Lymphœducts* and of *Wood,* successively, from year to year." He goes on to show how different is the annual growth in the case of different trees, and even of different years of the same tree. He understood also the distinction between Spring and Autumn wood; "on the inner *Verge* of every annual *Ring* of *Wood,* . . . the old *sap-Vessels* grow much more compact and close together." Grew tells us then the common opinion in his day was that "the *Barque* only surrounds the *Body,* as a *Scabbard* does a *Sword*"; and points out that this is a mistake, for they are truly continuous. "Now the reason why the Barque nevertheless slips so easily from the *Wood,* is plain, viz. Because most of the *Young Vessels* and *Parenchymatous Parts,* are there every year successively formed; that is betwixt the *Wood* and *Barque*; where the said *Parts* newly formed, are as tender, as the tenderest *Vessels* in *Animals.*"

Grew had only an imperfect conception of the cellular structure of plants. The cells were not in his eyes the fundamental things; it was rather the fibres forming the cell walls which seemed to

him of the first importance. These fibres he believed to be
continuous from cell to cell, and he thought that some parts
of the plant consisted of fibres alone, not woven together to form
" bladders" or cells. His elaborate comparison with "pillow
lace" gives the clearest idea of his mental picture of plant
tissues. "The most unfeigned and proper resemblance we can
at present, make of the whole *Body* of a *Plant*, is, To a piece
of *fine Bone-Lace*, when the *Women* are working it upon the
Cushion, For the *Pith*, *Insertions*, and *Parenchyma* of the *Barque*,
are all extream Fine and Perfect *Lace-Work*: the *Fibres* of the
Pith running *Horizontally*, as do the *Threds* in a Piece of *Lace*;
and bounding the several *Bladders* of the *Pith* and *Barque*, as
the *Threds* do the several *Holes* of the *Lace*; and making up the
Insertions without *Bladders*, or with very small ones, as the same
Threds likewise do the *close* Parts of the *Lace*, which they call the
Cloth-Work. And lastly, both the *Lignous* and *Aer-Vessels*, stand
all *Perpendicular*, and so cross to the *Horizontal Fibres* of all the
said *Parenchymous Parts*; even as in a Piece of *Lace* upon the
Cushion, the *Pins* do to the *Threds*. The *Pins* being also con-
ceived to be *Tubular*, and prolonged to any length ; and the same
Lace-Work to be wrought many Thousands of times over and over
again, to any thickness or hight, according to the hight of any
Plant."

It is for his feeling for morphology that Nehemiah Grew
seems to me to have been chiefly remarkable, as witnessed, for
instance, by his lucid accounts of the nature of bulbs, and of
thorns. In "all Bulbous Roots . . . the *strings* only, are absolute *Roots*; the *Bulb*, actually containing those *Parts*, which
springing up, make the *Leaves* or *Body*; and is, as it were, a
Great *Bud* under ground." The Thorns of the Hawthorn "are
constituted of all the same substantial *Parts* whereof the *Germen*
or *Bud* it self (is), and in a like proportion : which also in their
Infancy are set with the resemblances of divers minute *Leaves*."
Tendrils, or as he calls them "Claspers," are regarded as having
the natures of stems and roots compounded together, which
he ingeniously observes is shown by "their Circumvolutions,
wherein they often mutually ascend and descend." He has got
hold of the generalisation that the stems of twining plants are
apt to have their vascular system concentrated about the central
axis. "As to their *Spiral Motion*, it is to be noted; That the
Wood of all *Convolvula's* or *Winder's*, stands more close and

round together in or near the Center, thereby making a round, and slender *Trunk*. To the end, it may be more tractable, to the power of the external *Motor*, what ever that be: also more secure from breaking by its winding *motion*." The winding of some plants in one direction and some in the other he surmises to be due respectively to the influence of the sun and the moon. The following passage seems to be an anticipation of modern work on the nutation of non-climbers [cf. Darwin's *Power of Movement in Plants*]. "The *Convolution* of *Plants*, hath been observed only in those that Climb. But it seems probable, that many others do also *wind*; in which, the main *stalk*, is as the *Axis* to the *Branches* round about. Of which number, I conceive, are all those whose *Roots* are twisted; . . . whether it be so, or not, the Experiment may easily be made by tying a *Thred* upon any of the *Branches*; setting down the respect it then hath to any Quarter in the *Heavens*: for, if it shall appear in two or three Months, to have changed its Situation towards some other Quarter; it is a certain proof hereof."

Dimorphic leaves interested Nehemiah Grew a good deal; for instance, those of the "Little Bell" (Harebell). His explanation in this case is that the radical leaves were formed in the seed, whereas the upper later-formed leaves have been fed with a different sap supplied direct by the root. He enters into a detailed account of methods of vernation and bud-protection, which shows careful and extensive observation. He notes, for instance, the mucilage which occurs in "the first *spring-leaves* of all kinds of *Docks*; betwixt the *leaves* and the *Veil* wherein they are involved." The æstivation of flowers appeals to him no less than the vernation of leaves. In the Poppy, for instance, the petals "are cramb'd up within the *Empalement* by hundreds of little *Wrinckles* or *Puckers*; as if Three or Four fine *Cambrick Handcherchifs* were thrust into ones Pocket."

The anthers he finds contain certain powders "which as they start out, and stand betwixt the two lips of each *Cleft*, have some resemblance to the common Sculpture of *Pomegranate* with its *Seeds* looking out at the *Cleft* of its Rind." The use of the "Attire" (andrœcium) he says is "for Ornament and Distinction to us, and for *Food* to other *Animals*. . . . We must not think, that *God Almighty* hath left any of the whole Family of his Creatures unprovided for; but as the Great Master, some where or other carveth out to all; and

that for a great number of these little Folk, He hath stored up
their peculiar provisions in the *Attires* of *Flowers*, each *Flower*
thus becoming their Lodging and their Dining-Room, both in
one." He goes on however to say that this is merely a
secondary use, and that it was suggested to him in conversation
by Sir Thomas Millington that the stamens were male organs.
To this he agrees, but adds that the andrœcium also serves
to draw off some of the redundant sap, so that only the purest
goes to the seed.

Grew gives an excellent account of seeds and fruits,
drawing special attention to the " Branchery " or vascular
system ,in the latter. His account of the Poppy capsule is a
good example of his way of attacking the difficulties of botanical
description. " The *Poppy-Head*, is a little *Dove-Coat*; divided
by Eight or Ten *Partitions*, into so many *Stalls*. On both
sides the *Partitions*, hang a most numerous *Brood* of *Seeds*.
. . . as it dries, it gradually opens at the *Top*, into several
Windows, one for every *Stall*: which are all covered with a
very fair *Canopy*. . . . As the several *Windows*, serve to let
in *Aer*, for the drying of the *Seeds*, after their full Growth:
So the *Canopy* over them, serves to keep out *Rain*. For here,
the *case* not cleaving down the *side*, as it usually doth ; should
the Rain get in, it would stand in it, as in a *Pot*, and so
rot the seeds. And as the *Canopy* serves to preserve the
seeds ; so the several *Partitions*, or *Walls*, for their better
Stowage. For by an easie survey of this little piece of
Ground, it is plain, that as they stand on both *Sides* every
Wall, there is as much more Ground for them to stand upon,
as if there were no parting *Walls*, but the *seeds* stuck all
round about upon the *Ambit* or *Sides* of the *Case*; or upon
a great *Bed* or *Placenta* within it, as in *Hyoscyamus, Anagallis*,
etc., where there is a less numerous brood." Amongst Seed-
cases Grew naturally includes the sporangium of the fern, which
he describes with great clearness, recognising that the annulus
is a " sturdy Tendon or Spring " by means of which the seed is
flung abroad.

The morphological part of the book is illustrated by a
large number of engravings, many of them remarkably good.
I can only draw attention to a very few of the points which
they bring out. Stomates are shown (though not in detail);
annual rings in stem and roots ; the pith diaphragms of the

Walnut; various types of vernation; the form of pollen grains, and so on. But perhaps the most remarkable drawing is that illustrating the anatomy of a Corin branch. A segment of the stem is shown on a large scale, cut so as to expose both radial and transverse sections, and with part of the cortex removed, so as to reveal the rays and wood as seen in tangential section.

The subject of the tastes, scents, and odours of plants seems to have exercised Nehemiah Grew a good deal. Tastes, he tells us, should be distinguished in their degrees, and these "may be extended . . . with easie distinction from *One* to *Five*: So the *Root* of *Sorrel*, is Bitter in the *first*: of *Dock*, in the *second*; of *Dog-Rose*, in the *third*; of *Dandelyon*, in the *fourth*; of *Gentian*, in the *fifth*. . . . All kinds of *Tastes*, in all their Degrees, and in differing Numbers, may be variously Compounded together: . . . in *Aloes, Bitter* and *Sweet*; the one in the *fifth*, the other in the *first Degree*; as upon an unprejudiced tryal may be perceived." It is a matter for thankfulness that the power of drawing these subtle distinctions is no longer recognised as an essential qualification in a botanist! Grew pours scorn upon the methods adopted by the uninitiated in the attempt to produce variations in flower colour. To do this, he says, "by putting the *Colour* desired in the *Flower*, into the *Body* or *Root* of the *Plant*, is vainly talked of by some; being such a piece of cunning, as for the obteining a painted face, to eat a good store of *white* and *Red Lead*."

Grew by no means confined himself to morphology, but carried out some physiological experiments and suggested many others. For instance, he placed a plant upside down in a box of mould with its shoot sticking out of a hole in the bottom, and found that the stem curved up, showing that the upward growth of this organ does not depend merely on a desire to be in the air. He held very definite views as to the ascent of the sap; and the "ladder hypothesis," a theory closely resembling his, had some vogue at a later date. He believed the rise of the sap was caused by a joint action of parenchyma and wood, the sap ascending a certain distance in the wood, and then a certain distance in the parenchyma, then back to the wood, and so on in a zigzag.

A good deal of the *Anatomy of Plants* is taken up with rather mystical discussions about the "saline principles" of plants, the dependence of the form of the leaf on certain intersecting circles, of flowers on intersecting spheres, and so on, which have rather tended to obscure the excellence of the purely morphological parts. For it is in his feeling for *form* that Nehemiah Grew seems to excel. The fact that the microscope was a comparatively new invention was very much in his favour; for not being perfectly accustomed to it, he was in the habit of perpetually checking his work by reference to naked-eye observations—a habit which unfortunately botanists of the present day tend to lose!

In conclusion I should like to quote Nehemiah Grew's own apology for the magnitude of the task he had undertaken: " The Way is long and dark: and as Travellers sometimes amongst Mountains, by gaining the top of one, are so far from their Journeys end; that they only come to see another lies before them: so the Way of *Nature*, is so impervious, and, as I may say, down Hill and up Hill, that how far soever we go, yet the surmounting of one difficulty, is wont still to give us the prospect of another. . . . A War is not to be quitted for the hazards which attend it; nor the *Councils* of *Princes* broken up, because those that sit at them, have not the Spirit of Prophecy, as well as of Wisdom. To conclude, if but little should be effected, yet to design more, can do us no harm: For although a Man shall never be able to hit *Stars* by shooting at them; yet he shall come much nearer to them, than another that throws at Apples."

THE UTILISATION OF PROTEIDS IN THE ANIMAL

BY F. G. HOPKINS, D.Sc., F.R.S.

Reader in Chemical Physiology at the University of Cambridge

As an essential part of the all-important results of photosynthesis in the green plant, there arise from inorganic precursors those complex nitrogenous substances, the proteids or proteins, which form the chemical basis of every living cell, animal or vegetable.

For its supply of these the animal is ultimately dependent upon the plant, and cannot maintain its tissue equilibrium without consuming pre-formed proteid as part of its dietary. How precisely the ingested proteid is utilised; in what form the dead proteid of the food reaches the living tissues; whether it all becomes part of the living tissues before it can be broken down and energy extracted from it; what, in any case, are the precise chemical changes associated with the liberation of its energy—all these are questions which it has proved very difficult to study experimentally, and they still await final solution. They are related, however, to problems which concern a branch of physiology now at a very interesting stage of its development, and they yield a case in which opinion is at the moment emerging from the influence of older teachings. It is justifiable therefore to consider here the present attitude of physiologists towards them.

The study of animal metabolism reached its scientific phase under the influence of Liebig, and the questions which concern us were perhaps first clearly stated, as they were certainly first clearly answered, rightly or wrongly, by him. To Liebig, indeed, the fundamental aspects of proteid metabolism in the animal appeared to suggest no great difficulties, and his answers to the questions partake of the lusty self-confidence which was characteristic of biological science in his day. Deceived by the fact that the percentage analysis of any one proteid (its content, that is, of carbon, hydrogen, nitrogen, and oxygen) differed so little from that of any other, Liebig conceived of a practical chemical identity throughout the group of proteid substances. He held therefore that the proteid of the diet remained to all intents and purposes unaltered, digestive processes notwithstanding, as it slipped from the alimentary canal into its place in the living tissue, taking there a position made vacant by the previous breakdown of the tissue proteid, or leading, in other cases, to growth. While, to Liebig, this was the only fate conceivable for proteid eaten by an animal—to enter the tissues as intact proteid—his views were equally clear as to its subsequent destiny. Proteid was the sole source of mechanical energy in the body, and was only broken down—could only be broken down—when such mechanical energy was being liberated. The proteid contained in a muscle could undergo metabolism only when the muscle was actually contracting. Thus Liebig came sharply to distinguish between the part played by proteid and that played by fats and carbohydrates. These could be oxidised at any time, and supplied heat, not mechanical energy, for the body. Proteid could supply both, but the former only after being broken down in the course of tissue activity.

I am tempted to quote two paragraphs from Liebig's Lectures on Animal Chemistry, published in 1842, not only to show how clearly defined were his views on the points referred to, but to give a base from which to measure our present divergence of view. Referring to the proofs of identity in the percentage composition of vegetable and animal proteids, he says : " How beautifully and admirably simple, with the aid of these discoveries, appears the process of nutrition in animals, the formation of their organs, in which vitality chiefly resides ! Those vegetable principles which, in animals, are used to form blood, contain the chief constituents of blood—fibrin and albumen—ready formed, as far as regards their composition." And to illustrate his belief in the locking up in the tissues of all proteid consumed, and of its stability therein in the absence of mechanical activity, we have the following—" Man when confined to animal food, respires, like the carnivora, at the expense of the matters produced by the metamorphosis of organised tissues ; and just as the lion, tiger, hyæna, in the cages of a menagerie, are compelled to accelerate the waste of the organised tissues by incessant motion, in order to furnish the matter necessary for respiration, so, the savage, for the very same object, is forced to make the most laborious exertions,

and go through a vast amount of muscular exercise. He is compelled to consume force merely in order to supply matter for respiration."

This remarkable conception of Liebig's, that proteid is only metabolised during the mechanical activity of organs was in reality disproved during his lifetime, for quantitative studies of the course and rate of excretion showed that the elimination of nitrogen from the body during any given period is closely and immediately related to the contemporary ingestion of proteid, and much less closely, if at all, to the muscular work being done at the time. But to quote Liebig is not often to quote effete opinion, and his doctrines regarding the utilisation of proteid influenced physiological thought long after experiment had proved much of them to be faulty. The direct utilisation of proteid, indeed, was in its essence hardly questioned by physiologists till lately. But Liebig's teaching has probably influenced medical and semi-popular opinion still more ; and has been at the bottom of the idea that the value of a dietary to an active animal necessarily increases with the amount of proteid contained in it. It is true that this last opinion in a modified form is still held by no less an authority than Pflüger, and it must be referred to again later.

In any case, Liebig's teaching as to the simplicity of the phenomena involved in proteid assimilation is no longer to be accepted ; his conception of the proteid we eat entering the tissues as intact proteid is being replaced by a belief that the process of assimilation is much more subtle and elaborate, and his view that the whole of the proteid absorbed from the bowel is, *stricto sensu*, assimilated, appears to be giving way to an understanding that the organism deals with the material in a highly discriminative and selective manner, true assimilation of proteid being confined to a small portion of what is actually consumed.

The first task to be undertaken here is to show how our views have become thus modified.

While Liebig, Mulder, and the earlier workers did not perhaps fully realise it, a recognition of the complexity of the proteid molecule has long existed ; but it remained for quite recent analytical studies to demonstrate fully how great is this complexity. Recent work has not perhaps led to any exaggeration of our ideas as to the magnitude of the molecule, but has shown how heterogeneous are the details of structure within it. When the proteid is broken down into the stable and, comparatively, very small molecular groups, which, in the original complex, are associated together by atomic linkages labile to hydrolysis; when, that is, the proteid is fully digested by a suitable succession of enzymes, or by boiling with water in the presence of strong acids, a crowd of substances, chemically characterised as amino-acids, arises. Of these derivatives some twenty have already been recognised as arising from any one proteid, and we have good evidence for the belief that the tale of them is by no means complete.

It is not purposed, nor is it necessary, to enter at all fully into chemical details here; but it is well to illustrate, by indicating the constitution of some of the typical amino-acids, how heterogeneous is the molecule from which they are derived.

The simplest is *glycocoll*, or amino-acetic acid, CH_2NH_2COOH. The aliphatic compound which, from typical proteids, arises in greatest quantity is *leucine*, an amino-caproic acid, $\begin{smallmatrix} CH_3 \\ CH_3 \end{smallmatrix}>CH.CH_2.CHNH_2.COOH$. A dibasic acid of the aliphatic type is *glutaminic acid*, $COOH.CH_2.CH_2.CHNH_2.COOH$. Hydroxy-acids of the type of *serine*, $CH_2.OH.CH.NH_2.COOH$, also appear, and diamino-acids in which the NH_2 (amino) group occurs twice. Of aromatic substances (containing the ring structure in their molecule) we may instance the long familiar *tyrosine*, $HO\langle\ \rangle CH_2.CHNH_2.COOH$, and, as one containing its nitrogen in the ring, *prolin*, or pyrrolidine carboxylic acid,

$$\begin{array}{l} CH_2-CHCOOH \\ |\quad\ >NH \\ CH_2-CH_2 \end{array}$$

Lastly, we have in *tryptophane* a double ring compound the structure of which is, in all probability,

$$\begin{array}{l} -C-CH_2=CHNH_2.COOH \\ \quad \equiv CH \\ -NH \end{array}$$

These substances and their numerous congeners are condensed together to form the complex proteid molecule.[1] Space

---

[1] Those unaccustomed to the structural formulæ of organic chemistry need do no more than recognise the heterogeneity which these substances exhibit. Other proteid derivatives are alanine, phenyl alanine, aspartic acid, cystine (which contains the sulphur of the proteid), arginine, lysine, pistidine, glucosamine, etc.

does not permit that the question as to the nature of the linkages or the relation of any one amino-acid to others should be dealt with here. Current researches, and especially the investigations being carried out in the laboratory of Emil Fisher, are yielding us important information on points such as these, but a discussion of them may be here dispensed with.

It is easy to recognise that the heterogeneity displayed by a large molecule, such as that we must attribute to proteids, offers abundant opportunity for variation, and a recognition of such complexity raises at once the question as to whether all proteids are, after all, so nearly identical with each other as Liebig believed. Recent research indicates clearly that they are not.

While it is even possible that proteids of diverse origin may display differences in the actual permutations and combinations—in the sorting, so to speak—of their constituent amino-acids, we have, at present, no evidence to prove this; but that the constitution of one proteid may differ quantitatively from that of others is now certain. We may sufficiently illustrate this significant fact, by indicating the differences which have been found between certain prominent dietetic proteids and the proteids of the blood. Thus, comparing the chief proteid of milk, caseinogen, with the last mentioned, we find that it yields, when similarly treated, little more than half the amount of leucine; caseinogen gives about 10 per cent., the blood proteids about 20 per cent. The proportionate amount of cystine in the milk proteid is also decidedly less, while its content of glutaminic acid is notably greater. The proteids of wheat are astonishingly rich in glutaminic acid, one of them (gliadine) yielding nearly five times as much as that got from blood proteids [Osborne and Harris]. The mixture of proteids which we consume in white bread would probably yield at least four times as much glutaminic acid as would an equal weight of the proteids in our blood. Blood is comparatively rich in tryptophane, whereas a proteid which forms a large proportion of the widely used foodstuff, maize, contains none at all.

Such differences might be further illustrated, but the facts given will suffice. They show fundamental variations in the constitution of proteids which the elementary analyses relied upon by Liebig did not bring to light. Having realised such differences, it is no longer possible to conceive, with Liebig, that the proteid eaten goes intact to the tissues ; to believe that vegetables " contain the chief constituents of blood . . . ready formed as regards their composition."

In considering the question further we have, indeed, to recognise more subtle distinctions. It is, at least, probable that the proteids of one species of animal differ from those of others, even as regards homologous tissues ; that the blood, say, of the horse is not, in respect of its proteids, the same as that of the ox ;· and that each of these bloods differs still more from that of man. Evidence is accumulating to show that there is a chemistry of species ; that chemical differences in the protoplasm underlie morphological differences, and that these differences concern, in particular, the complex and easily variable proteids. We see, therefore, that the assimilation of the very diverse nitrogenous materials eaten by animals must be a complex matter, involving, in no small degree, processes of remoulding and modification in order that the normal composition and specificity of the tissue proteids may appear.

.That a proteid which differs from blood proteids in its fundamental quantitative constitution in the sense discussed above does not enter the blood stream of an animal consuming it with any trace of such deep-seated differences remaining, is strongly suggested by the results of an experiment carried out by Abderhalden and Samuely. These observers bled a horse till the amount of circulating blood proteids was greatly reduced, and then, having withheld food for a time, they fed it upon large quantities of gliadine, the wheat proteid which, as we have seen, contains five times as much glutaminic acid as do the normal blood proteids. After the horse had been kept upon the gliadine, sufficient blood was again withdrawn to permit a further analysis of its proteids. These still exhibited precisely their normal amount of glutaminic acid ; their specific constitution being unaffected by the absorption of the digested gliadine.

It is perhaps an open question, as will be seen, whether an ingested proteid enters the blood at all in the form of intact proteid molecules. If it does, the experiment just described seems to indicate that the remoulding and modification spoken of above, occurs at the very first stage of assimilation, and is brought about in the active cellular mechanism of the intestinal walls.

But in what form does ingested proteid leave the cavity of the alimentary canal? How is it presented to the absorbing mechanism? Owing to the conditions which prevail in the animal, the endeavour to obtain an experimental answer to this question has proved extraordinarily difficult; and the increasing ingenuity displayed recently in attacking it has only emphasised the elusive nature of the problem. We have no complete answer, but evidence has accumulated which is extremely suggestive, and it points to a conclusion very different from that arrived at by earlier workers. So far from the proteid being (as Liebig thought) scarcely altered in its essential composition during physiological digestion, it is probably broken down in the completest way, and leaves the gut lumen in the form of the liberated individual amino-acids.

A priori considerations have earlier stood in the way of accepting this conclusion—considerations which we now see have no weight. In the first place, a belief, due originally to the influence of Liebig, that the animal body has very small power of determining chemical synthesis, made such a conception of proteid metabolism unacceptable, for synthesis in the tissues would have to follow such a breakdown in the bowel. But increased knowledge has made us well able to credit the body with such synthetic powers as these. The animal cannot emulate the primary magic of the green plant, which, by the aid of the sun's energy, calls forth organic material with high potential energy from inorganic stuff containing minimal potential energy; but once this step has been taken, the animal cell seems to have a wide range of synthetic capacities. Another apparent objection had teleological bearings. Such decomposition of proteid in the gut would be "a waste of chemical potential energy which would serve no purpose" [Bunge]. But we now know that hydrolytic breakdown involves only minimal changes in potential energy, and the mixture of amino-acids, derived from a proteid, contains well-nigh the whole of the energy of the original substance [Rubner, Loewi].

Many considerations now prepare us indeed for a ready belief in the completeness of the breakdown.

We have come first of all to realise more fully how efficient is the organisation of proteoclastic enzymes in the alimentary tract. We know that the pepsin of the gastric juice does not merely depolymerise the proteid molecule, but breaks it down into heterogeneous fractions [Pick]. We know that the activised trypsin of the pancreatic juice is capable by itself of producing a more complete breakdown than was earlier thought [Kutscher], and the physiological succession of peptic action, followed by tryptic action, is an ordered arrangement making for still greater efficiency in the process of hydrolysis [Fisher and Abderhalden]. Lastly, we know that the intestinal cell contains, and to some extent secretes, a ferment—erepsin—which has a potent influence in promoting hydrolysis at its final stages [Cohnheim]. The existence of such an array of proteoclastic agents might conceivably be explained on other grounds; but a frank recognition of the facts certainly leads to the view that a more or less complete breakdown of our nitrogenous foodstuffs in the gut is a physiological necessity.

Again, if the full significance of the specificity of tissue proteids has been grasped, implying as it does the necessity of a re-moulding of the dietetic proteid, it will explain why the intestinal breakdown should be profound—why something more is required of the gut than the mere production of soluble absorbable material. There has to be a change in the architecture of the proteid, and we may here use an analogy, and urge that if, for instance, a gothic cathedral had to be constructed out of a classic temple, the latter would need first to be disintegrated well-nigh into its constituent stones.[1] In the simpler case of carbohydrate metabolism, the vegetable starch has to be broken down into sugar before it receives the stamp of animal life, and is resynthesised into glycogen; the vegetable proteid, on the other hand, has to obtain not merely the stamp of the animal, but that of the species.

Even in unicellular animals a mechanism for the breakdown of proteid before assimilation is always present, and in some species of amœba it has been shown that ferments of the tryptic, or more destructive type, are present [H. Mouton], and during the digestive processes of other protozoa the reaction of the fluid in the vacuoles also indicates that a tryptic ferment

[1] This analogy, or something like it, has now been frequently used, and the idea it illustrates is common property; but, to the best of my belief, it was first employed in this connection by the late Professor Huppert of Prague, who luminously wrote upon the importance of chemical differentiation in animal species, long before the subject received general attention (*Ueber die Erhaltung der Arteigenschaften*, Prag, 1896).

is at work [Greenwood and Saunders]. Significant, too, as Kutscher and Seeman have pointed out, is the circumstance that, in plants, the reserve proteid—stored in seeds, tubers, and bulbs—is utilised by the germinating organism only after complete breakdown, whereas simple solution would suffice for mere transport.

Such considerations as these should not bias us into a too ready acceptance of the experimental evidence bearing on what actually is formed in the intestine, which is admittedly still imperfect. Yet all the indications of modern research are in favour of the belief that intact proteid molecules do not normally enter the paths of assimilation. Thus Kutscher and Seeman, confirming the less complete observations of Kühne and Lea, found that the material passed through an opening made in the middle of the small intestine of a dog, in course of full proteid digestion, contained amino-acids in quantity. They failed to detect them in the blood, as have all observers ; but this is no puzzle to any one acquainted with the conditions of the circulation, and certainly does not in itself necessitate a belief in the reconstitution of proteid from amino-acids during their passage through the gut wall. The quite recent experiments of Leathes and Cathcart show, indeed, that during absorption of proteid material from the gut, non-proteid nitrogenous substances are found in the blood in a definitely increased amount.

Confirmatory evidence for the probability of a complete breakdown of proteid during digestion, is offered from another side. An interesting line of experimentation initiated by Loewi has shown that an animal maintains its normal equilibrium when fed with no proteid other than what has been previously digested as completely as possible. Whatever may be the amount of intact proteid necessary for presentation to the tissues, animals so fed must obtain it by a resynthesis, for their food contained none.

But certain very cogent facts have led recent thought to go even further than has yet been indicated in departure from the teaching of Liebig, and have suggested the view that a large proportion of the amino-acids liberated in digestion may suffer yet further breakdown before entering the domain of internal metabolism as strictly defined. It is a fact long familiar, but always puzzling, that the rate of excretion of nitrogen from the body seems to be determined by the amount of proteid eaten rather than by the presumptive needs of the tissues for a supply of energy. An increased elimination rapidly follows consumption, and well-nigh the whole of the nitrogen ingested in the form of proteid appears in the urine within twenty-four hours.

The excretion is equally complete if the amount of proteid taken be increased greatly, and it goes on just the same if the consumer remains completely at rest on a diet amply sufficient for hard work. The adult human body does not store extra nitrogen (or stores it only under very exceptional circumstances), and so, clearly, does not retain any excess of proteid. But, if so, it would seem that proteid as a source of energy suffers a great disadvantage when compared with fats or carbohydrates. These can be stored and used at convenience; the utilisation of proteid, on the other hand, is apparently compelled by its mere arrival, and cannot await the needs and conveniences of the body.

It is the great merit of a theory recently advanced by Folin that it enables us to escape from this paradoxical view. In the light of Folin's theory the rate of nitrogen excretion through the kidney shows itself, not as a measure of the rate in which energy is being extracted from proteids, but as an indicator of the activity of a process whereby the nitrogen is immediately removed from the amino-acids formed in the bowel, while their non-nitrogenous remainders are used subsequently and at convenience. Such removal of nitrogen occurs without practical loss of energy.

It was shown, two years ago, by Lang in Hofmeister's laboratory, that the tissues, generally, have the power of removing nitrogen from amino-acids in the form of ammonia; and it has since become clear that the direct removal of the amino group (NH_2) as ammonia from various substances is a characteristic chemical event in the body. It may occur by a simple process of reduction, or by one of hydrolysis, as when alanine (amino-propionic acid) is converted into lactic acid,

$$CH_3CHNH_2COOH + H_2O = CH_3CHOHCOOH + NH_3.$$

Now, during proteid digestion ammonia is undoubtedly increased in the portal blood [Nencki and Zaleski; Folin[2]], and being carried to the liver becomes the precursor of urea, which is rapidly eliminated. Folin's own very elaborate studies of human urine have shown that the nitrogen elimination which varies so intimately with the amount of proteid ingested concerns. alone that which is in the form of urea. Other nitrogenous. substances which we excrete in much smaller quantities—such as creatinine, uric acid, and ammonia—are much more stable in their output. The urea, according to Folin, is, in the main, the result of that preliminary treatment of the diet proteid which we have discussed—it is of " exogenous " origin ; the other substances arise from the living tissues, and their amount is a measure of tissue breakdown—their origin is " endogenous." I am at one with Dr. Noel Paton, and differ from its author in feeling that the chief support of Folin's wholly acceptable theory is not derived from these considerations, but rather from its. satisfactory dissociation of the phenomenon of nitrogen elimi- nation from what is, in a stricter sense, the true period of the utilisation of the proteid.

It must of course not be supposed that all the amino-acids. derived from a meal of proteid are thus further disintegrated, with loss of their nitrogen, before entering the more intimate domain of tissue metabolism, though the proportion under average conditions may be large. However great may be the stability which we ascribe to healthy tissues, we must. recognise that their repair in some degree is always necessary, and the materials for the re-formation of proteid must always. be reaching them. It is still unsettled whether or not fresh synthesis of proteid occurs in the intestinal wall. If it does, some part of the blood proteids must represent transport material. More likely free amino-acids are transported, as such, and it is only the analytical difficulty which prevents their identification in the blood.

It is interesting to remember that the proteids of one tissue differ from those of another, and the individual tissue has, like the whole animal, a need for preserving its specific stamp. If material for repair reaches it in the form of proteid in the blood, then each tissue must deal in its own way with this common. supply, and a second breakdown, preceding synthesis, would seem to be necessary. For the determination of this we should look to the activity of those intracellular ferments which are to be discussed immediately. If the free amino-acids circulate, material for selective synthesis is directly presented to the cell.

The considerations so far presented will show that our knowledge of assimilative processes is yet far from complete. They will show also, however, how far we have travelled from the original position of Liebig. Instead of the diet proteid marching intact into the tissue cell, it seems probable that the animal body treats the proteid presented to it as the plant treats its reserve proteid, that is, as a convenient store of amino-acids, on the liberation of which depends the selective utilisation which is so necessary.

If our knowledge of the assimilation of the nitrogenous foodstuffs is yet very far from complete, we know less about what happens to proteids in the tissues themselves. In spite of some difference of opinion on the point, recent research seems to me to emphasise more than ever the view that when food supply and oxygen supply are normal, the nitrogenous basis of living tissues exhibits marked stability. At the same time, however, it is, of course, certain that degradation of the tissue proteids is always going on. Upon what lines and as the result of what processes does this occur?

The most enlightening research carried out in this connection during recent years has concerned itself with what is called "autolysis" (auto-digestion) of the tissues. When an organ removed from the body is placed under antiseptic conditions, chemical processes within it do not cease. Breakdown continues or is exaggerated [Salkowski]. If we chose aseptic rather than antiseptic conditions, these processes of the surviving life of the tissue are yet much more in evidence. The antiseptic, meant to prevent the intrusion of bacteria, also interferes with the activity of the normal agents acting in the tissue, but when we avoid contamination by aseptic removal of the organ to be studied, the survival processes are found to be rapid and important [Jacoby, Magnus-Levy, and many others].

Among other changes it is found that during such survival life the proteids present in the organ concerned undergo digestion and yield products for the most part identical with those formed by ordinary digestion in the bowel. The tissue is said to undergo autolysis. We recognise, indeed, that every cell has the power of producing within itself proteoclastic (as well as many other) enzymes.

It is difficult to avoid the belief that such intracellular ferments play their parts in the normal physiological life of the cell, and that a process of hydrolysis, due to their action, precedes or accompanies other changes in the proteids when tissue breakdown is, for any reason, determined. It must be clearly understood, however, that if the cell extracts energy from its own proteid, as proteid, the hydrolytic breakdown in question is incapable of yielding this energy, which must be extracted upon different lines, or arise from other changes accompanying or succeeding hydrolysis. It may well be believed that the products of autolysis suffer the same fate during utilisation as that suffered by amino-acids arriving from the bowel, since their nature is similar.

Indeed, there is one interesting aspect of metabolism which the recognition of the existence of intracellular ferments has undoubtedly illuminated—that constituted by the events of starvation. It is well known that the wasting which occurs in the absence of food falls but little upon organs the functions of which are absolutely necessary to the carrying on of life. Thus the heart and nervous system may lose but 3 per cent. of their weight when the muscles lose over 30 per cent. Since autolysis normally yields products similar to those formed from foodstuffs in the intestine, we realise that autolysis of the less necessary tissues may, in hunger, maintain uniform conditions of nutrition for the more necessary tissues [Leathes]. This would hold, not only as regards the maintenance of structural integrity, as in the case, say, of the heart; but also in the maintenance of the supply of nitrogenous precursors of secretions essential to life, as in the case of, say, the activities of the adrenal gland. This last instance will repay some closer attention. The body exhibits in a very striking way a correlation of function by chemical means. The normal course of activity in any one organ or tissue may be wholly dependent upon the circulation of chemical substances prepared specifically by quite other organs or tissues. The substances concerned in these phe-nomena—"chemical messengers which, speeding from cell to cell along the blood stream . . . co-ordinate the activities and growth of different parts of the body"—have been called "hormones" by Prof. Starling. Among these hormones is one (adrenaline), prepared by the suprarenal glands, the circulation of which is absolutely essential to the maintenance of normal blood-pressure and the visceral functions.

Now adrenaline happens to be the only hormone of known

chemical constitution. It is a nitrogenous substance and therefore arises in some way from proteid. It is an aromatic substance,
and therefore probably arises from one or more of the aromatic
nuclei of the proteid molecule. Whatever the chemical potentialities of such cells as those of the suprarenal gland, we are well
justified in believing that they need some special chemical substance from which to elaborate the final specific product. It is
extremely probable that adrenaline is formed from one of the
aromatic amino-acids derived from proteid. Now a constant
supply of adrenaline is absolutely necessary for the body. This
being so, be it noted that the 7 grammes, or thereabout, of
suprarenal tissue which are present in the human body will,
on a liberal estimate, contain no more than some 200 mg. of total
aromatic substances derivable from the whole of the gland
proteids. But in starvation the medulla of the gland undergoes
comparatively little wastage, and 1 per cent. of the total gland
tissue is the very most that would waste in a day. To these
considerations we might add others, for it is much more likely,
in my own belief, that some one, and not all of the aromatics
derivable from proteid can serve as a basis for the elaboration of
adrenaline, and, lastly, it is unlikely that the elaboration from
the available precursor would be complete. If, then, the assumed
limitations in the chemical potentialities of the glands hold good,[1]
it will be seen that the available supply of material within the
gland itself, for the daily production of adrenaline, is almost
infinitesimal, and though the active substance is produced in
very small amount, its continued supply would undoubtedly
call for assistance from other tissues; while to supply enough
of the particular amino-acid required, and to secure its duly
reaching the gland, a not inconsiderable amount of proteid must
be broken down. Similar considerations may apply to the

[1] Some, impressed by the wide range of chemical capacities displayed by
the body, may prefer to believe that such a substance as adrenaline may be
prepared by the gland *de novo*, starting from indifferent nitrogenous precursors;
if this were the case, the quantitative considerations offered above would lose their
weight. But I believe that to attribute such indefinite powers to an animal
cell when simpler chemical possibilities are available is of the nature of resort
to vitalism, and is against what is so far known of metabolism. French workers
have offered evidence, which is by no means conclusive, that tryptophane is a
precursor of adrenaline. However this may be, some experiments indicating
that an individual amino-acid may be essential for special purposes in the body,
other than tissue repair, will shortly be published from the Cambridge laboratories
by Miss E. G. Willcock.

maintenance of the supply of numerous other hormones. In my opinion, such a call, made by special organs · upon the tissues generally, may necessitate a (hydrolytic) nitrogenous breakdown during hunger quite as much as the maintenance of the weight of the heart, nervous system, etc.; and it would go on even while a store of fat and carbohydrate is still available for energy supply. I have raised this point here because it seems to me to illustrate anew the interdependence of all tissues, and strengthens one's belief in the physiological importance of autolysis. Such phenomena, though exaggerated in actual starvation, may well occur under normal conditions to a less degree. Supply from the gut is not always contemporaneous with sudden needs, and a general readiness on the part of the whole body to supply nutrition to any one part under stress of circumstances may be secured by the existence of autolytic ferments.

If intracellular ferments function physiologically in this sort of way, it becomes a question as to how their activity is controlled. Schryver has found that acidity accelerates the course of autolysis, while alkalinity diminishes its rate. He advances the view that the ammonia made available during the digestion of food-proteids in the gut maintains tissue alkalinity, while in its absence, as during starvation, the acids produced by other metabolic processes in tissue cells activise the ferments, and so determine autolysis. But there is no evidence, so long as oxidation is efficient, that the muscles of a starving animal are more acid than those of a well-fed one, and it is difficult to see how this intestinal supply of ammonia can affect the reaction of other tissues than the liver, in which it is converted into urea, and we know that the administration of fats and carbohydrates, alone, can reduce, if it cannot prevent, tissue breakdown.

The exact relation of autolytic ferments to metabolic processes cannot be regarded as settled; but their discovery has added interest to the subject, and will continue to suggest fresh lines of experimental attack.

Much of what has been written about proteid metabolism in past years has been on the lines of a controversy in which the protagonists were Voit and Pflüger. It was outside the scope and intention of this brief review to deal with such a historical controversy; but the fact that the high authority of Professor Pflüger is still on the side of a conception which in some

respects is nearer to the views of Liebig than to those which we
have described as based upon recent experimental work, renders
it necessary to make momentary reference to this controversy
before closing.

According to Voit's teaching the greater part of the proteid
which is absorbed from the bowel is too rapidly metabolised for
us to believe that it has ever formed part of living structures in
the body. (On this point, it will have been seen, recent views
are entirely in harmony with those of Voit.) The structural
materials are relatively stable, and their breakdown and reconstruction are slow processes. The proteid so rapidly dealt
with is oxidised under the influence of, but not as part of, the
living bioplasm.

Pflüger, on the other hand, holds that only what is " living"
can be metabolised in the body. For him it is the incoming
dead proteid that is stable, and only when it has undergone
certain intramolecular changes, which fit it to become the basis
of living bioplasm, does it show the characteristic instability that
leads to metabolism. All the proteid eaten becomes tissue-
proteid before it suffers change.

The phenomenon which may be called the fundamental
phenomenon of proteid metabolism, the rapid elimination of
ingested nitrogen—the apparent dependence of proteid utilisation
upon proteid supply—was much in favour of Voit's arguments,
and has always been hard to reconcile with Pflüger's view.
The latter himself, however, explains it logically enough in
this way. Proteid food stands nearest to the living material
of the body, and is to be looked upon as the preferential food
of the cell for all purposes. So long as enough of it is supplied,
the tissues use it to the neglect of fats and carbohydrates, which
may thus be stored. Only when so much proteid is available
in the day's supply as to be more than enough for *all* the
purposes of the body should we expect to find it stored, and
such an amount could not be eaten by man, owing to the
organisation of his digestive system. This is a logical conception; but such preference on the part of the cell is almost
impossible to prove experimentally, and its probability is much
lessened if we deny that all the nitrogen of ingested proteid is
built up into the bioplasm before it is excreted. Remembering
all that has been said as to the wide diversity of the proteids
eaten, and believing that the tissues retain a mean composition

proper to the species, I do not see how we can fail to deny this; while our developing conception of intracellular ferments acting as weapons of the cell, and accounting for its dynamics, makes the hypothesis of a difference of constitution between unstable living and stable dead proteid less and less necessary.

If we accept Folin's conceptions, we are at once at a distance from Pflüger's standpoint, and we are also well removed from that of Voit. The modern theory explains better than Voit's what is puzzling in the elimination of nitrogen, and so far as it impinges on Voit's teaching it removes from it what was vague.

If the views which are tempting us just now justify themselves, and if the organism really uses the dietetic proteid in such a partial and selective manner, it seems to many that we shall be compelled to divest the proteid of some of its dignity as the pre-eminent foodstuff. Such views predispose us to accept the results of experiments like those of Professor Chittenden, which have become so familiar, indicating that the optimum consumption of proteid is a good deal less than we should have thought possible some years ago. This is a matter of practical importance, and we must give such indications full consideration. At the same time we must be cautious in coming to a conclusion while our ignorance of detail is so great. Proteid as a source of energy may show itself to be of no higher order than fats or carbo-hydrates, and the amount actually needed for tissue repair may be small; but the demands of the body are complex, and the constituents of proteid may have uses which come under neither of these heads. What is the optimum supply for such purposes we cannot even yet be said to know.

REFERENCES

ABDERHALDEN and SAMUELY, *Zeitsch. f. physiol. Chem.* xlvi. 193 (1905).

BUNGE, *Text-Book of Physiological Chemistry.* 2nd ed. of Translation by F. A. Starling, p. 168.

COHNHEIM, *Zeitsch. f. physiol. Chem.* xxxiii. 451 (1901).

FISHER and ABDERHALDEN, *Zeitsch. f. physiol. Chem.* xxxix. 81 (1903).

FOLIN, *Amer. Journ. Physiol.* xiii. 117 (1905).

FOLIN[2], *Zeitsch. f. physiol. Chem.* xxxvii. 174 (1902).

GREENWOOD and SAUNDERS, *Journ. Physiol.* xvi. 441 (1894).

JACOBY, *Zeitsch. f. physiol. Chem.* xxx. (1900).

KUTSCHER, *Zeitsch. f. physiol. Chem.* xxxix. 155 (1897).

—— and SEEMAN, *Zeitsch. f. physiol. Chem.* xxxiv. 528 (1902).

LANG, *Hofmeister's Beiträge*, v. 321 (1904).

LEATHES, *Journ. Physiol.* xxviii. 365 (1902).

—— and CATHCART, *Journ. Physiol.* xxxiii. 462 (1906).

LIEBIG, *Animal Chemistry*, trs. Gregory. 2nd ed. 1842, pp. 48, 77.

LOEWI, *Arch. f. exper. Pathol. u. Pharm.* xlviii. 303 (1902).

MOUTON, *Compt. rend. de la Soc. de Biol.* liii. 801 (1901).

NENCKI and ZALESKI, *Zeitsch. f. physiol. Chem.* xxxiii. 206 (1901).

OSBORNE and HARRIS, *Amer. Journ. Physiol.* xiii. 35 (1905).

PICK, *Zeitsch. f. physiol. Chem.* xxviii. 219 (1899).

NOEL PATON, *Journ. Physiol.* xxxiii. 1 (1905).

RUBNER, quoted by Loewi, *loc. cit.*

SALKOWSKI, *Zeitsch. f. physiol. Chem.* xiii. 506 (1889).

SCHRYVER, *Biochemical Journal*, i. 123 (1906).

Printed by Hazell, Watson & Viney, Ld., London and Aylesbury.

THE PHYSICAL BASIS OF LIFE

By W. B. HARDY, M.A., F.R.S.

Fellow of Gonville and Caius College, Cambridge

In a famous lay sermon on the Physical Basis of Life, written nine years after the publication of *The Origin of Species*, Huxley writes as follows :

> When hydrogen and oxygen are mixed in a certain proportion and an electric spark is passed through them, they disappear, and a quantity of water, equal in weight to the sum of their weights, appears in their place. There is not the slightest parity between the passive and active powers of the water and those of the oxygen and hydrogen which have given rise to it. At 32°, and far below that temperature, oxygen and hydrogen are elastic gaseous bodies whose particles tend to rush away from one another with great force. Water at the same temperature is a strong though brittle solid, whose particles tend to cohere into definite geometrical shapes. . . .
>
> Nevertheless, we call these and many other strange phenomena the properties of the water, and we do not hesitate to believe that in some way or another they result from the properties of the component elements of water. We do not assume that a something called "aquosity" entered into and took possession of the oxide of hydrogen as soon as it was formed, and then guided the aqueous particles to their places on the facets of the crystal or amongst the leaflets of the hoar frost. On the contrary, we live in the hope and in the faith that by the advance of molecular physics we shall by-and-by be able to see our way as clearly from the constituents of water to the properties of water as we are now able to deduce the operations of a watch from the form of its parts and the manner in which they are put together. . . .
>
> If the properties of water may be properly said to result from the nature and disposition of its component molecules, I can find no intelligible ground for refusing to say that the properties of protoplasm result from the nature and disposition of its molecules.
>
> But I bid you beware that, in accepting these conclusions, you are placing your feet on the first rung of a ladder which, in most people's estimation, is the reverse of Jacob's, and leads to

the antipodes of heaven. It may seem a small thing to admit that the dull vital actions of a fungus or a foraminifer are the properties of their protoplasm, and are the direct results of the nature of the matter of which they are composed. But if, as I have endeavoured to prove to you, their protoplasm is essentially identical with, and most readily converted into, that of any animal, I can discover no logical halting-place between the admission that such is the case and the further concession that all vital action may, with equal propriety, be said to be the result of the molecular forces of the protoplasm which displays it. And if so, it must be true, in the same sense and to the same extent, that the thoughts to which I am now giving utterance, and your thoughts regarding them, are the expression of molecular changes in that matter of life which is the source of our other vital phenomena.

This uncompromising, virile attitude towards this most difficult and stupendous problem of science is characteristic both of the man and of the time. Huxley wrote in 1868 at the zenith of a period of strenuous intellectual life without doubt unsurpassed in the history of the world. The strong new wine of scientific discovery was running in men's veins.

A mere chronological table of the chief scientific events shows how fast was the growth. In the forties the labours of Joule provided a basis for the conception of the conservation of energy which at a step unified all the sciences. In the forties, too, the unification of the biological sciences was begun by the recognition of the cell as the unit of all life, and of the glutinous sarcode as its physical basis, and was crowned by the publication of *The Origin of Species* in 1859, which gave force and authority to the older doctrine of the continuous development and progression of life.

The spirit of the age was one of conflict, and men's minds were tuned by it to Pisgah-like visions of the country to be conquered. The ideal of the new learning was the unity of all knowledge, its quest the establishment of a scheme of things animate and inanimate which should show them, linked together, without break, in orderly progress from the simple to the complex, from the lower to the higher, and its duty the warfare against a piecemeal and partial outlook of separate creation and catastrophic change. For the new learning no one did battle more strenuously than did Huxley.

The doctrine of the unity of knowledge and experience is not an easy one ; it is justified even now rather by the steady trend of science than by its completed demonstrations. Knowledge may be seen to be growing from the sides of many a chasm like the two arms of a cantilever, and we believe that human industry and the human intellect will one day complete a bridge across which all may pass in safety. But in the meantime there are grave signs in the scientific and semi-scientific literature of the day of a growing impatience with the rate of progress, which, on the one hand, ensures ready and uncritical acceptance of the crude attempts of an amateur biologist to make living matter, and, on the other, breeds a feebler purpose which seeks an unhealthy opiate in " vitalism" or some other " ism" of like nature.

Nearly forty years of vigorous scientific work have elapsed since Huxley wrote, and it is still possible for the vitalist to assert that no single vital process can be completely expressed in terms of physics and chemistry, that is, of motion and of matter. The biologist is reproached, for instance, with the undoubted fact that the power which a living cell has of selecting certain chemical substances and of rejecting others cannot yet be explained by, and indeed in some ways seems to contradict, the known laws of molecular physics.

To this reproach I would reply after the fashion of Socrates, and with the same purpose, by a question.

Here are two pairs of gases, one of hydrogen and oxygen, the other of hydrogen and chlorine. I burn the members of each pair together, and from the one pair I get water, a fluid odour-less, innocuous, and of relatively slight chemical activity, while from the other I get hydrochloric acid gas, acrid, poisonous, and of the highest chemical activity. Now, the molecules of those three gases have certain inalienable properties, an invariable weight, a fixed capacity for electricity. They perform move-ments the harmonic periods of which are so fixed that apparent departures from them have been used to detect and measure the velocity of approach of a star towards the earth. I ask the chemist or molecular physicist to explain the amazingly divergent properties of the compound in terms of the properties of the component gases. I ask him to do what over-hasty people, forgetful of the extreme youth, the paucity in years of human knowledge, ask the biologist to do with respect to living matter, and the reply is that the question is unanswerable.

It cannot be sufficiently insisted upon that in many regions not the simplest more advance has been made towards a material explanation of vital phenomena than towards a solution of the simple question why one pair of gases should combine to form a fluid, while another pair combine to form a gas.

An unanswerable question concerning the elements of natural knowledge is a sharp reminder of our ignorance, and such a reminder is needed to curb the spiritual arrogance which in our time has brought this greatest of all mysteries, the relation of living to non-living matter, to the temples of vulgar credulity, and has prostituted it to the purposes of common charlatans and impostors.

Since Huxley wrote, our knowledge of the physical basis of life has developed in many directions. The properties of secretion and absorption, of contractility and irritability, have been studied in great detail. The classical fields of physiology, the detailed investigation of form, and the anatomy of function have been continuously worked. But the greatest advance has come in the domain of chemical physiology. Ten years ago this was a scientific No Man's Land, despised by the pure chemist and traversed only in a distrustful, amateurish way by the physiologist. Now that is changed; on the one hand a race of physiologists has sprung up who are at the same time expert chemists, on the other one sees a pure chemist, Emil Fischer, of Berlin, bending all the resources of his great laboratory in men and materials to the central chemical problem of living matter, the chemical structure of proteid.

The few pages at my disposal would not hold even a *catalogue raisonnée* of the new departures. Therefore it is necessary to select a few problems, and for purposes of contrast I choose not the new but the old, which were agitated half a century ago.

At the outset, however, it is necessary to state certain elementary facts—that there is a unit of living matter called the cell, which everywhere and in all places has recognisably the same structure; and that all forms of life are divisible into two divisions: those in which the individual and the cell are coterminous—the simple-celled forms; and those more complex and larger types in which the individual is a cell complex—the multicellular forms. The former are probably the more numerous, but they escape notice by reason of their small size, which is imposed upon them by a law, wellnigh without exception, which must strike very deeply into the nature of living matter—namely, that no single unit, no cell, that is, can increase to more than microscopical dimensions. When it reaches the limit of size it becomes unstable, a field of force of a peculiar and special nature is formed within it, and by this field of force the cell is presently rent in twain.

The basis of this curious limitation of size is not far to seek. Living matter is composed of very large molecules, and substances so built possess certain special properties which mark them off from simpler substances. To them the name of colloids is given, after the type of the class the jellies. Now, a jelly is a curious half-way house between the solid and the liquid states. Like a solid, it is capable of retaining differences of state, it is rarely of uniform character throughout. The rate of relaxation, as Clerk Maxwell called it, of jellies is slow, much slower than that of simple liquids, much faster than that of true solids. Combined with this characteristic inertia, however, is a degree of molecular mobility sufficient for chemical changes of great velocity. A jelly is in this way a meeting-place of extremes, and this it is which enables the colloidal state to manifest life.

Consider now a small free cell, an infusorian swimming in a wayside pool. It displays many activities, it digests in this region of its living body, it maintains a store of starch in that region, in the movements of its parts there is diversity. Both its chemical and physical characters betoken a complexity which show it to be not a homogeneous droplet, but, in spite of its minute size (less than $\frac{1}{100}$th inch), to be heterogeneous. It has a structure, an architecture, the coarser features of which we can decipher with the aid of the microscope.

It is only in the colloidal state that we could have within so small a space so great a diversity of matter, and such differences of chemical potential as must exist to support the multifarious activities of the living cell, combined with the molecular mobility necessary to give chemical change free play. At the same time this capacity for maintaining differences of state imposes limitations, one of which is that of size. Large molecules can move in the substance of the cell scarcely at all. Therefore, when the size exceeds a certain critical limit, the dynamical balance fails, and internal strains appear of a magnitude great enough to tear the cell apart. On this blending of opposites, on the curious combination of inertia and chemical mobility in the colloidal state, is reared the whole fabric of the dynamics of living matter.

Each living cell is a machine; it breathes, taking in oxygen; it feeds, and the food is burnt by the oxygen to chemically simpler bodies. The living cell, like the gas engine, can tap the stores of chemical energy—and, like the gas engine, it is an internal combustion engine. Now in a power station where electricity is being produced to run a score of trams, there' is a steady hum or drone, the varying pitch of which marks the speed of the engine. To the engineer in charge, from long habit, that varying sound speaks of events happening in remote parts of the system. A glance at the clock, and he will tell you that the sound is falling because the engine is adjusting itself to the increased load due to such and such a tram breasting such and such a hill. In the same way, watching the movement of a living cell under the microscope, if we were sufficiently skilled we could refer the continual change in the rate and direction of its movement to temporary inequalities of temperature, of lighting or of chemical composition, etc., in the water in which it lives. If we resort to experiment, the effects are obvious: an electric shock causes the irregular Amœba to come to rest as a sphere, a trace of acid slows its movements, of alkali accelerates them.

These things—the electric shock, the acid, or the alkali—are what the biologist calls "stimuli," and by varying their nature or intensity he can control the activity of living matter to a very remarkable extent.

Let us return for a moment to the Amœba. We watch it crawling amid sand, fragments of decayed leaves, and living diatoms, and we notice that of the particles which it eats some are nutritious food, some are innutritious and absolutely useless. But we also notice that there is a decided balance in favour of the nutritious particle. Like Autolycus, it is a picker-up of unconsidered trifles, guided by a decided preference for things useful to itself. Therefore, the tiny animal manifests discrimination or choice—imperfect, no doubt, but clearly recognisable. And the choice is beneficial; it contains an element of purpose.

Watch an Amœba long enough, and it will be seen to divide into two, and these again into two, so forming successive generations the individuals of which resemble one another. This labile, creeping fragment of jelly has recognisable form, and zoologists classify the various forms in so many species, each of which breeds true. They manifest that property of living matter called heredity.

Lastly, the individual Amœba is in incessant movement, and with each successive generation there is growth of individuals and increase in the mass of living matter. Now, to these three features, *choice and purpose*, *growth*, and *heredity*, I propose to confine myself, and I will consider them separately and in the order named.

THE FACULTY OF CHOICE

In one of Jules Verne's books, which at one time or another held most of us in thrall, there is an account of a submarine vessel which for a long time was conjectured to be some mighty marine monster. Now, I want you to put yourself in a similar position with respect to a model Whitehead torpedo, to consider yourself as meeting one of these for the first time and studying its movements under the impression that it is a living being. The torpedo must be without its charge, otherwise your experiments would come to an abrupt end; for I wish you to consider yourselves as inquiring why this curious beast should always swim at the same depth. Push it down, pull it to the surface, it would presently be swimming again as many inches below the surface as before, and you would say to yourselves, Why on earth does it *choose* to swim there?

Instead of a model torpedo, here in a drop of fluid are countless thousands of the most minute forms of life, each actively darting hither and thither, each so small that 5,000 would make only one large Amœba. Into that drop you introduce two fine capillaries, the one filled with very dilute acid, the other with very dilute alkali, and in no very long time you will find that the Vibrios have collected in a mass at one or other of the tubes—probably the acid tube. If you followed ordinary usage, you would express the result again in terms of choice by saying that the animals are attracted to the acid and are repelled from the alkali.

Both cases illustrate the difficulty in freeing the imagination from the tyranny of the counters it employs. The first case, that of the torpedo, has served its purpose as an illustration, and it interests us no longer. Let us see whether a purely mechanical conception will explain the second.

The acid and alkali diffusing out of the tubes destroy the

uniformity of the water, so that, starting from one tube and moving to the other, one passes through gradually diminishing acidity, through neutrality, to a region of gradually increasing alkalinity, which reaches a maximum at the orifice of the other tube. The medium between the tubes, therefore, is accurately graded in composition.

Now let us see the effect of a trace of acid and alkali upon these Vibrios. It is not always the same; it depends upon the particular forms we are examining. I choose the case where acid slows the movéments, alkali increases them. Each individual Vibrio, as we watch it, is seen to move in an erratic and irregular orbit, so erratic that we can consider it as completely irregular.

The problem now becomes a simple mathematical one. Given a number of particles, each moving in an irregular orbit, and uniformly distributed throughout a homogeneous medium. The medium ceases to be homogeneous and is changed so that in one region the mean velocity of each of the particles is augmented, in another it is diminished. What will be the effect upon the distribution of the particles? The answer is, that they will collect where the motion is slowest.

We now try the experiment, and we find that the Vibrios do collect where their motion is most slowed—namely, in the region of maximal acidity. And they do not swim directly there; they, as it were, settle out in that region, as the hypothesis demands.

The influence of a chemically heterogeneous medium upon the free cells living in it is called "chemiotaxis." I have analysed a simple case, but it would take a session's lectures to follow out the application of the principle to biological problems. It explains great regions of disease, it has even been applied to the workings of the nervous system. At one sweep it embraces the directive effects of the surrounding medium upon the movements of free cells, in the waters of the earth and in the bodies of animals and of plants.

The choice of food particles, the discrimination manifested by Amœba, is the chemiotactic response of its irregularly flowing protoplasm to the chemical atmosphere, if I may so put it, of the food particle. At the Mint a chance collection of sovereigns are presented to a certain machine, and it sorts them into those of full weight and those of short weight. A chance collection of particles are presented to Amœba, and it sorts them very imperfectly into those which modify the incessant streaming of its protoplasm so that they become engulphed, and those which do not so modify the streaming. The element of choice, or, as we may now put it, the directive influence of the surroundings, is much less perfect here than in the case of the Vibrios, because the oscillations—the movements of the protoplasm on which it operates—are both less in extent and much more regular than are the movements of the Vibrios.

In the choice of food particles, and in the sorting of the coins, there is an end to be served. Looked at from this point of view, chemiotaxis sometimes presents novel features. *Amœba proteus*, large and slow-moving, frequently captures an active ciliate called *Colpidium*. Observers describe the capture as being due to the *Colpidium* swimming as though attracted into the pseudopodial jaws, whence it makes no efforts to escape. Here the element of purpose, looked at from the standpoint of *Colpidium*, is that of a Christmas ox marching to the kitchen to be converted into beef-steaks.

The directive effect of the medium upon a free cell is usually more complex than in the case considered. *Opalina* is a large ciliate which in a uniform medium swims straight forward, owing to the movement of the cilia or vibratile hairs which cover its surface. The movement in this case starts at the front end of the animal and sweeps back as a wave like the wave over a cornfield. In a heterogeneous medium the movement starts excentrically, the waves sweep obliquely down the animal, and the direction of motion changes. The net result again is the same—the animal ceases to be distributed evenly when the water ceases to be of uniform composition.

The next example raises the question of choice to a higher level. It brings into the response of the animal its previous history. We will take the simplest case, as it is offered by Opalina. This animal is parasitic in the intestine of the Frog, and it thrives in a very slightly acid medium. But its attraction to acid is not an inalienable quality. Glut it with acid, soak it in dilute acid for an hour, and it now collects in a region of alkali; bathe it for an hour in very dilute alkali, and its chemiotactic response is once more changed—it collects about the acid.

The mechanism underlying this change of response must be patent to every chemist. There are many substances whose chemical and physical characters are completely reversed by change from a trace of acid to a trace of alkali, or *vice versa*. Amongst these substances, and markedly possessed of this character, are the chemical substances, called proteids, of which all living matter is composed. The varied response to acid or alkali may unquestionably be traced in the first instance to the directive influence of the amphoteric proteid on the surface energy of the animal and upon the train of chemical events in its interior.

A parallel differential response is furnished by *Stentor*—a large, trumpet-shaped animalcule—which fixes itself by its foot to some solid object. Touched on one side by a fine glass hair very lightly, it bends towards the hair; touched more heavily, it bends *away*. Therefore there is a touch of a certain strength which produces no response. To a series of touches regularly repeated it gives the following responses. At first it simply bends away, then it contracts right down on to its foot; if that does not get rid of the irritant, it looses its hold and swims away.

These responses have been analysed with great care in order to elucidate the underlying mechanism. I can stop only to point out the element of purpose. In order to get rid of an irritation, a certain movement is tried; it fails, another movement is tried; it fails, and a third movement is tried.

Now, it must be clearly understood that organisation will account for these phenomena. Quite as remarkable a series of responses, each in turn designed to get rid of an irritant, can be obtained from a Frog which has been deprived of its brain, and therefore presumably lacks both consciousness and intelligence. And step by step, as organisation advances, the response gains in complexity, until the human imagination is unable to unravel the chain of cause and effect. But the biologist is cognisant of no break in the series from the choice of a Vibrio, which can be analysed algebraically, to the choice of a child between two toys.

The faculty, clearly seen in the case of *Stentor*, of storing impressions, so that the response to any particular stimulus is in part conditioned by the stimuli which have preceded it, is a familiar property of living matter, and also of matter in the colloidal state. The molecular state of a jelly is not fixed by the conditions of the moment. Just as a piece of wrought iron has properties different from those of cast iron, so the circumstances which attend the making of a jelly—temperature, concentration, and the like—confer on it an internal structure which controls its properties for years to come. Each jelly, therefore, has an individuality due to the record which it bears of its past.

Take another case. A vertical rod of wax is bent, first north, then south, then east, then west, and so on. Left to itself, it will quietly work out these movements in the reverse order. It bends first west, then east, then south, then north, and so on. The molecular structure of the wax is such as to preserve a record not only of the fact that it has been moved, but also of the number, direction, and order of the several movements.

The Faculty of Growth

If it be true, as some chemists think, that in the process of oxidation there are always two processes more or less concurrent—the first one of synthesis, in which bodies of increased chemical complexity are formed by the union of the oxygen and the combustible substance; the second one of analysis, which supervenes only when the synthetic products reach a degree of complexity where they are unstable at the particular temperature and pressure—then, considered in a general way, the processes of assimilation and growth of living matter are exceptional only in the prominence and permanence of the synthetic stage. The living cell, on this view, is like the flame in being an oxygen vortex; it is unlike it in the extraordinary latency or delay in the advent of the analytic processes.

The peculiar feature of the living cell, however, considered as a machine, lies in the fact that, of the total amount of energy which it acquires, a fraction is retained and devoted to the increase of its own substance. In other words, it grows. After a while it divides, and the daughter cells are like itself, so that there is not only the power of increasing the bulk of living matter by growth, but also a directive faculty called heredity, which constrains the new living matter, made from non-living matter, into the pattern of the old. The problems of growth and multiplication can be reduced to their simplest terms only in the case of minute forms like Amoeba, each of which is at once a single cell and an individual. Each individual amongst the higher forms of life is built of countless cells, all of which, with one or two exceptions, are predestined to death. The exceptions—the true immortals—are chosen from the germ cells. When, however, an Amoeba multiplies, it divides bodily into two, and by this simple process a new generation is formed. Clearly, as Weissmann first pointed out, in such cases death intrudes only in the guise of accident.

Now, the conditions of life of these simplest forms are by no means simple. Make an infusion of hay in boiling water, and let it cool. In the course of a day or so it will be found to be swarming with rod-like bacteria (*Bacterium subtilis*), engaged in feeding on the organic matter dissolved out of the hay. A few days later numbers of an actively mobile slipper-shaped animal, called *Paramœcium caudatum*, make their appearance to actively swallow and digest the bacteria; and so the round goes on. The bacteria and the Paramœcia alike have developed from wind-carried spores. Therefore in the natural life of these creatures are periods of physiological activity alternating with periods when life seems to be completely dormant—periods which follow one another according to no regular sequence, but in consequence of chance rainfall and of drought, when the inhabitants of the dried-up pool are caught up and carried away as dust.

Watch any chance collection of Paramœcia, and individuals will be seen not only to divide, but occasionally to fuse. Two individuals swim together, adhere closely, and effect an extensive interchange of substance. This is the process of conjugation: it is the first beginning of sexual reproduction. It is followed by increased physiological activity, increased rate of growth and of multiplication. If we could follow this mating process fully, if our imagination could grasp the events which lead up to it and the effects which follow, we should see in it the response of life to the flux of cosmical energy, just as the oscillations of a particle in Brownian movement are the response to the flux of electrical potential. This is no careless phrase: it is sober truth, for the air currents carry and mix spores from far-distant places which have, therefore, had different life-histories. They have lived in waters draining different kinds of soil, and therefore chemically different, the balance of sunshine and shade has been different, and the wind capriciously sows these spores from north, south, east, and west in the pot of hay tea. There they become active, and mate in conjugation, but not fortuitously. Guided by chemiotaxis, unlikes meet and fuse, just as do unlike cells when an ovum and a spermatozoon fuse, and the fusion of a pair of unlike individuals results in a thorough reorganisation, a fresh make-up, of the living matter of each. The continuance of the race depends upon the change of environment, upon the alternation of periods of activity with periods of dormancy, and upon the fusion of unlike individuals; and for these the sequence of natural phenomena, of summer and winter, of sunshine and shade, provides.

Now, the problem of growth is this. Suppose we eliminate these factors; suppose we isolate a pure strain of Paramœcia and keep them abundantly supplied with food, will the race continue to flourish and grow indefinitely, or will it attain old age and die off? The problem is far-reaching. It touches the simple questions of function, of digestion and assimilation, on the one side; while on the other it is concerned with the limitations of heredity in moulding successive generations after the common type. Three workers have attacked it with conspicuous success, Maupas,[1] Calkin,[2] and Woodworth,[3] and in each case the experimental method was the same.

Maupas was the first. He isolated individual Paramœcia under normal and healthy conditions—namely, in hay infusion containing the bacteria on which they feed, which was changed daily. Each individual was the starting-point in a sequence of generations, there being, on the average, two generations in three days. The rate of division was recorded, and the records furnished the basis for a curve of vitality.

The experiment established two points, the first being the presence of fluctuations of vitality of fairly regular character— "rhythms," they have been called. The curve alternately rises and falls, and each complete "rhythm"—a rise and a fall, that is—lasts about a month. The second is that the curve, as a whole, steadily falls, each successive rise in vitality is a little less than its predecessor, each depression a little lower, until— about the 170th generation—the period of old age, of senile decay, is reached, and the race dies out.

There the matter was allowed to rest, until fresh experiments were prompted by a remarkable observation made by Prof. Loeb. He found that the unfertilised eggs of sea-urchins could be made to develop by immersing them for a few hours in sea-water containing a higher percentage of salt than ordinary. If the eggs could be artificially aroused, why not the senile

[1] *Arch. d. Zool. Exp.* 1889 (2) vii.   [2] *Arch. f. Protistenkunde*, i. 1902.
[3] *Journ. of Exp. Zool.* ii. 1905.

Paramœcia? So argued Prof. Calkin. Therefore, when the period of decay had arrived and the individuals were dying off rapidly, he tried placing them in various infusions. Vegetable infusions were without effect, but infusions of animal tissues, and particularly beef extract, gave the required result. The rate of growth and of reproduction reached the normal level, and death ceased. Senile decay had given way to artificial rejuvenation. Instead of 170 generations being the limit, by stimulation in the periods of depression Calkin succeeded in carrying a race to the 740th generation, and Woodworth to the 860th generation, when the individuals were still healthy and fully active. The living matter of these cells without doubt is potentially immortal!

Consider for a moment what incredible chemical activity and stability of character these figures imply. If it was possible to preserve alive all the individuals, then at the 900th generation we should have a number which would need a row of some hundreds of figures to express. The parent cell would have produced the 900th power of 2 individuals like itself. The increase in the bulk of active living matter which would have been formed from non-living had there been space enough and food enough is not less wonderful. At the 350th generation it would have the dimensions of a sphere larger than the known universe![1] And the surface of the sphere would be growing outwards at the rate of miles a second. Nor is this all, for in addition to the enormous chemical activity implied by a rate of growth which would, if unchecked, produce a mass of living matter larger than the known universe in less than two years, there has been throughout continuous expenditure of energy on incessant and active movement. These animals have been watched continuously for five days, and throughout that time they were ceaselessly moving!

The recurring periods of depression show that in the living machine repair is not complete, and that after a time it will, if left to itself, cease working. With the condition of ill-repair there is associated a feature of singular interest. Woodworth specially draws attention to the fact that in the periods of depressed vitality the transmission of characters is imperfect. The moulding power of heredity fails, and many "monsters" are born.

[1] I owe this rough calculation to my friend Mr. Punnett.

Rejuvenescence can be brought about by a great variety of media, by extracts of muscle, of brain, of pancreas, by simple salts, by alcohol even. It is not food, but a marked and abrupt change of state that is needed—a stimulant, in fact—and beef extract produces its effect not *quâ* food, but as a stimulant pure and simple. Senile decay is due to monotony, under the influence of which the vital potential wears out!

The action of alcohol is remarkable. It was added to the water in which the animals lived so that they were always immersed in a solution of 1 part of spirit in 5,000 to 10,000 of water. In the effect produced there is the touch of nature which makes the whole world kin. The periods of depression were wiped out. The curve of vitality no longer showed the ominous recurrent depressions. At the same time the rate of growth and division—that is to say, the physiological activity— was increased by as much as 30 per cent.

Something of the same effect is produced by strychnine, but there is a remarkable and significant difference in the fundamental action of the two drugs, for whereas the beneficial effect of alcohol endures after the drug ceases to be administered, that of strychnine does not. Alcohol, as Calkin says, in spite of the prodigiously increased rate of living, "exacts no physiological usury," it is beneficial in its after effects. Strychnine is harmful in its after effects; the onset of decay and death is hastened.

What significance are we to attach to artificial rejuvenescence? There are two possibilities. The chemical agent employed may either add something which is missing or diminished in the chemical make-up of the protoplasm, or it may restore a physical state. The former implies that the chemistry of the growth process is imperfect: the process of converting non-living to living matter is subject to inaccuracies—inconceivably small it is true, since they need to be magnified to the 170th power of 2 before they destroy the working of the machine, but cumulative from generation to generation. I incline to think that senile decay is due not so much to such a chemical insufficiency as to the wearing out of a physical state, of a "potential."

Consider a special case. Thirty minutes' immersion of an individual Paramœcium in very dilute solution (1 part of salt in 1,000) of potassium phosphate was found to restore vitality, and

the effect persisted for 282 generations.   Now, in this case the
restoration and maintenance of the "vital potential," as Calkin
calls it, cannot be due to the presence in the individuals of a
trace of the salt, for each generation would halve the amount,
so that as early as the twentieth generation less than a
millionth part would be left for each individual.   One is
therefore driven to believe that the salt acts by restoring a
state which, in the absence of natural or artificial rejuvenescence,
wears out in about 170 generations.

The continual flux of energy and of matter which seems to
be necessary to the maintenance of life implies a high degree of
molecular mobility.   It is possible that living matter, like all
other forms of matter, tends to come into equilibrium with its
surroundings, and to attain a condition of too great stability.
To restore it the living substance needs stimulating at intervals,
just as a coherer needs tapping after each electric wave has
passed, in order to restore its particles to the non-conducting
position.   These are vague possibilities, but physical science
furnishes a case so suggestively akin to artificial rejuvenescence
as to merit description.

Matter is composed of molecules which in the liquid or solid
state are attracted to one another by forces of prodigious power.
Each molecule in the interior of a mass is pulled on the average
equally in all directions.   But consider the surface layer, a film
only a few molecules deep.   There the intermolecular forces are
necessarily to a great extent unbalanced, with the result that this
surface film acts something like a stretched elastic skin—it tries
always to compress the mass to the smallest possible dimension.
This, however, is not the only feature of the surface layer.   It is
also the layer which is in contact with adjacent masses of matter,
gas, liquid, or solid, as the case may be.

Now, masses of matter which do not mix when in contact,
and which therefore are defined by a surface of separation, are
rarely, perhaps never, without influence upon one another.
Interaction of the surface layer takes place, so that the balance
of molecular forces is modified, incomplete chemical reactions
occur, and a condition of molecular stress is produced which,
amongst other things, is manifested by the development of
electrical charges.

These molecular events on surfaces are very potent; they
can produce effects which are impossible and even inconceivable

in matter in bulk. It is, for instance, not only possible, but probable, that in the surface layers the conditions may sometimes be such as to associate decrease of volume with decrease of pressure, a relation so subversive of ordinary experience as to be unthinkable. In the surface layer a gas may be condensed to the liquid state when far above its critical temperature and below its critical pressure. Chemical changes occur or are suspended under conditions of temperature and pressure totally unlike those controlling the same changes in masses of matter. Concentration, electric conductivity, all physical properties in fact, become abnormal, therefore when the surface energy forms a large fraction of the total molecular energy, as in films, or fluid in fine capillaries, ordinary chemical or physical knowledge fails us.

Now, there is no lack of evidence to prove that the lifelike characteristics of colloidal matter, its capacity for storing impressions, the elusiveness of its chemical and physical states, are due to the fact that an exceptionally large fraction of its energy is in the form of surface energy.

There is also direct and unmistakable evidence in the nature of the effect of various salts upon the heart-beat, and in the optical characters of thin films, that living matter also contains a very large proportion of surface energy per unit of mass, and the curious and extreme physical and chemical powers which it manifests are without doubt largely due to this cause. Now, it is just in experiments on surface energy that one finds a case analogous to the effect of the salt in bringing about rejuvenescence of senile protoplasm, or in awaking the dormant powers of an unfertilised egg.

It has been shown recently by a French physicist, M. Perrin,[1] that by the use of minute amounts of salts one can give to the surface energy of a solid a certain direction—one can fix in the surface layer certain qualities which, for instance, define the electric properties of the surface. The effect once produced, no amount of washing will undo it; the salt can be removed, the effect remains. So far as we know, in the absence of active chemical intervention it will endure for all time, always exerting a directive influence upon the molecular events in its neighbourhood. In these experiments there is, it seems to me, a real clue to the nature of the phenomena of rejuvenescence.

[1] *Journ. d. Chem. Physique*, ii. p. 61, and iii. p. 50.

HEREDITY

On the earth are some half-million different species of animals and plants, each of which breeds true in virtue of what we call heredity. Each species therefore represents a strain or line of descent of living matter always growing, dividing, and increasing in mass, like the little Paramœcia we have already considered, each striving to occupy the whole earth, and restrained in the attempt only by the accident of death.

The strains of living matter are separated from one another by a wide gulf which we do not know how to bridge. Change of state seems to be without effect. Continuous supplies of the richest food will not convert a strain of dwarfs into giants. In the solemn words of the Burial Service, " All flesh is not the same flesh, but there is one kind of flesh of men, another flesh of beasts, another of fishes, and another of birds."

The nature of these differences in the kinds of living matter and their mastery so that we may be able to control them is without doubt the most difficult and the most important problem which science has attempted to solve : most difficult because it deals with a form of matter much more complex than any which the chemist or physicist so far has considered ; more important because on the solution of this problem depends the possibility of removing practical medicine, politics, and morality from the domain of empiricism and tradition to that of rational co-ordinate knowledge.

To speak of a strain breeding true is a bald way of describing a force so potent as heredity, so impish in its eccentricities. On the Antarctic ice there abound a race of birds called penguins. They have never seen a tree since they first were penguins ; they do not fly, for their wings have been reduced to small flat paddles with which they swim. The bird cannot tuck its head under its wing, because the wings are too shrunken ; but still, in mute worship and touching fidelity to its forbears of thousands of years ago, when it composes itself to sleep each individual bends round its head and tucks the tip of the beak—it is all it can do, poor thing !— under the dwarfed wing.

This lingering instinct, this obsession by the great past, is like a whale dreaming of the green fields in which his forefathers browsed ! Now, each individual penguin starts life as a microscopic fragment of living matter, a single cell, so wonderfully compounded, so cunningly devised, as to enshrine without loss all the diverse qualities and powers which the word "penguin" connotes, down to the trivial detail I have described! There is little wonder that the naturalists of half a century ago gave the problem up in despair. There is cause for wonder and for congratulation that, impelled by the divine dipsomania for research, knowledge has moved so far as to make a beginning in the assaults.

Given a fulcrum, anything can be moved. The necessary fulcrum was found when attention was directed not, as in Huxley's time, to the more obvious resemblances between the different kinds of protoplasm, but to the less obvious differences. The microscope for the most part fails us here, in the first place because the discrimination between different kinds of matter by the agency of sight is possible only when there are associated differences in optical properties, and when there is the possibility of getting a clear image. Now, living matter is singularly free from definite optical differences, and it has the optical characters of ground glass. Therefore, the ultimate refinements of microscopic vision are for the most part wasted upon it. The dead cell exhibits remarkable structural details, but in the act of death there is of necessity a redistribution of matter which obscures and defaces the finer details of the real living structure, and replaces them by structure which is formed in the process of dying. For the material basis of the difference in the strains of living matter we have to look below the limits of microscopic vision, below the limits even of the living molecule, to the chemical molecule of which that living matter is built up.

The nearest chemical approach to living matter is the proteid, the chemical substance of which all protoplasm is, water excepted, chiefly composed. Now, the fulcrum I spoke of, or, better, the thought which loosed the fetters of imagination, was the appreciation of the significance of the fact that proteids chemically are not all alike, and that the strains of living matter differ from one another in the kinds of proteid of which they are built up—that is to say, in their ultimate chemical constitution.

All proteids are not the same proteids; there are proteids of men, others of beasts, others of fishes, and others of birds!

The nature of the differences leads us to a real picture of the underlying differences between the kinds of protoplasm. The tide of thought of the older observers was fettered by the act that all proteids have about the same atomic composition. The biologist of to-day owes his emancipation to the chemical discovery that the properties of a complex substance are defined not so much by the kind of atoms or number of atoms of which it is built, as by the arrangement of those atoms in space.

Here is a simple and startling case. The molecules of two chemical substances, benzonitrile and phenylisocyanide, are composed of seven atoms of carbon, five of hydrogen, and one of nitrogen:

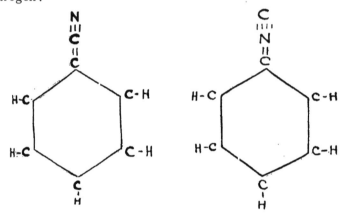

BENZONITRILE   PHENYLISOCYANIDE

There is a small difference in the arrangement of these atoms which is illustrated by the diagram. Now, what are the properties of these two substances? They are as unlike as possible. The first is a harmless fluid with an aromatic smell of bitter almonds. The second is very poisonous and offensive.

A vivid impression in regard to the odour of the iso-cyanides may be produced by the following experiment. In a test-tube bring together a *little* chloroform, aniline, and alcoholic potash. The reaction takes place at once. *It is better to perform the experiment out of doors and in such a place that the tube with its contents can be thrown away without molesting any one.*

In the building of a complex molecule one has atoms gathered together to form groups, these to form larger groups, and the whole structure is arranged on a fundamental plan or style, like, for instance, the ring of carbon atoms in the two substances just mentioned. There is, therefore, a molecular architecture, and, as in ordinary architecture, there are differences of style and of general plan, Gothic, Norman, etc., with endless variety in detail. Amongst the recognised styles of molecular architecture is the proteid style, and the qualities common to all forms of life are based ultimately upon the essential features of that style, while the differences between one kind of living matter and another are the expression of the differences in detail—the omission of this group, the addition of that.

Some of the atomic groups which find a place in the proteid molecule are readily recognisable by chemical tests—one of these groups occurs as a separate chemical substance called *Tryptophane.* It shows a vivid purple colour with sulphuric acid and reduced oxalic acid. Here are solutions of two proteids, one from maize seeds, the other from the white of egg, the one of the vegetable strain, the other of the animal strain. The former lacks, the latter possesses this group.

Now, in order to represent the great varieties of living matter the proteid molecules must be capable of very many variations of structure. That is, after all, mainly a question of size—the larger it is the greater the possibility of variations in detail; and as the molecule of proteid seems to contain from ten to thirty thousand atoms, whereas the most complex molecule known to the organic chemist contains less than a hundred, there is no lack in this respect.

Proteids unquestionably are the material basis of life, but when isolated after the death of the cell they are not living. They are chemically stable bodies. They show no signs of the characteristic chemical flux. It is therefore conjectured on experimental grounds that the living molecule is built up of proteid molecules, that it is so complex, so huge, as to include as units of its structure even such large molecules as these. But when such very large molecules enter into chemical combination with one another, whether by reason of the great magnitude of the masses of matter in each in relation to the magnitude of the directive forces, or because the molecules themselves, owing to their great size, to a certain extent cease to be molecules at all in the physical sense, and possess the properties of matter in mass, it is at any rate certain that in their chemical combinations they cease to follow the law of definite combining weights which is the basis of chemistry. The quantity of the substance A which will combine with a fixed quantity of the substance B is determined not only by the chemical nature of A and of B, but also by the chance conditions of temperature and concentration of the moment. This class of chemical compounds is within limits continuously adjustable to changes in its surroundings, while at the same time it resists those changes by reason of its inertia. Here is a real adumbration in non-living matter of the chemical flux which is the abiding characteristic of the matter of life.

The biologist speaks of these molecular complexes as *molecules*, and in that he is wrong in so far as the word implies a *defined* structure, a chemical unit. The biogen, or chemical unit of living matter, is not a fixed unit like the molecule of dead proteid ; it is an average state. That we know from the chemical phenomena of living matter.

Why should this be ? Consider what must happen if you make the atomic building much larger than it already is in the molecule of dead proteid. You already have a molecule so large as to be liable to fracture on mere mechanical agitation. A molecule composed of fifty proteid molecules would cease to be a molecule in the physical sense : it would be matter in mass, defined by a surface; it would break up the waves of light, so radiant energy would profoundly affect it.

In a mass so large, a portion of the energy would of necessity be in a borderland between what we call osmotic energy and surface energy, the fraction in the one state or the other being determined from moment to moment by the changing relations with the enveloping matter. If the chemical structure was such as to produce a shape other than a sphere, surface energy would tend to produce chemical rearrangements, and the opposing play of these forces might result in oscillations of form which would reflect the irregular flux of cosmical forces just as does the particle in Brownian movement. The chemical relations of such a mass would be defined in the first instance by the surface layer, but any simple chemical event on the surface would be likely to fire a train of events leading to an eruption like a sun-spot on the sun.

It is not, I think, difficult on these lines to conceive of a substance the chemical units of which could maintain themselves only in virtue of a continual flux of matter and energy—only, that is, as an average state; but it is certain that to develop the hypothesis we need what has not, so far as I know, yet been begun —namely, a kinetic theory of those intramolecular relations of atoms which are statically expressed by the geometrical methods of stereo-chemistry. The living cell, like a gas engine at work, is a chemical vortex, and there is no hope of analysing the motions of its parts so long as we are limited to statical methods.

In the history of the study of heredity there is a note of tragedy. In the early days of last century Lamarck began the revolt against the dogma of the immutability of species, which culminated in 1859 in the publication of *The Origin of Species*. Between Lamarck and Darwin, however, stand a scanty band of men forgotten by all but a few specialists, who strove by experiments in cross-fertilisation to pierce the mystery of heredity. Amongst them, and the last of the line, was a monk of the Abbey of Brünn, one Gregor Mendel, who in 1865 communicated to the Brünn Natural History Society the results of eight years devoted to experiments with peas, under the modest title of *Experiments in Plant Hybridisation*.

The fate of Darwin's work is known to every one : how "it was considered a decidedly dangerous book by old ladies of both sexes," and how, "overflowing the narrow bounds of scientific circles, it divided with Italy and the Volunteers the attention of general society." The fate of Father Mendel's work was different. For the rest of the century it lay completely forgotten and buried in the annals of the little local society. But when it was rediscovered in 1900 by Professor de Vries, of Amsterdam, it was at once realised by the very few competent to judge that the pursuit of a hobby in the abbey garden had led to a theory of the nature and workings of heredity so clear and complete as to leave to others only the application of principles and the amplification of details.

To find an achievement parallel to Mendel's, in the difficulty of the problem attacked and the all-embracing nature of the solution reached, one has to turn to Willard Gibbs's clean sweep of the domain of chemical equilibrium. But the author of the Phase Rule lived to see the work rediscovered—again by a Professor of the University of Amsterdam—and become the inspiration of a cloud of workers in all lands. The Mendelian laws of heredity, established twenty years earlier, are only now beginning to bear fruit, twenty years after Mendel's death.

The magnitude of Mendel's achievement can be appreciated by calling to mind the acute intellects which have been foiled by the problem. For a century the study of heredity has remained a repellent mass of statistics, with scarcely more discernible order than might be found in any chance collection of facts ; and of the would-be student it might be said, " Quæsivit cælo lumen ingemuitque reperta." And for half of that century there has lain hidden a solution of the riddle which brings these facts into an order so straightforward that a child might learn it.

We should have nothing to do with the Mendelian laws here were it not that they have given singular meaning and interest to certain details of cell structure which before were a mere collection of unintelligent facts. To take things in their proper sequence I will first state the laws of inheritance so far as they concern us, and then consider the structural characters which seem to be their material basis.

The first Mendelian principle which concerns us is this : that what is transmitted from generation to generation may be analysed into certain qualities or characters—constant characters, as Mendel calls them—each of which is a unit in heredity, each of which, therefore, is capable of independent trans- mission. Thus in peas are length of stem, character of inflorescence, colour of seeds, flavour, and so on. Underlying these characters—each of which is capable of being picked out or put back by a breeder, forming a substrate on which they are erected—there would seem to be a basal character which is inalienable, and which the breeder cannot, at present at any rate, touch. Thus, in the case of peas, what is of necessity transmitted is the fundamental qualities of " plant " as opposed to " animal," and of " pea " as opposed to other plants. To proceed in Mr. Bateson's words :

These [unit] qualities or characters whose transmission in heredity is examined are found to be distributed among the germ cells, or gametes, as they are called, according to a definite system. This system is such that these characters are treated by the cell divisions (from which the gametes result) as existing in pairs, each member of a pair being alternative to the other in the composition of the germ. Now, as every zygote—that is, any ordinary animal or plant—is formed by the union of two gametes [in the process of sexual fertilisation], it may either be made by the union of two gametes bearing similar members of any pair, say two blacks or two whites, . . . or the gametes from which it originates may be bearers of the dissimilar characters, say a black and a white. [In the first case, no matter what its parents or their pedigrees may have been, the zygote breeds true indefinitely, unless some fresh variation occurs.]

If, however, the zygote be gametrically cross-bred, its gametes [or germ cells] in their formation separate the pair of characters again, so that each gamete contains only one character of each pair. At least one cell division in the process of gametogenesis is therefore a differentiating or segregating division, out of which each gamete comes sensibly pure in respect of the unit characters it carries, exactly as if it had not been formed by a cross-bed zygote at all.

For our purposes this may be reduced to three propositions: (1) that inheritance consists in the transmission of independent characters, of which each race or species possesses a definite number; (2) that these characters form pairs of opposites or alternatives; (3) that in the formation of the germ cells these characters are sorted out and distributed so that no germ cell carries both members of a pair. Can any material basis be found for these? To this we will now turn.

Five years ago it is doubtful whether there existed in the whole domain of science such a charnel-house of dead facts as in that of the science of cell structure. Thirty years of active study of animal and plant cells prepared for microscopical examination in various ways had resulted in the accumulation of a multitude of details respecting the structure of the cell nucleus and of the extraordinary way in which it behaves in cell division, and especially in those cell divisions which produce the germ cell. It was known that from the characteristic substance of the nucleus—which stains very deeply with aniline dyes, and hence is called chromatin—a continuous thread is spun as the first step in cell division, and that this thread of chromatin splits across into rods called chromosomes, each of which again splits, this time not across, but lengthwise, so as to form two "daughter" chromosomes, which, under the influence of a peculiar field of force formed in the substance of the cell, move away from one another and gather at the opposite poles of the spindle-shaped field, there to fuse and form the two nuclei of the "daughter" cells.

A further very significant and curious fact was known— namely, that the number of chromosomes formed in the process is not a chance one, but, in the first place, it is always an even number, and, in the second place, each species of animal or plant has a characteristic number. In the division of the cells of the human body, for instance, there are formed thirty-two chromosomes. But to these and many other similar facts no significance could be attached, beyond the obvious one that the nuclear substance is not divided grossly to form a new cell generation, but distributed by a complex and minutely detailed process of subdivision and segregation.

Not only were the facts of nuclear division without significance, but the presence of the nucleus itself seemed to be meaningless. The contractility of the muscle cell, the conductivity of the nerve cell, the chemical activities of the gland cell, reside in the cell body, and not, save perhaps in the last case, at all in the nucleus. Throughout cell life it lies to all appearance an inert mass, which becomes active only in the process of cell division. And yet actual experiments on enucleated fragments of Protozoon cells and on the nerve cells of higher forms had proved that in the absence of the nucleus the cell body cannot live. True, it carries on all the life functions for a while, but it seems to have lost with the nucleus the power of growth and of repair.

The last six years have witnessed the rehabilitation of the nucleus, and biologists now see in it the seat of that influence which directs the formative process by which living matter is produced from non-living matter, and controls the distribution of characters in heredity.

The actual agent in the latter process seems to be the chromosome, and the material basis of the limitation in the number of characters transmitted from generation to generation, and of their definiteness lies in the restriction of the number of chromosomes to a definite number for each species of animal or plant. The chromosomes are not fragments of nuclear substances of accidental composition, nor are they all alike. On the contrary, the probability is that they are unlike, and possessed of a high degree of individuality.

The material process which underlies the segregation of characters in the germ cell and the fusion of characters in pure and cross-bred zygotes can also be followed in the peculiar features of the cell divisions which form the male and female gametes.

As I have already said, each species has a characteristic

number of chromosomes, but in the cell divisions which form ova or spermatozoa, ovule or pollen grain, this number is halved, so that each spermatozoon or ovum receives only half the proper number. In this sense, therefore, the germ cell is only half a cell. When two germ cells fuse in the act of fertilisation, the full number of chromosomes is restored. Thus, to choose an instance, the full number of chromosomes which make up a nucleus in a cell of the human body, no matter where it be placed, is 32; in the formation of the spermatozoon or ovum, however, there is a redistribution of chromatin, so that each receives only 16 chromosomes. When a spermatozoon fuses with an ovum, a zygote with the full number, 32, is formed.

The chromosomes therefore are the elements, the organs, as it were, of heredity. They have individuality, the limitations of which are not yet known. Each bears a unit character or a group of unit characters. The evidence for the individuality of the chromosomes is very remarkable.

Fundulus and *Menidia* are two fishes belonging to separate orders. Each has 36 chromosomes, but the chromosomes of the former are so much longer than those of the latter as to be readily distinguishable. Moenkhaus[1] crossed these two forms, and traced the fusion of the long and short chromosomes in the formation of the hybrid zygote. But when the zygote prepared to divide, the paternal and maternal elements segregated and formed two groups of chromosomes, the one of long and the other of short chromosomes, and in each segment division the paternal long and the maternal short chromosomes reappeared and acted independently. Another case has been furnished by Wilson.[2] In certain groups of insects there is among the chromosomes of the male cells one distinguished by its small size. The total number of chromosomes, instead of being even, is odd: there are thirteen, and this small chromosome is the thirteenth. It is, in point of fact, only half a chromosome, therefore when each of the others divides into two, it does not divide, but passes bodily to one or other of the two new cells. In this way two different kinds of spermatozoa are formed—those which possess the odd half chromosome, and those in which it is missing—and they are formed in equal numbers. Now, ova fertilised by the former grow into females, those

---

[1] *Amer. J. Anat.* iii. 1904.
[2] *Journ. of Exp. Zool.* iii. 1906.

which are fertilised by the latter grow into males. Therefore this particular chromosome is the carrier of the sex character.

I have stated the theory of the mechanism of heredity as it seems to be developing. A word of caution, however, is necessary. It is quite possible that we are attaching too much importance to the chromosome simply because, owing to the affinity of its substance for dyes, we can follow it in the phases of cell history. The rest of the nucleus and cell body does not happen to show such constant affinities, and therefore the sense of sight yields no evidence as to their action in cell division. Yet, so far as we know, the same detailed processes of synthesis and analysis which we can follow in the chromatin substance may divide the units of the rest of the cell in cell division, and guide the half of each unit to its allotted place in the architecture of the new cells.

The observations of Conklin[1] upon a curious Ascidian egg makes this even probable. The body of this egg is built of five kinds of protoplasm recognisably different to sight during life. These are (1) deep yellow, (2) light yellow, (3) light grey, (4) slate grey, (5) clear transparent. Each of these has a separate history: the deep yellow protoplasm makes the muscular system, the light grey the brain, the clear transparent the skin, and so on. This egg therefore is a mosaic, an architecture of different kinds of living matter, which we can detect and follow owing to associated optical differences. Had these been absent, we should have known as little of the architecture of this egg as we know of that of eggs in general.

The independent transmission of characters, and the presence in the germ cells of different kinds of living matter, are indisputable. They lead us, however, to a riddle which I leave to my readers to solve as they will. We are driven to believe that in the material make-up of any race there are several kinds of living matter which cannot be changed the one into the other, and of which some will mix, others will not or cannot mix. These materials, bricks, as it were, in the building, are transmitted from generation to generation by the agency of the germ cells, which therefore are heterogeneous structures.[2] Now, the

[1] *Journ. of Exp. Zool.* ii. 1905.

[2] The beginnings of the science of their architecture is to be found in the last report of Mr. Bateson and Mr. Punnett to the Committee of the Royal Society on Evolution.

doctrine of the direct transmission of the various living substances employed in the make-up of the individual lands us in this difficulty. The fertilised egg has all the material necessary for the make-up, therefore it can, and does, develop into an adult. The generative cells also possess amongst them all the necessary material. Therefore amongst the earlier generations of cells produced by the growth and division of the fertilised ovum that cell or those cells which will form the generative organs must contain all the substances. But direct experiment contradicts this conclusion. Possibly in the very first cleavage, certainly in the second cleavage of the egg, there is a distribution of material amongst the two or four cells such that each one lacks something in the general make-up, and therefore can, and will, grow only to an imperfect monster if isolated. But one of those four incomplete cells will give rise, amongst other things, to the generative organs, each cell of which, in the first instance, is complete. Therefore, as we may "neither confound the persons nor divide the substance," we seem to be in a region of incomprehensibles.

"Just as that normal truth to type," says Bateson, "which we call heredity is in its simplest elements only an expression of that qualitative symmetry characteristic of all non-differentiating cell divisions, so is genetic variation the expression of a qualitative asymmetry beginning in gameto-genesis [the genesis, that is, of the germ cells]. Variation is a novel cell division. . . . What is the cause of variation?" Cross-breeding—that is, the union of unlike germ cells—may modify the character units. So, too, apparently may the long-continued *absence* of cross-breeding. It has been noticed in the cycles of a pure strain of Paramœcium that the periods of depressed vitality are also periods when the directive force of heredity is weakened. The individuals of successive generations show great departures from the normal type, and monsters are of frequent occurrence. With the lowered rate of growth, the lowered "vitality," as we call it, for want of a more precise word, there is associated a lowered degree of fixity of type.

SOME WORLD'S WEATHER PROBLEMS

By WILLIAM J. S. LOCKYER, M.A., PH.D., F.R.A.S.

Ce que nous connaissons est peu de chose, mais ce que nous ignorons est immense.—LAPLACE.

A POPULAR writer once said, " The earth we inhabit is surrounded by an atmosphere of air, the height of which is known to be at least forty-five miles. It presses upon the earth with a weight equal, at the level of the sea, to about fifteen pounds on every square inch of surface. As we ascend high mountains, this weight becomes less; as we go down into deep mines, it becomes sensibly greater. We breathe this atmospheric air, and without it we could not live many moments. It floats around the earth, being in perpetual motion; and, according to the swiftness with which it moves, it produces gentle breezes, high winds, or terrible tornadoes."

In these few words we have a very good general definition of the aerial envelope surrounding our earth, and it is this atmosphere and some of its movements which form the subject of the present article.

Many of us, I think, are rather apt to consider the height of our atmosphere in relation to the size of the earth greater than it really is. Although there is no real limit to its boundary, we must still remember that the loftiest mountain peak rides above about three-quarters of its total amount (by weight), while the highest altitudes permanently inhabited by man have still about half the atmosphere below them. If therefore we were to draw a circle to represent the figure of the earth, we should only have to increase the radius by about one-thousandth part in order to include half the atmosphere. As the circumference of the earth is about 24,000 miles, it will be seen therefore that the greater aerial movements must take place, for the most part, in a direction parallel to the earth's surface, since the depth of the atmosphere is, comparatively speaking, so restricted.

Even at the break of the twentieth century it is nevertheless astonishing how little is known about the general movements of this immense ocean of air. This perhaps is more our misfortune than our fault, because we live at the base of this aerial ocean, like flat-fish live at the bottom of the sea. It is only very occasionally that we have been able to gather information about the upper air currents which play such an important part in the circulation of our air. From the observation of the movements of very elevated clouds, and the drift of fine dust ejected into the air by volcanic outbursts, such as Krakatoa, some notion has been obtained of the directions of movements of the higher currents.

The time has now fortunately arrived when soundings in the air can be made nearly as easily as those in the ocean of water. This is not accomplished by employing manned balloons, for the heights to which these can attain are very restricted. Modern air-sounding machines are of two kinds, and are known as *ballons sondres* (free balloons) and *kites*. The former are small hydrogen balloons launched into the air, with their compact set of self-registering meteorological instruments, and allowed to wander where the aerial currents please to take them. Kites, on the other hand, are not quite such free agents, for, like the ocean-sounding machine of Lord Kelvin, they are controlled by pianoforte wire, operated by a winding engine.

In the hands of Messrs. Laurence Rotch, Helm Clayton, Tesserenc de Bord, Assmann, Dines, and others, great heights have already been reached by these means; and it is possible now to obtain a large amount of very valuable data to a distance of about 14 kilometres, or about 9 miles, from the earth's surface.

Perhaps one of the greatest advances in meteorology during the present century will be the knowledge gained by the sounding of the air by these and possibly other means; and since this method of research is only in its infancy, very much greater elevations will no doubt be brought within the range of routine investigation.

For the present, therefore, we must still be content, like flat-fish, to consider only the thin stratum in our immediate neighbourhood, and for a time to make the best use of the meteorological data gathered in this, the bottom of the aerial ocean.

As in many other walks of life, it is a good thing to utilise what you have got. Even if the meteorological data which we possess are restricted mainly to this bottom atmospheric layer, even if they do not extend over a great number of consecutive years, and, lastly, even if they are not homogeneously distributed over the earth's surface, they should be co-ordinated and discussed, and, if possible, working hypotheses based on them. Year by year more facts will be gleaned, and the hypotheses will be strengthened or rejected according as the new observations corroborate or render them untenable.

The beginning of the present century marks an important step in the progress of meteorology, an advance which is necessarily the result of the steady accumulation of facts by workers scattered all over the world. Up to about the end of the last century, each of the national meteorological institutions of all countries had been busy in organising, extending their services, and in gathering into their own particular net meteorological observations. They studied the changes which were occurring in their own immediate areas, and by means of these data improved their forecasts as to the kind of weather which might be expected on the following day, week, month, or season—as the case may be. In fact, meteorology was then "national" or "parochial."

Some of these institutions had already found that their own areas were too limited in extent to give them the necessary data for successful forecasting work, so they entered into a mutual compact with neighbouring countries for the exchange of certain pieces of meteorological information. From "national," meteorology then became, so to say, "continental."

The present stage of meteorological investigation has, in the last few years, indicated that even this mutual help of neighbouring countries, each working for its own immediate ends, is not sufficient. It is imperative not only to know what is occurring in the neighbouring countries, but also in the Antipodes, the most distant part of the world possible. In fact, the new meteorology—namely, "World Meteorology"—has dawned.

In the present article it is proposed to lay before the reader a general idea of the steps which have led to this broad-minded view of the meteorological problem, and, if possible, to point out the general trend of research in the future.

It may be mentioned here, however, that the remarks which follow refer altogether to the prospects of defining the character of weather during a season or a year, and not to such short intervals of time as a day, a week, or a month. It is of far greater consequence to know whether a summer will be fine or wet, or a winter mild or cold, than to be informed that the day after to-morrow will be wet.

Every one is acquainted with the fact that not only in these but in other lands the weather is not the same every year. Sometimes there is a great abundance of rain, while at other times we have very little. Some winters are impressed on our memory by the abundance of snow, while for several years in succession snow is conspicuous by its absence. Each country is more or less visited at times with floods, droughts, famines, or such-like misfortunes, and the names of India and Australia are at once brought up in the mind.

Now, without going into too great detail, it may be said that all these conditions are produced by changes in the intensity or direction, or both, of the main currents in our atmosphere, and, as I have stated above, we are only acquainted, like flat-fish, with the currents at the bottom of this atmosphere. We know, however, that these currents are dependent for the main part on the distribution of the atmosphere over the earth's surface, and we are further familiar with the fact that this distribution is not homogeneous, for by means of our barometers we are able to weigh, in inches of mercury, vertical columns of the air, and these weights are far from being equal.

Observations of barometric pressure seem therefore to be at the base of all weather changes, and the statements here made will be for the main part restricted to this meteorological element. The reader must, however, be reminded that it is really temperature which dominates the atmospheric circulation, but the changes of this important element are rendered so complicated by local causes, that for simplicity all reference to it will be omitted here.

To describe now some of the steps which have led up to this new aspect of meteorology, reference must first be made to the work of H. F. Blanford, who, in 1879, was the Meteorological Reporter to the Government of India. Blanford, from a discussion of the variations of barometric pressure over many years, found that there existed a kind of see-saw of

14

pressure between Siberia and Indo-Malaysia and Australia. In other words, when in some years the pressure in Siberia was extra high, that in Indo-Malaysia and Australia was correspondingly low. More recently the great Swedish meteorologist Hildebrandsson discovered that there were several regions which behaved in this see-saw manner with each other, and such localities he termed " centres of action" of the atmosphere. Thus the Azores and Iceland behaved in this manner, and also Siberia and Alaska, Tahiti and Tierra del Fuego, India and Siberia, Greenland and Key West, and Buenos Ayres and Sydney (Australia).

The present writer, with Sir Norman Lockyer, carried the investigation still further forward, and examined the pressure changes at ninety-five stations scattered all over the globe. The result of this inquiry led to the discovery that there really existed only one large see-saw, and this between nearly antipodal parts of the earth! In fact, it seemed that the Indian region was about the centre of one area, while the Argentine and Chili formed the centre of the antipodal region. By classifying the different types of pressure changes according as they resembled the two main and opposite types, namely, Bombay (India) and Cordoba (Argentine), it became possible to trace out a boundary line separating these two large areas. It was at once seen that each of the regions which Blanford and Hildebrandsson had pointed out as behaving in a reverse manner to each other were situated on opposite sides of this boundary line, and should therefore exhibit opposite pressure changes, according to the nature of this large see-saw. A similar classification of pressure types was undertaken by Professor Bigelow, of the United States Weather Bureau, and he corroborated almost exactly this world-wide barometric surge.

The accompanying map (fig. 1) shows the world divided into these two large regions. The continuous and dotted curved lines represent iso-phase conditions in the eastern and western hemispheres respectively, and the boundary between them is shown by the heavy lines. The see-saw regions referred to by Blanford and Hildebrandsson are connected by straight lines, which, as will be seen, all, with one exception—namely, that between Tahiti and Tierra del Fuego, which has been here omitted—cross the boundary line.

Now what does the presence of this see-saw inform us?

The reader must in the first instance be reminded that the amount of atmosphere surrounding our globe may be regarded as an invariable or constant quantity. The reduction of it, therefore, in any one part of the world must be counterbalanced simultaneously by a corresponding increase in another region. Now this large barometric see-saw tells us that such a transference of air is really in operation, and the direction of this exchange is from east to west and from west to east alternately, and not between the equatorial and polar regions.

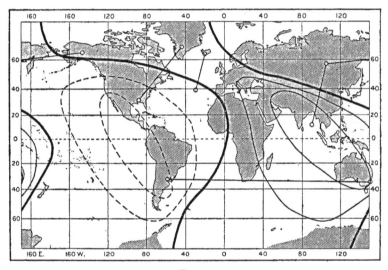

FIG. 1.

Map to show the central areas and boundaries (thick lines) of the two extensive regions the pressures over which behave in a see-saw manner. Previously known see-saws for special areas are joined by continuous straight lines.

It is difficult at the present time to give even an approximate idea of the amount of air that is being moved backwards and forwards; but the quantity must be very considerable, since it can so easily be detected by the changes of the barometer. In the first instance it was thought that possibly we might here be in the presence of a wave of high and low pressure travelling round the world; but further investigation showed that this was not the case, but that simply a huge to-and-fro action, world-wide in extent, was occurring. The reader may next naturally ask whether the existence of this barometric see-saw is going to help meteorology? The reply is that it will help this science very materially.

In the first instance meteorologists have, for the most part, been led to consider atmospheric changes as occurring mainly between the equator and the poles; but here we have a distinct east and west action taking place. This must in the future be legislated for.

Again, as we are in the presence of air being transferred from one half of the world to the other half, along a parallel of latitude, the places in each of these halves must be closely allied meteorologically. In fact we have here possibly an important clue to the close connection between the meteorological behaviour of regions which are widely separated. For instance, it is only quite recently that Sir John Eliot pointed out that the drought in the Indian region during the years 1895 to 1902 was a more or less general meteorological feature of the whole area, including Abyssinia, East and South Africa, Persia, Baluchistan, Afghanistan, and probably Tibet, and the greater part or the whole of Australia. A glance at the map here given shows that all these localities fall in the eastern hemisphere portion of the see-saw, so that they should be affected similarly.

Now a very important point in relation to this transference of air from one hemisphere to the other is the interval of time occupied by each of these barometric surges. One to-and-fro motion occupies nearly, but not quite, four years on the average; but the intervals are not all equal, so that this value is only approximately true.

It must be understood, however, that this particular four-year variation applies only to the more central parts of the two reciprocating areas, such as, for instance, those enclosed in the smaller dotted and continuous lines shown in the map to which reference has previously been made. Outside of these the variation seems to be shortened as regards duration, and in such cases as the British Isles, Europe, Canada, United States, etc., the changes exhibit more of a three than a four-year variation. In these instances also the intervals are not perfectly regular.

There is, I think now, little doubt that this pressure oscillation dominates, and is therefore responsible for, the very different types of weather that are experienced at any one place

in the course of three or four years. This alternate trans-
portation of air from the east to the west, and vice versa, must
affect to some degree, also alternately, the existing currents
of air. In tropical regions, such as India, for instance, the south-
west monsoon—the rain-bearing current which mainly fertilises
that vast country—must suffer some change, and it only requires
a slight deflection from its mean course to prevent this land from
receiving during its summer months its main water supply
requirements.

Even countries which are not in the zone of steady trade or
monsoon winds, but whose weather is the resultant of the
passage of cyclones and anticyclones—eddies in the general
atmospheric circulation—must also be affected.

The middle and southern portion of Australia, for instance,
is in the path of great air eddies of high pressure—or "anti-
cycles" as they are called—travelling from west to east, while
New Zealand, farther to the south, is in the track of a series
of cyclones—areas of low pressure—which are trying to wedge
themselves, sometimes with success, between the Australian
anticyclones.

Slight changes in the paths or intensity of these air systems
which must and do occur in consequence of the alternate excess
and deficiency of pressure caused by this barometric see-saw, alter
very considerably the distribution of rainfall over this large tract
of land in that part of the world.

Similar changes occur in the British Isles. The paths of low
pressure areas are sometimes to the north of Scotland, while at
other times they are, on the average, situated well across the
middle of England.

In the former case they are pushed to these higher latitudes by
the high pressure areas traversing the Continent, since cyclones
always move on the fringe of such high pressure systems. In
the latter case, the course of the high pressure systems being
situated farther to the south, the cyclones follow suit.

This movement of the tracks is also bound up with the
world-wide barometric see-saw; Great Britain being, however,
so far removed from the main central regions of action—namely,
those enclosed by the smaller oval curves shown on the map—
presents consecutive changes of three years' rather than four
years' duration.

While it is easy to represent with considerable accuracy the changes of air pressure from year to year over a large area, by means of a curve constructed from the readings of one barometer, it is not generally possible to determine accurately the rainfall changes from the records of one rain-gauge. The amount of rain collected in a rain-gauge is so dependent on its position and the local configuration of the land, that, in order to form some idea of the variation from year to year over a large area, the data gathered from numerous well-scattered instruments have to be combined. ⁻ Granting, therefore, that barometric curves represent the pressure facts over large stretches of country, much more accurately than rainfall curves illustrate the actual variations of rainfall, the reader's attention may be drawn to the set of pressure and rainfall curves brought together in the accompanying figure (fig. 2).

Now the rainfall curves here shown represent the variations from year to year at places where a reduction of pressure corresponds to an increase of rainfall. In other words, for those comparatively small areas over which the fall of rain is here taken into account, the greater the pressure the less the rainfall, and the lower the pressure the greater the rainfall. The function of an excess of high pressure, then, in those areas, is to obstruct as much as possible the passage of the rain-bearing air currents, which results either in the deflection of the moist winds from these regions or in a great reduction in their normal strength. If the pressure be abnormally low, we are presented with the most favourable conditions for great rainfall. In the curves here shown, then, those illustrating the pressure changes are inverted, so that their highest and lowest points will synchronise with those representing the times of greatest and least rainfall. The peaks in the pressure curves indicate, therefore, the years when the pressure was very low; and it will be seen that they correspond remarkably closely with the highest points of the rainfall curves, which pick out the wettest years for the regions investigated.

To illustrate the generality of this behaviour between pressure and rainfall for widely separated areas, typical instances are given for India, Australia, and England. Further, in order not to present rainfall curves which may only show fortuitous agreements with the inverted pressure curves, two absolutely independent sets of rainfall curves are shown with each pressure curve.

In the case of India, for instance, the area called Konkan and Ghats, on the west coast, is represented by the mean readings of seven rain-gauges; while the Deccan, another

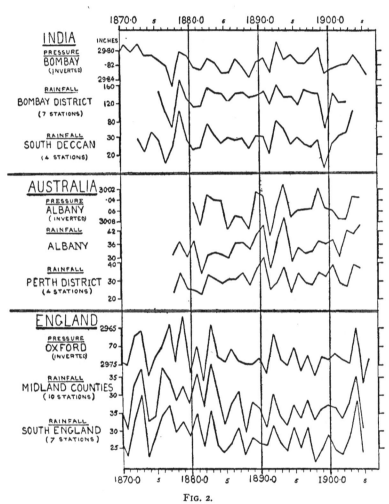

Fig. 2.

Curves to illustrate the close association of increase of rainfall with decrease of pressure for areas widely separated. (The peaks of the pressure and rainfall curves denote excess low pressure and abundant rainfall respectively.)

region, is the mean value from four other stations. The way all these curves change together demonstrates conclusively the very rigid connection between these two meteorological elements.

In no less degree is this the case for the south-west coast of Australia, for Albany (one station) and Perth district (four stations) show distinct rain beats in those years when the pressure over this region and the whole of Australia was deficient.

In our own islands the same story holds good, only, as I have previously pointed out, the duration of those changes is not the same as in the two areas which have just been considered. The pressure changes are here compared with two districts, termed by the Meteorological Office "Midland Counties" and "England, South." The curves here shown represent the rainfall variations as deduced from ten rain-gauges in the former and seven in the latter district. The very close correspondence of these curves with that for pressure at Oxford—a representative of English pressure variation from year to year—endorses for this region the deduction pointed out above.

Now the reader must not imagine that in every part of the world a decrease of pressure means an increase of rainfall. It can, and does, often happen that in some regions excess of high pressure means excess of rainfall. This is brought about by the fact that this pressure condition may be very favourable for the particular rain-bearing air current for some areas. In discussing, therefore, the rainfall of any part of the world in relation to pressure, this fact has to be borne in mind. It is on account of this, among other reasons, that the rainfall problem is so much more complicated than that of the pressure one; and the reader will, I think, agree with me that it becomes imperative to thoroughly understand the latter before proper consideration can be given to the former.

The existence of these alternate high and low pressure conditions will call for a considerable amount of work on the part of the meteorological representatives of different areas. It will be their duty to study the effects of these changes in the rainfall of their immediate localities, and find out how the "normal" conditions—conditions, by-the-bye, which seldom actually exist—are thereby influenced. The old saying—

> Be it fine or be it wet,
> The weather 'll always pay its debt—

has therefore so far a great amount of truth in it, and, as will be seen further on, its application to weather changes of longer duration is equally appropriate.

Enough perhaps has been written about this oscillatory pressure change to give the reader an insight into its bearing on the progress of meteorology during the present century.

Its detection opens up many fields for future research. Not the least of these will be the determination of the height above the earth's surface to which it can be traced. The highest station at which it has been found so far is Leh, in India, which stands at an elevation of 11,503 ft. above sea level. From an atmospheric point of view, this is not a great altitude; but it may be possible, when systematic kite flying is a matter of daily routine, to examine their records from this special point of view.

A piece of work which undoubtedly calls for immediate attention is the thorough examination of the pressure changes at every station which possesses a series of barometric observations. The map accompanying this article is only a very rough approximation of the prevailing conditions gathered from a discussion of ninety-five stations distributed over the globe. The paucity of observations in some regions, especially the oceans, is sufficient in itself to indicate the tentative nature of the pressure distribution here displayed; but the map will serve its purpose if it indicates, as I think it does, the regions where observations are most wanted. We are here truly in the domain of World Meteorology, for an island, ever so remote from the mainland, becomes a unit of the first importance in the determination of the extent and intensity of these pressure changes. Islands may not be ideal places for meteorological observers to live on, but they are perfect stations for studying the movements of the atmospheric currents, which are there unhampered, and therefore not rendered so complex by the presence of large tracts of land or lofty mountains.

Any one who has examined a long series of barometric observations will have noticed that, in addition to the short waves of fluctuation just described, there are others of much longer duration. Those changes are not due to any defects in the measuring instruments, but are actual variations due to our atmosphere. The reader is probably better acquainted with somewhat similar variations in the case of rainfall; but, as I have previously stated, rainfall being an effect of pressure, the pressure changes must first be examined to understand more fully those fluctuations recorded by rain-gauges.

So far as the investigations with pressure observations has gone at present, there are two conclusions which I think are now beyond criticism. The first is, that all over the world changes of long duration are in operation; and secondly, these changes are not all alike either in intensity or time of duration.

In spite of these marked differences, there seems nevertheless to be an underlying connection between them all. It is not therefore without, but rather within, the bounds of possibility, that the time will come when such apparent divergences will all be found to be very closely allied to each other, and be the natural resultants of one or more primary world atmospheric fluctuations.

Just as we have been able to place before the reader a map, showing a first approximate relationship between the barometric changes of short duration extending over the whole of the earth's surface, so during the present century it will be possible to make a similar survey for those air fluctuations which require more than three or four years to complete a pulsation.

The only reason why such a chart is not available to-day is that the length of time over which observations extend for regions well distributed is not yet sufficiently great.

Our appetites have, however, been wetted by numerous investigations, which are quite sufficient to indicate what a promising field of inquiry lies before us.

We know, for instance, that in the Indian area there is a gradual rise and fall of pressure, a complete alternation which occupies about eleven years. The same area shows distinct traces also of a variation which takes more than three times this length of time to go through a complete swing.

In Australia this eleven-year variation is also in operation, but so far as can at present be determined, it appears to be so modified that it presents a fluctuation of about nineteen years' duration.

Similar to this in length of time, but with a distinct phase difference—that is, the epochs of maxima or minima do not take place simultaneously—is a variation in progress in the southern part of South America. There we have nearly, but not quite, an opposite state of things to that occurring in Australia. Our own islands are not even free from these long barometric swings, for here we can trace a variation which seems to be of about thirty-five years in length. By a continuous and steady accumulation of more data, the time will eventually arrive when a more accurate knowledge of the lengths of these various changes will be acquired; but at present we are, so to speak, practically groping our way in the dark. There is no question, however, about the importance of gaining as much knowledge as possible about the lengths, phases, and relative intensities of these changes of long duration. Even with regard to these, compensating effects must be taking place in different parts of the world, a deficiency in one region corresponding to a surplus somewhere else. The behaviour of such pressure changes in a very remote region from us may therefore be of very great concern to us in trying to look into the past and future of these sequences of changes. In fact, it seems impossible for the investigator to take too broad a view of the domain of meteorology.

In this article an attempt has been made, and I hope with some success, to acquaint the reader with some of the surges which are taking place in this atmosphere of ours, and which are initially responsible for the variations of weather from year to year and from decade to decade. We have, however, left out of consideration altogether the prime mover and originator of all these varied pressure permutations. In fact, we might be describing the play of *Hamlet*, omitting the rôle of the Prince of Denmark! The sun—

> Great source of day! best image here below
> Of thy Creator, ever pouring wide,
> From world to world, the vital ocean round,
> On Nature write with every beam His praise.
>
> THOMSON'S *Hymn*—

as Professor Young writes in his excellent manual of Astronomy, " is the nearest of the stars—a hot, self-luminous globe, enormous as compared with the earth and moon, though probably only of medium size compared with other stars; but, to the earth and the other planets which circle around it, it is the most magnificent and important of all the heavenly bodies. Its attraction controls their motions, and its rays supply the energy which maintains every form of activity upon their surfaces." As a child of the sun, therefore, and bathed continually in his rays, the earth should respond to all his varied moods if they are emitted into space and represented by either an increase or reduction of radiation of sufficient magnitude. How can these moods be detected? Has it been possible to measure any variation in the sun's radiating power? In spite of the rapid strides made in the designing and construction of instruments of precision, I do not think it can be said for certain that any changes that have been recorded can be positively attributed to the variation of solar radiation. Our atmosphere, the very movements and changes of which we wish to investigate, is responsible for this deficiency of our knowledge, for it masks or mitigates all effects which, if it were absent, might be recorded.

By carefully studying the sun, and observing the many phenomena rendered visible by means of telescopes and spectroscopes, it has been found that the sun appears very much more disturbed at some times than at others. Failing absolute measurements of solar radiation, we are therefore driven to employ deductions. Thus, for instance, we know that sunspots are indications of solar disturbance; so we can say that the greater the number or size of the spots the larger will be the areas of disturbance. From this it is deduced that the radiation must be greater, and consequently the sun hotter. It has been found further that there is a periodicity in spot activity; that is, during some years only a few spots are visible on the solar disc, while in other years they are very numerous. In fact, they are most or least numerous about every eleven years. We are thus led finally to infer that the sun becomes hotter and colder alternately every eleven years.

From sunspot and other solar data it has also been concluded that changes which extend over about four years and thirty-five years are also in operation.

We are thus brought into the presence of three variations in solar activity, the oscillations of which cover approximately four, eleven, and thirty-five years each.

It is needless to ask the reader whether we have in our atmosphere any changes which correspond to these solar fluctuations, for he must now be quite familiar with them.

The fact that the only three solar pulses known should have their terrestrial equivalents in our atmosphere is, I think, a very strong case in favour of the solar origin of these barometric surges.

There seems reason also to believe that, if these solar

changes are of sufficient magnitude to originate such large fluctuations in the terrestrial atmosphere, they ought to be capable of direct measurement by instrumental means at no very distant date.

In stepping from the earth to the sun, to look for the origin of the changes in the air around us, the domain of World Meteorology is extended by about 96,000,000 of miles.

The meteorologist of the twentieth century must no longer limit his vision to the height of the most elevated cirrus cloud, but, like the astronomer, must keep in touch with all the manifold changes which are for ever taking place in the incandescent envelope of our central luminary.

> O glittering host! O golden line!
> I would I had an angel's ken,
> Your deepest secrets to divine,
> And read your mysteries to men.

THE ORIGIN OF GYMNOSPERMS

By E. A. NEWELL ARBER, M.A., F.L.S., F.G.S.

PROBABLY no aspect of botanical research is held to be of greater interest than that which tends to throw light on the ancestry of some race or group of living plants. The study of fossil plant-remains possesses a peculiar fascination in this respect. Although attempts have been made, from a detailed comparison of the organs and the structure of recent plants, to obtain clues as to their lines of descent, it is but rarely, and to a limited extent, that success has been achieved. The tendency to segregate living plants into self-contained groups, by the erection of isolated phyla for their reception—*e.g.* Coniferales, Cycadales, and the like—as well as the lack of definite evidence as to their probable ancestry, are proofs that for help in these matters we must seek elsewhere. Happily, within the last few decades, Palæobotany has come to the rescue. The study of Palæozoic fossil plants, the possibilities of which in this direction were foreseen by Brongniart, and Williamson, among others who bore the brunt of the early reconnaissance or pioneer work in this domain during the past century, has proved especially fruitful in this respect. But the greatest achievement to be recorded, so far, is the clear light which such researches have thrown on the origin of the plants known to botanists as the Gymnosperms. This race includes the four phyla—Coniferales, Cycadales, Ginkgoales, and Gnetales—all of which are still represented, in greater or less degree, in the vegetation of the world at the present day. As opposed to the Angiosperms, the outstanding feature of this group is found in the fact that the seed is naked; carpels, which in the Angiosperms envelop the seed, being absent. While the solution of the problem of the ancestry of the Angiosperms themselves remains for the future, that relating to the Gymnosperms has now been solved. It is proposed to consider briefly here the bearing of a series of remarkable, but very recent, researches, which have proved so fruitful in this respect.

It has been known for nearly a century that seeds, similar in organisation to those of modern Gymnosperms, occur in rocks which are among the oldest in which fossil plants are found. Thus the race of Gymnospermous seed-plants is geologically ancient. Curiously enough, as will be seen at a later stage in this consideration, one of the results of modern work has been a tendency to push back, in the scale of geological time, the appearance of the naked seed-bearing habit to an even earlier period. This particular form of heterospory is found to be incredibly old in a geological sense—far more ancient, in fact, than those rocks in which the earliest known fossil plants are found at present. Yet, while a primitive type of seed-plant is still quite unknown to us in the fossil state, a study of the oldest seed-plants with which we are acquainted has afforded distinct evidence as to the ancestors from which some, at least, of them were derived.

The particular members of the Gymnospermous stock, whose line of descent has now become clear, are the Cycads. We know that the modern Cycads,[1] of the Tropics or sub-Tropics of both the Old and New Worlds, are but survivals of what was once a great group, holding a position in Mesozoic times similar to that which the Dicotyledonous Flowering Plants hold in the vegetation of to-day. This group has now been traced back to a more ancient type of seed-bearer, long extinct, and this in turn to a still older stock, which was homosporous, and essentially Cryptogamic, as opposed to seed-bearing.

A glance at the fossil plants of Carboniferous age preserved in our larger museums, or even the casual collection of specimens on the waste heaps of some colliery, will serve to impress at least one fact. It will be found that impressions of fronds—such as those known technically as *Sphenopteris*, *Neuropteris*, *Alethopteris*, etc., all of which closely recall the leaves of the modern ferns — were not only extremely abundant, but greatly varied, in Carboniferous times. The casual observer perhaps would be startled to learn that the majority of such fossils, so very fern-like in every aspect, were not ferns at all. Such, however, is the fact. Not only

---

[1] The present-day Cycads are included in about nine genera, and from seventy-five to a hundred species.

were they not ferns, as even specialists in fossil botany fully believed until recently, but they had attained to the very different rank or status of seed-plants. In brief, the discovery of this significant fact forms the greatest contribution to fossil botany within modern times. We know now beyond doubt that most of these fern-like plants of Carboniferous age were Gymnosperms, and not Cryptogams.

It would take too long here, nor would it be advisable, to try to illustrate even the fundamental botanical differences between the seed-plants, of which the Gymnosperms are one group, and the Cryptogams. A rough analogy may, however, serve to give some conception of the gap which separates these two races of the Vegetable Kingdom. In the animal world, living and extinct, two great subdivisions are easily distinguished—the Vertebrata, or animals having a " back-bone," and the Invertebrata, which do not possess a spinal column. Partly on these grounds, but in reality for many other reasons, the Vertebrata are generally regarded as having attained to a far higher position in the scale of morphological development and complexity than the Invertebrata.

In the Vegetable Kingdom there may be found a superficial analogy. The seed-plants (*Spermophyta*) correspond roughly to the Vertebrata among animals, while those plants—including the Algæ, Mosses, and Ferns—which possess the type of reproduction termed Cryptogamic, lie roughly parallel to the Invertebrata. In both kingdoms there are further subdivisions, or races—the Reptiles, Birds, Fishes, and Mammals among Vertebrata, and the Gymnosperms and Angiosperms among Spermophytes. The highest plants—the Angiosperms, or Flowering Plants—correspond to the Mammals among Vertebrata.

Now, if a fossil had been discovered, let us say resembling externally a Crustacean—the group of Invertebrata to which the Lobsters and Crabs belong—which in addition to a tough outer skeleton (*exoskeleton*) possessed a bony spinal column, the main conclusion that an animal had been found, closely resembling an Invertebrate, but which was in reality a Vertebrate, would be of great interest to zoologists, and to the scientific world in general. In the case of these seed-bearing, fern-like plants, this is roughly what has happened. What was formerly regarded as a Fern has proved to be a Gymnosperm, two stages, let us say equivalent, in imagination, to the Crustacea among Invertebrata, and to

Fishes or Birds among Vertebrata. This illustration must not, of course, be taken literally. It may, however, serve to give to those who are not specially familiar with the characters of the subdivisions of the Vegetable Kingdom, some idea of the enormous gap which the botanist perceives between the two main subdivisions of that kingdom, the Cryptogamic and the Spermophytic; hence the interest of the present subject, the origin of certain seed-plants from ancestors which did not bear seeds.

But to return to our Carboniferous fern-like fossils. The first suspicion that in these cases we may be dealing with plants, which are not really ferns at all, suggested itself to the Austrian palæobotanist Stur in 1883. The fact that a fern-like fructification had never been observed on the fronds of any of the thousands of examples of *Alethopteris* or *Neuropteris* which had been collected at one time or another, led Stur[1] to the conclusion that whatever these plants may have been, they were certainly not ferns. But for twenty years afterwards no definite evidence was forthcoming as to the precise nature of the fructifications of these plants.

Meanwhile progress was being made in what seemed to be other directions, though, in respect at least to one group of fossils, it subsequently proved that this apparently new direction was in reality identical with the old.

The study of Carboniferous petrifactions, *i.e.* plant remains in which the internal structure is preserved, begun in this country by Witham and Binney, but finding its greatest exponent in Williamson, had already commenced to make rapid progress. Among the members of the Carboniferous vegetation investigated by Williamson,[2] and subsequently by that botanist in conjunction with Dr. Scott,[3] were certain fossils known as *Lyginodendron* and *Heterangium*, which possess a peculiar type of structure, combining features common both to living ferns and also to living Cycads. Other genera also possessing this peculiarity had been discovered by Continental workers, as well as by British investigators.[4] By the year 1898 it had become clear that there existed a considerable plexus of such plants in Carboniferous and Permian times, and for this group Professor Potonié,[5] of Berlin, proposed the appropriate name, Cycadofilices, thus indicating the synthetic nature of their morphology.

[1] Stur, 1883, p. 638.　　[2] Williamson, 1873.　　[3] Williamson and Scott, 1895.
[4] Scott, 1899.　　　　　[5] Potonié, 1897, p. 160.

During the progress of this work it had been gradually discovered that many of the types of fern-like foliage to which I have already alluded, especially the frond-genera *Alethopteris*, *Neuropteris*, and some Sphenopterids, belonged in reality to these same plants, the Cycadofilices. Thus, for some years past, it has been apparent that Stur was correct in his view that such fronds were not the leaves of plants which were really Ferns, in the sense that we imply when we speak of the recent Ferns.

Nothing, however, had so far been discovered as to the nature of the fructification of any of the Cycadofilices. It was not until 1903 that we possessed any definite information on this point. The real initiation of the whole of our now comparatively advanced knowledge of the ancestry of the Gymnosperms, as well as of the reproductive organs of this group, dates from that year, when the researches of Prof. Oliver and Dr. Scott on this subject were first made public. Prof. Oliver had been engaged for some time previously upon an investigation of the petrified seeds of the Carboniferous and Permian periods. Among these was one, previously named by Williamson *Lagenostoma Loxmai*, which was reinvestigated by Prof. Oliver conjointly with Dr. Scott. The main result of this work was that these authors [1] were able to recognise in this seed the female fructification of *Lyginodendron oldhamium*, a member of the Cycadofilices, possessing highly compound, fern-like foliage of the *Sphenopteris* type.

The seed itself is a highly evolved structure, agreeing in certain respects with that of living Cycads, more especially in the possession of a well-developed pollen-chamber, and pollen-collecting mechanism, but in many of its features much more complex than the seed of any recent Gymnosperm.

It may be of interest to point out that this seed has not, even to this day, been found attached to *Lyginodendron*. Yet its attribution, although indirect, is none the less certain. The seed itself was enclosed in an envelope, known as the cupule, the whole resembling somewhat a hazel nut surrounded by its husk. The cupule was a lobed structure, studded with peculiar glandular hairs. These glands agree exactly with those long known to occur on the stem, petioles, and even the leaflets of *Lyginodendron*, but have not been found on any other

[1] Oliver and Scott, 1903, 1904.

fossil plants of Palæozoic age. It was chiefly, but not entirely, on the identity of these glands, that Prof. Oliver and Dr. Scott were able to refer *Lagenostoma Lomaxi* to *Lyginodendron*.

Although the seed-bearing habit in this particular instance is as yet unknown, in the case of another *Lagenostoma* (*L. Sinclairi*), described more recently,[1] the seeds, also enclosed in cupules, have been found attached to a highly branched structure, which is interpreted as a frond with reduced lamina. In fact, in all the examples among these fern-like plants, in which the seed-bearing habit has been ascertained, the seeds were borne on fronds, either little modified and similar to the sterile foliage, or on fronds with reduced blade or lamina.[2] Thus these ancient seed-bearing, fern-like plants must have presented, in the lax manner in which the fructification was borne, a striking contrast to almost all the other great Palæozoic groups of plants, where the sporangia were aggregated into cones. If we could see to-day a landscape as it was in Carboniferous times, we should no doubt be struck by the diversity of cone-bearing plants as compared with those in which the fructification was to be found on the fronds, in much the same manner as sporangia occur on the fronds of living ferns. Only here, in most cases, but not in all, the sporangia were heterosporous, the female or megasporangium being of that particular nature which we call a seed.

Since 1903, as we shall see, several other discoveries relating to the nature of the female fructifications of members of the Cycadofilices have followed rapidly on the heels of the initial attribution of the seed *Lagenostoma* to *Lyginodendron*. We may, however, before considering these, complete the story so far as it relates to that genus.

It is only quite recently that we have come to know anything of the male organs of the Cycadofilices. Yet, curiously enough, the first, and at present the only discovery in this direction, also relates to this same plant *Lyginodendron*. Mr. Kidston[3] showed last year (1905) that a particular form of sporangial aggregation or sorus, borne on a frond in which the blade or lamina was somewhat reduced and already well known as a detached fossil under the name *Crossotheca*, was in reality the male fructification of *Lyginodendron*. He had

[1] Arber, 1905[1] ; Scott and Arber, 1905.
[2] Arber, 1905[2].   [3] Kidston, 1905, 1906.

the good fortune to obtain fronds showing both the reduced, fertile pinnules, and the ordinary, sterile foliage of *Lyginodendron*, an exceedingly rare circumstance among fossil plants. On the whole these male organs of *Lyginodendron* are almost startling in their simplicity, and in their obvious resemblance to the fructification of certain isosporous ferns. Surprise at this circumstance is only increased when we reflect that the corresponding female organ of the same plant is of an exceedingly advanced type, as has been already intimated. Hence, if the female fructification points strongly in the direction of the Cycads, the discovery of the male organs serves but to forge another link in the chain which connects the Cycadofilices with the Ferns : links which have been already recognised in the general habit of the foliage, and in certain features of the anatomy of the stem. In other words, the Cycadofilices are synthetic in their fructification, as in their anatomy.

In the case of the Crossothecas described by Mr. Kidston the structure is to some extent preserved. There is, further, a possibility that some sporangia, occurring in petrified material, which were discussed by Miss Benson[1] in 1904, under the name *Telangium Scotti*,[2] may also eventually prove to be the male organs of a Pteridosperm. If this is the case, their structure lends further support to the conclusion as to their fern-like nature.

Within a few months of the initial discovery by Prof. Oliver and Dr. Scott, their main conclusion that some at least of the Cycadofilices were seed-bearing plants was confirmed in a striking manner by Mr. Kidston.[3] At the close of 1903 Mr. Kidston obtained some ironstone nodules from South Staffordshire containing plant impressions, which proved to be of exceptional interest. Some of them showed a large seed, about 3 cm. long, identified by that author as a *Rhabdocarpus*, to the stalk of which were still attached typical leaflets of *Neuropteris heterophylla*, Brongn. Now, the *Neuropteris* type of foliage, although almost always occurring detached like other fossil fronds, had been known for some years previously to belong to the stem *Medullosa*, another typical genus of the Cycadofilices. Thus *Medullosa* has also proved to be a seed-

---

[1] Benson, 1904.
[2] Mr. Kidston regards *Telangium* as quite distinct from *Crossotheca*.
[3] Kidston, 1903.

bearing plant. There is also some evidence which appears to relate to the male organs[1] of this genus, but it is not entirely satisfactory, and so need not be discussed here.

Of the three great British genera of Cycadofilices, two are known now to be seed-bearers. The third, *Heterangium*,[2] perhaps the most interesting of all in this connection on account of its still closer relationship to the Ferns, remains for the future. Should it eventually be found to have also attained to the seed-bearing status, the case will be even more complete.

It is obvious that some new name is necessary to denote this group. Prof. Oliver and Dr. Scott,[3] recognising in 1903 that *Lyginodendron* would not be found eventually to stand alone among the Cycadofilices as a seed plant, proposed the name *Pteridospermeæ* for this race, " embracing those Palæozoic plants with the habit and much of the internal organisation of Ferns, which were reproduced by means of seeds." This suggestion has already met with wide acceptance.

At present, and for some time to come, these discoveries will stand as the greatest contribution of the British School of Palæobotany. Their wide-reaching importance has been quickly realised, and special accounts of this progress have already been published in France,[4] Germany,[5] Austria,[6] Sweden,[7] and the United States.[8]

More recently, valuable contributions on the same subject have been made both in the United States and in France. These further discoveries concern genera in which the internal structure is at present unknown, and which therefore have not been so far recognised as synthetic types. They are, however, all plants possessing the fern-like habit of foliage, and thus would seem to fall within the limits of the Pteridospermeæ, now that they have proved to be seed-plants.

In 1904 Mr. David White[9] announced the discovery of

---

[1] Kidston, 1903.

[2] For a general account of the structure of the Cycadofilices, see Scott, 1900, chapters x. and xi.

[3] Oliver and Scott, 1904, p. 239.

[4] Zeiller, 1904, 1905.

[5] Oliver, 1905.

[6] Scott, 1906.

[7] Nathorst, 1906.

[8] Berry, 1904[1], 1904[2]; Coulter, 1904; Ward, 1904; White, 1904[2].

[9] White, 1904[1].

specimens from Virginia, belonging to a typical Lower Carboniferous frond-genus, *Aneimites*, better known in this country under the name *Adiantites*, which bore seeds on the fronds. In this case the lamina of the fertile frond was reduced, as compared with that of the sterile, and the seeds appear to have had a distinct marginal wing. Thus seed-bearing fern-like plants have been traced back so far as Lower Carboniferous times.

An equally interesting discovery, and one perhaps of greater significance at the present time, was made by the veteran French Palæobotanist, M. Grand'Eury,[1] in the following year (1905). The frond-genus shown, in this instance, to belong to a seed-bearing plant, is that known as *Pecopteris*, which is admittedly very fern-like. It occurs commonly in the higher zones of the Upper Carboniferous and in the Permian. Probably of all the fern-like genera found in these rocks, *Pecopteris* was the one which was least suspected of being a Pteridosperm, for many species have been known for a long time in the fertile state, and, so far, the fructification has invariably proved to consist of sporangia, which may be closely compared with those of a family of living ferns, the *Marattiaceæ*. Yet M. Grand'Eury has shown that at least one Pecopterid, *P. Pluckeneti*, Schl., bore seeds, and that the seeds were attached to the lower surface of a fertile frond,[2] which scarcely differed as regards the lamina from the sterile leaf. It is true that this species is not a very typical member of the genus. But it seems quite possible that other Pecopterids may have borne seeds, and that some of the fructifications, at present regarded as sporangia, comparable to the organs of the homosporous ferns which they so closely resemble, may have been in reality the male organs of Pteridosperms. However, for information on this point we must look to the future.

This completes the list of Pteridosperms which have been discovered during the last three years. There are, however, many genera which at present rest under the strong suspicion of having also been members of this race, and not ferns, but in these cases the material at present available is not sufficient to place the matter beyond doubt. In most of these good evidence of continuity between the seed and the frond is at present lacking.

[1] Grand'Eury, 1905.   [2] See Zeiller (1905), fig. 7 on p. 725.

One of the earliest instances in which there appeared to be some grounds for attributing a seed to a certain frond was discussed by Wild[1] in 1900. This investigator showed that it is exceedingly probable that the well-known seed, *Trigonocarpus*, really belonged to a *Medullosa* with the *Alethopteris*-type of frond, not only on the grounds of constant association, but on the agreement of certain anatomical features common to the seed and the vegetative organs. This attribution still appears to be extremely probable, but definite evidence of the fact has not so far been forthcoming.

M. Grand'Eury,[2] to whose work reference has been made, has supported this conclusion, and has also shown that equally strong evidence, so far as constant association can be regarded as having weight in this connection, exists for referring the Upper Carboniferous or Permian genera *Odontopteris*, *Linopteris*, and *Callipteridum* to the Pteridosperms. In a more recent paper the same author[3] deals with further data, also derived from association, affecting the frond-genus, *Sphenopteris* and the seed, *Codonospermum*.

M. Grand'Eury has also called attention to the structural diversity of the Carboniferous seeds which may probably be attributed to the Pteridosperms. The investigation of certain Palæozoic seeds showing structure, at the hands of Prof. Oliver,[4] has also led to the same conclusion.

I have dealt with these recent discoveries at some length here, for this work is of extreme importance as supplying the missing link connecting at least one group of the Gymnosperms with Fern-like ancestors. The Pteridosperms may be regarded as standing, roughly, half-way between the Ferns and the Cycads, though in reality they themselves have crossed the borderland, and are obviously on the Gymnospermic side. Thus the views of Dr. Scott and others, who for some time previously have urged the conclusion that the Cycads, the most primitive of Gymnosperms still existing, were derived from the Ferns, have received full confirmation. Dr. Scott[5] has pointed out that practically every step in the line of descent of the Cycads can now be

---

[1] Wild, 1900 ; see also Scott and Maslen, 1906.
[2] Grand'Eury, 1904[1], 1904[2].
[3] *Ibid.* 1906.
[4] Oliver, 1903, 1904[1], 1904[2].
[5] Scott, 1903, 1905[1], 1905[2], 1906.

traced, beginning in some distant geological period with a homosporous ancestor which, although quite unknown at present as a fossil, must have been essentially similar to a modern fern.

As a corollary of these discoveries, a discussion on the origin of the seed itself has naturally arisen. Unfortunately in this respect, the seeds of the Pteridosperms have all proved to be highly evolved and complicated structures, and thus, in the absence of features which might be considered primitive, they hardly advance our ideas very definitely. Certain considerations arising out of recent work have, however, been put forward by Prof. Oliver,[1] Miss Benson,[2] and Miss Stopes.[3] The subject has also been fully discussed within the last few months by Prof. Oliver,[4] in his presidential address to the Botanical Section of the British Association at York.

With regard to the ancestry of other groups of the Gymnosperms, such as the Coniferales and Ginkgoales, the evidence derived from a study of fossil botany is by no means so clear at present as in the case of the Cycads. This, no doubt, is chiefly due to the fact that plant remains showing anatomical structure are unfortunately very rare in the Mesozoic rocks.

Both the Ginkgoales and the Coniferales begin to appear before the close of the Palæozoic period. The former have but one representative living to-day—the Maidenhair Tree (*Ginkgo biloba*, L.) of China and Japan, now almost extinct except where cultivated. In the Mesozoic rocks we find abundant evidence that this group, now reduced to a bare unit, formed at that time a highly characteristic factor in the flora. A study of the living plant has shown that in *Ginkgo* we have again a synthetic type, in that it combines characters common to both Cycads and Ferns[5] on the one hand and to Cycads and Conifers on the other.[6] So far, however, the study of fossil botany has not added materially to what can be ascertained as to the ancestry of this group from the still surviving member.

When we turn to the Coniferales, the group which includes the Pines and Firs, the fossil evidence is more valuable in this connection. Yet botanists are not, so far, agreed as to whether the Coniferæ as a whole are monophyletic, *i.e.* derived from a

[1] Oliver, 1902, 1903.
[2] Benson, 1904.
[3] Stopes, 1905.
[4] Oliver, 1906.
[5] Seward and Gowan, 1900, p. 146.
[6] Scott, 1900, p. 521.

single, pre-existing race, or polyphyletic, from several distinct ancestors.[1]

Those who hold the former view probably constitute a majority at the present time. So far as the fossil evidence is concerned, they are inclined to regard all the modern families of Coniferæ as derived from a very ancient, and now long extinct group, the Cordaitales. This race of seed-bearing plants flourished in Palæozoic times, before the earliest appearance of the Coniferæ, and is now well known—thanks to the researches of Grand'Eury and Renault. *Cordaites* itself is the typical example. This plant was a tall tree, bearing a crown of branches above, which, in turn, bore long, strap-shaped leaves. The veining of the leaf was parallel, thus strongly recalling that of a Monocotyledon, a comparison which, however, is quite without any value as a clue to affinity. Among the leaves were borne catkin-like bodies, the inflorescences, consisting of male and female flowers. The seeds were naked, and altogether agree closely with those of the Cycads.

As regards the vegetative characters, "the Cordaiteæ hold the balance very evenly between Cycads and Conifers."[2] The female inflorescence recalls that of a Conifer, but does not resemble a Cycadean cone in the least. The male is quite unlike anything found among modern Gymnosperms. It is in this group, in many respects a less specialised one than any Gymnosperm now existing, that those who hold to the Monophyletic theory seek for the origin of the Coniferales. At any rate, the study of fossil botany has produced evidence of an ancient group of Gymnosperms, neither identical with the Conifers nor with the Cycads, but combining certain characters common to both ; and the existence of this group must ever continue as a factor to be taken into account in any discussion of the origin of the Coniferæ.

In this connection it may be pointed out that, fortunately, we have already a clue as to the possible ancestry of the Cordaitales themselves, thanks to the researches of Dr. Scott.[3] While, in the internal structure of the stem of many of their later representatives in geological time, we find a condition of affairs almost identical with that of an *Araucaria* among living Conifers, yet some of the earlier members possessed,

---

[1] Oliver, Arber, Scott, and Seward, 1906.
[2] Scott, 1900, p. 441.   [3] Scott, 1902.

in addition, the peculiar type of primary wood, which is characteristic of certain Pteridospermeæ. Thus the Cordaiteæ combine characters common to the Cycadofilices and Coniferæ, and, if special importance is attached to the value of anatomical evidence as a clue to affinity, as is held in many quarters, the ancestry of the Coniferæ, through the Cordaitales, may be traced back ultimately to the Fern alliance, by means of the connection between the Cordaitales and the Pteridospermeæ.[1]

An important fact, derived from a study of fossil plants, is that the *Araucarieæ*, to which family belongs the well-known tree, commonly cultivated in this country, the " Monkey-Puzzle " (*Araucaria imbricata*, Pav.), is a very ancient, if not the most ancient, family of Coniferæ. Mr. Seward[2] and Miss Ford, in a critical study of the Araucarieæ, recent and extinct, which has just been completed, have pointed out that certain morphological features, especially in connection with the reproductive organs of these plants, agree with others to be found in the Lycopod group. They have advanced arguments " in favour of the view that this group of Gymnosperms, unlike the Cycadales, was probably derived from Lycopodiaceous ancestors," and again, that "the general consent which has deservedly been given to the view that the Cycadales and Filicales are intimately connected by descent, may have the effect of inducing an attitude too prone to overestimate the value of the arguments advanced in support of an extension of the idea of a filicinean ancestry to other sections of the Gymnosperms."[3] Thus, the view that the Coniferæ may have had a direct connection with the Lycopods in past times, although by no means a new one, has received further acceptance quite recently. By those who hold to it, the Coniferæ can scarcely be regarded as other than polyphyletic. Mr. Seward and Miss Ford have indeed drawn " attention to the various characters in which the Araucarieæ differ from other members of the Coniferales, and suggest the advisability of giving more definite expression to their somewhat isolated position by substituting the designation Araucariales for Araucarieæ."[4]

The fact that some of the Palæozoic Lycopods, *e.g. Lepidocarpon*, are now known to have borne seeds, a recent discovery which we

[1] Scott, 1905[1], 1906.
[2] Seward and Ford, 1906.
[3] Seward and Ford, 1906, p. 164.
[4] *Ibid.* p. 164.

owe to Dr. Scott,[1] is, no doubt, regarded as lending support to the view that some of the Coniferæ may have been derived from Lycopods.

In conclusion, it has been shown that the study of fossil botany has thrown light on the ancestry of at least one group of Gymnosperms, the Cycads. These plants, still surviving as mere remnants of what was once a great race in past times, were descended, ultimately, from the Ferns. The intermediate stage, represented by the Pteridospermeæ is now known.

As regards the Coniferæ, two views are held. Those who regard the group as monophyletic, trace them back also to the Ferns, through the Cordaitales. Others derive some of them from Lycopodiaceous ancestors, and hold that the Coniferæ as a whole were polyphyletic.

BIBLIOGRAPHY

ARBER, E. A. N. (1905[1]), On some new species of *Lagenostoma*, a type of Pteridospermous Seed from the Coal Measures, *Proc. Roy. Soc.* Ser. B. vol. lxxvi. p. 245, 1905.

—— (1905[2]), The Seed-bearing Habit in the Lyginodendreæ, *Proc. Cambridge Phil. Soc.* vol. xiii. pt. iii. p. 158, 1905.

—— *see* Scott and Arber (1905).

BENSON, M. (1902), A New Lycopodiaceous Seed-like Organ, *New Phytologist*, vol. i. p. 58, 1902.

—— (1904), *Telangium Scotti*, a new species of *Telangium* (*Calymmatotheca*), showing structure, *Ann. Bot.* vol. xviii. p. 161, 1904.

BERRY, E. W. (1904[1]), A Notable Paleobotanical Discovery, *Science*, N.S. vol. xx. pp. 56 and 86, 1904.

—— (1904[2]) Recent Contributions to our Knowledge of Paleozoic Seed-plants, *Torreya*, vol. iv. p. 185, 1904.

COULTER, J. M. (1904), Pteridospermaphyta, *Science*, N.S. vol. xx. p. 149, 1904.

FORD (S. O.), *see* Seward and Ford (1906).

GOWAN (J.), *see* Seward and Gowan (1900).

GRAND'EURY, C. (1904[1]), Sur les graines des Néuroptéridées, *Compt. Rend. Acad. Sci.* vol. cxxxix. p. 23, 1904.

—— (1904[2]), Sur les graines des Néuroptéridées, *Compt. Rend. Acad. Sci.* vol. cxxxix. p. 782, 1904.

—— (1905), Sur les graines trouvées attachées au *Pecopteris pluckeneti*, Schlot, *Compt. Rend. Acad. Sci.* vol. cxl. p. 920, 1905.

—— (1906) Sur les graines de *Sphenopteris*, sur l'attribution des *Codonospermum* et sur l'extrême variété des "graines de fougères," *Compt. Rend. Acad. Sci.* vol. cxli. p. 812, 1906.

KIDSTON, R. (1903), On the Fructification of *Neuropteris heterophylla*, Brongniart, *Proc. Roy. Soc.* vol. lxxii. p. 487, 1903, and *Phil. Trans. Roy. Soc.* Ser. B. vol. cxcvii. p. 1, 1904.

[1] Scott, 1901 ; see also Benson, 1902.

KIDSTON, R. (1905) Preliminary Note on the Occurrence of Microsporangia in Organic Connection with the Foliage of *Lyginodendron, Proc. Roy. Soc.* Ser. B. vol. lxxvi. p. 358, 1905.

—— (1906), On the Microsporangia of the Pteridosperms, *Proc. Roy. Soc.* Ser. B. vol. lxxvii. p. 161, 1906, and *Phil. Trans. Roy. Soc.* (in press).

MASLEN (A. J.), *see* Scott and Maslen (1906).

NATHORST, A. G. (1906). De äldsta fröväxterna. En ny klass inom växtriket, *Fauna och Flora*, vol. i. heft 1, p. 30, 1906.

OLIVER, F. W. (1902), On a Vascular Sporangium from the Stephanian of Grand'Croix, *New Phytologist*, vol. i. p. 60, 1902.

—— (1903) The ovules of the older Gymnosperms, *Ann. Bot.* vol. xvii. p. 451, 1903.

—— (1904[1]), On the structure and affinities of *Stephanospermum*, Brongniart, a genus of fossil Gymnosperm seeds, *Trans. Linn. Soc.* Ser. 2 Bot., vol. vi. p. 361, 1904.

—— (1904[2]), Notes on *Trigonocarpus* Brongn. and *Polylophospermum*, Brongn. two genera of Palæozoic seeds, *New Phytologist*, vol. iii. p. 96, 1904.

—— (1905) Über die neuentdeckten Samen der Steinkohlenfarne, *Biolog. Centralblatt*, vol. xxv. p. 401, 1905.

—— (1906), The Seed, a Chapter in Evolution, Pres. Address, Sect. K, *Brit. Assoc. Rep.* Brit. Assoc. York (1906), in press.

OLIVER, F. W., ARBER, E. A. N., SCOTT, D. H., and SEWARD, A. C., Discussion on "The Origin of Gymnosperms," Linnean Soc., March 5 and May 3, 1906. See *New Phytologist*, vol. v. pp. 68 and 141, 1906.

OLIVER, F. W., and SCOTT, D. H. (1903), On *Lagenostoma Lomaxi*, the Seed of *Lyginodendron, Proc. Roy. Soc.* vol. lxxi. p. 477, 1903.

—— —— (1904) On the Structure of the Palæozoic Seed, *Lagenostoma lomaxi*, with a statement of the evidence upon which it is referred to *Lyginodendron*, *Phil. Trans. Roy. Soc.* Ser. B. vol. cxcvii. p. 193, 1904.

POTONIÉ, H. (1897), Lehrbuch der Pflanzenpalæontologie, Berlin, 1897-99.

SCOTT, D. H. (1899) On the Structure and Affinities of Fossil Plants from the Palæozoic rocks. III. On *Medullosa anglica*, a new representative of the Cycadofilices, *Phil. Trans. Roy. Soc.* Ser. B, vol. cxci. p. 81, 1899.

—— (1900), Studies in Fossil Botany, London, 1900.

—— (1901) On the Structure and Affinities of Fossil Plants from the Palæozoic Rocks. IV. The Seed-like Fructification of *Lepidocarpon*, a genus of Lycopodiaceous Cones from the Carboniferous Formation, *Phil. Trans. Roy. Soc.* Ser. B. vol. cxciv. p. 291, 1901.

—— (1902), On the Primary Structure of Certain Palæozoic Stems with the Dadoxylon Type of Wood, *Trans. Roy. Soc. Edinburgh*, vol. xl. pt. ii. p. 331, 1902.

—— (1903), The Origin of Seed-bearing Plants. A Lecture delivered before the Royal Institution of Great Britain, May 15th, 1903.

—— (1905[1]), The Early History of Seed-bearing Plants, as recorded in the Carboniferous Flora (The Wilde Lecture), *Mem. and Proc. Manchester Lit. and Phil. Soc.* vol. xlix. pt. iii. Mem. 12, 1905.

—— (1905[2]), What were the Carboniferous Ferns? *Journ. Roy. Micr. Soc.* for 1905, p. 137, 1905.

—— (1905[3]) The Sporangia of *Stauropteris oldhamia*, Binney (*Rachiopteris oldhamia* Will.), *New Phytologist*, vol. iv. p. 114, 1905.

SCOTT, D. H. (1906), The Fern-like Seed-Plants of the Carboniferous Flora, *Résult. scient. Congress intern. Bot. Wien.* (1905), p. 270, 1906.
—— *see* Oliver and Scott (1903) (1904).
—— *see* Williamson and Scott (1895).
SCOTT, D. H., and ARBER, E. A. N. (1905), On Some New Lagenostomas, *Rep. Brit. Assoc. Cambridge* (1904), p. 778, 1905.
SCOTT, D. H., and MASLEN, A. J. (1906), Note on the Structure of *Trigonocarpon olivæforme*, *Ann. Bot.* vol. xx. p. 109, 1906.
SEWARD, A. C., and GOWAN, J. (1900), The Maidenhair Tree (*Ginkgo biloba*, L.) *Ann. Bot.* vol. xiv. p. 109, 1900.
SEWARD, A. C., and FORD, S. O. (1906), The Araucarieæ, Recent and Extinct, *Proc. Roy. Soc.* Ser. B, vol. lxxvii. p. 163, 1906 ; *Phil. Trans. Roy. Soc.* Ser. B, vol. cxcviii. p. 305, 1906.
STOPES, M. C. (1905), On the Double Nature of the Cycadean Integument, *Ann. Bot.* vol. xix. p. 561, 1905.
STUR, D. (1883), Zur Morphologie und Systematik der Kulm und Karbonfarne, *Sitzungsber. K. Acad. Wissen. Wien.* vol. lxxxviii. p. 633, 1883.
WARD, L. F. (1904), I. The Pteridospermaphyta. II. Palæozoic Seed Plants, *Science*, vol. xx. pp. 25, 279, 1904.
WHITE, D. (1904¹), The Seeds of *Aneimites*, *Smithsonian Miscel. Collect.* vol. xlvii. pt. iii. p. 322, 1904.
—— (1904²), Fossil Plants of the Group Cycadofilices, *Smithsonian Miscel. Collect.* vol. xlvii. pt. iii. p. 377, 1904.
WILD, G. (1900), On New and Interesting Features in *Trigonocarpon olivæforme*, *Trans. Geol. Soc. Manchester*, vol. xxvi. p. 434, 1900.
WILLIAMSON, W. C. (1873), On the Organization of the Fossil Plants of the Coal Measures. Part IV. *Dictyoxylon, Lyginodendron*, and *Heterangium*. *Phil. Trans. Roy. Soc.* vol. clxiii. p. 377, 1873.
WILLIAMSON, W. C., and SCOTT, D. H. (1895), Further Observations on the Organization of the Fossil Plants of the Coal Measures. Part III. *Lyginodendron* and *Heterangium*. *Phil. Trans. Roy. Soc.* Ser. B, vol. clxxxvi, p. 703, 1895.
ZEILLER, R. (1904), Observations au sujet du Mode de Fructification des Cycadofilicinées, *Compt. Rend. Acad. Sci.* vol. cxxxviii. p. 663, 1904.
—— (1905), Une nouvelle classe de Gymnospermes : les Ptéridospermées, *Rev. génér. Sci.* vol. xvi. p. 718, 1905.

SCIENCE IN MEDICINE:

BEING AN ACCOUNT OF SOME RECENT ADVANCES IN THE TREATMENT OF DISEASE FOLLOWING ON THE DISCOVERY OF OPSONINS

By A. C. INMAN, M.A., M.B. (Oxon.)

Introduction—Chemical Antiseptics—Surgery—The Anti-bacterial substances of the blood—Bacteriolysins—Bactericidal substances—Agglutinins—Opsonins—Practical Adaptation to Medicine—Passive Immunity—Serum-therapy—Active Immunity—Inoculation of bacterial vaccines—The Vaccine—The Opsonic Index—Classification of diseases in which vaccination is applicable : I. Local—II. Local with general disturbances—Pulmonary Tuberculosis—III. Septicæmias—Cancer—Difficulties of Inoculation—Reasons—Their treatment—Sero-diagnosis—Conclusion.

MEDICINE has always depended on science for advancement; and there are few branches of science to which it does not owe a deep debt of gratitude. Without the microscope some of the most potent causes of disease must have remained for ever unknown; as far back as 1680 it enabled Leeuwenhoek to publish an exhaustive treatise on the organisms of putrefaction, and two centuries later it was mainly owing to the high degree of efficiency to which this instrument had been brought that Pasteur was able to lay the foundations of modern bacteriology. Since that date medicine has entered upon what has been appropriately called the bacteriological era; and this, short though it has been, has already witnessed a complete revolution in the treatment of the large class of diseases known as bacterial infections. Moreover, it must be felt by most observers that further advances will be made in this direction; that with improved methods and an increasing insight into physiology and pathology, a still larger number of diseases will be traced to a bacterial origin, and will receive a corresponding and consequently a more effective treatment.

It is the object of this paper to examine the recent advances in our knowledge of the rôle played by the blood in bacterial infections. Before doing this it will be well to make a brief reference to some of the agents which have been of service in the battle with the germs of disease.

Chemical Antiseptics.—In the first place there are the chemical antiseptics, such as carbolic acid, corrosive sublimate, iodoform, etc.   First introduced into the practice of surgery by Lord Lister, they have widened the scope of operations to an enormous extent.   But it is important to realise the action of these potent chemical substances, and the object we have in view in using them.   As time goes on it is becoming apparent that aseptic surgery is gradually ousting antiseptic surgery from the field; and it is very doubtful whether the practice of using antiseptics in the dressing of septic wounds is advisable, since the action of a large number of them, at any rate, is in direct antagonism to the action of the protective substances in the blood, which must be regarded as the prime factors in the art of healing.

A further limitation to their use in the treatment of bacterial infections arises from the fact that, however readily they may be applied to a local lesion, such as an abscess or a wound, their application in generalised infections has been a matter of extreme difficulty.   The administration of the coal-tar derivatives and of such preparations as the sulphocarbolate of soda by the mouth has been rewarded with but scant success, and the intra-venous injections of the preparations of formaline, etc., have proved themselves too uncertain and too dangerous to warrant their adoption in practice.

Success, however, in the treatment of two diseases is to be credited to chemical antiseptics.   In the case of Malaria the use of quinine has been attended by brilliant results, and no one would doubt, bearing in mind its method of application, that it actually destroys the malaria parasites which are circulating in the blood.   The other disease is Syphilis, in the treatment of which mercury has been so effective.   The recent work of Metchnikoff and his fellow-workers at the Pasteur Institute on the relationship of the *Spirochæta pallida* to syphilis, and on the effect of mercurial inunction after inoculation of the syphilitic virus, is of great importance and interest in this connection.[1] Similar success was hoped for in the treatment of acute rheumatism with salicylic acid or its salts, but this has unfortunately not been realised.   Although these preparations are of the utmost

[1] *Harben Lectures*, 1906.

value in the treatment of rheumatism as regards alleviation of pain and swelling, and the reduction of fever, it is unfortunately not uncommon for a patient to develop some of the most serious sequelæ of acute rheumatism, such as endocarditis, pericarditis or pleurisy, whilst fully under the influence of the drug.

Surgery.—Another weapon which must be mentioned, if it be only to do homage to the importance and extent of its work and to the extraordinary skill which has been attained in its use, is the surgeon's knife.   With respect, however, to the diseases under consideration, those of bacterial origin, it is probable that recourse to its aid will be less frequent in the future; at any rate every effort will be made by rational and scientific means to place the diseased body in a position to exert its own resources against the invading micro-organisms, and not until these have failed will the knife be called into use. Further, these means may be called upon with advantage to act in conjunction with surgical procedures, and the latter may often be helped by them to a happy result.

Anti-bacterial substances in the blood.—The blood, then, contains substances which are antagonistic to bacterial life; that this is so is now a matter of certainty.   The body, that is to say, is normally endowed with a certain amount of destructive power against all micro-organisms; and in immunisation (that is, artificial increase in resistance to bacterial infection) more of these substances are called into existence and conveyed into the blood from the body tissues.   The blood, therefore, may be said to be strongly antiseptic, and it is only possible for the most virulent types of bacteria to exist in it, and these rarely in any numbers for any length of time.

The study of the effect the blood exerts on bacteria has gone on continuously throughout the bacteriological era, and it has long been a matter of controversy whether the blood serum or the white corpuscles are the agents in destroying invading micro-organisms.

A series of researches inaugurated by the well-known paper by Nuttall[1] emanating from Flügge's laboratory showed that the blood serum could, by itself, kill certain bacteria.   These researches led to Buchner formulating his Alexic theory of immunity, which attributes the protection of the body against micro-organisms to the presence in the serum of certain chemical

---

[1] *Leitschrift f. Hygiene,* 1888.

bodies (alexines) which have an affinity for, and are able to destroy, bacteria.

Metchnikoff[1] and the French school stoutly upheld the theory that the bacteria are destroyed by being eaten up by the phagocytic white cells—the macrophages and microphages —and did not attribute any of the action to the serum. Though in the light of recent work they have had to modify their views to a considerable extent, they still consider the white blood cells to be the essentially important agents, and attribute any action of the serum to the fact that it contains substances set free from the cells. But, granting the presence of these bodies in the serum, their origin is still a matter of doubt. Metchnikoff considers them to be products of the phagocytes, whilst many followers of Buchner in Germany deny the cells any part in their manufacture. Without pursuing this difficult and complex question any further, we will briefly touch on those substances in the serum, as yet discovered, which have anti-bacterial properties, and then consider how they may be used in the treatment of bacterial infections.

Contemporaneous with the inquiry into the origin of the protective substances of the blood, further varieties of anti-bacterial substances were discovered.

Bacteriolysins.—Pfeiffer, working on cholera, was the first to show in an unmistakable way that cholera bacilli introduced in enormous quantities into the peritoneal cavity of immunised animals were in the space of a few minutes entirely dissolved. These substances, whose action is thus demonstrated, are technically known as Bacteriolysins.

Agglutinins.—These were first discovered by Durham working, in Grüber's laboratory, on typhoid in 1895. They cause the bacteria to swell up, to lose their motility, and finally to aggregate into clumps. This does not take place in the case of all bacteria, and for all practical purposes is limited to cases of typhoid and Malta fever when running a favourable course. Widal and Grünbaum applied the experiment to clinical medicine, and since then the agglutination test has proved of great value in the diagnosis of both typhoid and Malta fevers. The degree of agglutination may further be taken as a useful guide as to the patient's progress, at all events as regards the manufacture of protective substances.

[1] *Leçons sur la Pathologie comparée de l'inflammation,* 1892.

16

Opsonins.—These substances, discovered recently by Wright and Douglas,[1] are chemical bodies which act upon the offending micro-organisms and render them capable of being ingested by the white corpuscles. They do not kill the bacteria, but only prepare them for ingestion by the phagocytes.

This research, while confirming the important part played by the phagocytic cells of the blood in bacterial infections, has shown that the bacteria must be acted on by the opsonins before they can be ingested ·and ultimately destroyed by these cells.

The opsonins are thermolabile substances, being destroyed by exposure for ten minutes to a temperature of 60° C., and also by sunlight. In contrast to the bactericidal substances and agglutinins, which have been demonstrated only in connection with isolated species of organisms, opsonins exist for each species of bacteria, and they can be severally extracted from the serum by a process of saturation; they are "specific for the different strains of bacteria."[2] The discovery of these substances, and the technique by which they can be quantitatively estimated to an accurate degree in a small sample of the blood taken from the finger, has led to an extremely important research, which has thrown light on many dark problems of medicine.

Immunity, using this term as the equivalent of increased resistance of an organism to bacterial infection, can be achieved in two ways. To one of these Ehrlich has given the name "active immunity," to the other "passive immunity."

Active immunity may be defined as the increased resistance to bacterial infection which is obtained by the inoculation of bacterial vaccines. Passive immunity may be defined as the condition of increased resistance which is obtained by transferring to an organism protective substances elaborated in the organism of another animal which has been actively immunised.

Of these two kinds of immunisation, passive immunity, although dependent upon active immunity, was the first as regards its application to medicine. For this reason serum-therapy will be dealt with first, and vaccine treatment will receive consideration later on in this paper.

Serum-therapy.—Anti-sera are of two kinds: (*a*) Antitoxic, (*b*) Anti-bacterial. The former, of which the diphtheria and

[1] *Proc. Roy. Soc.*, vol. 72, 1903.
[2] Bulloch and Western, *Proc. Roy. Soc.*, 1906.

tetanus antitoxins are examples, possess the power of neutral-
ising poisons, and have met with great success in the treatment
of diphtheria, and with some success in that of tetanus; further,
both are standardised so as to enable the physician to regulate
the dose at will.

The anti-bacterial sera, on the other hand, in spite of an
extended trial, have practically met with no success; the cause
of this must now be inquired into. Wright,[1] in a paper read
before the Chelsea Clinical Society, explains their failure in
this way. In anti-bacterial serum-therapy the physician proposes to administer to the patient protective substances elaborated in the organism of an animal which has been vicariously
inoculated with the appropriate bacterial vaccine. If the serum
used is to be of benefit to the patient it is essential that it should
contain a certain quantum of anti-bacterial substances. As we
saw in the case of the anti-diphtheria and anti-tetanus sera the
antitoxic value is measured, and they have been found to contain
a large number of anti-bodies. But in the case of other sera
no such examination is instituted, and in consequence it is not
known whether these preparations have any anti-bacterial properties at all. It is assumed in their manufacture that the
animal which is being inoculated will respond to each inoculation
by the elaboration of more anti-bacterial substances. This
assumption is based on analogy with the manufacture of the
anti-diphtheritic and anti-tetanic sera, but that this is not
justified is proved by the fact that every animal inoculated
against the diphtheria bacillus does not elaborate anti-bodies;
some only do so to a limited degree, and some die during the
inoculations. It is therefore essential that all sera should be
subjected to a searching examination; the proper course for
the bacteriologist is "to verify everything, to measure everything, and to declare in each case the results of his measurement."
After the inoculation of a bacterial vaccine into an animal the
immediate effect is a lowering of the resistance of the blood,
and this is followed by an increase in resistance above the
normal. If this animal is bled when it has completely recovered
from the effects of the inoculation, it is probable that its serum
contains a certain quantum of anti-bacterial substances—but it
is inexcusable to assume that this is so without measurement;
for even if, in response to the infection, the elaboration of pro-

[1] *Clin. Journ.*, May 16, 1906.

tective substances has taken place successfully, it cannot be known, without measurement, whether a superabundance of these substances has been made, and if so, whether they are sufficient to be able to stand the dilution of the total volume of serum. The success obtained by all anti-sera—excepting, of course, the diphtheria and tetanus antitoxins—has been most disappointing; and in spite of the enormous amount of work bestowed upon serum-therapy, it has not advanced at all in its application to practical medicine.

But imagine that the blood has been drawn from the experimental animal before it has been able to elaborate sufficient anti-substances to neutralise the dose of poison administered to it; that is to say, that the blood has been drawn off during the negative phase. Under these circumstances it is quite comprehensible that the serum thus drawn off may actually contain some of the bacterial poison. If this serum be now placed in bottles and labelled *anti-serum*, "terminological inexactitude" would hardly describe it. Not that in every case the patient would be the loser by these methods. Undesirable consequences have followed the administration of such sera containing very virulent bacterial toxins; but, on the other hand, if the poison be not in excess good may follow evil, and a rapid cure be effected after a severe negative phase. Chantemesse[1] has, during the past five years, been treating cases of typhoid fever with an "anti-typhoid" serum. His results have been brilliant; but the fact that the dose administered is only a fraction of a cubic centimetre, that a rise of temperature, associated with enlargement of the spleen, takes place, and that this "phase of reaction" lasts from four to five days, and is followed by a phase of defervescence and amelioration of the general condition—these facts have led to the suggestion that this serum is really a vaccine in disguise.

Serum-therapy has other disadvantages which should be briefly mentioned. The administration of sera very frequently leads to a series of symptoms to which the Germans have given the name "Serum-Krankheit." Urticaria and joint pains are the most prominent features. Studying the coagulability of the blood, Wright has shown that these symptoms are considerably lessened in such cases, and suggested that calcium salts, which he had found increased the coagulability, might be used in the

[1] *Presse Médicale*, Feb. 1906.

treatment of this condition. A recent clinical paper by Netter confirms the value of calcium as a prophylactic for these undesirable complications.

Briefly it may be said, as regards the use of anti-bacterial sera in the treatment of bacterial infections, that only too frequently there has been a tendency to guess at the bacterial nature of some cases on rule-of-thumb knowledge, with the result that a so-called anti-serum, boldly professing to counter-act the class of bacteria hypothetically assumed to be the offenders, is injected into the unfortunate patients in conjectural doses with mechanical regularity.

Active Immunity.—When a bacterial vaccine is introduced into an animal the immediate effect on the blood is a lowering of its resisting power to the micro-organism in question. This phase is followed by one during which protective substances

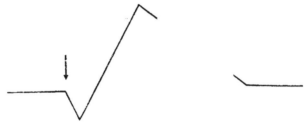

FIG. I.—An Inoculation Curve.

are elaborated in response to the infection, and the resisting power of the blood becomes greater than the normal resistance. A gradual return to the normal then occurs (fig. 1). This train of events, which follows upon a single inoculation, was first fully described by Wright, and was built up on the study of the effects of the inoculation of bacterial vaccine made first in connection with typhoid inoculation. To the various phases this author has given the names "the negative phase," "the positive phase," and "the phase of maintained high level."

Next as regards a series of inoculations. It is possible to give subsequent inoculations during any of these three phases, and the result in each case will be different and must be described :

(1) Inoculation during the negative phase. In this case a cumulative action in the direction of the negative phase is produced (fig. 2).

(2) Inoculation during the positive phase; in this case an indefinite rise in the resistance of the blood should be theoretically produced. This is, unfortunately, not the case in practice, and whereas a cumulative action in the direction of the negative phase is only too readily achieved, it is generally found

FIG. 2.—Re-inoculation during the Negative Phase.

impossible to produce a cumulative action in the direction of the positive phase (fig. 3).

(3) If the subsequent inoculations take place as the effect of the previous inoculation is passing off, the next inoculation may act as an independent event, and this is the ordinary

FIG. 3.—Ideal Inoculation.

result achieved when animals are allowed to recover completely between the successive inoculations (fig. 4).

The practical import of these experiments will be readily understood. If inoculation is resorted to in a haphazard manner, it is possible that each dose will be given during a negative phase, and the patient's resistance driven down and down with disastrous results. The ideal inoculation—a cumulative action in the direction of the positive phase—is unfortunately hardly ever achieved, so that it is most convenient to attempt to keep the patient's resistance above the

FIG. 4.—Correct Inoculation.

normal by allowing each inoculation to act as an independent event.

Now, to be able to treat any bacterial disease with success, there are certain things of which we must take cognisance. If possible, we must know (1) the particular bacterial agent causing the disease; (2) something which will remove or counteract the offending agent; (3) the "dose" of the remedy employed; and (4) we must be able to follow the

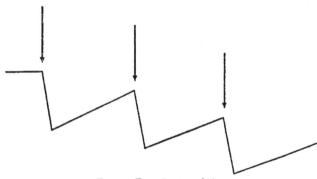

FIG. 5.—Excessive Inoculation.

effect of the drug, especially if, as is certainly the case in the majority of instances, the remedy be not devoid of danger.

(1) As regards the causative agent, we are now able in a very large number of cases to say what bacteria are present. The bacteriological examination of sputum, urine, pus, effusions, etc., both by cultural methods and by microscopical inspection, has met with a very large amount of success, and in competent hands

is of the utmost value. More recently, even in the graver general infections—the septicæmias—by actual bacteriological examination of the blood obtained by venesection it is possible to determine what bacterium is the cause of offence in each case.

Further, inferential diagnosis may be made by the examination of the blood serum or lymph—as, for instance, by means of the Widal-Grünbaum test. It will be seen later that the examination of the blood serum, lymph, effusion, etc., as regards their opsonic power, may also be of great value in this connection.

(2) *The Vaccine.*—The therapeutic agent is a vaccine—the bacterium in the suspension being the one corresponding to the causative agent of the disease under treatment. If necessary— and this is so in many cases, especially when dealing with bacteria of the Colon group and the Streptococci—the vaccine may be prepared from the actual offending bacteria. Cultures taken from the site of infection are made on a suitable medium, and the resulting growth is washed off and suspended in normal salt solution. This emulsion is then poured into a test tube, hermetically sealed off in the flame, and then sterilised by heat. The temperature employed must be sufficient to kill the bacteria, but only just sufficient for this purpose, an important consideration in the preparation of a vaccine being to alter as little as possible the chemical composition of the bacterial protoplasm. In the case of most of the commoner pathogenic bacteria a temperature of 60° C. for one hour is enough. A water bath is the form of steriliser employed. After sterilisation some of the vaccine is withdrawn, planted out on culture media, and incubated for 24 hours. If these media remain sterile at the end of this time the vaccine may be regarded as sterile, though for precaution against subsequent contamination $\frac{1}{4}$ per cent. of Lysol or some such antiseptic is added.

(3) Before giving the inoculation it is essential to know how much is being given, and for this reason the vaccine has to be standardised. This is accomplished by counting the number of bacteria in the emulsion we have made. At first this seems almost impossible, but Wright has practically solved the difficulty in this way. "Availing himself of the fact that the normal blood contains about 5,000,000 red cells per cubic millimetre (the actual number can, of course, be determined by making a red

count), he compares an equal part of blood with the bacterial emulsion which is to be counted. These are mixed together and a stained film is made. Then, by counting a large number of red blood-corpuscles and the number of bacteria seen whilst counting them, a comparison between the two is easily instituted. For instance, if 1,000 bacteria are seen whilst counting 500 red cells, the bacterial emulsion contains twice as many bacteria as the blood does red blood-corpuscles. We will suppose that the red blood-count proved to be normal—that is to say, that the individual had 5,000,000 red cells in 1 c.mm. of his blood—then the emulsion under investigation contains 10,000,000 bacteria per c.m. In this way we can regulate our doses of vaccine almost as accurately as we can our dose of alkaloid or salt." [1]

An alternative to counting the number of bacteria in a vaccine is to weigh them. This method is used in the preparation of Koch's New Tuberculin (T. R.) The growth of tubercle bacilli is suspended in alcohol and put into cylinders with a number of porcelain marbles. The cylinders are rolled round and round, and in this way the culture is broken up. The alcohol is then allowed to evaporate, and there remains a powder, which consists of crushed tubercle bacilli. This powder is weighed, and 10 mgm. are suspended in 1 c.c. of salt solution.

(4) *The Opsonic Index.*—We must have the means of determining an appropriate dose. The discovery of Opsonins and the method of estimating them quantitatively has given us a means of doing this, a dose being good or bad according as it increases the opsonins in the blood or no.

The method of estimating the opsonic power of any given blood consists in bringing together white blood-corpuscles, bacteria suspended in salt solution, and the blood serum, and allowing them to remain together for a certain time at the temperature of the body.

A solution containing 1·5 per cent. sodium citrate and ·85 per cent. sodium chloride is prepared and sterilised. The finger is then pricked with a sterile needle and a bandage wound round the finger above the punctures, so that the return of blood from the distal portion of the finger is prevented. The blood is then allowed to flow into a test tube containing the above solution.

[1] *Treatment*, May, 1906.

The citrate prevents clotting of the blood and the isotonic salt solution does not allow hæmolysis to take place. About twenty drops of blood are thus prepared. The mixture is then centrifugalised, when it will be found that the corpuscles have been driven to the bottom whilst the plasma remains in the supernatant fluid. The latter is syphoned off and the corpuscles are again washed in physiological salt solution. A second centrifuge again drives the corpuscles to the bottom and leaves the clear salt solution above. If this is syphoned off the tube only contains washed human corpuscles, and it is found that the white corpuscles occupy the upper layers whilst the majority of the reds are at the bottom ; so that if the tube is well sloped on its side, and some of the " cream " be drawn up into a pipette, a large number of white corpuscles are available. The bacteria are obtained by taking a loopful of growth off an agar slope and mixing it up in salt solution (in some cases a ·85 per cent., in others a 1·5 per cent., solution is used). An emulsion of bacteria is thus obtained.

The blood serum is readily obtained by pricking the finger and drawing a small quantity into a capillary pipette or a small capsule. The blood thus obtained is allowed to clot, and as the clot contracts the clear serum is expressed. An equal volume of the blood-cream, the bacterial emulsion, and the serum are drawn up into a capillary pipette and intimately mixed together. The pipette is sealed off in the flame and the mixture is placed in an incubator, kept at body temperature (37° C.) for fifteen minutes. At the end of this time the pipette is taken out, the end broken off, and the mixture blown on to a clean slide, where it is again well mixed. A sample of the mixture is then placed on another slide and a blood film prepared. This is stained with Leishman's dye and is ready for microscopic examination. We will suppose that this experiment has been conducted with the serum obtained from a healthy man; in actual practice the serum obtained from a number of normal men is " pooled " together, equal quantities of each contributing to the whole. The experiment is now repeated in exactly the same way, but the serum used in this case is obtained from the patient under treatment. We now have two blood films to examine.

Using a $\frac{1}{12}$th oil-immersion lens for the purpose it is seen that the phagocytic white cells have eaten up a certain number of the bacteria, and by counting, say, fifty consecutive poly-morphonuclear neutrophils, and at the same time the number of bacteria ingested by those cells, it is possible to work out the average number of bacteria ingested per cell. We will suppose that in the first film—that is, the one obtained after using normal serum—this average works out at 9·5 bacteria per cell, and that in the second film, obtained with the patient's serum, it is only 8·5 per cell.

Then in this case the patient's serum has not such a high opsonic power as that of the normal serum. Taking the normal " index " as unity, and comparing the patient's average with the normal average, the patient's opsonic "index" is in this case 0·89.

Thus: Normal serum=9·5. As for purposes of comparison normal serum is always taken as unity, then in the same ratio patient's serum=$\frac{8·5}{9·5}$=0·89. This is spoken of as the patient's opsonic index, and it can be taken as a guide as to the richness of his blood in these protective substances.

Having inquired into the nature of opsonins as regards the part played by them in immunity, and having briefly reviewed the method employed for their detection and measurement, it is necessary to consider the class of cases in which therapeutic inoculation is applicable. For this purpose the classification suggested by Wright is the most convenient:

(1) Localised bacterial infections.

(2) Localised infections associated with constitutional dis-turbance.

(3) Septicæmias.

Class 1.—In the case of localised infections the bacteria have found a suitable soil for growth ; and it may be taken generally that the less well supplied with blood any particular part of the body is the more likely is it to be a favourite site for infection. In these cases, therefore, therapeutic inoculation offers the best chance of success. Auto-inoculation is hardly taking place at all, and in most cases there is nothing to prevent the elaboration of enough protective substances, in response to vaccination, to cure the disease. Carbuncles, boils, etc., rapidly disappear under this treatment. Early pulmonary tuberculosis should probably be included under this category, and it is very likely that in such cases treatment with a tubercle vaccine would be rewarded with great success.

Class 2.—Constitutional disturbance accompanying a localised bacterial infection may be taken as evidence of auto-inoculation from the site of infection; and it should be remembered that this auto-inoculation is with a living microbe, and therefore one which is capable of multiplication. Under these circumstances the examination of the blood of a patient as regards its opsonic power reveals the fact that he is living " in a succession of negative and positive phases "; that is to say, the resistance of the blood is lowered as an immediate result of the infection, and then an increase in protective substances above the normal is elaborated in response to the infection. The more advanced cases of pulmonary tuberculosis are essentially of this type. That this fact complicates matters as regards therapeutic inoculation will be readily understood. Two courses may be followed :

(*a*) The patient may be put under the condition least suitable for auto-inoculation, *i.e.* at rest, and not until auto-inoculation has been put a stop to as far as possible are inoculations employed.

(*b*) A daily examination of the blood may be instituted, and only very small doses of vaccine injected at opportune moments, according as that examination shows evidence of recent auto-inoculation or no.

It will strike any one who has thought on this matter that clinical experience has led to the adoption of probably the most satisfactory method of treatment without thoroughly understanding why it so acted. The patient suffering from an acute bacterial infection has always been most rigidly kept in bed, and has been placed upon a very low diet which consists chiefly of milk—a study of the diet sheet of any of our hospitals will always be found to contain a fever diet, which often only consists of milk; and in the more chronic infection by the tubercle bacillus a full diet with extra milk (3 to 5 pints per diem) is the rule. Now by these means the blood and lymph streams are reduced to their minimum flow. The blood is at its maximum of coagulability, and the least amount of lymph exudes through the walls of the blood-vessels. In fact, those circumstances exist which are least suitable for auto-inoculation, and by these means the infection is converted as far as possible into a local one. And if the blood be examined during this treatment as regards its opsonic power it will be found that the fluctation gradually gives way to a steady line.

Pulmonary Tuberculosis.—The above line of treatment associated with the all-important fresh air has formed the basis of the successful sanatorium treatment of pulmonary tuberculosis. It is not, however, uncommon in these institutions to find, after a patient has undergone a long course of rest and forced feeding which has resulted in the temperature remaining normal and in a considerable gain in body weight, that when he is allowed to get up the temperature chart again shows an evening rise, indicating that the disease is still active. In these cases, again, if the blood be examined as regards its opsonic power, it will be found that the succession of negative and positive phases is re-established. Koch's old Tuberculin, first available in 1890, was extensively tried, but had, at all events in this country, fallen into disrepute. The chief cause of its failure was that the method of administration was wrong. Anything between a milligram and ten milligrams was given at intervals of two or three days, and it has been found that such doses reduced the resistance of the body, as evidenced by severe constitutional disturbances. The effect of administering it in that manner was, in the light of modern knowledge, to produce a cumulative action in the direction of the negative phase. The resistance of the patient was driven lower and lower, and when it is remembered that at the same time in many cases the patient was inoculating himself with living bacilli, it is not surprising to find that this method of treatment did not meet with success, but often, unfortunately, with disastrous results. It has now been established that $\frac{1}{10000}$th to $\frac{1}{800}$th of a mgm. is the dose of tuberculin which offers the best chance of success. Even the administration of this dose is followed by a slight negative phase, but not of a sufficient degree or duration to do any harm. The effect of a single inoculation would seem to last for about ten days when dealing with a patient who is not inoculating himself. It is still a question as to how far inoculation should be undertaken during active tuberculosis. If possible it would seem to be wisest, by fresh air, rest, and forced feeding, to localise the infection as far as possible, before beginning treatment with tuberculin. This method has been carried out by Turban in Davos for a considerable number of years, though he has been giving large but infrequent doses, with great success as regards the small number of recurrences. But the bacteriology of acute pulmonary tuberculosis, associated as it so often is with other organisms (mixed infections), still requires much investigation; and the effect of these associated micro-organisms on the blood also requires detailed study before much success is to be expected from the use of tuberculin in the treatment of acute pulmonary tuberculosis.

Class 3.—The acute bacterial infections, such as septicæmia, ulcerative endocarditis, etc., seemed to offer few possibilities for the exploitation of vaccine treatment. It does not, however, appear in the light of recent events that their exhibition in these cases will be given up without a struggle. The all-important thing is that frequent, even daily, examinations of the blood must be instituted during treatment. It is only by these means that we can follow what is going on; that is to say, know to what extent the patient is inoculating himself. Next in importance is the administration of such small doses of vaccine that the negative phase may be made as slight as possible, or may, if possible, be avoided altogether.

It is difficult, without an intimate knowledge of the subject, to conceive how the injection of dead bacteria into a patient who is already inoculating himself to an excessive degree can be expected to be of advantage to that patient. Wright suggests the following as an explanation of the mechanism: " In the case where bacteria are, as in the septicæmic conditions, found in the blood-stream, or in organs standing in direct relation with this, the bacterial derivatives are of necessity diluted by the whole volume of the blood and lymph before they can come into application upon the tissues in which, we may take it, the machinery for the elaboration of protective substances is located. In conformity with this great dilution of the bacterial derivatives, a comparatively speaking ineffective immunising stimulus will here be administered. In contrast with this, where a bacterial vaccine is inoculated directly into the tissues the bacterial products will come into application upon these in very concentrated form, calling forth a correspondingly larger production of protective substances."

Cancer.—One group of diseases remains for consideration, though it is only in the light of very recent work that it is necessary to mention it in connection with bacterial disease. Every attempt has been made to trace cancer to a microbic origin, but no reliable work has, as yet, succeeded in proving this. It may be said that to-day the research into the origin of cancer is proceeding along two main lines. First we have the grafting experiments of Ehrlich, secondly the investigation along bacteriological lines.

As regards Ehrlich's work, one point of importance as regards treatment is the fact that he has succeeded in immunising mice against virulent tumours by previously inoculating them with less virulent ones. He has thus shown that a high degree of immunity against cancers may be achieved by the inoculation of tumours.

With regard to the bacterial research, the work of Doyen seems to promise something of practical utility. This observer has recovered a micro-organism, which he has called the *Micrococcus neoformans*, from a very large percentage of tumours. An attempt was made to furnish an anti-serum for this bacterium, but without success. In fact, it would appear likely, from some results published, that this serum possessed in some cases definite toxic properties. As has been shown in a previous paragraph, such haphazard preparations are always likely to fail. Karwacki, continuing along the same lines of research, confirmed the presence of this micro-organism, and contends, from the fact that the blood of cancer patients has a characteristic agglutinating action upon the bacterium, that it plays an important part in connection with the disease.

Jacob and Geets, pursuing the same subject but using Wright's method, examined the blood of a number of cancer patients as regards its opsonic power to the *M. neoformans*, and it was found to be low. Further, when these patients were inoculated with a "neoformans vaccine" the opsonic power of the blood was raised considerably above the normal. As regards results it is as yet too early to say anything definite. It is not intended to convey the idea that this work even suggests a microbic origin of cancer, but the frequency with which the *M. neoformans* has been recovered from tumours, and the behaviour of the blood in relation to it, are indications that this coccus plays a not unimportant part in the life-history of cancers.

These few notes on cancer will serve to indicate that interesting and important work is being done in connection with this disease, that it is being approached from the scientific side, and good results may be expected.

Difficulties of Inoculation.—It will be gathered from what has been said that therapeutic inoculation of bacterial vaccines has a very wide range of applicability; it has been indicated that this range is extending as our knowledge of disease increases, and it has been suggested that more diseases will come under its scope in the near future. From the nature of things such a method cannot always be successful. Some cases will not respond to inoculation, and although it is possible by vaccination to produce in them a negative phase, the desired positive phase does not follow. Further, it must not be thought that a high opsonic power of the blood necessarily spells " cure," for a patient may actually die with a high index. This is more especially the case in the septicæmias, the action of the bacterial poison on the tissue elements, the heart, and central nervous system causing death by disturbance of a vital function; in these cases a blood rich in protective substances will naturally be of little avail. In some cases the cause of failure of therapeutic inoculation is quite obscure; in others, Wright has suggested that one of the obstacles to success lies in a deficient blood and lymph supply to the affected part.

The methods of combating these causes of failure must now be inquired into.

The production of a local hyperæmia—" drawing the blood to the part "—by means of hot fomentations has been the staple treatment for local bacterial infections from time immemorial, and its efficacy, which has stood the test of time, is doubtless due to the fact that the protective substances of the blood are thus brought to bear against the invading organisms. The X-rays, massage, radiant light, etc., are more modern methods of accomplishing the same thing. Bier [1] has for several years been treating local bacterial infections with great success. His method consists in producing a passive hyperæmia at the site of infection for a certain time, then allowing the blood thus confined to the affected spot to flow on, and imprisoning a further supply of fresh blood for another period. By this means of "intermittent" hyperæmia the protective substances of the blood are enabled to exert their effect upon the bacteria in more concentrated form. Further, the same surgeon uses various devices, such as suction pumps, to draw the blood to the surface of the skin, and instead of widely lancing boils, etc., he punctures them and intermittently draws blood to the surface by means of these pumps.

[1] *Hyperämie als Heilmittel*, 1906.

Were this method combined with the inoculation of bacterial vaccines, and a blood rich in protective substances brought to the site of infection, its value should be considerably increased. There are, therefore, at our disposal many methods of drawing blood to the surface, and thereby concentrating its anti-bacterial properties on local bacterial infections in these parts. In the case where after operation a sinus has formed, and the wound is not completely healed, the situation is probably as follows: the walls of such a sinus consist of coagulated lymph, which furnishes a splendid medium for bacteria to grow on ; moreover, this clot is almost without blood supply, so that the bacteria meet with little opposition. Antiseptics are used to wash it out frequently and to plug it, but unless every single bacterium be killed, in a few hours there are thousands more. Wright has suggested that if a solution of sodium citrate ($1 \cdot 5$ per cent.) is introduced into the sinus, the coagulum of lymph is dissolved ; and if to this a solution of dextrose (12 per cent.) is added, serum will be drawn through the wall of the sinus by a process of osmosis, and thus the anti-bacterial properties of the blood can be utilised. This is a scientific application of another very old household remedy, the outcome presumably of accident or instinct and long experience—namely, the soap and sugar poultice,[1] the soap exerting the same decalcifying action on the blood as the citrate, and the sugar " drawing " the lymph to the surface.

Sero-diagnosis.—It will also be understood from what has been said that the examination of the blood, as regards its opsonic power, may be used not only as a guide in treatment, but as an aid in diagnosis. This means of diagnosis—comparable to the Widal test now so well known in clinical medicine—has already been fully investigated as regards the diagnosis of Tubercle.[2] The following are the main results obtained in this research.

1. If the opsonic index of a patient suffering from a local bacterial infection which suggests tuberculosis is persistently low to the tubercle bacillus, the lesion in question may be taken to be of a tubercular nature.

2. If on a single examination of the blood the opsonic index is found to be high, and on a succession of examinations is found to be continually fluctuating, then, if the patient is suffering from

---

[1] *Med. Chi. Trans.*, vol. 89.   [2] *Proc. Roy. Soc.*, 1906.

an illness suggesting tuberculosis, it may be inferred that the patient is suffering from tuberculosis which is active or has recently been so.

3. A persistently normal index to the tubercle bacillus would be taken as evidence against the case being of a tubercular nature.

An attempt has been made in this paper to show the reader one of the lines along which our knowledge of bacterial infections has advanced, and to indicate to him how our power of control over these infections has recently been increased. But this has not been the only object in view. The intention has been to signify that medicine is on the threshold of a new era in which science will hold the chief place ; that it is from the laboratory that further advances in medical knowledge are to be expected. The great problems still remain unsolved; we need mention only consumption and cancer in proof of this ; but careful scientific researches are being made in many places and in many directions, and the results already attained promise well for the future.

The chemist, the physiologist, the pathologist (the latter nowadays almost synonymous with bacteriologist) will be the physicians of the future, and it is to their efforts that men may look with increasing hope for new and greater victories over the terrible power of disease.

THE QUANTITATIVE CLASSIFICATION ON IGNEOUS ROCKS

By JOHN W. EVANS, D.Sc., F.G.S.

Imperial Institute, London

IN the course of my work on the crystalline rocks of the cataracts of the Rio Madeira and the adjoining rivers, I thought it would be of interest to ascertain the position of such of the rocks as had been analysed, in the quantitative classification of igneous rocks which has recently been brought forward [1] by four of the leading petrologists of the United States, and almost universally accepted in that country.

As I have given some attention to the details of this classification, and as it is still comparatively little known, it may be useful to give a brief account of its main features, to show how it is applied in concrete cases, and to state—for what they may be worth—the conclusions to which I have come as to its claims for adoption beyond the limits of the land that gave it birth.

It is impossible not to sympathise with the ideal which the authors of this bold attempt at reform set before them. Every petrologist deplores the confusion that reigns in the classification and nomenclature of igneous rocks. It is true that the unprecedented advances that have been made in the last thirty years in our knowledge of their composition and structure, origin and relations, have not been without result. The old point of view that attached an illogical importance to geological age and stratigraphical relations has been to a large extent discarded; but although the recognition of the importance of chemical and mineral composition and of structure as the only legitimate bases of classification marks an important step in advance, yet there has been—and still is—such lack of agree-

---

[1] *Quantitative Classification of Igneous Rocks*, by Whitman Cross, Joseph P. Iddings, Louis V. Pirrson, and Henry S. Washington (Chicago, 1903; also published in vol. x. of *The Journal of Geology*, 1902, p. 555); *Chemical Analyses of Igneous Rocks published from 1884 to 1900*, by Henry S. Washington (Washington, 1903); *Chemical Composition of Igneous Rocks expressed by means of Diagrams*, by Joseph P. Iddings (Washington, 1903).

ment as to the characters which should distinguish the different classes and the nomenclature which should be employed, that those who wish to keep in touch with the world's petrological literature must make themselves acquainted with some half-dozen systems of classification differing from one another in many important points, and none of them entirely satisfactory.[1]

In almost all cases the main rock types are now defined by the conjunction of certain mineral and structural characters, though by some authorities considerations of structure are still subordinated to stratigraphical relations. But while in this country most petrologists take the mineral composition as the basis of the larger divisions, others give this place to the structural or stratigraphical characters.

Chemical composition, on the other hand, has up to the present played a comparatively small part in any system of classification, in spite of the fact that in the early days of geology Élie de Beaumont divided igneous rocks into acid, containing more than 65 per cent. of silica ; intermediate, with from 55 to 65 per cent., and basic, with 40 to 55 per cent. ; and that a similar division has been made by many subsequent writers, though somewhat different percentages may have been substituted, or two groups have taken the place of three.

The importance of a knowledge of the chemical composition has long been recognised, and is now appreciated more than ever ; but it is still the exception for chemical analysis to be resorted to. The complete analysis of an igneous rock is a long and complex operation, and few geologists possess the technical skill and laboratory appliances to carry it out themselves. Professional analysts, too, however expert they may be in other directions, have rarely the special training in rock analysis which is necessary if satisfactory results are to be obtained. Workers have therefore, as a rule, contented themselves with the microscopical examination, and as a result the prevailing systems of classification are based on mineralogical and structural characters, and only indirectly and approximately express the chemical composition.

While, however, the investigation of the chemical composition

---

[1] A full, though by no means exhaustive, account of the different attempts to classify rocks, from the time of Linnæus to the present day, will be found in the form of an introductory review to the *Quantitative Classification of Igneous Rocks* cited above.

of a rock has almost always been quantitative in character, the mineralogical examination is rarely anything but qualitative; and, as a natural consequence, in all classifications based on mineral characters, quantitative considerations have been almost entirely disregarded. As long as the minerals which form part of the definition of a rock are present, they may vary in amount indefinitely without in most cases affecting its nomenclature or its place in classification.

It was with the object of remedying this particular defect, and at the same time replacing the prevailing discordance with uniformity and simplicity, that Professor Iddings and his coadjutors devoted themselves for ten years to the elaboration of their quantitative classification of igneous rocks.

They recognised at once that by far the most important character of an igneous rock was its chemical composition.

"The chemical composition of the rock," we read on p. 108 of the *Quantitative Classification*, "is its most fundamental character, being a quality inherent in the magma before its solidification, and is therefore of greatest importance for its correlation with other rocks."

"All rocks of like chemical composition should be classed together, and degrees of similarity should be expressed by the relative positions or values of the systematic divisions of the classification."

"The mineral and textural characters, being dependent largely on external conditions attending rock consolidation, are to be regarded as of subsidiary importance in classification, but should receive due recognition in the system."

Again, on p. 110: "Structure or texture is now known to depend so largely on variable conditions attending the consolidation of magmas, that it can no longer be given the prominent rôle hitherto assigned to it. Chemical and mineral composition then remain as those characters of igneous rocks most available for their classification. Of these, it is to be noted that while the two are most intimately related, the former is more fundamental, since it pertains to a magma which may consolidate as a glass, or become a holocrystalline rock, and in the latter case the mineral constitution varies with attendant conditions."[1]

[1] Cf. Prof. Bonney's Presidential Address, *Quart. Journ. Geol. Soc.*, vol. xli. (1885) p. 66.

But they were not sufficiently consistent to carry out their classification on chemical lines.

"While the chemical composition of igneous rocks is their most fundamental characteristic, it is known that there is an absence of stoichiometric proportions among the chemical elements or components" and "an absence of chemical division lines or of groups or clusters of similar combinations of elements"; this, and the fact that "all holocrystalline, and many of the partially crystalline rocks, derive their most obvious characters from the mineral particles composing them," led to the conclusion that "the systematic classification should be constructed, as far as may be, by the use of mineralogical data in one form or other," and it was finally decided "to treat the chemical composition of rocks in terms of minerals, and to make the basis of primary subdivisions chemico-mineralogical."

As, however, magmas having the same chemical composition may crystallise out in different mineral aggregates, or may consolidate as a glass, they devised a method of calculating from the chemical analysis of a rock, a standard mineral composition called the "norm," which might in some cases practically coincide with the actual mineral composition or "mode," but, as a rule, differed more or less widely from it. Finally, from the minerals of this ideal rock, and the proportions in which they occurred, the position of the rock in the quantitative classification was determined.

OUTLINES OF THE QUANTITATIVE CLASSIFICATION

(1) *The Chemical Composition*

The starting point of the new quantitative classification is the chemical composition of the rock, which may be ascertained either from a chemical analysis or calculated from a quantitative estimation of the minerals present. The latter operation is best carried out by the linear traversing method of Rosiwal, which gives the volumetric proportions.[1] By multiplying these by the specific gravities we get the gravimetric proportions, and thence, with a knowledge of the chemical composition of the different minerals, that of the rock can be obtained. Those

[1] See Rosiwal, *Verh. Wien Geol. Reichs-Anst.*, vol. xxxii. (1898) p. 143; *Quant. Class., supra cit.*, 1903 ed., p. 204; Joseph P. Iddings, *Journal of Geology*, vol. xii. (1904) p. 225; Ira A. Williams, *American Geologist*, vol. xxxv. (1905) p. 34.

substances which, like quartz, have a constant composition present no difficulties. In other cases the composition of the mineral can be ascertained from the study of the optical characters; and when this fails, it is assumed to be the same as in other rocks of a similar character.

The whole process will be found worked out in detail in the case of a microgranite from the island just below Santo Antonio, Rio Madeira, Brazil, in my paper on the rocks of the cataracts of the River Madeira and the adjoining portions of the Beni and Mamoré.[1]

(2) *The Calculation of the Norm*

The chemical composition having been determined, the calculation of the theoretical mineral composition or norm may be proceeded with.

By dividing the percentages of the different chemical constituents (which are, as usual in mineral analyses, given in the form of oxides) by their molecular weights, the molecular proportions are obtained. The molecules are then distributed according to fixed rules to form the minerals of the ideal rock or norm. Minerals whose composition is too complex, or which are otherwise unsuitable, are excluded: for instance, the only silicates containing alumina are the simple felspars (orthoclase, albite, and anorthite) and the felspathoids or "lenads," as they are called by the authors of this classification.

Subject to these restrictions, the method of allotment of the molecules is intended to express, as far as possible, the natural affinities and preferences among the chemical constituents of igneous rocks. The actual rules are necessarily somewhat complex, and are too lengthy for repetition. They will be most easily understood if I show how they are applied in a particular case, that of the microgranite already referred to.

In the following table the first column gives the results of the analysis by Mr. G. S. Blake, and the second the molecular proportions obtained by dividing by the molecular weights; the third column shows the manner in which these molecules are distributed or allotted; and the fourth, the number of the "mineral molecules" (viz. molecules representing minerals) that result.

[1] *Quart. Journ. Geol. Soc.* vol. lxii. (1906) p. 116.

Percentage Composition. Per cent.	Molecular Proportions.	Allotment.	Mineral Molecules.
		⎧ ·0196 anorthite . .	·0098
		⎪ ·0089 orthosilicates .	—
		⎪ ·3222 orthoclase . .	·0537
SiO$_2$ . . . 73·96 . . 1·2327	.	⎨ ·3438 albite . . .	·0573
		⎪ ·0088 orthosilicates to	
		⎪ metasilicates .	·0177
		⎩ ·5294 quartz . . .	·5294
TiO$_2$ . . . Trace			
		⎧ ·0537 orthoclase . .	—
Al$_2$O$_3$ . . . 13·10 . . ·1284	.	⎨ ·0573 albite . . .	—
		⎪ ·0098 anorthite . .	—
		⎩ ·0076 corundum . .	·0076
Fe$_2$O$_3$ . . . 0·74 . . ·0046	.	·0046 magnetite . .	·0046
FeO . . . 1·28 . . ·0178 ⎫		⎧ ·0046 magnetite . .	—
	⎬ ·0184	⎨ ·0006 pyrite . . .	·0006
MnO . . . 0·04 . . ·0006 ⎭		⎩ ·0132 metasilicate .	—
MgO . . . 0·18 . . ·0045	.	·0045 metasilicate .	—
BaO . . . 0·06 . . ·0004 ⎫		⎧ ·0024 apatite . . .	—
SrO . . . 0·13 . . ·0013 ⎬ ·0142		⎨ ·0020 fluor . . .	·0020
CaO . . . 0·70 . . ·0125 ⎭		⎩ ·0098 anorthite . .	—
KO . . . 5·05 . . ·0537	.	·0537 orthoclase . .	—
Na$_2$O . . . 3·55 . . ·0573	.	·0573 albite . . .	—
P$_2$O$_5$ . . . 0·102 . . ·0007	.	·0007 apatite . . .	·0007[1]
S . . . 0·041 . . ·0013	.	·0013 pyrite . . .	—
F . . . 0·084 . . ·0044	.	⎰ ·0005 apatite . . .	—
		⎱ ·0039 fluor . . .	—
99·017	1·5202	1·5202	

Less O
replaced
by F and S   ·046

       98·971

H$_2$O above 100° C 0·93
H$_2$O at 100° C   0·23

      100·13

This distribution is carried out in the manner following :—

In the first place sufficient lime and fluorine are allotted to phosphorus pentoxide to form apatite, $3(CaO)_3P_2O_5 + CaF_2$. As much lime is then allotted to the remaining fluorine as will form the mineral fluor, while the sulphur and a part of the ferrous iron are united to form pyrite.

After these minor matters are disposed of, the alumina is allotted to the potash, soda, and remaining lime to form the groups $K_2O . Al_2O_3$, $Na_2O . Al_2O_3$, and $CaO . Al_2O_3$ respectively. The excess of alumina is represented as forming corundum, although that mineral is, as a matter of fact, not present in the rock. To the ferric oxide the same molecular amount of ferrous oxide is added

[1] Measured in molecules of P_2O_5.

to form magnetite. Silica is now allotted first to the group $CaO.Al_2O_3$ to form anorthite, next to the remaining ferrous oxide and the magnesia to form the orthosilicates $(FeO)_2SiO_2$ and $(MgO)_2SiO_2$, and then to the $K_2O.Al_2O_3$ and $Na_2O.Al_2O_3$ to form orthoclase and albite. Sufficient silica now remains to raise the orthosilicates of iron and magnesia to metasilicates $FeO.SiO_2$ and $MgO.SiO_2$, and finally an uncombined residue is left over to form quartz.

In the next table the first column gives the proportions of the different mineral molecules, and the second column the actual percentage weights of the minerals obtained by multiplying by the mineral-molecular weights.

	Molecular Proportions of minerals.		Mineral Molecular weights.		Percentages by weight.		
Quartz	·5294	×	60	=	31·76		
Orthoclase . .	·0537	×	556	=	29·86	felspars 62·71	salic 95·25
Albite . . .	·0573	×	524	=	30·03		
Anorthite .	·0098	×	(278)	=	2·82		
Corundum . .	·0076	×	102	=	0·78		
Ferrous metasilicate	·0132	×	(132)	=	1·74	metasilicates 2·19	
Magnesium metasilicate .	·0045	×	100	=	0·45	(hypersthene)	femic 3·73
Apatite (measured in molecules of P_2O_5) .	·0007	×	336	=	0·24		
Fluor . . .	·0020	×	78	=	0·16	oxides, etc. 1·54	
Magnetite . .	·0046	×	232	=	1·07		
Pyrite	·0006	×	120	=	0·07		

Total (exclusive of moisture and combined
water) . . . . . .   98·98

The anorthite includes a certain amount of strontium and barium, and the ferrous metasilicate some manganese metasilicate. These have been allowed for in the above calculations.

The close approximation of the total of the minerals of the norm (98·98) to the original total of the analysis (less combined water and moisture), 98·971, is strong evidence that the operation has been correctly carried out. The authors of the system only calculate the number of molecules to the third place. This saves trouble, but the final figures are then only approximate, especially in the case of the felspars whose mineral-molecular weights are very high: 556 in the case of orthoclase.

If there is not enough alumina to satisfy both the alkalies and the lime, the latter has to go short, and is subsequently allotted silica to form either Wollastonite, $CaO.SiO_2$, or Akermanite, $(CaO)_4 (SiO_2)_3$, according to the amount of silica available.

When the silicate is insufficient to convert all the alkali-alumina groups to felspars, it is divided up among felspars and felspathoids according to definite rules which have their foundation in petrological experience. Other contingencies of a similar nature may arise, but are provided for by the scheme, which has been elaborated in detail with the greatest care and exactness.

The operations in the calculation of the norm are both interesting and instructive, but they are to a large extent unreal, and cannot compare in value and scientific character with that of calculating from the analysis the proportions of the minerals which have been actually ascertained to be present by microscopical examination,[1] and verifying the result by means of a quantitative determination by the linear method.

(3) *The Classes*

The primary divisions into which igneous rocks are classified in this system are dependent on the separation of the rock-forming minerals which may be present in the norm into two main groups, the salic and femic minerals. The former consist of quartz, zircon, corundum, the felspars and the "lenads," viz. felspathoids, while all the remainder are classed as femic. They include the silicates of iron, manganese, magnesia, lime, and the alkalies (but not of course the felspars or felspathoids), also the metallic oxides and apatite, fluor, pyrite, and other minerals of a similar character.

[1] The minerals present in the microgranite consisted of quartz, orthoclase and microcline, plagioclase (partly microperthitic, partly independent), biotite, fluor, apatite, and pyrite. The proportions in which these were present were calculated from the analysis to be as follows :—

	Per cent.
Quartz	32·90
Orthoclase and microcline	27·58
Albite } plagioclase allied to albite	{ 29·92
Anorthite }	{ 2·76
Biotite	4·28
Apatite	0·24
Fluor	0·13
Magnetite	0·72
Pyrite	0·07
Excess alumina	0·46
Excess water	1·05
	100·11

The excess of alumina and part of the water may represent aluminium hydrates formed in the course of incipient lateritisation. These figures do not agree very well with the results which I actually obtained by the linear method, for reasons that it is needless to enter upon here.

If the ratio of the total weight of the salic minerals to that of the femic minerals be greater than 7 to 1, the rock falls within the persalane class; if it fall short of this, but is more than 5 to 3, it is in the class of the dosalanes. If the ratio lie between $\frac{5}{3}$ and $\frac{3}{5}$, the rock is a salfemane, if between $\frac{3}{5}$ and $\frac{1}{7}$, a dofemane; and if less than $\frac{1}{7}$, a perfemane.

(4) The Orders

Each of these five classes [1] is again divided into orders. The subdivision into orders of the first three classes—the persalanes, dosalanes, and salfemanes—depends on the proportion of felspars to quartz on the one hand and felspathoids on the other. As is well known, these do not occur together. There are nine orders in each class, as shown by the following table, where Q denotes quartz, F felspar, and L lenad or felspathoid.

1. $\frac{Q}{F} > \frac{7}{1}$ perquaric. 4. $\frac{Q}{F} < \frac{3}{5} > \frac{1}{7}$ quardofelic. 7. $\frac{L}{F} < \frac{5}{3} > \frac{3}{5}$ lenfelic.

2. $\frac{Q}{F} < \frac{7}{1} > \frac{5}{3}$ doquaric. 5. $\frac{Q\ or\ L}{F} < \frac{1}{7}$ perfelic. 8. $\frac{L}{F} < \frac{7}{1} > \frac{5}{3}$ dolenic.

3. $\frac{Q}{F} < \frac{5}{3} > \frac{3}{5}$ quarfelic. 6. $\frac{L}{F} < \frac{3}{5} > \frac{1}{7}$ lendofelic. 9. $\frac{L}{F} > \frac{7}{1}$ perlenic.

The dofemane and perfemane classes are each divided into orders [2] according to the proportions of femic silicates to non-silicates.

Most of the orders have received special names ending in "are," and based on the names of countries or of states. For instance, the quardofelic order of the persalanes is known as Britannare, the perfelic order of the dosalanes as Germanare. [3]

(5) The Rangs and Sub-rangs

Each of the orders of the first three classes is subdivided into five minor groups, known as "rangs," according to the

[1] There are also sub-classes, which are only of importance in the case of rocks with a considerable amount of corundum (see *post*, p. 271) and zircon (in the first three classes), or apatite, fluor, etc. (in the last two).

[2] These are again split up into sections of orders according to the ratio of the more acid to the less acid femic silicates, or into sub-orders according to the ratio of the iron ores to the titanium minerals.

[3] The sections of orders and sub-orders have similar names, the former ending in -iare, the latter in -ore.

molecular proportion, in the salic minerals, of the alkalies to the lime.

1. $\dfrac{K_2O + Na_2O}{CaO} > \dfrac{7}{1}$ peralkalic.

3. $\dfrac{K_2O + Na_2O}{CaO} < \dfrac{5}{3} > \dfrac{3}{5}$ alkalicalcic.

2. $\dfrac{K_2O + Na_2O}{CaO} < \dfrac{7}{1} > \dfrac{5}{3}$ domalkalic.

4. $\dfrac{K_2O + Na_2O}{CaO} < \dfrac{3}{5} > \dfrac{1}{7}$ docalcic.

5. $\dfrac{K_2O + Na_2O}{CaO} < \dfrac{1}{7}$ percalcic.

Each of these rangs is again split up in exactly similar fashion into five sub-rangs, according to the molecular proportion of the salic potash to the salic soda. In the last two classes the rangs and sub-rangs are based on the molecular proportions of the bases in the femic minerals.[1]

The division into classes and orders[2] is based on *gravimetric* calculations—that is to say, it is determined by the ratio of the total weights of the groups of minerals on which the classification depends; but in forming the smaller divisions, the rangs and sub-rangs, it is the *molecular* proportion of certain chemical constituents that is considered. It is difficult to find a logical justification for this fundamental change in the basis of classification. Again, it is not the whole of the chemical constituents in question that determines the classification, but only so much as forms part of the salic or femic minerals, as the case may be. For instance, the lime in the femic minerals has no influence on the division into rangs in the first three classes, only that in the salic minerals being taken into consideration, while in the subdivisions of the other two classes only the femic lime is considered.

Both rangs and sub-rangs receive distinctive names taken from localities where typical rocks are met with; the former end in -ase, the latter in -ose. It is the name of the sub-rang which appears to be generally employed when referring to a rock.[3]

As specimens of the nomenclature, it may be mentioned that Essexose is the dosodic sub-rang of Essexase, which is the domalkalic rang of Norgare, the lendofelic order of the class

[1] There are also sections of rangs and of sub-rangs with which we need not concern ourselves here.

[2] As well as that into sections of orders and sub-orders in the last two classes.

[3] See note, p. 269.

Dosalane; while Amiatose is the sodipotassic sub-rang of Coloradase, which is the alkalicalcic rang of Britannare, the quardofelic order of the Persalanes.[1]

It is unnecessary to deal here with the still smaller divisions into grades and sub-grades, which are carried out on somewhat similar lines to those of larger size, as they appear to be seldom if ever employed, or with the adjectives which have been devised to indicate the manner in which the " mode " or actual mineral composition differs from the " norm " or ideal composition, for these are matters of entirely subsidiary importance. Nor is this the place to describe the nomenclature to be employed to express the degree of crystallisation, the size of the crystals, and the character of their development; but I may say that the terms selected appear to be well chosen, and might be conveniently adopted, whatever may be the fate of the quantitative classification itself.

The principles of the classification will, however, be more easily understood when they are applied to a particular rock, such as the microgranite already referred to.

As a result of the calculations that have been made, it will be seen that the gravimetric proportion of salic to femic minerals in the norm or ideal mineral composition of this rock is 95·25 to 3·73, which is more than 7 to 1. The rock is therefore a persalane. The quartz in the norm amounts to 31·76 per cent. and the felspars to 62·71, which is less than three-fifths and more than one-seventh. The rock, therefore, belongs to the quardofelic order of the persalane class—an order which, as already stated, has received the name of Britannare.

The proportion of the molecules of the alkalies, which are in this case all salic, to those of the salic lime is 1110 to 98, which is more than 7 to 1. The rock is therefore peralkalic, and belongs to the Liparase rang of the order. Again, the proportion of the potash to the soda molecules is 537 to 573, which is less than 5 to 3 and more than 3 to 5. This places the rock in the sodipotassic sub-rang, which is known as Liparose.

The calculation of the norm seems long and complicated, but it can, with practice and the help of the tables included in

---

[1] In the last two classes the matter becomes more complicated. For instance, Marquettose is the domagnesic sub-rang of Marquettiase, the permiric section of Texase, which is the permirlic rang of Texiare, which again is the pyrolic section of Scotare, the dopolic order of the class Dofemane.

the *Quantitative Classification of Igneous Rocks*, 1903 ed., be rapidly carried out.

Dr. Washington has had forms printed on cards to facilitate the calculations required for determining the place of a rock in the classification. On one side is the analysis, with the molecular proportions to three places of decimals; on the other, the calculations and classificatory results. He very kindly sent me some of them filled in with the particulars of the rocks from the River Madeira. The results he obtained for the rock already referred to were as follows:

		Percentages by weight.		
Quartz		31·26		
Orthoclase		30·02		
Albite		29·87	63·50	95·17
Anorthite		3·61		
Corundum		·41		
Hypersthene	{FeO.SiO₂ / MgO·SiO₂}	2·48		3·64
Magnetite, etc.		1·16		
				98·81

This is practically the same result as that given above, though by a curious accident the orthoclase and albite are almost exactly interchanged.

Results of the Classification

Rocks classed as Liparose by Dr. Washington [1] had been previously described as granite, granitite, hornblende granite, soda granite, aplitic granite, aplite (haplite), lithionite granite, quartz syenite, quartz lindoite, granophyre, porphyry, granite-porphyry, Rapakivi granite and granite-porphyry, quartz porphyry, quartz tourmaline porphyry, quartz syenite-porphyry, syenite-porphyry, felsite-porphyry, bostonite, keratophyre, quartz keratophyre, rhyolite, rhyolitic glass, obsidian, rhyolite-obsidian, pitchstone, trachyte, quartz pantellarite, nevadite, liparite, dacite, andesite-obsidian, andesite (decomposed), comendite, paisanite, alaskite, and arfvedsonite grorudite.

This long list of names is accounted for to some extent by the differences of structure and mineral composition resulting from varying conditions of consolidation. The indefiniteness, confusion, and redundancy of modern nomenclature are also

[1] See *Chemical Analyses of Igneous Rocks, supra cit.*, pp. 144-53.

responsible for much; but the following table, giving the varying percentage of the different chemical constituents, compiled from Dr. Washington's Analyses, shows that the new classification is also at fault:

Molecular proportions.

SiO_2	.	.	.	.	77·61 to 66·28	.	.	.	1·294 to 1·105
Al_2O_3	.	.	.	.	18·41 to 10·84	.	.	.	·180 to ·106
Fe_2O_3	.	.	.	.	5·31 to nil	.	.	.	·033 to nil
FeO	.	.	.	.	4·76 to nil	.	.	.	·066 to nil
MgO	.	.	.	.	0·84 to nil	.	.	.	·021 to nil
CaO	.	.	.	.	1·42 to trace	.	.	.	·025 to trace
K_2O	.	.	.	.	6·77 to 3·70	.	.	.	·072 to ·039
Na_2O	.	.	.	.	5·67 to 2·56	.	.	.	·091 to ·041
$KO_2 + Na_2O$	.	.	.	.	.	.	.	.	·149 to ·102

This shows a wide range for what is, after all, a very minor subdivision of the classification. The silica has a variation of more than 11 per cent. and the other constituents show wide variations.

Other sub-rangs give a similar result. Toscanose, the sodi-potassic sub-rang of the domalkalic rang of the quardofelic order of the persalanes, includes rocks variously described as granites, syenites, and diorites, besides their hemi-crystalline and glassy representatives, and has a range of silica from 76·48 to 62·20, alumina from 20·82 to 12·04, potash from 7·09 to 3·28, and the alkali molecules from ·141 to ·079.

Andose, the dosodic sub-rang of the alkalicalcic rang of the perfelic order of the dosalanes, has a percentage of silica varying from 58·20 to 44·85, and includes rocks whose descriptions range from tonalite and quartz diorite to gabbro.

These examples are sufficient to demonstrate that these small groups have individually no very definite chemical significance.

If we turn to the larger divisions, we find that they also do not correspond to any clearly characterised variations in chemical composition.

The class persalane includes[1] a quartz porphyry with a silica percentage of 80·99 and an anorthosite with one of 45·78. The extreme limits are, however, much farther apart, and would include pegmatites with over 85 per cent. and intrusive quartz veins with practically 100 per cent. of silica.[2] In the other direction, the analysis of a rock consisting mainly of corundum, spinel, and anorthite, described by Morozevicz[3] under the

[1] Washington, *op. cit.*, pp. 122, 206.
[2] Iddings, *op. cit.*, p. 64.
[3] *Tscherm. Min. Pet. Mitth.*, vol. xviii. (1898) p. 212 ; Washington, *op. cit.*, p. 217.

name of 'Kyschtymite, shows a silica percentage as low as 16·80, and yet it is a persalane. Indeed, rocks consisting of practically pure corundum, such as have been described from India,[1] would still be persalanes, though they are almost free from silica. The range of the alumina might theoretically be as great as that of the silica. Apparently Kyschtymite has about 70 per cent., and the Indian rocks nearly 100, while intrusive quartz has none. The alkalies, too, may exceed 21 per cent., as in the soda of a rock such as is known to exist composed almost entirely of nepheline, and may fall to zero.

The rocks of this class have therefore nothing in common except the fact that they contain less than 12·5 per cent. of ferromagnesian silicates and other minerals classed as femic.

In the other classes there is a similar wide range of silica percentage: in the dosalanes from 75·40 to 38·11; in the salfemanes from 73·60 to 36·36; in the dofemanes from 53·05 to 31·59 and even 9·79 in the case of certain iron ores which are believed to be of igneous origin; and in the perfemanes from 55·52 in a lherzolite to 0·76 in an iron ore.[2] The limits between which the silica might possibly vary would, I need not say, be still wider. The other oxides have ranges almost as great.

Even the total amount of iron, magnesium, and calcium oxides, which is the most constant character of the classes, shows wide variations. The persalane analyses range from 1·05 to 22·07 per cent., and might theoretically reach 30 per cent. In the dosalanes the percentage varies from 2·44 to 30·03, and might be as high as 50. In the salfemanes it ranges from 13·21 to 39·87 per cent., with a higher limit of 70; in the dofemanes, from 33·32 to 65·90, with a possible percentage of 90; while in the perfemanes it varies between 41·16 and 72·55, and might rise to 100 per cent.[3] These figures will not be materially

---

[1] See F. R. Mallet, *Records Geol. Surv. India* (1872), vol. v. p. 20 and vol. vi. (1873) p. 43 ; and *The Manual of the Geological Survey of India, Mineralogy,* (Calcutta, 1887) p. 48 ; and Prof. J. W. Judd, " On some Simple Massive Minerals (Crystalline Rocks) from India and Australia," *Mineralogical Magazine,* vol. xi. (1895) p. 56. At the Imperial Institute there are specimens of microcrystalline corundum from the Malay Peninsula which might also be classed as a rock. It is not clear, however, whether these rocks are of igneous origin.

[2] See Washington, *op. cit.,* pp. 218, 306, 310, 350, 354, 364 ; Iddings, *Chem. Comp., supra cit.,* pp. 59, 62.

[3] See Washington, *op. cit.,* pp. 142, 206, 218, 292, 312, 338, 354, 364, 366, and 368.

modified by the inclusion of titanium oxide, which would merely raise the upper limit. There are, however, good reasons for placing it beside silica for classificatory purposes. The total of the silica, alumina, and alkalies, the essential chemical constituents of the more acid rocks is approximately complementary to that of the iron, magnesium, and calcium oxides, and varies in a similar fashion.

The reason for these wide variations in the chemical composition of the classes is not far to seek. They must be attributed to the system by which the oxides are combined to form the minerals of the norm, and these are divided arbitrarily between the salic and femic groups. The result is that a considerable amount of silica is combined with the ferrous, manganous, magnesium, and calcium oxides, and counts on the femic side, while a large percentage of lime is often transferred in anorthite in the opposite direction.

The amount of silica carried to the femic side by the metallic bases is not proportional to the amount in which they are present. In acid rocks they usually form more acid silicates, and give a femic character to a larger proportion of silica than in basic rocks. The position of lime is peculiar. If there be enough alumina to form anorthite, the lime as well as the alumina and silica combined with it are salic, but if no alumina be available, both lime and silica are femic. Alumina has therefore an influence on the classification of rocks far beyond the amount in which it is present, for it may change from one side to the other lime and silica equal to 1·7 of its own weight.

The state of oxidation of the iron also materially influences the result; which is obviously undesirable, in view of the fact that it is not always easy to ascertain the extent of the oxidation that existed at the time of the consolidation of the rock. In the calculation of the ideal constitution or norm, the ferric oxide is either combined with ferrous oxide to form magnetite or, if there is not enough ferrous oxide, remains uncombined as hæmatite.[1] If, therefore, there is much ferric oxide, a large proportion, or, it may be, the whole, of the iron takes no silica with it to the femic side, while if the iron be all present as ferrous oxide it will take over from ·42 to ·83 of its own weight of silica, so that the femic total will be considerably increased.

[1] With the exception of any that may be needed to unite with soda and silica to form "acmite" molecules.

In the orders there is less variation in the silica percentage. It is, however, still very considerable in many cases. In the analyses of the rocks included in Canadare, the perfelic order of persalanes, the percentage of silica varies from 68·01 to 45·78, and the possible range is from nearly 73 to less than 38. In Germanare, the corresponding order of the dosalanes, it varies from 64·46 to 42·92; and in Scotare, the dopolic order of the dofemanes, from 53·05 to 33·84.[1]

The variation in the composition of the orders of the classes which are divided according to the proportion of quartz to felspars, and of felspars to lenads or felspathoids, is explained to a large extent by the fact that the alkali felspars with six molecules of silica, and anorthite with only two, have exactly the same classificatory value. In addition to this, in those orders where the lenads have their share in the classification, the position of rocks rich in soda is thrown out of relation to that of those rich in potash by the fact that in the former nepheline, with only two molecules of silica, has the same value as leucite, with four such molecules, has in the latter.

In the more basic classes the division into orders is based on the proportion of femic silicates to non-silicates which are mainly oxides, and this depends largely on the state of oxidation of the iron, which ought not to be relied on for classificatory purposes.

In all the classes, the varying amounts of alumina, which transfer lime and silica one way or the other between femic silicates and felspars, effect considerable and unreasonable changes in the delimitation of the orders.

In the rangs the silica percentage varies practically to the same extent as in the orders, for the lines of division do not directly depend on the amount of silica present. In the more acid classes the limits of the rangs in each order are, as we have seen, determined by the ratio of salic alkalies to salic lime, not that of the total alkalies to the total lime, which may be widely different. The distinction between total and salic alkali is of no great importance, as it is only when the percentage of alkali is very high or that of alumina exceptionally low that any alkali passes over to form femic silicates; but the salic lime depends on the amount of anorthite, and that is determined by the alumina which is left when the alkalies are satisfied. If this be

[1] See Washington, *op. cit.*, and Iddings, *Chem. Comp.*, p. 56.

small, the salic lime is diminished, and the ratio of salic alkali to salic lime is correspondingly increased, and vice versa.

The rangs of the first three classes being divided into sub-rangs on the basis of the ratio between the salic potash and salic soda, the variations in chemical composition remain in almost all other respects as great as in the rangs, the chief exception being the sub-rangs of the orders in whose determination the lenads play a part. Here the variation caused by the different silica percentages of leucite and nepheline practically disappears when the ratio of the salic potash and soda becomes fairly constant.[1]

Throughout the classification, both in the large and small divisions, the extent of the silica variation is very striking, and though it may not be relatively greater than that of the other oxides, yet it has on the whole a greater influence on the character and mineral composition of the rock.

It is sufficiently apparent that the lines of division of the Quantitative Classification do not stand in any logical relation to the chemical composition, and that it does not satisfy the condition laid down by the authors that " all rocks of like chemical composition should be classed together, and degrees of similarity should be expressed by the relative positions or values of the systematic divisions of the classification."

The principle of employing larger groups than the oxides as units of classification undoubtedly finds considerable support from the phenomena attending the differentiation and consolidation of igneous rocks.[2] A scientific classification should, as far as possible, express the origin and genetic relations of the divisions and subdivisions to which it gives rise. Igneous rocks, being formed by the consolidation of magmas, should be classified with reference to the laws in accordance with which igneous magmas are differentiated, either while still in the liquid state or in the course of consolidation.[3]

These laws are still imperfectly known, but there is no

[1] It is for this reason, I presume, that the sub-rangs are most frequently used in nomenclature.

[2] Iddings, *The Chemical Composition of Igneous Rocks*, p. 69.

[3] While it is probable that the majority of igneous magmas are the result of the differentiation of the original molten substance of the earth, there are some rocks which appear to be formed from magmas of an unusual type, due to the absorption of solid material by a liquid magma. These would have to be specially dealt with in any scientific classification.

department of petrology in which so much progress has been
made in recent years, or in which there is such promise of still
further advances in the immediate future.[1] Yet, although the
actual phenomena that attend the consolidation of igneous rocks
are still obscure, there seems little doubt that in some cases,
at least, differentiation takes place in a partially crystallised
magma by the straining out, for some reason, of the still liquid
residue from the crystals that have already formed. The former
is usually eutectic in character, and therefore more or less definite
in chemical composition. The different minerals that have crystallised out and the liquid residue may therefore be considered, for
some purposes, units of differentiation; but these are, in many
respects, widely different from the mineral molecules that play
their part in the quantitative system of classification, and which
have been chosen because they lend themselves to a comparatively simple, but arbitrary, resolution into mineral groups.

But the differentiation which appears to occur in a magma
that is still entirely liquid is at least equally important, and
in this case, too, there is reason to believe the oxides do not,
in all cases at least, act as independent units, but are associated
in larger molecules.[2] Alumina, for instance, may form with
the alkalies or lime compounds analogous to the spinels, and
these again appear to combine under certain circumstances to
form the simple felspars, or leucite or nepheline. There are,
however, no grounds for supposing that the minerals that
ultimately separate out necessarily represent the previous
molecular condition of the magma.[3] At present our information

[1] The whole subject has been reviewed by Mr. J. J. H. Teall in his Presidential
Address to the Geological Society (*Quart. Journ. Geol. Soc.*, vol. lvii. [1901]
pp. lxx to lxxxvi), and by Prof. H. A. Miers in his opening address in Section
C of the British Assciation in 1905. See also J. H. L. Vogt, " Physikalisch-
chemische Gesetze der Krystallisationsfolge in Eruptivgesteinen," *Tscherm. Min.
Pet. Mitt.*, vol. xxiv. (1906) p. 437.

[2] See, *inter alia*, Rosenbusch, *Elemente der Gesteinslehre*, 2nd edition, Stuttgart
(1901), pp. 186-95.

[3] The fact that felspar crystals, having a composition intermediate between
those of albite and anorthite, crystallise out from a magma does not prove that
molecules, with the same complex composition, existed in the magma. What
evidence there is tells in the opposite direction, for the albite and anorthite
molecules appear to differentiate quite independently. But while the authors
of the quantitative classification are quite right in considering only the simple
felspars as units of classification, there appears to be no good reason for assuming,
as they do, that albite and anorthite are equivalent molecules which play nearly
the same part in the process of differentiation.

is too scanty for us to dogmatise, but in this case also it is impossible to believe that the units selected for the quantitative classification correspond in any but the most imperfect manner to those actually present in molten magmas, which probably vary in molecular composition within wide limits under different conditions of temperature and of pressure.

But even if the elements of differentiation were sufficiently permanent to be taken as units of classification, there would be no justification for the arbitrary division of group from group by lines drawn at arithmetical intervals, which can correspond to nothing that has occurred in the evolution and differentiation of igneous rocks.

These purely arbitrary boundaries would in many cases separate a common and simple rock-type into two or more portions, with different names and positions in the classification. Indeed, there would be few rock masses sufficiently uniform to be included under one name, for a slight increase in the proportion of one mineral to another would frequently be sufficient to transfer the rock from one order, or even class, to another.

The authors defend the position they have taken up by the contention that rocks pass into each other by imperceptible gradations, and claim that there are no natural groups that could be adopted in classification.

This view cannot, I believe, be supported. Rocks do not differentiate alike in all directions. There are certain associations of oxides which increase or decrease together, so that in the extremes at least definite types must become prominent; and while it is true that intermediate links occur which bridge over the interval between rocks of extreme types, they are not, there is every reason to believe, distributed with anything like uniformity.

Prof. Iddings has constructed diagrams in which each rock is represented by a point whose horizontal co-ordinate is proportional to the percentage of silica and whose vertical co-ordinate to the ratio of the total alkali molecules to those of silica. In these diagrams he has included all the satisfactory analyses available so as to indicate the frequency of different types as far as concerns the amounts of silica and the alkalies. A brief examination of these diagrams shows that there are definite regions where the dots representing analyses are closer together than in the intermediate areas, and we have no right to assume, with Prof. Iddings, that this is due to the limited number of rocks that have yet been analysed, and that as the number of analyses of igneous rocks, selected without bias in one direction or another, is indefinitely increased, the points representing them will tend to distribute themselves in such a manner as to approximate to perfect uniformity.

As a matter of fact, analyses are not selected impartially from all occurrences of igneous rocks. It is not the most widely developed rocks that are the most frequently analysed. They belong, as a general rule, to well-known types that are easily recognised in hand specimens or thin sections. Analysis is, in their case, often considered unnecessary, and when it is resorted to, the number of analyses is not in proportion to the extent to which the rocks occur. It is of the exceptional rock which presents difficulties in determination, that the worker most frequently obtains an analysis to verify his determinations and to illustrate his description. Such a rock is often of very local occurrence and so inconstant in character, that it may well be considered worthy of more than one analysis. It is therefore inevitable that a tabulation of analyses gives an altogether false idea of the frequency of occurrence of different types, and such a diagram as is above described exhibits a much greater uniformity of distribution than actually exists.

If we could adequately represent the quantitative occurrence of rocks of different chemical types by points in a diagram, we should find, there can be little doubt, that the whole resembled a complex star cluster, which throws out arms in this direction or that, and contains within its boundaries condensations of various shapes separated by regions more sparsely occupied. Unfortunately the value of the method of representing rocks by means of points in a diagram is seriously diminished by the fact that whether the simple oxides or the larger molecules be taken as units, the number of variants far exceeds the dimensions which are available. It is only by a kind of projection that the different components, which would require a space of at least six or seven dimensions for even approximately complete expression, can find their places in the two dimensions of a diagram. The variants must be reduced to two by a process of combination or exclusion, or both.[1] The result is, so to

---

[1] One more variant may be introduced into plane diagrams by the use of the triangular diagrams of Ozann and others, but these are somewhat more difficult to construct and to read.

speak, a foreshortening by which rocks that differ considerably in composition may be represented by the same point, so that the natural groupings are more or less obscured. The difficulty may to some extent be solved by constructing a number of diagrams which, though based on the same general principle, have the components differently selected or combined, and correspond to projections of solid figures on different planes.

The natural groupings which future research will, I believe, map out, must stand in close relation to the principles on which the separation of rock magmas takes place, and will guide us to a simple and natural classification of igneous rocks which, while essentially chemical, will indicate as far as possible the origin and genetic relations of the different rocks with which it deals.

The task of framing such a classification must be postponed to a future day—which may not, however, be far distant—when the principles of the segregation of igneous magmas are better understood than they are at present.

In the meantime it would be the wisest course to employ a system of classification on a mineral basis, which, like that outlined by Professor Bonney,[1] takes the holocrystalline rocks as the types, and considers the less completely developed or the glassy forms with the same chemical composition as modifications of them.

The general adoption of the quantitative system of classification which has been described in the present paper would, I fear, tend to obscure the true issues, and to obstruct rather than to facilitate the ultimate evolution of a natural division and nomenclature of igneous rocks; but petrologists will always owe a debt of gratitude to its authors for making the first definite reconstructive movement towards the radical reform of the present system, or rather want of system.

There is something to be said for a purely chemical classification, but it is undesirable to introduce a number of divisions and names that must soon be discarded, and a systematic use of diagrams showing the chemical composition would serve every purpose of a detailed classification on the same lines.

In any case, if we wish to facilitate the introduction of a scientific classification, and desire that the results of our work should ultimately find their natural place amongst its divisions,

[1] *Op. cit.*, pp. 67-78.

we must not be content to label the rock with the name of one of our present groups, or enumerate the minerals which are present and the characters they present. We must make a quantitative determination of the relative amounts of the minerals, and an optical determination of their composition, so far as the present state of our knowledge permits, and must finally, from these data, calculate the chemical composition of the rock, verifying it whenever circumstances permit by a chemical analysis. Even where a glassy magma exists we may be able to form some idea of its composition by the specific gravity and refractive index, assisted to a limited extent by the colour, magnetic susceptibility in strong fields, fusibility, and the simpler chemical characters.

Wherever the chemical composition of a rock has been determined, a careful comparison should be made with other rocks having a similar composition, and the differences that present themselves in mineral composition and structure carefully studied and if possible explained.

But it is not only quantitative determination of the minerals that is important. An estimate, as close as circumstances permit, should only be made of the amount in which the different rock types are developed in each district, and their relations with one another in space should be carefully studied.

By these means workers in the field and with the microscope may co-operate with those engaged in experimental research on the fusion, differentiation, and solidification of silicates, in tracing out the true history of igneous rocks and the physical principles in accordance with which they have come into existence. It will only be when these aims have been at length accomplished that we shall be in a position to frame a truly scientific and universally acceptable system of classification of igneous rocks.

THE NATURE OF ENZYME ACTION

By W. M. BAYLISS, M.A., D.Sc., F.R.S.

Assistant Professor of Physiology in University College, London

	PAGE
CATALYSIS	281
REVERSIBILITY	284
VELOCITY OF REACTION	289
Effect of Concentration of Enzyme.	
Effect of Temperature.	
Accelerators and Retarders.	
CO-ENZYMES	297
ZYMOGENS	298
CHEMICAL AND PHYSICAL PROPERTIES OF ENZYMES . . .	298
MODE OF ACTION	299
COMPLEX SYSTEMS	303
CONCLUSION	305
LITERATURE	305

CATALYSIS

ALL investigators who have had occasion to consider the nature of the chemical changes taking place in living organisms must have been struck with the comparative ease with which highly stable bodies are split up under these conditions. Sugar is oxidised to carbon dioxide and water, egg-white is hydrolysed to amino-acids and other simple bodies. Under ordinary laboratory conditions, strong acids and high temperature or similar powerful agents are necessary to effect such changes. Schönbein may be mentioned as one of those who early called attention to this circumstance.

Now although, at first sight, the powers referred to may be regarded as a distinguishing characteristic of systems which are called "living," it must be remembered that there is a large and increasing class of phenomena, known to chemists as "catalytic," which manifest properties in many ways analogous. Oxygen and hydrogen, for instance, combine at ordinary temperatures so slowly that the production of water is not to be detected, and the application of the high temperature of a flame or electric spark is needed. But the presence of a minute

quantity of finely divided platinum is sufficient to cause combination to take place at ordinary temperatures. Again, the
oxidations produced by hydrogen peroxide proceed in many
cases at a very slow rate, which can be enormously increased
by the addition of traces of iron or manganese.

What, then, are the characteristics of such phenomena?
And are there to be found in the living organism bodies acting
in a similar way to the platinum or iron in the above cases?

The essential property of a catalyst is that given in the
well-known definition by Ostwald, in which it is stated to be
a body which accelerates (or in some cases retards) a chemical
reaction. The original definition also states that the chemical
reaction in question is one that, left to itself, proceeds at an
extremely slow rate. But it is obvious that it is theoretically
immaterial what the original velocity of the reaction is: any
foreign substance that changes the rate of the reaction is a
catalyst. At the same time the practically important cases are
those where the reaction is almost or quite inappreciable without
the assistance of a catalyst. A further characteristic of these
bodies is that they do not appear in the final products of the
reaction and usually are to be found unaltered at the end.
There are, however, exceptions to this last statement, as we
shall see later, owing to the instability of the catalyst under
the conditions of the reaction.

At the outset it is important to be quite clear as to the
distinction between what may be called "trigger-action" and
catalysis. Let us suppose a block of some solid material resting
at the top of an inclined plane, which has a somewhat rough
surface, and is at such an angle that the block would, if left
to itself, slide down very slowly. This is prevented by a stop
of some kind which can be removed, say by pulling a trigger.
Now it is plain that it makes no difference either to the rate
at which the block slides down or to the energy developed in
its fall whether the trigger is very easily worked or is stiff
and needs the expenditure of much energy to move it. But let
us suppose that the block is slowly sliding and that we apply
a catalyst, in the shape of grease, to the inclined plane. The
block will now move faster and the rate will be proportional
to the amount of grease used, although the energy developed
will, as before, be unaltered. Here we see illustrated two
additional properties of catalysts—viz., that the amount of

increase of velocity of reaction is, within limits of course, proportional to the concentration of the catalyst, and also that no energy is added to the system by its presence. The fact that when the falling block has reached the bottom of the slope the energy developed by its fall is the same whether it fell rapidly or slowly calls attention to yet another important property of catalysts, which consists in the fact that a small quantity will, in the end, produce as much effect as a larger one, provided that it be given a longer time to act, on the understanding that during this longer time the catalyst itself remains intact.

As a rule these bodies are very active, so that it is usually taken as one of their characteristics that they act in extremely low concentrations compared to those of the bodies acted upon.

When we now turn again to the living organism, with the properties of catalysts in our minds, we note that bodies answering to this description have for many years been prepared from various cells and tissues. To mention one or two only, we have diastase from malt, which converts starch to maltose; pepsin from the stomach, which splits up various proteins into peptones; peroxydase, which separates active oxygen from various organic peroxides; and so on. These various preparations were originally called "ferments," but owing to a certain confusion between the organism producing certain changes— yeast, for example—and the chemical substance by whose agency the changes are produced, it was suggested by Kühne to call these chemical agents "enzymes," expressing the fact that invertase—one of the earliest known—was contained in yeast (ἐν ζύμη). The name has come into general use, though "ferment" is also frequently used.

"Enzymes," then, are the organic catalysts met with in the living cell. They have certain properties in addition to those already mentioned, one of which is that of being destroyed by a temperature of from 50 to 70° C., probably due to their being colloidal in nature. In contradistinction to the changes produced by living protoplasm, their activities are not prevented by the presence of antiseptics, such as chloroform or toluol, although in certain cases these substances have a more or less retarding action.

Although, as yet, none of the enzymes has been prepared synthetically, and their only source is the living organism, it seems to me unscientific to include this condition in their definition. Oppenheimer gives practically Ostwald's definition of a catalyst, as referred to above, with the addition of the proviso that an enzyme is a "catalytically active substance which is produced by living cells." But the origin of a substance does not concern us when investigating its properties, and, as Bredig points out, this fact does not come into consideration any more than the fact of the heart being composed of living cells affects the discussion of its properties as a pumping-engine.

As catalysts, then, the function of enzymes is to accelerate the velocity of reactions, so that the study of enzyme action in the main becomes one of velocity of reaction. Our attention must, accordingly, be directed first of all to this question. But, as will be seen, it is necessary, before we can profitably discuss the general laws of reactions as catalysed by enzymes, to consider shortly how far these reactions are to be regarded as reversible. In other words, is a synthetic process capable of acceleration by the same catalyst that accelerates the corresponding analytic one? and in what way is the position of equilibrium affected by the presence of a catalyst?

REVERSIBILITY

If we add ethyl acetate to water we find, on examination of the mixture after some time, that, in addition to ethyl acetate and water, there are present the products of hydrolysis of the former, viz. ethyl alcohol and acetic acid. By analysis of the solution at various intervals of time, we find, moreover, that a state of equilibrium is arrived at. In this condition, provided that no change takes place in other conditions, such as temperature, the relative proportions of ethyl acetate, acetic acid, and alcohol are always the same. Again, suppose that we start from the other end, so to speak, and take acetic acid and alcohol, instead of ethyl acetate, we find that the same relative proportion of the three bodies is present when equilibrium is established. We have to deal, therefore, with two opposite reactions. A little consideration will show that, as Van 't Hoff points out, these two reactions must be regarded as both proceeding simultaneously, but at varying rates in proportion to the concentration of the reacting bodies, until in the equilibrium position they possess the same velocity. In fact, suppose we start with ethyl acetate in water, for a moment the reaction between ethyl alcohol and acetic acid will be absent, but as soon as traces of these bodies are formed by splitting of the ester the opposite reaction will commence, although at first slowly.

Now the hydrolysis of ethyl acetate is accelerated by the presence of various catalysts, such as acid (H^+ ions), and the enzyme lipase. Is the opposite (synthetic) reaction also accelerated by these bodies?

Since the equilibrium position is given by the equality of the two opposite velocities, it is plain that if one only of these is changed the equilibrium position must alter. So that, if it is found experimentally that catalysts do not affect the final equilibrium position, it follows that both the opposite reactions must be correspondingly changed.

If the view taken in the introductory remarks as to the action of catalysts is correct, and that no energy is added or taken away from a reacting system by the presence therein of a catalyst, it would seem, _à priori_, necessary that the position of equilibrium should be unaffected. There is, moreover, direct evidence that such is the case. The question has been thoroughly investigated in Ostwald's laboratory by Koelichen in the case of this polymerisation of acetone by bases, and by Turbaba in that of the equilibrium between aldehyde and paraldehyde. It was found in both cases that the equilibrium position was unaffected by catalysts. Naturally the time taken to attain thereto varied considerably, according to the catalyst used.

Many cases are known where a catalyst which accelerates a particular reaction also accelerates the opposite one, when acting on the products of the former. The instance already given of ethyl acetate is one of these; hydrochloric acid not only accelerates the hydrolysis of the ester, but also its formation when acting on a mixture of acetic acid and alcohol.

But the problem which concerns us here is, do the colloidal organic catalysts, the enzymes, fall into line with the simpler inorganic ones in this respect? Practically this resolves itself into this question, does an enzyme under one set of conditions accelerate a particular reaction and under another set of conditions accelerate the opposite reaction? Does lipase, for example, behave like hydrochloric acid in the case just mentioned?

In certain cases synthetic actions of this kind have already been observed. I may refer to the production of ethyl butyrate by the action of lipase on butyric acid and ethyl alcohol in the experiments of Kastle and Loevenhart, of amygdalin from glucose and mandelic-nitrile-glucoside by the maltase of yeast (Emmerling), of salicin by emulsin, and probably of saccharose by invertase from their respective products of hydrolysis (Visser), of a disaccharide from glucose by maltase (Croft Hill), and of isolactose from galactose and glucose by lactase (Emil Fischer and Frankland Armstrong). There is some evidence also of a similar action in the case of the enzymes which act upon proteins.

We may take it, then, that, so far as investigated from this point of view, enzymes have shown themselves to be capable of accelerating both of the opposing components in reversible reactions.

There are, at the same time, certain cases which need a little more consideration; but before proceeding further it will simplify matters if a few words be said as to the terminology of the subject. It was suggested by Duclaux that the termination "-ase" should be taken as expressing an enzyme, and that this termination should be added to the body acted on by the enzyme—e.g. lactase is the enzyme hydrolysing lactose. It is, of course, inconvenient to displace some of the old-established names, such as "pepsin" and "trypsin," but, as far as possible, Duclaux's recommendation should be acted upon. Again, an English name for the substance split by the enzyme is badly wanted. "Substrate" is used by the Germans, and, in default of a better, it will be used in the present article. "Hydrolyte" is used by some authors; but this name does not include cases where the action is not one of hydrolysis but of intra-molecular splitting or oxidation.

It has been the custom to speak of an enzyme which attacks starch or protein, for example, as "amylolytic" or "proteolytic" respectively; but, as Prof. Armstrong has pointed out, these names are incorrectly formed. "Amylolytic," in analogy with "electrolytic," should mean a decomposition by means of starch, and, to avoid this confusion, Prof. Armstrong advises the use of the termination "-clastic" instead of "-lytic" in speaking of enzyme action.

After this digression I now return to the discussion of some recent work on the synthetic action of enzymes. It will be noticed that, in the list of cases given above, isolactose and not lactose is stated to be formed by the action of lactase on glucose and galactose. Accordingly, the sugar synthesised by the enzyme is not the one which it hydrolyses, but the optical isomer of the latter. Similarly, Croft Hill found that maltose is only produced in small amount by the action of maltase on glucose, and that a new disaccharide (revertose) makes its appearance. Frankland Armstrong states that this revertose is identical with isomaltose, as Emmerling had also concluded. Frankland Armstrong is, indeed, of the opinion that the rule is that with a particular enzyme that body is synthesised which is *incapable* of being hydrolysed back by the enzyme in question.

It will be obvious that this conception involves many difficulties. To follow the evidence on which it rests, and which consists in the mode of action of maltase and emulsin respectively on glucose, a brief explanation of the stereochemical relations of the glucosides is necessary. Glucose itself is now looked upon as having the structure of a glucoside—*i.e.* the internal anhydride structure of a lactone. It therefore exists in two optically isomeric forms known as a- and β-glucose. In water these two forms, which differ in their rotatory power, exist together in equilibrium; according to Tanret, in 10 per cent. solution there is 3·7 per cent. of the a-form and 6·3 per cent. of the β-form. When a solution of glucose in methyl-alcohol, in which also it exists in both a and β forms, is acted on by hydrochloric acid, two stereo-isomeric methyl-glucosides are formed, corresponding to the a- and β-glucoses. Of these the a-compound alone is hydrolysed by the maltase of yeast, and the β-compound alone by emulsin—the enzyme found so frequently in the higher plants, and which hydrolyses so many of the natural glucosides, such as amygdalin. These latter, then, naturally must be regarded as β-glucosides, while maltose (glucose-glucoside) is an a-glucoside; isomaltose is presumably the glucose-β-glucoside.

If we take a solution of glucose, therefore, and act upon it by maltose, we should expect, from all that we know of reversible reactions, that a certain amount of maltose would be synthesised; and this, in fact, is what Croft Hill originally believed to take place. Later he found that only a small part of the bi-hexose consisted of maltose, the remainder being, as he thought, a new sugar, revertose. Moreover, as already mentioned, Frankland Armstrong thinks it doubtful whether maltose is produced at all by the action of maltase, and that revertose is really isomaltose. Emulsin, on the other hand, which does not hydrolyse maltose, does produce this sugar, though apparently in very small quantity.

If we accept this point of view, we are landed in many difficulties, as well as having to reject many experimental results of other observers. Croft Hill, for example, found that the same equilibrium point was reached whether he started from glucose or from a mixture consisting chiefly of maltose—viz. for a 40 per cent. solution, equilibrium was arrived at when the percentage of maltose was 15 and that of glucose 85. This would seem impossible if the synthetic product were isomaltose, on which the enzyme present has no action. Since there is no agent capable of hydrolysing the isomaltose formed, there is no reason why the glucose should not be completely transformed to this disaccharide. Again, as Croft Hill showed, the relative percentages of glucose and "maltose" varied according to the concentration of the original solution taken; in a 10 per cent. solution, for example, the percentage of maltose is only 5 at the equilibrium point. If, therefore, we were to take a 40 per cent. solution of glucose and act upon it with maltase until the 15 per cent. of the synthetic disaccharide has been produced and then dilute to four times its volume, if the synthetic sugar formed were isomaltose it should remain unchanged, whereas, if maltose, it would, of course, be hydrolysed back to glucose by the enzyme present. This experiment has been performed by Croft Hill with the result of hydrolysis back to glucose, according to the latter of the two above alternatives.

The results of Visser, who found that emulsin synthesised salicin, the same glucoside which it hydrolyses, and that invertase is able to synthesise cane-sugar, are also at variance with Frankland Armstrong's results.

It must be admitted, therefore, that further evidence is necessary before we can accept this point of view of the synthesis by enzymes of bodies which they do not hydrolyse.

Croft Hill seems to suggest the possibility of more than one enzyme being present in the preparations used. This suggestion must be kept in mind, especially as the methods used by Frankland Armstrong depend on the destruction of glucose or maltose by fermentation with living yeasts. The importance of the synthetic aspect of enzyme action has been pointed out by Croft Hill. The various enzymes present in living cells may indeed be responsible for the building-up processes taking place therein. Take such a case as the formation of starch from sugar in the green leaf. Even supposing that less than 99 per cent. of the glucose or maltose produced in photosynthesis is converted by the intracellular amylase into starch, the fact that starch is insoluble and is at once deposited out of the reacting system renders possible a considerable amount of conversion in a reasonably short time.

Before passing on to the subject of the next section it is advisable to refer to another kind of equilibrium, which is of a somewhat more complex nature than that already discussed, and which is met with in certain instances, especially where the products of the reaction have a specific retarding action on the enzyme. Such a case is that of amygdalin and emulsin, as investigated by Tammann. These cases differ from what may be called "true" chemical equilibrium in that the enzyme itself forms part of the system when in equilibrium. This is shown by the fact that addition of more enzyme to a solution which has arrived at a stationary condition causes further progress of the reaction. It is not correct to call these cases "false" equilibrium, since they show all the properties of a true equilibrium, such as alteration by adding more substrate, removing products of reaction, dilution or concentration or change of temperature, with also that characteristic already referred to, viz. additional change by adding more enzyme. It is in such instances that the question of compounds between enzyme and substrate or products of reaction comes more prominently under our notice but this aspect of the phenomena will be discussed later.

VELOCITY OF REACTION

Practically, all enzyme actions with which we have to deal are unimolecular, or, in other words, their velocity may be represented by the rate of change of one molecule into two or more. The great majority of them, in fact, consist in the splitting up of a molecule by introducing into it the elements of water.

To simplify matters we will commence by neglecting the

reversible nature of these reactions and assume that they proceed to completion.

In such cases the rate at which the change proceeds will be, at any moment, proportional to the amount of unaltered substrate present at that moment. A simple application of the calculus shows us that the time course of the reaction will therefore form a logarithmic curve, when put into the graphic form.

What is the nature of the corresponding curve when enzyme actions are concerned? Several cases have already been investigated; but it can only be regarded as somewhat unfortunate that the first of such reactions to be completely analysed, chiefly by Victor Henri, turns out to be an anomalous one. I refer to the action of invertase on cane-sugar. When the values of the velocity constant in this instance are calculated by application of the unimolecular formula to the experimental data, it is found that the values so found steadily *increase*, instead of remaining the same all through. In nearly all other examples worked out up to the present time the corresponding values steadily *decrease*. Such are maltase and lactase (E. F. Armstrong), emulsin (Tammann), lipase (Kastle), and trypsin (by myself). Amylase is stated by some to follow the same law as invertase.

In the greater number of cases there are, then, some influences at work which either diminish the amount of enzyme acting or in some way make it less active. The chief cause seems to be the increasing concentration of the products of reaction, since by adding these, or certain of them, it is found that the reaction is slowed. There are several possible ways in which this effect may be produced. In the first place there is some evidence that there is a kind of combination taking place between the enzyme and certain of the products of its action; this would result in a part of the enzyme being withdrawn from participation in the reaction. In the second place, as insisted on by Moore, the reversible nature of these reactions has not been sufficiently taken into account; even where the reaction appears to proceed to completion it is probable that this means that the equilibrium point is very near one end, and the tendency for the reaction to run in the opposite direction will show itself more and more as this equilibrium point is approached. In the cases of invertase and emulsin investigated by Visser from this point of view, it was found that the occurrence of the reverse reaction was capable of accounting for the greater part of the deviation from the unimolecular logarithmic law, so that much more regular values were found for the velocity constant when calculated from expressions including terms taking this reverse reaction into account. It did not seem, nevertheless, that the whole of the falling-off from the logarithmic curve could be explained in this way. Visser introduces the idea of "intensity" of action of the enzyme which is in some way diminished as the reaction proceeds. In some instances, in which the enzyme is one sensitive to changes of chemical reaction, it seems possible that the influence of the products of reaction may be partly accounted for in this manner. Trypsin is very sensitive to alkali, and the various amino-acids formed by its action may, by combining with alkali, diminish the intensity of the tryptic action. This is a point requiring investigation.

The consideration just mentioned leads us back to the peculiarity of the velocity of reaction of invertase. This is found neither to proceed in accordance with the strict logarithmic law nor to fall away from it, but, on the contrary, to become more rapid than the law requires. In other words, the activity of the enzyme appears to be *increased* as the reaction proceeds. Two suggestions have been made to account for the similar state of affairs when cane-sugar is inverted simply by the action of water at 100°. Kullgren has shown that an acid is produced, and as we know that the reaction is catalysed by H^+ ions, this appears sufficient to account for the experimental results. Now it is possible that the hydrolysis produced by invertase may also be accompanied by the formation of acid, and, as the activity of this enzyme is increased considerably by H^+ ions in moderate concentration, the explanation of Victor Henri's results may be found here. There is also another possibility suggested by Mellor and Bradshaw in connection with inversion of cane-sugar by water at 100°—namely, that the glucose and fructose as first formed may be in those modifications which have the higher rotatory powers, and that only at a later period, when equilibrium is attained, is the normal rotation of glucose arrived at. If this is the case with invertase action, it is obvious that if readings are taken of the change as it progresses in the tube of the polarimeter itself, these readings will indicate a higher concentration of inverted sugar than really exists. Both of the possibilities mentioned are capable of experimental test.

On the whole, then, we may take it that when proper account is taken of the reversible nature of enzyme action and of the action of some of the products of the reaction on the activity of the enzyme itself (positive or negative auto-catalysis), the unimolecular logarithmic formula, deduced from the law of mass action, satisfactorily applies to the phenomena in question.

When we proceed to investigate the rate of change more in detail, we find that this simple law does not apply to the initial and final stages of the reaction—at all events, in the cases which have been investigated up to the present time. (I refer to those of lactase and maltase as investigated by Frankland Armstrong, and of trypsin as investigated by myself.) Although the greater part of the reaction follows the logarithmic law, with the qualifications above referred to, the initial stage and the final stage follow a linear course. In other words, it is found that equal amounts of the substrate are hydrolysed in equal times, notwithstanding the change in its concentration which is taking place. As Frankland Armstrong explains this result, it is due to the combination which takes place between the enzyme and the substrate as a necessary preliminary to the hydrolysis of the latter. As to the nature of this combination and the evidence for its existence, more will be said later; for the present we may assume that some kind of intimate association does take place. When, therefore, at the commencement of the reaction, the concentration of the substrate is considerably in excess of that of the enzyme, practically the whole of the latter is in combination with substrate; and since it will not be set free to attack a further quantity of substrate until the first is hydrolysed, it is obvious that as long as the substrate is in considerable excess equal amounts of it will be split in equal times. As the concentration of the substrate diminishes, the amount at any time in active association with enzyme will be proportional to this concentration, so that the rate of change will follow the law of mass action. In the final stage the amount of change is again equal in equal times, due now to the enzyme being in excess, so that it is able to attack the whole of the substrate available.

Moore has pointed out that, by taking into consideration the reversibility of the reaction and also the alteration of the intensity of the action of the enzyme, due to some effect of certain products of the reaction, it is possible to frame a mathematical expression which not only includes the above three stages of the hydrolytic part of the reaction, but also a stage of zero change at the equilibrium point and a reversed velocity beyond this point.¯ The expression is undoubtedly a complex one, although capable of simplification in particular cases. It is, moreover, satisfactory to find that such an equation can be formed on the basis of experimental data.

Concentration of Enzyme.

When we come to consider the various influences that are capable of changing the rate of a reaction as catalysed by enzymes, the first point that suggests itself is, how is this velocity related to the amount of enzyme present? We have already seen that in the case of inorganic catalysts it has been shown that the final equilibrium point is unaffected by the amount of catalyst present. The time taken to reach this point is alone affected. With regard to enzymes, it has been shown by Croft Hill that the same law applies to the case of maltase, and by Visser that it applies to invertase and emulsin.

Various statements have been made from time to time with respect to the relation of the concentration of enzyme to the rate of the change produced by it. Indeed, when we consider the different states of affairs in the three stages of the reaction, we understand the cause of the disagreement amongst observers as to this question. The relation cannot, in fact, be the same at all stages of the reaction. In the first stage, so long as the enzyme is in small proportion to the substrate, the rate of change is in direct proportion to the amount of enzyme present; in the second (logarithmic) stage the ratio is some exponential function of the enzyme concentration, so that when this is doubled, for example, the rate of change is not also doubled, but is usually multiplied by about $2^{\frac{1}{5}}$ to $2^{\frac{1}{2}}$; whereas in the last stage, in which the enzyme is in excess, it is clear that the rate of change will be independent of the amount of enzyme present. These facts come out quite unmistakably in the cases in which they have been looked for, such as lactase (E. F. Armstrong, trypsin (the present author), and others. The bearing of the exponential function of the second stage on the nature of the combination˙ between enzyme and substrate will be seen later.

Effect of Temperature

Increase of temperature is well known to have a considerable effect on the rate of ordinary chemical reactions, the usual factor being between 2 and 4 for 10°: for example, the velocity at 40° is from 2 to 4 times that at 30°. In the case of enzymes the corresponding factor has been determined for emulsin by Tammann to be 7·14 between 60 and 70°, for trypsin by myself to be 5·3 between 20 and 30°, and for the catalase of blood by Senter to be 1·5 between 0 and 10° C. But whereas the curve expressing this fact becomes steeper as the temperature rises in the case of ordinary chemical reactions, it is found that enzymes show a phenomenon known as the "optimum" temperature— viz. a particular temperature at which during a given time a greater amount of substrate is acted upon than at any temperature either above or below this. To understand the meaning of the "optimum" it is sufficient to bear in mind two facts— viz. that enzymes are rapidly destroyed at high temperatures, probably because of their colloidal character; and that, in order to determine the degree of their activity, it is necessary to allow the action to proceed for a considerable time at the raised temperature. This being so, we are unable to observe directly the rate of change at the beginning of the action at any new temperature, and at the higher temperatures, before sufficient change has occurred to enable measurements to be made, a considerable part of the enzyme has been destroyed. This is shown very distinctly by the observations of Frost Blackman on the respiration of plants. If curves are drawn expressing the evolution of carbon dioxide at various temperatures and at various times after the commencement of exposure to these temperatures, and these curves are continued so as to show the rate at the first moment of action of the particular temperature, it is found that the points so fixed lie on a curve exactly like that of Van 't Hoff for ordinary chemical reactions, so that the higher the temperature the greater the initial rate of change, and the optimum temperature is merely an expression of the fact that at a certain temperature the increased velocity due to this raised temperature is more than sufficient to counteract the rapid destruction of the enzyme. It follows also from these experiments that the apparent optimum temperature will vary considerably, according to the time which has elapsed between the commencement of the exposure to the temperature and the period at which the observation is made.

The fact that during the brief life of an enzyme at a high temperature its activity is so enormously increased makes caution necessary in experimental work. It is sometimes the practice, in order to stop the action of an enzyme at a given moment, to raise the temperature of the reacting mixture as rapidly as possible to 100° or thereabouts. This cannot be done instantaneously, and during the necessary interval of time the enzyme becomes exceedingly active, so that considerable further change may take place. That such change, in point of fact, does occur, was found by Delezenne in the case of papain, and by myself in the case of trypsin. This method of stopping further action is, therefore, inadmissible in accurate investigations. It is better to freeze solid as rapidly as possible, or, when such admixture is immaterial to subsequent work, to add some substance which stops the action of the enzyme, such as alkali to solutions containing pepsin or invertase.

From the point of view of the theory of enzyme actions, the large temperature coefficient has some importance. As pointed out by Senter, it shows that the interpretation of the form of the velocity curve as due to diffusion in a heterogeneous system, on the lines of the theory of Nernst, does not agree with the experimental facts. The temperature coefficient of diffusion processes is low, whereas that of enzyme actions is unusually high.

Accelerators and Retarders

Of Ehrlich's two classes of bodies acting on the living cell enzymes belong to that one which consists of bodies of high molecular weight, like the foodstuffs and bacterial toxins, which are directly assimilated by the protoplasm and built up into its giant molecule. The other class—to which drugs belong, as well as the various chemical messengers, or hormones, produced by the organism—are of simpler chemical constitution, and act by the physico-chemical characters of their molecule as a whole. The first class of bodies alone gives rise, when injected into the living organism, to the production of antagonistic bodies—anti-toxins or anti-enzymes, etc. Of these latter several are known to be normally present in the blood, such as anti-trypsin and anti-rennet. Others have been produced by hypodermic injection of the respective enzymes.   Anti-trypsin has also been shown by Weinland to exist in the bodies of intestinal worms, which are thus protected from the action of the pancreatic juice. It is stated by this observer that the intestinal epithelium also contains a similar substance; but recent work by Hamill has failed to confirm this statement, although the mucous membrane of the stomach was found to contain an anti-pepsin, as affirmed by Weinland.   These anti-bodies appear not to destroy the enzyme, since a mixture of trypsin and anti-trypsin, at first inactive, slowly recovers its proteoclastic power, apparently by gradual disappearance of the anti-body.   The mode of combination of toxin and anti-toxin has been shown by Craw to be of the nature of adsorption, the characteristic of which process is that it partakes of some of the properties of physical union and some of those of chemical combination—it is, in fact, as Ostwald puts it, a combination in varying proportion.   The dyeing of cotton by Congo-red is such a process.   If two pieces of cotton of the same size are dyed in solutions of Congo-red, one of which is twice the strength of the other, it is found that the cotton does not take up equal quantities of dye from both, as it would if a true chemical compound were formed, nor does it take up twice the amount from the stronger solution, as it would if the process were purely physical, such as solid solution; it is found, in fact, that relatively more is taken up from the more dilute solution; so that the amount taken up from that of twice the strength is not double that taken up from the weaker, but less than this— indeed, $\times 2^{\frac{1}{x}}$, where x is greater than 1 and usually less than 2.

It must suffice here to refer to the action of various electrolytes, especially that of OH^- and H^+ ions, to which various enzymes are differently sensitive.   Pepsin is paralysed by OH^- ions, while being increased in activity by H^+ ions.   It appears probable, from the researches of the present writer, that the action of neutral salts may chiefly consist in facilitating or otherwise the mutual adsorption of the enzyme and substrate.

The enzymes connected with oxidation processes are extremely sensitive to traces of certain metallic salts.   Laccase has been shown by Bertrand to be considerably increased in activity by the presence of manganese.   A similar relation of tyrosinase to iron has been pointed out by Miss Durham.   It is not certain whether these enzymes are unable to exert this activity in the absence of the metals referred to; if this is the case, manganese and iron would be for them "co-enzymes," to which we may now turn our attention.

CO-ENZYMES

It was noticed by Magnus that an extract of liver containing lipase became inactive on dialysis. On investigation it was found that the activity was restored by adding a portion of a boiled extract of liver, or even a similar extract after proteins had been precipitated by uranyl-acetate. The activating body is soluble in absolute alcohol but not in ether, and is not present in the ash of liver. It is, therefore, a non-colloidal substance, but not inorganic. What may be called, then, the lipoclastic system of the liver consists of two components—the one, destroyed by boiling, may be regarded as the enzyme proper, but it is inactive except in the presence of the dialysable body, the "co-enzyme."[1]

A similar state of affairs has been shown by Harden and Young to exist in the case of the alcoholic enzyme of yeast, zymase. The press juice of yeast may be separated by dialysis, or by filtration through gelatin, into two parts, each by itself inactive, but becoming active when mixed. The filtrate or dialysate always contains phosphates, and it was found that the addition of phosphate to the inactive residue had a similar activating effect to that of the filtrate itself. A remarkable fact is that the amount of carbon dioxide (and alcohol) produced is proportional to the amount of phosphate added—viz. one molecule of carbon dioxide for each atom of phosphorus. The authors are not prepared as yet to state whether phosphate is the only co-enzyme concerned.

Certain experiments of Cohnheim indicate that the pancreas produces a co-enzyme for the glycolytic oxidase of muscle. The difficulty in the investigation of this phenomenon is that the pancreatic extract in 90 per cent. alcohol, which was added to watery extracts of muscle, caused an inhibition of sugar oxidation, if present even in slight excess. This may, perhaps, be the reason why other observers have not been able to confirm Cohnheim's results. The matter cannot be looked upon as decided at present, but it appears from Cohnheim's

[1] According to recent work (Loevenhart, "Proc. Am. Physiolog. Soc.," in *Amer. Journ. of Physiology*, vol. xv. p. xxvii. 1906), the co-enzyme in this case is bile salts.

data that there is something in his pancreatic extracts which considerably increases the destruction of sugar in muscle extracts, when added to these in a certain proportion.

ZYMOGENS

Since all enzymes are products of cell activity, it follows that, in all probability, some intermediate stage of their formation from cell protoplasm may be separated from the cells producing them. Such a body is the mother-substance of pepsin, prepared from the gastric mucous membrane by Langley and Edkins. This "pepsinogen" is itself inactive, but is transformed into active pepsin by the action of acid. The similar body formed by the pancreatic cells (trypsinogen) is secreted as such, and remains inactive until it comes into contact with a specific enzyme (enterokinase), found by Pawlow and his pupils in the small intestine. This enterokinase is secreted by the epithelium of the small intestine, and converts the trypsinogen into active trypsin.

These zymogens must be carefully distinguished from the enzymes above mentioned as being inactive apart from their respective co-enzymes. In this latter case the process of activation is a reversible one, in the sense that an active system can be rendered inactive by removing the co-enzyme and restored to activity by replacing the latter. An active enzyme, however, once produced from its zymogen, cannot be reconverted into its precursor; the process appears to be one of hydrolysis, by which a new body is formed. The system of inactive enzyme and co-enzyme is rather an association of two different bodies, probably of the nature of adsorption.

CHEMICAL AND PHYSICAL PROPERTIES

Enzymes, as mentioned already, are colloids, and therefore exhibit the characteristic properties of this state of matter. Such are indiffusibility, precipitation by heat, greater or less instability, and probably an electric charge of their suspended particles.

It is impossible to make any definite statement as to their chemical nature, except of a negative character. The really active subtance is present in the usual preparations in so small an amount that it appears almost hopeless to discover its chemical constitution. As colloids they readily carry along with them, in a state of adsorption, constituents of the solutions from which they are precipitated. It is not surprising, then, to find that proteoclastic enzymes show certain reactions of proteins, invertase apparently contains carbohydrate, and so on. But when the solutions are more and more carefully purified it is found that, although powerfully active preparations can be obtained, these solutions show fewer and fewer characteristic reactions. Pekelharing's pepsin, for instance, showed only a minority of the usual protein tests, and was found to contain no phosphorus. This last fact definitely excludes the view taken by some that enzymes in general are nucleo-proteids.

At the same time it seems, from evidence to be mentioned immediately, that there is some very close similarity between a particular enzyme and its substrate, so that, when it is found that invertase cannot be prepared free from mannose, it may well be that this sugar is a necessary part of its molecule.

The facts referred to in the earlier pages of this article as to the relations between certain enzymes and optically isomeric substrates led Emil Fischer to put forward his famous simile of lock and key relationship.

In connection with this possible similarity of chemical structure it is of interest that the statement has recently been made that lipase is soluble in ether, and therefore probably of a fatty nature. The lipase of the cytoplasm of castor-oil seeds has also been shown by Nicloux to be destroyed by contact with water.

Mode of Action

It is evident from what has already been said that there is a very intimate association between an enzyme and its appropriate substrate. A few further facts pointing to some kind of combination may be mentioned here. Enzymes in presence of substrate or products of reaction are considerably less sensitive to the action of heat. It appears also that a particular enzyme has a special affinity for certain constituents of the substrate, as, for instance, invertase for fructose (E. F. Armstrong), trypsin (according to Emil Fischer) for tyrosin, leucin and, to a less degree, alanin. Now these bodies are precisely those which exercise the greatest inhibitory power on the velocity of the (hydrolytic) reaction ; in fact, glucose does not affect the action of invertase, whereas fructose has a considerable action of this kind. If an enzyme accelerates the reverse reaction, as well as the usual hydrolysis or splitting-up, it is easy to understand why it is capable of combining with products of reaction in addition to combining with the unaltered substrate; if we suppose that in order to exert its catalytic influence it must enter into some such intimate connection with the bodies to be acted upon, such "compounds" must be formed between enzyme and products before synthesis of the latter will occur.

Enzymes, being colloids, are particularly prone to form adsorption compounds, so that we naturally look for evidence as to the nature of the compounds in question. The fact that, at all events in the logarithmic stage of the reaction, increasing the concentration of the enzyme to say twice its value does not double the reaction velocity, but something less than this, is considerable evidence in this direction. A relationship of this nature is characteristic of adsorption phenomena. It also tends to confirm the view of Victor Henri that the enzyme is shared by the substrate and the water present in such a manner that the proportion taken up by the former is greater the lower the concentration of the enzyme. If we put: $a =$ the quantity of enzyme adsorbed from a solution of concentration $= 1$, and $b =$ that adsorbed from a similar solution of concentration $= 2$, then $a = b \times 2^{\frac{1}{x}}$, in which x is what is called the "proportionality factor," and is always greater than unity, which value it would have if the process were a purely physical one, such as solid solution, for example. The higher the value of x the more nearly the process approximates to a chemical combination, in which case it is obvious that $a = b$, and therefore $2^{\frac{1}{x}}$ in the above expression becomes unity or $x =$ infinity. In the case of trypsin x varies from 1·5 to 1·67, and according to Schütz and Borissow's "law of squares"—which is, however, merely a special case—$x = 2$. I find in some experiments made recently that in the case of Congo-red and paper x is usually less than 2, about 1·6 in fact, but it varies considerably according to the amount of electrolyte present. It would appear, then, that the union of enzyme and substrate follows a similar law to that of ordinary adsorption phenomena.

A particularly interesting case as regards the question before us is that of the lipase of the liver as investigated by Dakin.

It was found that this enzyme, acting on the optically inactive mandelic esters, hydrolysed more rapidly the dextro component than the lævo one. In this way the percentage of dextro-mandelic acid in the products of the reaction was greater at first than that of the lævo-mandelic acid; as the reaction proceeded the relative amounts became closer, until finally both were present in equal quantity and the acid was optically inactive. These facts can only be satisfactorily explained on the hypo-thesis that the enzyme is itself optically active and forms addition compounds with the ester. In the words of Dakin, "The dextro and lævo components of the inactive ester first combine with the enzyme, but the latter is assumed to be an optically active asymmetric substance, so that the rates of combination of the enzyme with the d. and l. esters are different. The second stage in the reaction consists in the hydrolysis of the complex molecules of (enzyme + ester). Since the complex molecule (enzyme + d. ester) would not be the optical opposite of (enzyme + l. ester), the rate of change in the two cases would again be different. Judging by analogy with other reactions one might anticipate that the complex molecule which is formed with the greater velocity would be more rapidly decomposed. In the present case it would appear that the dextro component of the inactive mandelic ester combines more readily than the lævo component with the enzyme, and that the complex molecules (d. ester + enzyme) are hydrolysed more rapidly than (l. ester + enzyme), so that if the hydrolysis be incomplete dextro acid is found in solution and the residual ester is lævo-rotatory."

We have as yet no definite information as to what happens after the combination between enzyme and substrate has taken place. In the case of inorganic catalysts the most acceptable explanation is that the intermediate compound is split up again with the formation of the particular products of the reaction and the liberation of the catalyst in its original form. The intermediate formation of ethyl-sulphuric acid in the production of ether and that of nitrosyl-sulphuric acid in the old chamber process of manufacture of sulphuric acid may be given as illustrations. Practical experience has found that in the latter process the nitric acid gradually disappears in the form of bye-products, so that the catalyst is not completely restored. This fact is of importance in connection with the view taken by some investigators that certain bodies, such as Buchner's zymase, are not enzymes, since there is a proportionality between the amount of enzyme and the total amount of change produced. This fact is probably to be explained by the disappearance of the enzyme during the reaction. A similar state of affairs is, indeed, to be seen in the case of trypsin, though in a less marked degree. It is found that small quantities of this enzyme will not effect the same amount of hydrolysis as larger quantities, even when allowed a very long time for the action. Experiment shows that trypsin does disappear slowly even in the presence of substrate or products.

Ostwald has insisted that, in order that the formation of an intermediate compound should be regarded as an adequate explanation, it is necessary to show that the two reactions, formation of intermediate compound and splitting-up of this compound, taken together progress at a velocity greater than that of the reaction without the presence of the catalyst. This proof has actually been afforded by Brode. When hydrogen peroxide acts upon hydriodic acid, this latter is decomposed at a certain rate, which rate is greatly accelerated by the presence of a trace of molybdic acid. Now Brode has been able to show that permolybdic acids are formed by action of the peroxide on molybdic acid and that these permolybdic acids, themselves produced at a considerable velocity, react on the hydriodic acid with very great velocity, molybdic acid being formed again, ready for a further activity. These two reactions, then, together progress at a greater rate than the action of the peroxide by itself upon the hydriodic acid.

When we consider the extraordinarily minute concentration in which enzymes exert their activity, there seems some justification for such a view as that of Arthus and de Jager—viz. that the properties known as those of enzymes are not associated with definite chemical individuals, but may be conferred, to a greater or less degree, on various kinds of bodies. In the process of purification of certain enzymes solutions have been obtained which, while very active, contain only an infinitesimal amount of solid matter. On the other hand Brode has shown that, in the reaction above described, the presence of 1 gram molecule of molybdic acid dissolved in 31,000,000 litres of water is capable of obvious catalytic action. In this case the catalyst is a definite chemical compound.

Space will not allow of the discussion of other theories of

enzyme action, such as the molecular vibration theory of Liebeg, or that of Barendrecht as to the "radiations" emitted by enzymes. I must content myself with calling to mind what is so well insisted on by Bredig—viz. that enzyme solutions are, as colloids, heterogeneous systems, and that it is necessary always to remember that surface-action and adsorption undoubtedly play a considerable part in the reactions taking place therein.

It may perhaps be objected that the view advocated in the preceding pages, that the combination between enzyme and substrate or products is to be looked upon as of the nature of adsorption, does not sufficiently take account of the specific nature of enzyme action. It may be said that caseinogen, for example, would be expected to form adsorption compounds indifferently with either trypsin or amylase. Adsorption, however, is not a purely physical process, and is undoubtedly more or less specific. The "adsorption affinity" of gelatin is considerably greater for acid-fuchsin than for Congo-red. But we know as yet too little about the essential nature of the process to warrant further discussion.

One more experimental result may be mentioned to conclude this section. It was found by Korschun, in investigating the relations between rennet and its anti-body, that by filtration through porous clay a solution of rennet could be separated into fractions which by appropriate dilution of the stronger fractions could be brought to the same strength as regards combination with the anti-body, but which differed considerably in their power of coagulating milk. In other words, the original solution appeared to contain a modified form of the enzyme analogous to Ehrlich's "toxoids"; that is, a part of the enzyme had lost its characteristic activity while still retaining its power of combining with the anti-body. I have myself met with some facts which point to the production of a similar modification of trypsin by warming to about 25° for a day or two; I have suggested calling these modified enzymes "zymoids." The facts afford support to Ehrlich's view that the combining power and the fermentative activity are functions of distinct side-chains.

COMPLEX SYSTEMS

It may be of interest if, in conclusion, I give a brief indication of some recent work on systems in which enzymes play an essential part and which simulate as a whole the properties of enzymes.

It has long been known that certain preparations can be obtained from tissues of both plants and animals which have the power of carrying active oxygen to certain oxidisable substances, such as guaiaconic acid, salicylic aldehyde, tyrosin, etc., and thereby causing the oxidation of these substances. These various phenomena were very obscure until the researches of Bach and Chodat threw light upon them. These observers show that there are three distinct classes of bodies involved. Firstly, organic peroxides of the type of substituted hydrogen peroxide; these are not enzymes, although they are decomposed by boiling. Secondly, peroxydases, enzymes which act upon the peroxides and produce active oxygen, which is capable of oxidising certain bodies when these are present. The system, peroxide and peroxydase, forms the oxidising agent, or oxydase as it was originally called. In addition to these two classes there is a third, also of enzyme nature, viz. the catalases, which decompose hydrogen peroxide with evolution of inactive or molecular oxygen and appear to act only on this particular peroxide. These are widespread in both plants and animals, but their function is not altogether clear, with the exception of one important case to be mentioned immediately. The "hæmase" of blood investigated by Senter is one of these catalases. Although the point of view taken by Bach and Chodat undoubtedly tends to clear up much obscurity, it does not explain all the facts known. The markedly specific nature of certain "oxydases," such as tyrosinase and aldehydase, is difficult to understand, and the mechanism by which the peroxide takes up atmospheric oxygen after being decomposed by the peroxydase, and thus reconstitutes itself, is as yet unknown.

Another complex system is that responsible for the activity of the green leaf in the forming of starch and oxygen from carbon dioxide and water. Important advances have recently been made in this region by Ussher and Priestley. They show that the system concerned consists of three partners—the protoplasm of the chlorophyll corpuscle, the chlorophyll itself, and a catalase. By means of the pigment, acting as both chemical and optical sensitiser, light energy is employed to cause reaction between carbon dioxide and water in such a manner that formaldehyde and hydrogen peroxide are formed. Now both these bodies are poisonous and, if allowed to accumulate, the reaction would soon come to an end. The formaldehyde is, however, rapidly polymerised by the protoplasm of the chloroplast, and the hydrogen peroxide is split up into oxygen and water by the catalase. We see then why the reaction as a whole does not occur in non-living preparations or extracts of green leaves. Formaldehyde is, indeed, produced in the presence of chloroform on exposure to light, but since no polymerisation occurs the chlorophyll is destroyed by it, and no further reaction is possible. The production of hydrogen peroxide and formaldehyde even takes place in light in leaves killed by boiling, and in this case, since the enzyme (catalase) is destroyed, as well as the protoplasm, the hydrogen peroxide also contributes to the destruction of the chlorophyll.

CONCLUSION

The living organism is enabled by the use of enzymes to bring about, under ordinary conditions of temperature and moderate concentrations of acid or alkali, many chemical reactions which would otherwise necessitate high temperature or powerful reagents. These enzymes are catalysts of a colloidal nature, and obey the usual laws of catalytic phenomena. Certain properties in which they differ from most inorganic catalysts are to be explained by this colloidal condition. One such property is destruction by heat and comparative instability as the temperature rises, thus affording an explanation of the so-called optimum temperature. The facility with which additive or adsorption compounds are formed with substrate or products is also due to the colloidal character of these enzymes.

LITERATURE

ARMSTRONG (E. Frankland), The Rate of the Change conditioned by Sucroclastic Enzymes, *Proc. Roy. Soc.* 73, p. 500, 1904.
—— The Influence of the Products of Change on the Rate of Change conditioned by Sucroclastic Enzymes, *Proc. Roy. Soc.* 73, p. 516, 1904.
—— Hydrolysis of Isomeric Glucosides, *Proc. Roy. Soc.* 74, p. 188, 1904.
—— The Synthetic Action of Acids contrasted with that of Enzymes, *Proc. Roy. Soc.* B. 76, p. 592, 1905.
ARMSTRONG (H. E.), The Terminology of Hydrolysis, especially as affected by Ferments, *Chem. Soc. Trans.* 1890, p. 528. Also footnote, *Proc. Roy. Soc.* 73, P. 500, 1904.

BACH and CHODAT, Various papers in *Ber. d. deutsch. Chem. Ges.* 36, 1903, and subsequently. Summary in *Arch. d. Sciences Phys. et Nat.* Geneva, tome xvii. p. 477, 1904.

BAYLISS, The Kinetics of Tryptic Action, *Arch. d. Sciences Biologiques*, tome xi. supplement, p. 261, St. Petersburg, 1904.

—— On Some Aspects of Adsorption Phenomena, *Biochem. Journal*, i. p. 175, Liverpool, 1906.

BAYLISS and STARLING, The Proteolytic Activities of the Pancreatic Juice, *Journal of Physiology*, xxx. p. 61, 1903.

BERTRAND, Sur l'intervention du Manganese dans les Oxydations Provoquées par la Laccase, *Compt. Rendus*, 124, p. 1032, 1897.

BLACKMAN (F. F.) Optima and Limiting Factors, *Annals of Botany*, 19, p. 281, 1905.

BREDIG, Die Elemente d. Chem. Kinetik mit Besond. Berücks. d. Katalyse u. d. Ferment-wirkung, *Ergebn. d. Physiologie*, I. i. p. 134, 1902.

BRODE, *Zeitsch. f. physik. Chemie*, 37, p. 257, 1901.

COHNHEIM, *Zeitsch. f. physiolog. Chemie*, 39, p. 396, 1903; 42, p. 401, 1904; and 47, p. 253, 1906.

COLE, *Journal of Physiology*, xxx. pp. 202 and 281, 1904.

CRAW, Physical Chemistry of the Toxin-Antitoxin Reaction, *Proc. Roy. Soc.* B. 76, p. 179, 1905.

DAKIN, Action of Lipase, *Journ. of Physiology*, xxx. p. 253, 1904.

DELEZENNE, On Papain, *C. R. Soc. de Biologie*, 1906.

DUCLAUX, *Microbiologie*, Paris, 1898.

DURHAM (F. M), Tyrosinase in Skin, *Proc. Roy. Soc.* 74, p. 310, 1904.

EMMERLING, Synthesis of Amygdalin, *Berichte d. deutsch. Chem. Ges.* 34, p. 3810, 1901.

FISCHER (EMIL), (Lock and Key Simile) *Ber. d. deutsch. Chem. Ges.* 27, p. 2992.

FISCHER und ABDERHALDEN, Ueber d. Verhalten verschied. Polypeptide Gegen Pankreassaft u.s.w. *Zeitsch. f. physiolog. Chemie*, 46, p. 52, 1905.

—— Ueber d. Verdauung einiger Eiweisskörper durch Pankreasfermente, *ibid.* 39, p. 81, 1903.

FISCHER und ARMSTRONG, Formation of Isolactose by Lactase, *Ber. d. deutsch. Chem. Ges.* 35, p. 3151, 1902.

FISCHER und BERGELL, *Ber. d. deutsch. Chem. Ges.* 36, p. 2503, 1903.

HARDEN and YOUNG, The Alcoholic Ferment of Yeast-Juice, *Proc. Roy. Soc.* 77 B. p. 405, 1906.

HENRI, VICTOR, Ueber d. Gesetz d. Wirkung d. Invertins, *Zeitsch. f. physik. Chemie*, 39, p. 194, 1901.

—— *Lois générales de l'Action des Diastases*, Paris, Hermann, 1903.

HILL, CROFT, *Journ. of Chem. Soc. Lond.* vol. 73, p. 634, 1898; *ibid.* 83, p. 578, 1903; and *Ber. d. deutsch. Chem. Ges.* 34, p. 1380, 1901.

HOFF, VAN 'T, *Vorlesungen über theor. und physik. Chemie*, i. 99, 2 Aufl. 1901.

KASTLE and LOEVENHART, *Amer. Chem. Journal*, 26, p. 533, 1901.

KOELICHEN, *Zeitsch. f. physik. Chemie*, 33, p. 149, 1900.

KORSCHUN, *Zeitsch. f. physiolog. Chemie*, 37, p. 366, 1903.

KÜHNE, (suggestion of name "Enzyme") *Heidelberg Unters.* i. p. 293, 1878.

KULLGREN, *Zeitsch. f. physik. Chemie*, 41, p. 407, 1902.

LANGLEY and EDKINS, *Journ. of Physiology*, vii. p. 371, 1886.

MAGNUS, *Zeitsch. f. physiolog. Chemie*, 42, p. 149, 1904.

MELLOR and BRADSHAW, *Zeitsch. f. physik. Chemie*, 48, p. 353, 1904.

MOORE, Article in *Recent Advances in Physiology*, edited by L. Hill, 1906.

NICLOUX, Series of notes in *Comptes Rendus de l'Académie*, tomes 138 and 139, 1904.

OSTWALD, *Lehrbuch d. All. Chemie*, 2 Aufl. ii. 2, p. 248, 1896-1902.

—— *Ueber Katalyse. Vortrag*, Leipzig, 1902.

PAWLOW and CHEPOWALNIKOW, *Gazette clinique de Botkin*, St. Petersburg, 1900.

PEKELHARING, Ueber eine neue Bereitungsweise d. Pepsins, *Zeitsch. f. physiolog. Chemie*, 22, p. 233, 1896.

SENTER, Das Wasserstoff-superoxyd-zersetzende Enzym d. Blutes, *Zeitsch. f. physik. Chemie*, 44, p. 257, 1903.

—— The Rôle of Diffusion in the Catalysis of Hydrogen Peroxide by Colloidal Platinum, *Proc. Roy. Soc.* 74, p. 566, 1905.

TAMMANN, *Zeitsch. f. physik. Chemie*, 3, p. 35, 1889; 18, p. 436, 1895; *Zeitsch. f. physiolog. Chemie*, 16, p. 281, 1892.

TURBABA, *Zeitsch. f. physik. Chemie*, 38, p. 505, 1901.

USSHER and PRIESTLEY, *Proc. Roy. Soc.* 77 B. p. 369, 1906.

VISSER, *Zeitsch. f. physik. Chemie*, 52, p. 257, 1905.

WEINLAND, *Zeitsch. f. Biologie*, 44, p. 1, 1903.

THE PROGRESS OF BOTANICAL AND AGRICULTURAL SCIENCE IN CEYLON

By J. C. WILLIS, Sc.D.

Director, Royal Botanic Gardens, Ceylon

THE tropics contain a vast number of species of plants of every kind, many more, generally speaking, than the temperate zones, and among these there are correspondingly more plants which in some way or other are useful to man.   But prior to the arrival of the European nations in the tropics there was but little, if any, intercourse between the different countries in that zone, and consequently the number of useful plants in any one country was but few.   How small the number could be is now hardly to be realised.   Thus Ceylon, to take a familiar example, owes practically all her useful crops, even rice, the national cultivation, coco-nuts, the commonest crop, and jak (*Artocarpus integrifolia*), the commonest fruit, to introductions from abroad; the only indigenous cultivation of importance is that of cinnamon.   It is difficult now to realise upon what plants the few natives of the island lived prior to the introduction of rice by Wi-jayo, the Sinhalese conqueror, in the sixth century B.C.

Not only were there but few useful plants in any one country, but agriculture, excepting in the Indian countries, was but little practised, and only so far as necessary for the actual subsistence of the people.   The principle was of the simplest—grow all you want, and consume all you grow.   There was but little internal trade in the products of the soil, and no export trade at all. With the arrival in the tropics of Europeans all this was changed. At first the white men—as has quite recently happened upon the West Coast of Africa—settled down at the river-mouths and upon the islands, *i.e.* in the places where means of transport were available, and began to trade.   Before long the insecurity of the traders, and the general inefficiency of the tropical natives, ed little by little to actual conquest of the countries, and with this the doom of the old style of agriculture in them was pronounced.

Once settled in the tropical countries in the capacity of masters, the white races proceeded to investigate the indigenous crops, and to realise that by their command of the means of transport they could introduce into any country many new crops which might thrive and prove useful to the inhabitants. Even the early Portuguese settlers and merchants realised this, and many plants, now common in the Eastern tropics, were introduced by them from the West Indies, and *vice versa*. Among other things which they brought to Ceylon, for example, are the guava (*Psidium guava*), the pine-apple, chillies (*Capsicum* spp.), coffee, the papaw (*Carica papaya*), the Malay apple (*Eugenia Jambos*), the rose-apple (*E. malaccensis*), the dhal or pigeon pea (*Cajanus indicus*), the horse-radish tree (*Moringa pterygosperma*), the source of oil of Ben, cotton, the custard-apple (*Anona reticulata*), and several other plants. All of these they had introduced before 1678.

The next comers to the tropics, the Dutch, continued the work of acclimatisation with great vigour, and first placed it upon a more scientific and practical basis by establishing botanic gardens in many of their tropical colonies. The object of these gardens was, to a very large extent, the introduction and acclimatisation of the useful plants of other countries. A further large number of useful plants were introduced into Ceylon during the rule of the Dutch, and many, both of these and of the plants introduced by the Portuguese, are now so common and so universally cultivated in the island that no one would suppose them to be other than native. As examples, take the papaw, the guava, the pine-apple, and the cassava (*Manihot utilissima*).

Still later came the English. At first they hardly realised the advantages to be gained from botanic gardens, and in Ceylon, for instance, they closed the old Dutch garden, and did not reopen it until 1810 (they having captured Ceylon in 1795), when, on the advice of Sir Joseph Banks, then President of the Royal Society, it was reopened in Colombo, under the charge of William Kerr (after whom is named the genus *Kerria*).

The ostensible object with which the English opened their tropical gardens was as much the investigation of the native flora of the countries in which they were situated, as the introduction of new and useful plants from abroad; and at first, owing to the absence of any central organisation, such as there is now in the Royal Botanic Gardens, Kew, and the absence

of any good method of transferring plants over long distances—
unless, as in a few cases, they would travel as seeds—they were
able to do but little in this latter direction.   The history of the
botanic gardens in Ceylon, which, both in the investigation of
the native flora and in the introduction of plants from abroad,
have been most efficient and successful, may therefore be taken
as typical, and will in general illustrate the history of all.

Started in Colombo in 1810, and afterwards transferred to
Kalutara on the south-west coast, the gardens did not begin their
really useful life till the final transfer was made, under the rule
of Alexander Moon, to their present site at Peradeniya, near
Kandy, in the mountain region in the central province of the
island.   This was effected in 1821, six years after the conquest
of the Kandyan kingdom by the English.   Prior to this period
the only works dealing with the flora of Ceylon were Hermann's
Musæum Zeylanicum (1717), Burmann's *Thesaurus Zeylanicus*
(1737), and Linnæus's *Flora Zeylanica* (1747).   How very incomplete our knowledge was of the flora may be seen from the fact
that the last-named book only contains 429 determined and 228
undetermined species.   Moon set vigorously to work to remedy
this state of affairs, and in 1824 published his *Catalogue of the
Indigenous and Exotic Plants of Ceylon*, which contains 1,127
species, or a trifle over one-third of the total flora of the island.

With the death of Moon, in 1825, work at the garden
languished for about twenty years, the place being under
the rule of " practical " gardeners, and gradually sinking into the
condition of a Government market garden, whose produce was
sold in Kandy.   This phase came to an end at about the same
period as the realisation in England of the fact that there must
be some central institution in that country, if acclimatisation of
plants in the tropics was to be properly attended to.   One
of the first appointments in the tropics made under the guidance
of Sir William Hooker at Kew was that of George Gardner, the
well-known Brazilian traveller, to the superintendence of the
Ceylon Botanic Gardens.   Unfortunately, Gardner died of
apoplexy within a few years, but he had already taken a great
step in advance : the market gardening work was given up, the
gardens, aided by Kew, had set about the introduction of useful
and valuable plants from other countries, and the investigation
of the Ceylon flora was being vigorously pushed on.

Gardner was succeeded by George Henry Kendrick Thwaites,

the discoverer of conjugation in the Diatoms, and author of other good work in microscopic botany. Thwaites was probably the best man who has ever been concerned with any English botanic garden in the tropics, and he devoted himself to his work, without once taking a holiday or leaving the island for a period of thirty-one years (1849-80). Dropping the study of his favourite Cryptogams, he at once began to work at the Ceylon flora, and with such success that during 1858-64 he was able, aided by Sir Joseph Hooker, to bring out the first flora of the island which had any approach to completeness, his *Enumeratio Plantarum Zeylaniæ*. This is not, it is true, a complete flora, even so far as it goes, for it only describes (in Latin) the new species, and is too technical for any but the botanist. Under the species which had already been described elsewhere, nothing is given but native names, and miscellaneous information. But by the publication of this book, the back of the work of preparing a Ceylon flora was broken, and it was merely a matter of further detailed work to prepare a complete flora. Thwaites himself intended to write a popular and complete flora of the island, but was prevented by pressure of work and advancing age.

Not only did Thwaites do yeoman work at the Ceylon flora, but he also devoted himself to the more practical side of his work with conspicuous success. Nearly all the plants which have proved so useful and valuable in Ceylon agriculture, except those previously introduced by the Dutch or Portuguese, were introduced during his tenure of office. We may best illustrate this statement by giving a few conspicuous examples.

The best known is, of course, cinchona, the source of the valuable alkaloid quinine. This was introduced from the Andes of Peru by a special expedition sent thither under the leadership of Mr. (now Sir) Clements Markham, and was established in the hill botanic garden at Hakgala, in Ceylon, in 1861. For a long time it was almost impossible to get any one to trouble about the plants, although they were given away; but as coffee cultivation began to fail, they were taken up, cautiously at first, but with a rush as soon as the profits realised by the pioneers became known. Rapidly the hills of Ceylon were covered with cinchona, and the export rose in a few years to fifteen million pounds, the price of quinine falling from 12s. to 1s. an ounce, and ruining the profitableness of the industry. The export declined as rapidly as it had risen, until now there is only a very trifling quantity sent out of the island.

Other importations made in Thwaites' time were the rubber-trees of South America. These were introduced in 1875-6. Ceãra rubber (*Manihot glaziovii*) became the source of a small industry in the early eighties; but was rapidly given up, as it did not prove profitable, and because tea was then coming in as the standard Ceylon crop. Para rubber (*Hevea brasiliensis*) did not seed so early or so freely, and was very slow in coming into cultivation. Until 1898 seed could only be obtained in small quantity, but since that time the planting has gone on very rapidly, until now there are probably 60,000 acres or more in Ceylon, and an equal quantity in the Federated Malay States, the profits of the early pioneers having proved to be enormous.

The great industry which has arisen in Ceylon in recent years is of course tea. This was introduced by the Botanic Gardens early in the last century, but was not touched until about 1870, and did not really begin to develop into an important industry until the return of the Commission sent to Assam to investigate methods, and provided with questions mainly by Dr. Thwaites. The Botanic Gardens, therefore, may claim to have had some considerable share in starting this industry also.

Thwaites retired in 1880, and was succeeded by Henry Trimen. The really important introductions of useful plants from abroad were by this time practically over, and Trimen hardly introduced anything in his sixteen years of work that is likely to prove of more than minor interest or value. But he devoted himself to spreading the use and cultivation of those useful plants which had already been introduced, more especially cinchona, cacao, and cardamoms, and to the writing of a complete flora of Ceylon, whose publication was begun in 1893. Unfortunately Trimen died in 1896, when only three volumes of the five of which the flora consists were published, but its completion was generously undertaken by the veteran Sir Joseph Hooker, and the last volume appeared in 1900.

In many respects the last twenty years of last century mark a check in the progress of the Ceylon Botanic Gardens, and it will be well, before resuming the history, to deal with what had been accomplished elsewhere. The other British gardens were in general (except Calcutta) simply inferior editions of those of Ceylon, and it is to the Dutch possessions that we must go for an illustrative example. In 1880 Dr. Melchior Treub was appointed to the direction of the famous Dutch colonial gardens at Buitenzorg in Java, which up to that time had been working on similar lines to those of Ceylon. Dr. Treub, unlike all previous directors, both in British and foreign gardens, was a "laboratory" botanist, and not a systematist.

With the advent of Dr. Treub, pure "laboratory" science commenced to have its innings in the tropics, both for itself, and for application to the needs of practical life, until now, if one may judge by the quality of the publications turned out by the Java department, the Dutch planter in that island is the most scientific agriculturist in the world. To leave pure science alone for a moment, one of the most noteworthy steps taken in Java was the scientific improvement of the bark of the cinchona-tree by careful selection, until now the best Java barks of *Cinchona Ledgeriana* contain as much as 17 per cent. of their weight of quinine, while the best Ceylon (unselected) barks contain only 8 per cent. This means that only half the weight of bark has to be grown and harvested in Java, as compared with Ceylon, to give an equal monetary return. Thanks largely to this work, Java has acquired a practical monopoly of the trade in cinchona, from which it scarcely seems likely that she will be easily ousted.

Other valuable work was done in grafting coffees, in the scientific improvement of sugar manufacture, and in many other lines, so that Java has remained, on the whole, a very prosperous colony, free from the terrible vicissitudes that have overtaken Ceylon and some other British possessions.

A laboratory for pure science was early established, and invitations to come and work in it issued to botanists in Europe and elsewhere, were responded to with considerable alacrity, especially as Dr. Treub himself, by turning out splendid work in several different departments of botanical science, showed what an opening there was for botanical work in the tropics.

The bulk of the botanical work done has been published in the well-known *Annales du Jardin botanique de Buitenzorg*, and a rough examination of this work will repay any one who is interested in this aspect of the subject. First-class work has throughout been turned out by Dr. Treub, and by a few distinguished visitors, such as Professors Goebel, Haberlandt, and others, but the work done by the ordinary run of visitors seems to show a distinct falling-off as time has passed by, or rather, the matter with which they have worked has been of less general interest. The earlier visitors were able to work at many "obvious" problems or phenomena, which stare in the face any one visiting a tropical garden, and yet whose solution exercised a considerable effect upon botany generally. Later visitors have found all these problems already worked at, and have had some trouble in finding any problems or phenomena which would afford the subjects of a few months' work.

The fact, in general, seems to be that the problems which can be solved by a visit of a few months to the tropics—except in the case of anatomical questions, the material for which can be collected in the tropics and taken home to Europe—are now but few and far between, and are not to be distinguished by the ordinary visitor, for want of the necessary preliminary acquaintance with the general life-phenomena of the tropics. The visitor who comes to the tropics for a short period may indeed benefit himself enormously, but he must not expect, unless he is a man of unusual penetration, to be able to do any very remarkable original work. These remarks are not to be understood as deprecating visits to the tropics, for the botanist who has never been there has not quite the true understanding of vegetable life, but as pointing out the reason why so many visitors somewhat disappoint expectations formed in Europe. A good many people in Europe have still the impression that a botanist has merely to go to the tropics immediately to see before him numerous problems or phenomena awaiting investigation, and to be solved in a short time.

What is wanted now, for the proper working out of the innumerable problems that remain for solution in the tropics, is that men shall spend sufficiently long there to become at home with the unwonted aspects of vegetable life, and to have the time to work at problems requiring long periods for their proper solution. The visitor who has only three or four months at his disposal should spend the bulk of it, not in doing research work in the laboratory, but in getting as much acquaintance and familiarity as he can with the many unfamiliar aspects under which vegetation presents itself in the hotter and damper regions of the globe.

To return now to the consideration of the Ceylon Botanic Gardens, it was in view of the considerations above set forth, that the first appointment made in enlarging upon modern lines the old department was that of a Scientific Assistant. This post can be held for any period not exceeding three years, and its occupant has to devote himself to research during that period. Similar appointments are in existence in Java. With a period like this at his disposal, the holder of the post can take up lines of work that would be impossible to a short-period visitor. As an indication of the kind of work that may be done, that carried on during the last eight years by the Ceylon incumbents may be mentioned. Mr. J. Parkin worked out the method of coagulating rubber into "biscuits," a method which is now worth many thousands of pounds per annum to the rubber planters of Ceylon and the Malay States. Mr. Herbert Wright wrote a monograph of the Ceylon species of *Diospyros*, hitherto very imperfectly known, and a full account of the periodicity of trees in Ceylon as to leaf-fall and renewal, work which it seems likely will lead to a reconsideration of many problems in physiology and fossil botany. Mr. R. H. Lock carried out some valuable experiments in breeding upon Mendelian lines—the first work of this kind to be done in the tropics, and one which has not only given valuable scientific results, but has produced some useful, new, and improved breeds. Mr. A. M. Smith, the present holder of the post, is carrying out valuable researches upon growth, which is by no means the same phenomenon in the tropics that it is in Europe.

During the last ten years the Ceylon Botanic Gardens have grown entirely beyond the old conception of botanic gardens, and have become practically a department of agriculture, though the name of "botanic gardens" is still retained, and the work of purely botanic gardens is still carried on among many new lines of work. The history of the transformation is interesting, and may be followed, as illustrating what is being considered in this paper. The gardens in Java commenced to grow in a similar way about 1885, and have lately been given the title of a department of agriculture. The great distinction between the two places is that the staff of the Ceylon department consists mainly of men who have to deal with all the different industries pursued in Ceylon, while that of the Java department consists mainly of men who are concerned with one industry only. Thus there are at Buitenzorg, and elsewhere in Java, laboratories devoted entirely to sugar, to tea, to coffee, to cinchona, and so on, while in Ceylon there are as yet no specialists devoted entirely to one industry. An agricultural department, similar to the Ceylon department, has been formed simultaneously in the West Indies, and within the last two years, stimulated by the success of these two institutions, departments of agriculture, on a larger or smaller scale, have been formed in practically every British tropical possession, even India.

The growth of the Ceylon department began in 1899, and in that year and the next two important appointments were made, those of an Entomologist and a Mycologist, for the study of the diseases caused in cultivated plants by insects and by fungi respectively. When these appointments were made, the general feeling of British planters was against owning up to any attack of disease on their plantations. If a man had such an attack, his idea was to keep it quiet, with the very common result that it spread badly over his own place, and got into his neighbours' plantations. The actual cause which stimulated the appointment of these officers was an outbreak of "cacao canker," a disease caused by a *Nectria* closely similar to that causing the apple canker of Europe. This disease began probably about 1894, and by 1897 was so rampant that it could no longer be concealed ; at that period there was a probability that Ceylon might lose her cacao industry altogether.

The newly appointed mycologist devoted much of his time to this disease, and worked out the methods of attacking it, with the result that those plantations on which his recommendations were carried out became gradually freed of the disease, and now nearly every cacao estate in Ceylon keeps it under by use of these methods, so that it is no longer looked upon as a serious menace to the cultivation.

Gradually the result of this work, and of similar work done by the Entomologist, has forced itself upon the minds of the Ceylon planters, and now on the majority of estates the greatest care is taken to look out for the first symptoms of disease, and to send at once to the Peradeniya staff for aid in combating it. So far has opinion turned in the opposite direction from that which it formerly occupied, that the newspapers of Ceylon are now almost the first to draw attention to any slight outbreak of disease, and to call for the help of the Mycologist or the Entomologist.

The passing of an ordinance (law) is under consideration in Ceylon, giving to the Governor power to proclaim any district as "infected" with some particular disease, and to appoint a board for that district, which shall have power to compel him to carry out the recommendations made by the Entomologist or Mycologist, when such recommendations have been approved by the larger Committee of Agricultural Experiments.

The next appointment to be made in Ceylon was that of an Agricultural Chemist. The work that lies before such a man scarcely needs detailed description; but some of the most important directions in which his work has been of great use to the people of Ceylon may be briefly mentioned. Thus he has taken up the question of the manuring of tea and other products, and shown people how to manure with the best results and at the lowest cost; prior to his work manuring was largely a matter of chance or luck. He has studied the distillation of camphor in Japan and elsewhere, and shown people in Ceylon how to distil cheaply, and how to get the largest percentage of camphor. He has investigated the manufacture of "oolong" teas in Formosa, and found out how to produce the peculiar flavour hitherto supposed to be a monopoly of such teas. Already Ceylon is making a considerable quantity of oolongs, and this may result in the capture of a large part of the American market, which prefers such teas. In many other ways his work has proved of the greatest value to the people of the Colony.

The appointment of the officers just mentioned of course involved the establishment of proper laboratories, and advantage was taken of their construction to make them large enough to accommodate visitors from abroad, as well as the staff of the department, so that now there is ample room at Peradeniya for any foreign visitors, even as many as five at one time.

To return to the consideration of the botanic gardens proper, and the lines along which their work has been extended and expanded to meet modern requirements. Their main object was always acclimatisation of plants from abroad, which of course involved the growth of these plants in the gardens, and the supply of small quantities of seeds or cuttings to those who

wished to try them further.   Plants which would not live in
the botanic gardens (which were established all over the island
in its different climates, hot and cold, wet and dry) could of
course not be acclimatised, and could never be of use in local
agriculture.   In the early days of the botanic gardens comparatively large areas could be given to individual plants ; but
as time went on, and the number of plants in the gardens
increased, this would necessarily become less and less possible,
even in the largest gardens, and at length the gardens must be
content with growing one or two specimens of each plant.
For the introduction of a new industry it is obvious that this
method can be of little use in a limited time unless the plants
can be easily multiplied.   Nor can real experiments upon the
best methods of cultivation be tried upon so small a scale.

To meet and overcome this difficulty the obvious best course
to pursue for the present was to open what may be called in
the American sense Experiment Stations, *i.e.* larger areas of
land entirely given up to a comparatively few commercial crops,
which could be studied with great care in all detail.   Two such
stations have been opened in Ceylon, and were practically the
first upon a large scale to be formed in any British colony.
Experiment Stations were already common in many other
places, but they were as a rule on a small scale, growing
perhaps one-quarter of an acre with this crop, one-half with
that, one-tenth with this manure, one-fifth with that ; and it
was in Ceylon that the opening of large stations was first
definitely put into practice, and larger areas devoted to the
different crops and experiments.   Thus the station at Peradeniya has twenty acres under tea, fifteen acres under lemon-
grass, and so on.

The preliminary trial of new plants is still done in the
botanic gardens as before ; but for the trial of anything likely
to be of importance in local agriculture, the venue of the experiments is as soon as possible transferred to the Experiment Station, and experiments are also made there with the
crops already in common cultivation.   This is a line of work
which the older botanic gardens were of course entirely un-
fitted for ; yet it is one which is already of great value, and is
increasing in importance every day.   The old day of "new
products" has practically gone by for most colonies and dependencies in the tropics.   No longer are there any products

which can be introduced into the country, and readily become the basis of new industries. In former days there were many such, which were in the hands either of the tropical or sub-tropical races of mankind, or which were simply collected from the jungles. Thus, to take only the industries which have at different times risen to prominence in Ceylon, coffee, tea, and cardamoms were cultivated only by the Chinese or other backward races, while cinchona and rubber were collected from the jungles. This is no longer the case, for everything of any serious value in the tropics is now in the hands of Europeans, Americans, or Japanese, and a fierce competition has to be faced by any one wishing to cultivate it.

This being so, it is evident that greater importance now attaches to the improvement of those industries which are already in any one country, and their conservation and expansion, than to the study of the comparatively minor chance of finding something that shall take their place if they fail. It is with this in view that the Experiment Stations of Ceylon were organised. They give a certain amount of attention to the trial of new products, *i.e.* products that are not as yet cultivated in the island upon any large scale, but their main care is devoted to the study of the existing crops of the country, the improvement of methods of cultivation, manuring, harvesting, and preparation of the produce for market.

Thus at the Experiment Station at Peradeniya, which lies in the "wet" zone of Ceylon and in the European planting districts, experiments are mainly devoted at present to cacao, rubber, tea, citronella, and ground-nuts, so far as old products are concerned, and to lemon-grass as a possible paying new product; while in the Experiment Station at Maha-iluppalama, which lies in the almost uninhabited "dry" zone of Ceylon, the experiments as yet have been mainly with cotton, a crop which is at present only slightly cultivated in Ceylon, and that in the poorest varieties; but which the experiments at Maha-iluppalama give reason to hope may prove to be a crop of considerable value to Ceylon, if the best kinds of cotton, the Sea Island and Egyptian, are cultivated.

Not only do the Experiment Stations work directly at the actual crops cultivated in the colony, but also at more general questions. Thus they have proved the great value to be derived from " green manuring " *i.e.* planting leguminous crops between the lines of the permanent cultivation, and hoeing them into the ground when they grow up. In this way the nitrogen contents of the soil, to say nothing of the humus, which is rarely present in great quantity in the tropics, may be considerably increased at small cost.

The Experiment Stations work not only at the cultivation of the various crops, but also at the best methods of harvesting them and of preparing their produce for the market. Thus in Ceylon much valuable work has been done upon the different ways of preparing the best quality of marketable rubber, upon the best ways of distilling oil of citronella or of lemon-grass, upon the best ways of fermenting and drying cacao, and similar problems.

The work of the Experiment Stations was at first looked at somewhat askance even by the European planting community, for whose benefit the establishments were mainly designed; but a change has come over this, and now the stations are visited almost daily by numbers of planters and other agri-culturists, who come to see what is going on, and to find out, in many instances, whether there is anything that they can themselves try with some prospect of success. In fact, the Experiment Stations are beginning to foster that most desirable spirit among the planters, the experimental habit of mind. A country in which the agriculturists are themselves ready and anxious to try experiments upon both old and new products, cannot fall very far behind in the race for agricultural prosperity.

Not only is it desirable that private individuals should try experiments upon their own land, but it is also of great advantage in another way. It is obvious that the results of the official experiments, say at Peradeniya, can only be of direct applicability to the soil and climate of that particular place, and to other places with similar soils and climates, while if they are applied directly, say to the Kelani Valley district, which has a more gravelly soil, and a much warmer and much wetter climate, they may fail utterly. Hence they require to be very carefully carried out and understood, and to apply them to other districts needs the personal attention of the actual officer who carried out the experiments, and who will therefore understand, as well as can be understood, the modifications that they may require before being applied to new districts. For this among other reasons, the Experiment Stations have started what we may term Co-operative Experiments. Thus, for example, to determine the effects upon the quality and quantity of tea produced, experiments with the various manures have been started upon no fewer than six selected estates at different elevations in Ceylon, and upon different types of soil; and other similar series of experiments are in progress. The actual work of the experiment, and the keeping of the records of the quantity of tea picked, or what not, is done by the superintendent of the estate, but the whole work is under the superintendence of the officer in charge of the Experiment Station, and all the returns are sent to him. In this way the work of the Experiment Station is, so to speak, multiplied, and increased in efficiency, by being made more directly applicable to many districts. As time goes on, probably this method of experimenting will become the common one, trial or control experiments being the main business of the Experiment Station proper.

It is obvious that in an Eastern country, with an unprogressive population, the work that has been outlined in the last few pages must mainly appeal to the European planter, but will not appeal to the ordinary villager. For one thing, all work of this kind requires capital, and the general tendency of all scientific improvement is, if anything, to increase the amount' of capital required; and the villager has not got capital at his disposal. To the native capitalist, who is every year going in for agriculture more and more, the work of the Entomologist, the Mycologist, the Chemist, and the superintendents of Experiment Stations appeals with almost as much force as to the European planter.

There is no doubt that agricultural progress is just as possible to the tropical villager as to his European or American prototype, but before it will be reasonably practicable he must have proper conditions as regards land, capital, transport, education, and labour. Land and labour have long been satisfactory enough for the purpose, transport is now provided to a moderate—though insufficient—extent, and capital and education are the main difficulties. We shall deal first with the latter.

The main differences between the West Indian and the Ceylon agricultural departments have been with regard to this. The former, having to deal mainly with a population of tropical

natives, has placed agricultural education first upon its programme; the latter, having first of all to deal with a large
population of the most energetic European planters in the
world, has placed it later.

There is no doubt that education, if properly conducted, is
one of the greatest agencies in agricultural progress. Not only
does it produce in the native new desires, and so tend to raise
the general standard of living in the country (and thus tend to
require more and more varied agriculture), but if rightly guided,
it may teach the boys to look to agriculture as a means of
livelihood, instead of, as has hitherto happened in most tropical
countries, tending to make them all look to the towns and to the
learned professions.

There is now very little doubt that direct teaching of
agriculture as such, in the primary schools, is undesirable.
The teacher, among other reasons, does not know enough
about it in the majority of cases to be able to do better
than the surrounding village population, and consequently
is·liable to criticism tending to reduce the value of his work.
What can be done, however, is that school gardens may be
opened, and in them the masters and pupils may learn to deal
with plants that are not as yet familiar in the villages. They
learn the general principles of agricultural and horticultural
work just as well, they tend to introduce " new products " to the
villages, and they do not lay themselves open to criticism in
the same way. The work should of course be under the
general supervision of trained agricultural teachers, who can go
round at frequent intervals and inspect the school gardens.
Deliberate and well-organised schemes of this kind have now
been in operation for some years in Ceylon and the West
Indies, at least.

Having passed through the primary school, the pupil may
receive more direct agricultural instruction in secondary schools
or in agricultural colleges, and this is already in full operation
in the West Indies. The old agricultural school in Ceylon was
closed some years ago, as it was found that its pupils went in
for anything rather than agriculture, and as yet no more modern
institution has been opened.

Before the poorer villager can do anything but cultivate his
old crops in the old way, he must be provided with some access
to small capital at a cheap rate. This in general must be

managed by some system of agricultural banks, perhaps most simply and efficiently by the Co-operative Credit Societies of the Raiffeisen type, which are now so common in Europe. Even these, however, are not altogether suited to the improvident tropical native, who tends to borrow money from them, not for agricultural purposes, but for weddings and other festivities. Probably the best method of helping him at first, and for a long time to come, will be by means of Co-operative Seed-supply Stores, a store being started in a village and advancing to the villagers the seeds they require on advantageous terms. Unless the society can be started by local capitalists, it will of course need to borrow money from the Government; but as soon as this is repaid, its further profits will go to its shareholders in the form of dividends, or perhaps better still, may be expended in purchasing seed of a better quality than that locally in use. Societies of this kind, as well as co-operative credit societies, are already in operation in Ceylon.

Provided that proper conditions obtain as to land, labour, transport, education, and capital, much may be done for the natives of a tropical country by the establishment of Agricultural Societies. There should in general be a central society, with as many local branches as possible, the central society publishing details of how to grow and dispose of produce, selling seeds at very low rates, and so on. The great thing in running a society of this kind is to have keen and enthusiastic officers in the central society, to keep up the interest and enthusiasm of the local branches. A society of this kind is now at work in Ceylon, and, when it has got over the difficulty of capital, will doubtless appeal to the villagers; at present its efforts mainly stimulate the local small capitalist. The society is, however, trying to get co-operative credit societies and co-operative seed-supplies started in the various places in which it has branches.

This must in general suffice as a sketch, necessarily brief and imperfect, of the progress in botanical and agricultural work that has gone on in the tropics. The serious student will at once notice that there has been but little said about scientific research, excepting in pure botany. As yet there is but little research carried on in the tropics on lines that do not promise immediate return, for in dealing with English people, at any rate, pure research must be floated in, so to speak, upon the top of research that has an immediate practical bearing.

Thus in Ceylon, while the work done at the Experiment Stations and elsewhere in the department is strictly scientific in regard to accuracy and method, the researches undertaken have always been such as afforded a reasonable prospect of some directly practical return within a moderate time. The results have been strikingly successful upon the whole, and the public have now acquired such a faith in the institution that there is comparatively little criticism of its work, and no repining, or attacks upon the conduct of the place, if work is undertaken which may require many years to yield a result. This is as it should be ; there is too much tendency among a certain class of scientific workers to regard work as of greater value the less it is applicable to the needs of practical life.

So far mention has only been made of the official institutions for the carrying on of scientific work, but the work done by true amateurs in the tropics is now of very slight importance, though in earlier times such people as Champion, Colonel and Mrs. Walker, Ferguson, and others did very valuable work in Ceylon, to take only one country out of the many. Now that there are first-class scientific institutions in the Philippine Islands, in Java, in Ceylon, in the West Indies, and elsewhere, it may be taken for granted that the scientific investigation of the innumerable problems of the tropics will go on much faster than hitherto, and produce great effects in the modification of the hitherto prevailing views in botany, if not in agriculture, which of course is an art of very different application in different climates.

STELAR THEORIES

By THOMAS G. HILL, A.R.C.S., F.L.S.

Lecturer on Biology, St. Thomas's Hospital

TEN years ago an article from the pen of Mr. A. G. Tansley, on the then current ideas regarding the nature of the vascular structures of plants, appeared in SCIENCE PROGRESS, and, inasmuch as during the last decade a large amount of work, directly bearing on the theory of the stele, has been published, no apology is needed for this present consideration.

It may at once be stated that what follows, necessarily brief and incomplete, is not intended for the botanist with special knowledge of the subject; rather, it is designed for those who require an introduction to a somewhat complicated chapter in modern botany.

If a transverse section of the stem of the sunflower be examined, there will be found a number of vascular strands arranged in a zone, and, if the stem be sufficiently young, separate one from the other. These strands, or bundles, are embedded in parenchyma which, on account of the disposition of the vascular bundles, is divided into two parts, the central medulla, or pith, and the peripheral cortex.

Immediately bounding the ring of fibro-vascular strands, on its outer side, the presence of a bundle-sheath, or endodermis, may be demonstrated.

Inasmuch as this brief account is purely descriptive, it may be taken as illustrative of the views held by botanists prior to 1886. In those days there was no attempt made to differentiate the constituent tissues of plants into morphological parts, after the manner obtaining in external morphology; the internal structures were treated rather from the physiological point of view. Thus, no essential difference was supposed to exist between the cortex and pith, the bundles were considered as separate structures, and the endodermis was regarded as being nothing more than a bundle-sheath; that is to say, it was not thought to have any special virtue in the delimitation of one tissue from another.

This older view, however, was replaced, for a time at any rate, and in many quarters, by the view first put forward by Van Tieghem (63, 69). He regarded the whole of the axial vascular system as a single structure, of which the bundles were parts.

Thus in the young sunflower stem, described above, the zone of vascular bundles, together with the pith and the interfascicular parenchyma or soft-walled tissue between the bundles, is morphologically distinct from the other tissues and forms a morphological unit, the stele, which is separated from the cortex by the endodermis, the innermost layer, according to Van Tieghem, of the cortex. Hence the central parenchymatous mass, enclosed by the vascular bundles, is part of the stele, and is thus morphologically distinct from the cortex.

The simplest possible case, according to this first theory of the stele, may be found, for example, in roots and in the stems of young ferns, where the endodermis immediately surrounds a solid vascular cylinder, or stele, which has for its external limit the pericycle. To this type of structure Van Tieghem gave the term monostele, and he derived all other vascular arrangements from it.

Thus a medullated monostele is formed by the aggregation of parenchyma in the centre of the original structure; the medulla, or pith, may extend outwards so as to break up the surrounding ring of vascular tissues into separate parts which are termed the bundles. This condition obtains in some ferns, e.g. Osmunda, and the majority of Dicotyledons.

It is to be observed that the term "bundle" is used by Van Tieghem in a more restricted sense than that employed by De Bary. Thus the root-structure referred to above was spoken of by De Bary as being a radial bundle, while Van Tieghem treats it as representing several bundles. The term "meristele," much used at the present time, is synonymous with the term "bundle" employed by Van Tieghem.

In other cases the endodermis and pericycle may extend inwards towards the pith, so as to enclose completely each meristele. There is now direct continuity between the medulla and the cortex through the medullary rays. The original stele is thus broken up, as it were, and ceases to exist as such. This condition is consequently termed astelic, and may be found in certain species of Equisetum, e.g. E. limosum, and in some aquatic plants, e.g. Nymphœa (fig. 1).

By the union of the meristeles of the astelic type, a continuous hollow vascular cylinder is reconstituted, which differs from the medullated monostele on account of the presence of an internal endodermis, and is termed gamodesmic. As an example the rhizome of *Equisetum sylvaticum* may be cited.

Thus far the theory accounts for the derivation of several distinct types of structure, viz. the medullated monostele, the astelic, and the gamodesmic conditions.

But there is an alternative line of development; the primitive monostele may undergo successive branching, so that the polystelic (or dialystelic) condition results, a very common

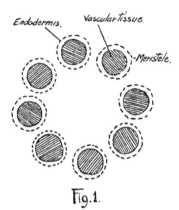

Fig. 1.

Diagram of a transverse section through the stem of an astelic plant.

arrangement to be met with in ferns, and which also occurs in certain Angiosperms, *e.g. Primula Auricula* (fig. 4, c).

In a transverse section of such a polystelic stem the appearance may be exactly like that obtaining in many astelic axes.

In much the same way as the meristeles unite to form a gamodesmic cylinder, so also may the separate steles of the polystelic phase unite together to form the gamostele, or, as it is frequently termed, the solenostele, *e.g. Marsilia* and *Loxsoma* (figs. 2 and 3).

In testing this theory two lines may be pursued:

1. Van Tieghem paid practically no attention to the ontogeny or development of the vascular structures. It is clear that the examination of the development of these tissues should show whether the polystele is formed by the successive branching of

the monostele; whether the gamodesmic system is arrived at by the reunion of separate meristeles; and so on.

2. It will be observed that the endodermis plays an important part, inasmuch as it is looked upon as the delimiting layer between the stele and the cortex. The question which arises is whether this endodermis has the morphological value attributed to it; if it has, then the theory may be justified.

Since the Ferns provide us with a greater diversity of vascular structure than the Phanerogams, it may be considered desirable first to follow the development in any convenient plant belonging to the former class.

A young plant of *Doodia aspera* R. Br. (see Chandler, 10) has a central cylinder consisting of a solid mass of vascular tissue, a rod of xylem or wood surrounded by phloem or bast, and

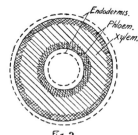

Fig. 2.

Solenostele. Diagram of a transverse section through an internode.

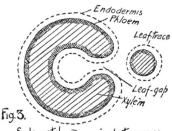

Fig. 3.

Solenostele. Diagram of a transverse section through a node.

the whole encircled by an endodermis. This type of structure has been termed the protostele. As development proceeds, a small collection of phloem elements appears in the centre of the xylem. The first leaf-trace bundle—concentric in nature—is formed by the outward passage of a portion of the vascular tissue; this leaves a gap in the central cylinder so that a transverse section through the node shows a horseshoe-shaped mass of xylem, and the internal and external phloem in continuity (fig. 4, A).

After the exit of the foliar strand, the leaf-gap is closed by the horns of the wood gradually coming together; hence there results a cylinder of xylem bounded both internally and externally by phloem, and the whole enclosed within the endodermis (fig. 4, B).

As the formation of successive leaf-traces is followed, it is found that the gap formed in the central cylinder by the outgoing foliar bundle persists, and becomes more obvious, the edges of the xylem being separated by parenchyma. This ground tissue is delimited from the vascular tissue by the endodermis, which is developed around the horns of the vascular tissues. At a still later stage a further elaboration obtains; the fundamental tissue encroaches, as it were, more and more towards the centre in such a way that ultimately the internal phloem forms a band lining the concavity of the xylem and enclosing the central mass of ground tissue, from which it is in turn separated by an inner endodermis (figs. 2 and 3).

During these progressive changes the vascular system has widened out; and further, as the leaf-traces become more numerous, the leaf-gaps overlap; hence, in any transverse section of a well-developed plant, the vascular system will be found to

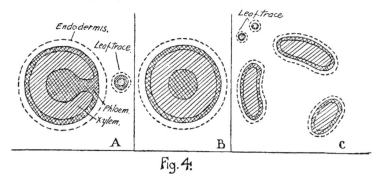

Fig. 4.

consist of a number of separate strands of xylem, each surrounded by a band of phloem and endodermis (fig. 4, c).

In other words, we have what—following Van Tieghem's hypothesis—would be termed the polystelic condition.

Doodia aspera may serve as an example for the development of the majority of the so-called polystelic ferns, although, it must be remarked, the type shows modifications in different plants. Thus, in some ferns, *e.g.* some species of *Schizœa*, it is found that the solid axial rod of xylem, which obtains in young plants, develops in its centre not phloem but parenchyma. In other cases, the appearance of the internal phloem may be delayed for some time after the first appearance of the central parenchyma, as in some species of *Angiopteris*, *Asplenium*, and *Nothochlœna*. Then, as regards the internal endodermis, the time when this tissue first makes its appearance also varies in different plants.

Turning to the Angiosperms, Gwynne-Vaughan (25) was one
of the first to test the validity of Van Tieghem's theory by the
examination of the development of the polystelic condition
obtaining in certain primulas.

The seedling of *Primula Auricula* is what may be termed
monostelic; which condition obtains until four or five leaves have
been formed, for the outgoing leaf-traces leave but a small gap
in the central cylinder. Afterwards it is found that these foliar
gaps are of some size, and the endodermis develops in the
concavity, thus separating the pith from the vascular tissue.
Before any one particular leaf-gap is repaired another is formed
by the giving off of another foliar bundle from the convex side
of the central stele; hence the latter becomes divided into two
portions. Further subdivision of the vascular zone takes place
in connection with the succeeding leaves; that is to say, a
polystelic condition is attained.

It is clear that, as Gwynne-Vaughan points out, the leaf-
traces are the all-important factors in determining the organisation
of the stele, a point overlooked by Van Tieghem.

The work of the former was carried much further by Jeffrey,
whose contributions entirely altered the aspect of the subject.
He examined various polystelic Angiosperms, viz. species of
Primula and *Gunnera* and *Parnassia palustris*. The comparison
of these with types of vascular structures of other Angiosperms
and Ferns led to conclusions, which will be dealt with later on,
at variance with the theory of Van Tieghem.

From the above account of the development of the vascular
systems of mature plants, it is obvious that the polystelic
condition does not arise by the successive *branching* of the
protostele, and the earlier work of Jeffrey (37) shows that not
only is the polystele not derived from the protostele in that
manner, but further that (1) the astelic type is not formed from
the medullated monostele by the separation of the latter into
bundles or meristeles, and the enclosure of each by an endodermis; for in young astelic axes, *e.g.* in that region of
seedlings situated immediately above the insertion of the seed-
leaves or cotyledons, there is a collateral stelar tube lined
internally and externally with an endodermis and having foliar
gaps; there is not a series of meristeles, as would be required
by Van Tieghem's hypothesis; (2) the three types, viz. polystelic, astelic, and medullated monostele, are modifications of one

type, which is the siphonostele, *i.e.* "the central cylinder is primitively a fibro-vascular tube with foliar lacunæ opposite the points of exit of the leaf-traces." He further remarks that "in the so-called polystelic modification, the central cylinder has internal as well as external phloem, and may be described consequently as amphiphloic. In the astelic type of axes so-called, the internal phloem is absent and the central cylinder is accordingly to be designated ectophloic. The medullated monostelic type of Van Tieghem is derived from the last-named by the degeneration of the internal phlœoterma or endodermis" (Jeffrey, 37, p. 38).

Thus, fig. 4 c represents the amphiphloic condition, inasmuch as the phloem completely surrounds each strand of xylem; the ectophloic type is similar, but the distribution of the phloem is not so extensive—instead of completely encircling the wood it is developed only on the outer margin of each strand.

The essential facts on which these conclusions are based have been amply verified by the work of Boodle (2—5), Brebner (8), Gwynne Vaughan (25 *et. seq.*), Farmer and Hill (20), Leclerc du Sablon (42), Tansley (57 *et. seq.*), and others.

The second factor which has an important bearing upon the original hypothesis is, whether the endodermis really is the innermost layer of the cortex.

Not a few instances have been noted where the endodermis and pericycle have seemingly arisen from the same initial layer of cells, and not only these two tissues, but others. Thus Boodle (3, p. 377) remarks: "In *Schizæa digitata* a comparison of transverse sections of the young stem at different stages of development points to a possibility that the endodermis, pericycle and phloem may be formed from the subdivision of a single layer." Again, Tansley and Chick (58) have shown that in the case of *Schizæa malaccana* the endodermis, pericycle, and the vascular ring all arise from a single initial layer.

These instances point to the fact that in many cases the endodermis is not the innermost cortical layer.

If a median longitudinal section through the stem-apex of a plant be examined, there may sometimes be made out three fairly distinct layers of cells: an outermost stratum consisting of a single layer of elements and termed the dermatogen; immediately below is a second series of cells, the periblem; and, finally, the central core, which is designated the plerome. These

three strata, during the growth of the plant, are in an active state
of division, and give origin to the tissues found in the mature
parts of the plant. Hanstein supposed that of these three layers,
or histogens, the dermatogen gave rise to the epidermis, the
periblem to the cortex, and the plerome to the central cylinder.
This is often spoken of as the theory of histogens.

Now if the cortex be formed from the periblem and the central
cylinder from the plerome, then the matter can be cleared up by
tracing the fate of these histogenic layers.

This important task was undertaken and acccomplished by .
Schoute (49), who discovered that it was the exception rather
than the rule for the endodermis to be absent. He found that
only seven plants out of four hundred Dicotyledons had no
endodermis. Thus, to a certain extent, the stelar theory may be
supported; but it must be pointed out that this tissue may only
occur in young stages of development. As regards the coincidence of the pericycle with the outermost layer of the plerome,
and of the endodermis with the innermost layer of the periblem,
Schoute has ascertained that this does not hold in the majority
of the cases examined. In some plants the plerome not only
gave rise to the pericycle, but also to the endodermis together
with some of the inner cortical elements; in other cases, the
plerome, as such, was non-existent.

Thus, incidentally, Hanstein's theory of histogens is destroyed, and, to the minds of some, the endodermis, considered as
a layer of morphological value, shares a similar fate.

Therefore it is clear that much of Van Tieghem's theory
is discredited; and although the majority of botanists have
discarded his hypothesis, taken as a whole, two important
conclusions have been retained in many quarters, viz.:

(1) That the stele, not the vascular bundle, is the unit of
structure.

(2) That a simple type of monostele (the protostele) is the
primitive form.

From these two ideas new views have arisen regarding the
stele.

Jeffrey (37, 38) considers that there are two types of central
cylinder, the protostele and the siphonostele, the former being
the more primitive. The siphonostele is tubular, has a medulla
derived from the cortical tissue, and has given successive origin,
by reduction, to the following modifications: the amphiphloic

siphonostele (with an internal endodermis and phloem as well as the normal outer phloem and endodermis); the ectophloic siphonostele (with external phloem and endodermis only); and finally to the condition obtaining in the majority of higher plants.

This theory may be illustrated by the consideration of the structure of certain Osmundaceæ.

In *Osmunda cinnamomea* the mature structure consists of a ring of vascular bundles enclosing a central mass of ground-tissue. Internal phloem, though not general, occurs here and there, more especially below the bifurcation of the stem; there may also be an internal endodermis. On the other hand, in *Osmunda regalis* and other plants of the Order, the internal phloem and endodermis are absent.

Hence, on the whole, the type of structure corresponds to Van Tieghem's medullated monostele; but, according to Jeffrey, the collateral bundles have arisen from concentric strands of vascular tissue, the inner phloem of which has become obliterated, evidence of this being supplied by the occasional appearance of internal phloem and endodermis in *Osmunda cinnamomea*.

This view is disagreed with by Boodle, Chandler, and others, chiefly on developmental grounds; thus Scott (53) remarks that "this theory of reduction is no doubt a tenable one, but as it receives no support from the development of the young plant, its basis is weak." On the other hand, the theory appears to be supported in particular cases, for Jeffrey has described the amphiphloic monostele as giving rise, through reduction, to the medullated monostele. For instance, in *Antrophyum semi-costatum*, *A. reticulatum*, *A. plantagineum*, *Vittaria elongata*, *Davallia stricta*, and *Adiantum pedatum* the vascular cylinders have lost entirely, or in part, the internal endodermis and phloem. On the other hand, the steles of *Vittaria lineata*, *Davallia plumosa*, and *D. fijiensis* are quite normal. Thus, in the plants first cited, a strong case may be argued for the medullated monostele being derived from the amphiphloic siphonostele, but it does not follow that the same holds good in all cases. As regards this last point Scott (53) remarks as follows: "An apparently protostelic structure is known to have arisen in certain cases (*e.g.* water-plants) by reduction, yet no one doubts that in other groups the protostele is really primitive."

The main features of Jeffrey's position are as follows:

(1) The endodermis has a morphological value.

(2) The majority of vascular arrangements are referable to the siphonostele.

(3) The pith is extra stelar, *i.e.* it is part of the ground or fundamental tissue, and has no separate morphological value.

Boodle, who has contributed largely to our knowledge of the anatomy of the Ferns, is opposed to the views just enumerated.

The main points in his position are as follows:

(1) The endodermis is not a dependable morphological limit.

(2) The medullated monostele and the solenostelic (and dialystelic) types have been derived from the protostele by the transformation of its central tissue into parenchyma, and into parenchyma together with phloem and endodermis respectively.

(3) The pith is thus stelar in origin.

The differences of opinion between these two observers are mainly concerned with the morphology of the endodermis and pith; and, indeed, it is around these two points that controversy centres. The majority agree as to the facts; it is in the interpretation of the facts that disagreement exists.

The Morphological Value of the Endodermis

It has been seen, from the work of Schoute and others, that, to the minds of many, considerable difficulties lie in the way of accepting the endodermis as a morphological layer— or as a layer which may be relied upon as delimiting the boundary of the cortex—owing to the fact that in many cases it is not the innermost layer of the cortex of Angiosperms.

In the case of the polystelic, or, better still, dictyostelic[1] plants, the apical region contains a number of strands, made up of many immature and dividing elements, showing no differentiation into definite tissues, and directly continuous with the fully developed steles. Taking the tissue between these immature and developing strands, where is the line of demarcation between periblem and plerome? Then again, the endodermis exhibits much variation in its appearance and non-appearance, in places where, theoretically, it should be absent and present respectively.

Lewis (43) found a well-marked endodermis enclosing a mass of cortical parenchyma in a root of *Ruscus*. In *Helminthostachys*

---

[1] A purely descriptive term suggested by Brebner (8) to replace the term polystele in order to avoid the theoretical connotation inseparable from the latter. He defines the dictyostele as "a vascular tube with large overlapping leaf-gaps, so that the whole structure becomes a network of vascular strands."

zeylanica (Farmer and Freeman, 19) this tissue does not appear in the stem of the young plant, while its appearance in the vascular strands of older plants is fitful and irregular.[2]   In *Botrychium lunaria* (Van Tieghem, 67) an external endodermis only is present in the young stems; hence it may be termed monostelic.   In slightly older stems an internal endodermis appears, and thus the structure becomes astelic, and ultimately gamodesmic.   In *Botrychium virginianum* (Jeffrey, 34) no internal endodermis is present, so that the stem is monostelic.   Hence, in closely related plants, a great difference in the structure— considered from the morphological point of view—results from the acceptance of the endodermis as a layer of morphological

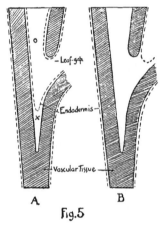

Fig. 5

value.   This same irregularity is exhibited in the Osmundaceæ (Faull, 21).   This difficulty was fully recognised by Strasburger, who suggested that the term "endodermis" should be used in the physiological sense, and restricted to that layer of cells which exhibits the well-known cuticularisation; while the term "phlœoterma" should be used to designate the band of tissue delimiting the cortex from the stele.   Here again arises a difficulty, for the only means of distinguishing the phlœoterma is that of position.

This difficulty associated with the endodermis may be illustrated by fig. 5, A, which represents a longitudinal section of a fern stem which has an internal endodermis originating in connection with the first leaf-gap.   Taking the region below

[2] See also Lang, *Annals of Botany*, xvi.

this gap (indicated by a X), the central parenchymatous tissue may be termed pith and may be considered to be stelar in origin. But above the gap (indicated by a circle) the central tissue would be morphologically distinct from the stele on account of the presence of the internal endodermis. Thus, in this upper region the pith would be absent unless there happened to be a band of parenchyma between the inner endodermis and the vascular tissue. On the other hand, another plant might be precisely similar excepting for the absence of the internal endodermis (fig. 5, B). So that, in this case, the pith at the base would be morphologically the same tissue as the central parenchymatous mass above the first leaf-gap.

On account of these facts, many have discarded the view that the endodermis is a tissue possessing any morphological signi-ficance; but, on the other hand, the older opinion is still held by some, for very little is known regarding the physiological rôle of the endodermis, and in the majority of cases the endodermis is present in the young stages, even when absent from the adult plants, having disappeared through reduction (Jeffrey).

It is possible that the endodermis may function as an air-tight layer of cells around the vascular tissue, separating it from air-containing intercellular spaces, and so preventing the access of air into the tracheæ.

Some such provision as this is apparently very necessary in the Ferns, for example, as, very generally, no secondary thickening takes place, and thus a tracheid rendered more or less functionless by the entry of air cannot be replaced. For this reason the endodermis completely surrounds the vascular tissue.

Drabble (17, p. 66) remarks that "It is with the xylem portion of the bundle that this [the above] function of the endodermis is generally assumed to be associated. This, however, is not very clear. In roots the endodermis may be provided with non-cuticularised 'passage-cells.' These cells occur opposite the xylem groups, while the cuticularised elements are usually continuous over the phloem regions. The frequent reduction of xylem in aquatic plants is not necessarily, or even generally, associated with the reduction in the endodermis."

He points out instances of the absence of the endodermis where, presumably, it should be present, and suggests "that the endodermis may find its function rather as a phloem sheath than a xylem sheath." This is, of course, possible, but it does not appear altogether probable.

In a great many roots, at any rate, the endodermis does form a continuous sheath around the vascular cylinder, and even in those cases where marked passage-cells are present (distinguishable by the fact that their basal walls are not thickened like the other cells of the endodermis), the radial walls are often cuticularised.[1] Then again, although the xylem of aquatic plants is much reduced it may not be entirely absent, and it performs its normal function of water-carriage. This fact was shown by Sauvageau[2] and, more recently, by Pond,[3] who demonstrated that the plants he experimented with, viz. *Vallisneria spiralis, Ranunculus aquatilis trichophyllus, Elodea canadensis, Myriophyllum spicatum,* and species of *Potamogeton* are dependent on the substratum for a considerable portion of their food-material, and that there is a current of water upwards from the roots.

Finally, the reason why the xylem requires protection against the access of air is clear. There frequently exists in the tracheæ a marked negative pressure ; should air gain an entry into these elements, while in this condition, the passage of water through them would be enormously impeded, if not entirely stopped, on account of the much greater resistance to the flow.[4]

On the other hand, the reason why the phloem elements should require the protection of an endodermis is not obvious. On the whole the anatomical evidence appears to support the first possibility, considered above, which is the view held by Strasburger, Boodle, and others. Jeffrey controverts this inasmuch as no experimental evidence has been brought forward.

Another use assigned to this tissue is connected with the reaction of certain parts of plants to the stimulus produced by weight, the consideration of which is outside the scope of the present paper.[5]

[1] The probability is not excluded that in those cases where the endodermis is greatly thickened both on its radial and basal walls, it may also perform a mechanical function.

[2] Sauvageau, C., *Sur les feuilles de quelques Monocotylédones aquatiques.* Paris 1891.

[3] Pond, Raymond H., "The Biological Relation of Aquatic Plants to the Substratum." *U.S. Fish Com. Rep.,* 1903.

[4] See Ewart, "The Ascent of Water in Trees," *Phil. Trans. Roy. Soc. Lond.* B. 198, 1905.

[5] See Darwin, Francis, Presidential address, British Association, Section K, Cambridge, 1904.

The Morphological Status of the Pith

In the majority of cases the first appearance of this tissue is an aggregation of parenchymatous elements within the hitherto solid rod of xylem of the protostele. This parenchyma may be directly continuous with the same tissue of the siphonostele. Now if the development of the vascular system stopped here there would probably be very little difficulty in accepting the pith as a morphological tissue ; but development goes further, the siphonostele gives rise to the dictyostele, which widens out considerably, enclosing in its network a mass of parenchyma— sometimes also sclerenchyma—which is indistinguishable from the more externally placed cortex. Where, in such cases, does the pith end and the cortex begin ? Again, in many cases, *e.g.* *Angiopteris*, *Marattia* (Farmer and Hill, 20), *Danæa* (Brebner, 8), *Matonia* (Tansley and Lulham, 61), and other plants, this " pith " is traversed by vascular strands. Further, vascular tissue may appear in the external ground tissue; Brebner (8) recorded a case in *Danæa* where a tracheid occurred in the ground tissue of the leaf-bases, which helps "to show that the line to be drawn between stelar and extrastelar tissue is not a hard and fast one." Hence, numerous difficulties have to be surmounted before many botanists can accept the tissue enclosed by the vascular cylinder, in all its varieties, as having any morphological significance.

Jeffrey argues that inasmuch as the internal phlœoterma, when present, is in direct continuity with the external phlœoterma[1] through the leaf-gaps, it is therefore of the same value morphologically, and hence it follows that the pith is not part of the stele, but represents a part of the cortical ground tissue; in other words, it is extrastelar.

Boodle (5, p. 533) considers that inasmuch as, in all probability, the solid protostele was the primitive type which gave rise to the more complicated forms of vascular systems, it follows that "whatever tissue is found within the xylem is presumably morphologically stelar. Assuming an exarch[2] protostele, a pith may have originated by incomplete differ-

---

[1] It should be mentioned that Jeffrey uses the term phlœoterma synonymously with endodermis.

[2] A term in common use to denote the position of the first-formed element, the protoxylem, of the wood. Thus, the protoxylem is exarch when it is situated in a peripheral position ; the metaxylem, or later-formed wood, is developed centripetally, and is thus more internally placed.

entiation of the xylem-mass. At any rate, if one regards the pith or other central tissue as having arisen in the first place by the transformation of potential tracheides into other tissue-elements, these latter should be treated morphologically as part of the stele." For these reasons, the pith is regarded as being stelar, and morphologically distinct from the cortex.

Gwynne-Vaughan (29, p. 737) does not consider that the inconstancy of the apical cells and meristems has any bearing on the main question. He remarks that "the consideration of the morphology of the stele can only begin when that region is definitely and satisfactorily delimited. It is not directly concerned with any question as to which particular segment it is in which the tangential wall appears that first of all delimits it. All that is required is that a definite delimitation should actually be possible at one point or another during the course of its development in a majority of cases sufficiently great to render the statement general."

He is of the opinion that it is better to consider the central parenchyma, or pith, as stelar, and to regard the internal endodermis as not being strictly homologous with the external endodermis. But inasmuch as the cortical origin of the central parenchyma is theoretically possible, it should not at once be rejected as inherently improbable.

Tansley and Lulham (61, p. 514) regard Jeffrey's view of the intrusion of the cortex into the stele as having "a merely metaphorical value."

It is pointed out that in order to show that the pith originated by an invasion of the cortex, it must be proved that the cortical elements actually pushed their way into the stele during the development, or that such had been the case during the course of descent. They consider that the pith "is morphologically an entirely new tissue, formed in the centre of the stele, in place of vascular tissue, which preceded it in ontogeny and phylogeny, but is always, in the Ferns, separated from the latter by endodermis."

As regards the difficulty connected with the occurrence of the internal commissural strands of vascular tissue in the Marattiaceæ and *Matonia*, Tansley and Lulham remark that, if their view of the pith be correct, then these accessory structures of such polycyclic ferns must also be regarded as new developments, which in part replace the pith in much the same manner as the pith replaced the central vascular tissue of the protostele.

Farmer and Hill (20), and Chandler (10) have expressed the opinion that, in view of the difficulties inseparable from the consideration of the pith as a morphological entity, it is better to confine attention to vascular and non-vascular tissue; or, in other words, to treat the matter from the physiological rather than from the morphological point of view. This opinion, it may be remarked, has not, on the whole, been accepted with favour; for although physiologically it is indisputable, it " radically ignores the morphological problems with which the anatomist is confronted when he is endeavouring to trace out the evolution of the tissues of vascular plants" (Tansley and Lulham, 61).

Thus we see that the stelar theory has undergone marked changes; indeed, there is in some quarters a return to the older views of De Bary and Sachs.

The physiological explanations for the variety of structures obtaining, and questions relating to phylogeny, together with certain other side issues, cannot here be dealt with; nor, indeed, is it desirable, for otherwise this introduction would become too complicated, and so defeat its object.

In conclusion, it must be borne in mind that although Van Tieghem's theory is not now accepted by the majority, there is no desire to underrate the great value of his work.

A theory, after all, is nothing more than a working hypothesis; and as such, few theories have been more productive than that of Van Tieghem. Without it, it is very doubtful whether the enormous advance in our knowledge of the structure of the Ferns during the last decade would have been so great.

LITERATURE

1. BELLI, Endoderma e periccilo nel G. Trifolium in rapporta colla teoria della stelia di V. Tieghem e Douliot, *Mem. Reale Accad. Sci.* Torino, 46, 1896.
2. BOODLE, Comparative anatomy of the Hymenophyllaceæ, Schizæaceæ, and Gleicheniaceæ. I. On the anatomy of the Hymenophyllaceæ, *Annals of Botany*, xiv.
3. —— *Id.* II. On the anatomy of the Schizæaceæ, *Ann. of Bot.* xiv.
4. —— *Id.* III. On the anatomy of the Gleicheniaceæ, *Ann. of Bot.* xv.
5. —— *Id.* IV. Further observations on Schizæa, *Ann. of Bot.* xvii.
6. —— On descriptions of vascular structures, *New Phytologist* ii.
7. —— Review: The anatomy of Palm roots, *New Phytol.* iv.

8. BREBNER, On the anatomy of Danæa and other Marattiaceæ, *Ann. of Bot.* xvi.

9. BRITTON and TAYLOR, Life-history of Schizæa pusilla, *Bull. Torrey Bot. Club*, xxviii.

10. CHANDLER, On the arrangement of the vascular strands in the "seedlings" of certain Leptosporangiate Ferns, *Ann. of Bot.* xix.

11. CHRYSLER, The development of the central cylinder of Araceæ and Liliaceæ, *Botanical Gazette* xxxviii.

12. CORMACK, On polystelic roots of certain Palms, *Trans. Linn. Soc.* Ser. 2, Bot. vol. v.

13. DANGEARD et BARBÉ, La polystélie dans le genre Pinguicula, *Bull. Soc. Bot. de France*, xxxiv.

14. DE BARY, Comparative anatomy of the Phanerogams and Ferns, 1884.

15. DRABBLE, On the anatomy of the roots of Palms, *Trans. Linn. Soc.* Ser. 2, Bot. vi.

16. —— A note on vascular tissue, *New Phytol.* iv.

17. —— The transition from stem to root in some Palm seedlings, *New Phytol.* v.

18. FARMER, On the embryogeny of Angiopteris evecta, *Ann. of Bot.* vi.

19. FARMER and FREEMAN, On the structure and affinities of Helminthostachys zeylanica, *Ann. of Bot.* xiii.

20. FARMER and HILL, On the arrangement and structure of the vascular strands in Angiopteris evecta and some other Marattiaceæ, *Ann. of Bot.* xvi.

21. FAULL, The anatomy of the Osmundaceæ, *Bot. Gaz.* xxxii.

22. FORD, The anatomy of Ceratopteris thalictroides, L., *Ann. of Bot.* xvi.

23. GWYNNE-VAUGHAN, A new case of polystely in Dicotyledons, *Ann. of Bot.* x.

24. —— On the arrangement of the vascular bundles in certain Nymphæaceæ, *Ann. of Bot.* x.

25. —— Polystely in the genus Primula, *Ann. of Bot.* xi.

26. —— Observations on the anatomy of Solenostelic Ferns. I. Loxsoma, *Ann. of Bot.* xv.

27. —— Remarks upon the nature of the stele of Equisetum, *Ann. of Bot.* xv.

28. —— Some observations upon the vascular anatomy of the Cyatheaceæ, *Ann. of Bot.* xv.

29. —— Observations on the anatomy of Solenostelic Ferns, Part II., *Ann. of Bot.* xvii.

30. —— On the possible existence of a fern stem having the form of a lattice-work tube, *New Phytol.* iv.

31. —— On the anatomy of Archangiopteris Henryi and other Marattiaceæ, *Ann. of Bot.* xix.

32. HOLLE, Ueber Bau und Entwickelung der Vegetationsorgane der Ophioglosseen, *Bot. Ztg.* 1875.

33. JANCEZEWSKI, Études morphol. sur le genre Anemone, *Rev. Gen. de Bot.* 1898,

34. JEFFREY, The gametophyte of Botrychium virginianum, *Trans. Canadian Inst.* v.

35. —— The development, structure, and affinities of the genus Equisetum, *Mem. Boston Soc. Nat. Hist.* v.

36. —— A theory of the morphology of Stelar structures, *Proc. Roy. Soc. Canada*, 1897.

37. —— The morphology of the central cylinder in the Angiosperms, *Trans. Canad. Inst.* 1900.

38. —— The structure and development of the stem in the Pteridophyta and Gymnosperms, *Phil. Trans. Roy. Soc. London*, Ser. B., vol. cxcv.

39. KAMIENSKI, Vergleichende Anatomie d. Primulaceen, 1878.
40. KÜHN, Untersuchungen ü. d. Anatomie der Marattiaceen u. Gefässkrypto-gamen, *Flora*, lxxii.
41. —— Ueber d. Anatomischen Bau von Danæa, *Flora*, lxxiii.
42. LECLERC DU SABLON, Recherches sur la formation de la tlge dans les Fougères, *Ann. Sci. Nat. Bot.* 7ᵉ sér. t. xi.
43. LEWIS, Formation of an irregular endodermis in the roots of Ruscus, sp. *Ann. of Bot.* xiv.
44. MERKER, Gunnera macrophylla, *Inaug. Diss. Marbourg*, 1888.
45. MOROT, Recherches sur le péricycle ou couche périphérique du cylindre central chez les phanérogames, *Ann. Sci. Nat. Bot.* 6ᵉ sér. 20.
46. PFITZER, Über d. Schutzscheide d. deut. Equisetaceen, *Jahrb. f. Wiss. Bot.* vi.
47. POIRAULT, Rech. anat. sur les Cryptogames vasculaires, *Ann. Sci. Nat. Bot.* 7 sér. t. xviii.
48. —— Sur l'Ophioglossum vulgatum, *Journ. de Bot.* vi.
49. SCHOUTE, Die Stelär-Theorie, Groningen, 1902.
50. SEWARD and DALE, The Structure and Affinities of Dipteris, *Phil. Trans. Roy. Soc. Lond.* B. cxci.
51. SEWARD and FORD, The anatomy of Todea, *Trans. Linn. Soc.* 2nd ser. vi.
52. SCOTT, On the origin of polystely in Dicotyledons, *Ann. of Bot.* v.
53. —— Prof. Jeffrey's theory of the stele, *New Phytol.* i.
54. SHOVE, On the structure of the stem of Angiopteris evecta, *Ann. of Bot.* xiv.
55. STRASBURGER, Über Bau u. verricht. d. Leitungsbahnen, Jena, 1891.
56. TANSLEY, The stelar theory, *Science Progress*, 1896.
57. TANSLEY and CHICK, Notes on the conducting-tissue system in Bryophyta, *Ann. of Bot.* xv.
58. —— —— On the anatomy of Schizæa malaccana, *Ann. of Bot.* xvii.
59. TANSLEY and LULHAM, On a new type of Fern stele, and its probable phylogenic relation, *Ann. of Bot.* xvi.
60. —— —— The vascular system of the rhizome and leaf-trace of Pteris aquilina, L., and Pteris incisa, Thunb. var. integrifolia, Beddome, *New Phytol.* iii.
61. —— —— A study of the vascular system of Matonia pectinata, *Ann. of Bot.* xix.
62. THOMAS, Some points in the anatomy of Acrostichum aureum, *New Phytol.* iv.
63. VAN TIEGHEM, Structure de la tige des primevères, *Bull. Soc. Bot.* xxxiii.
64. —— Sur la limite du cylindre central et de l'écorce dans les Cryptogames vasculaires, *Journ. de Bot.* ii.
65. —— Sur la dédoublement de l'endoderm, *Journ. de Bot.* ii.
66. —— Remarques sur la structure de la tige des Prêles, *Journ. de Bot.* iv.
67. —— Remarques sur la structure de la tige des Ophioglossées, *Journ. de Bot.* iv.
68. —— Traité de Botanique, Paris, 1891.
69. VAN TIEGHEM ET DOULIOT, Sur la polystélie, *Ann. Sci. Nat. Bot.* 7ᵉ sér. t. iii.
70. —— Sur les tiges à plusieures cylindres centraux, *Bull. Soc. Bot. de France*, xxxiii.
71. WIGGLESWORTH, Notes on the rhizome of Matonia pectinata, *New Phytol.* i.
72. ZANETTI, Das Leitungssystem im stamm von Osmunda regalis, *Bot. Ztg.* 1895.

ON A HILLSIDE IN DONEGAL : A GLIMPSE INTO THE GREAT EARTH-CALDRONS

By GRENVILLE A. J. COLE, F.G.S.

Professor of Geology in the Royal College of Science for Ireland

THE days for great sweeping surveys of the earth have long been over; and the *condottieri* of the geological armies, who would ride rough-shod across half a continent, lived to find local toilers rise against them, and to see weapons forged in nooks and crannies raised to bar their future passage. For a time, indeed, geological observation threatened to become lost in detail. The description of new fossil species seemed to go forward without regard to zonal grouping, or to the relations of faunas as a whole; while the neat labelling of a microscopic section became the logical end of an unnatural system of petrography. The signs of the times, after this diversion towards the histology of the subject, again indicate a desire to survey and correlate; and modern generalisation is certain to be wisely tempered by the advice of the specialists who now abound in every field.

Just as a lake may nurture a peculiar variety of trout, so some corner of the earth may have its own mineral associations; but many geological phenomena recur over such wide areas that we may fairly regard them as fundamental. Hence, when we sit down, for our personal edification, upon a bare and glaciated boss in Donegal, we may extend our view from the quartzite of Aghla and the granite of Glendowan to the crystalline crust at large, the field of many a tourney.

The rock around us seems gnarled and twisted; it is composed of interfolded sheets of differing constitution, some of which contain abundant dark and gleaming mica, while others are light in colour and rich in quartz and felspar. The rock is, in fact, a "gneiss," not merely with a general parallelism among its constituents, but with a well-marked banded structure. Here and there a sharply bounded band of quartz appears, with mica-schist on either side; the surfaces

of junction of these types of rock run parallel with the general
"flow-structure" of the gneiss, and the contrast between the
exceptionally micaceous and exceptionally quartzose portions
seems to imply some local differentiation in the mass.

Here, again, elongated blocks of hornblendic rock lie in the
gneiss, sometimes like pebbles, sometimes like very flat lenses ;
their direction of elongation is that of the prevalent structure
in the gneiss. At their margins they fade off into the gneiss
by the formation of layers of dark mica, whereby the more
felspathic rock around them becomes darker where they are
abundant. Epidote, sphene, and garnet are then found in it
as minerals of the junction-zones.

Cutting across the gnarled structure of the gneiss runs a
series of dykes and veins of pure granite, consisting mainly of
quartz and felspar. These clearly represent cracks that have
arisen subsequently to the crush or flow that produced the
gneiss. Molten material, capable of crystallising as granite,
oozed up into them from below, and the quartz and felspar
often separated out so slowly that coarse "graphic granite"
has been formed. At other times crystallisation is on a delicate
scale, and the material of the veins looks almost like sandstone.
The fact that this later granite contains red felspars has been
observed also in far-distant localities, but this as yet has no
known significance.

More important, perhaps, than what we have hitherto noticed
is the occurrence of a mass of limestone, quite large enough
to quarry into, enveloped as it were in the gneiss. The more
quartzose and felspathic bands of the gneiss appear as sheets
and veins cutting the limestone, and crystalline calcium-silicates
have freely arisen in the latter, as they would at the contact
with a once molten or "igneous" rock. Lime-garnets, brown
and nodular, with attempts at regular crystal-faces, are often conspicuous on the surfaces of junction. They may, indeed, be as
much as five centimetres across, and their size seems a measure
of the intensity of interaction, combined with the maintenance of
similar conditions throughout a long period of time.

These features may lead us to regard the gneiss itself as
a once molten mass that has become streaky during flow ;[1]

---

[1] *Cf.* the memorable views of G. P. Scrope, as far back as 1825 (*Considerations
on Volcanos*, p. 226); and of Charles Darwin, *Geological Observations on South
America* (1846), Minerva edition, p. 440.

but the view of the older geologists very generally was that the streakiness represented the remains of stratification, and that the gneiss arose from the crystallisation of the materials of normal sedimentary rocks. Consider our limestone bands, our quartzites, our darker patches of micaceous or hornblendic matter, all retaining a similar trend, or curving only with the curvings of the structure of the gneiss. Do not these represent residual masses that have escaped complete metamorphism? Do they not tell us of the obliteration of similar features in the main body of the gneiss? It has thus been suggested[1] that the blocks and balls of hornblendic rock embedded in a streaky gneiss record for us the presence of an old conglomerate which has become crystalline through sheer antiquity. Moreover, banded gneiss occurs again and again upon the margins of a normal granite mass. Is it not reasonable to suppose that the granite itself was once sedimentary, and represents to-day the extreme phase of alteration?

A third view to account for the streakiness of the gneiss has been generally held in recent years. Workers in regions where mountains have been reared by lateral pressures above the general level of the crust became impressed with the enormous forces that operated upon the central cores. The granite that is usually so prominent in these cores assumes a gneissose aspect, and is succeeded by fine-grained schists, as one passes from the centre towards the flanks of the mountain-chain. The intense crumpling, over-folding, and over-thrusting that can be traced among the stratified rocks of the foot-hills, led to the belief that the structures of the crystalline core could be explained by similar deformation. The microscope gave valuable aid in these inquiries, and proved that crushing and warping have frequently affected the crystalline constituents of schists and gneisses over very considerable areas. The parallel arrangement of the crystalline materials, the "foliation," in fact, was seen in many cases to be due to the action of pressure, and to be absolutely independent of any previous structure in the rock.

For instance, Judd writes in *The Student's Lyell* (1896, p. 549): "There can be little doubt that, in the great majority of cases, the schistose structure is an entirely superinduced one, and that foliation, like cleavage, must be referred to the

[1] Hardman, "On the Metamorphic Rocks of Counties Sligo and Leitrim, and the enclosed Minerals," *Sci. Proc. R. Dublin Soc.*, vol. iii. (1882), p. 358.

action of pressure, the planes of foliation being developed, like those of cleavage, at right angles to the direction in which the pressure is exerted."

Sometimes, under pressure, the larger crystals in a gneiss have become ground down to an ovoid form by the flow of smaller and powdered constituents round them. The coarser granites have been milled down,[1] as it were, and foliated gneisses, or even slate-like masses, have resulted. It has been asserted that development of mica along the surfaces of movement, has set up differences of constitution in different layers,[2] and a certain banded effect, resembling stratification, has resulted within rocks of truly igneous origin, as well as within altered sediments. At the same time, the intimate commingling of rocks of various kinds in the great earth-mill has constructed schists and gneisses out of very diverse materials. Here and there a fragment of sedimentary rock may have escaped destruction; elongated quartz-patches may similarly remain, as indications of the pebbles of some old conglomerate; but for the most part the origin of such schists and gneisses has become obscured by the breaking down of all their original structures.

Such, in a very few words, is the theory of dynamometamorphism, which has been employed to explain phenomena outwardly as different as the uniform cleavage of slate and the intense puckering of banded gneiss. Prof. Bonney[3] has consistently opposed the view that shearing by itself can produce a banded structure. He has, moreover, often laid stress upon the probability that sedimentation in very early days was so closely accompanied by metamorphic changes that the Archæan or most ancient rocks present certain distinct features that were impressed upon them practically at their birth. In such rocks, the foliation is held by Bonney to coincide with the original stratification. High temperatures prevailed, at the time of their deposition, comparatively near the surface of the globe, and differences in the constitution of certain layers of the sediments promoted a characteristic banding as the materials crystallised throughout the mass.

[1] Cf. Lapworth, *Intermediate Text-book of Geology* (1899), p. 124, etc.

[2] Lehmann, *Untersuchungen über die Entstehung der altkrystallinischen Schiefergesteine*, 1884, and other authors.

[3] E.g. Discourse on "The Foundation Stones of the Earth's Crust," British Association, 1888 (also in *Nature*, vol. xxxix. p. 89).

Van Hise, who does not pretend to have covered the wide field of foreign literature, even in his recent monumental work, supports Bonney's views as to the origin of banding in schists and gneisses of sedimentary origin.[1] On the other hand, "shearing" has been called in by many authors to account for the flow-structures in such gnarled rocks as that beneath our feet in Donegal, regardless of the curves and wrinklings that surely would have been obliterated or non-existent under any form of crushing combined with movement. It is time, therefore, to inquire further into the phenomena presented to us in the field.

In a large quarry opened in our gneissic series near Fintown, in Co. Donegal, we are introduced more clearly to the meaning of this particular "complex" of rocks. The foliation of the whole is here vertical—that is to say, crystalline rocks of different types run up side by side like walls, or like the uptilted beds of a stratified series. Here is a layer of grey gneiss rich in biotite, here a true mica-schist, here a band of uncrushed quartzite. On the north side of the quarry we have vertical bands of soft green shale, which prove to be decomposed basic igneous rocks; these are kept from complete denudation by bounding walls of the grey gneiss. So far, the gneiss, which looks like a granite with a general flow-structure, might be supposed to represent the most altered layers of a great sedimentary series.

But, higher up in the quarry, this grey gneiss cuts abruptly across all the bedded and foliated structure, and appears in a massive condition, with the jointing of an ordinary granite. Blocks of quartzite remain in it, here and there associated, as we noticed elsewhere, with clinging films of mica-schist; but the mica-schist seems mostly to have been destroyed. A huge block of diorite, into which the gneiss sends off veins, is seen to be entombed in the gneiss, which is here clearly an igneous mass, intrusive in and absorbing its surroundings.

We may be assured that the parallel structure in the lower part of the quarry is truly original and sedimentary, if we examine the vertical beds of quartzite and mica-schist—altered argillaceous rock—in the country at no great distance. Where granite invades these ancient strata, even their bedding some-

[1] "A Treatise on Metamorphism," *Monograph* xlvii. *U.S. Geological Survey* (1904), p. 762.

times disappears ; and it is difficult then to say how much of their crystalline constituents is original, and how much is derived from delicate interpenetration by the granite. The high degree of intermixture is astonishing when the microscope is brought into action, and the darkening of the granite near the contact is clearly connected with the absorption of material from the schists into which it has intruded.

Surely, then, where the same granite shows a banded structure parallel with the foliation of the adjacent schists, this may be, and very probably is, due to the presence of residual flakes of the schist, only partly "granitised," remaining along and near the contact-zone. If, moreover, as is here the case, the foliation of the schistose rocks represents the original bedding of a sedimentary series, the corresponding foliation in the granite is a relic of, and is induced by, the same original stratification. Whole masses of granite may thus invade and take the place of a pre-existing schistose series, without destroying all traces of the original structure of the district. This is the extreme case of the phenomenon of *lit par lit injection*, so ably described and explained by Michel Lévy,[1] and so adequately illustrated by Barrois, Lacroix, and other workers in French metamorphic areas, from Brittany to Mont Blanc and the Pyrenees. Lévy's own work on the granite of Flamanville[2] is a convincing exposition of the phenomena of marginal alteration and absorption; while in Great Britain, Miller,[3] in dealing with certain Scottish granites, has gone so far as to write that parts of these granites "are in fact pseudomorphs, or granite casts, preserving, as replacement structures, remains of the structure of the pre-existing rock."

Put in some such form as the above, and accompanied by demonstration of the phenomena in the field, it is easy enough to persuade the geological student, unacquainted as yet with the literature of metamorphism, of the correctness of Lévy's view in a wide series of examples. Take this embryo observer from Donegal, where the same type of granite-contact occurs over hundreds of miles of boundary, and set him in the county of

[1] "Sur l'origine des terrains cristallins primitifs," *Bull. Soc. géol. de France*, 1887.

[2] *Bulletin Carte géol. de France*, tome v. (1893).

[3] Quoted by Horne and Greenly, "On Foliated Granites and their Relations to the Crystalline Schists of E. Sutherland," *Quart. Journ. Geol. Soc. London*, vol. lii. (1896), p. 635.

Wicklow, above the romantic water-slide in Glenmacnass. He has already noticed the great curving joints of the granite bending down from the highland towards the foothills, which he now approaches, as if parallel with a surface of cooling, parallel, that is, with the margin of the intrusive mass. He now sees, running in much the same general direction through the granite, and parallel with the rock-surface that forms the water-slide, bands of dark micaceous material, which prepare him for encountering a contact with the schists. The granite is locally modified in structure and, clearly, in composition also. Looking farther down the valley, the student sees the walls formed of mica schist, the result of the contact-alteration of Silurian shales by the granite core of the Leinster Chain. The strike of the foliation of these schists, which is undoubtedly that of the original uptilted stratification, is parallel with the dark bands in the granite. Is, then, the student to conclude that the pressure that upraised the Silurian strata produced micaceous sheets in the granite parallel with the junction, and subsequent to the partial or complete consolidation of the mass; or that the dark bands are relics of the schists, and that the whole phenomena, the metamorphism of the Silurian beds and the modification of the granite into a handsome streaky gneiss, are manifestations of continuous action, and alike due to the conditions of intrusion?

J. B. Jukes has written truly of Co. Wicklow, "In the great majority of instances, the folia of the mica-schist, whether straight or puckered, are certainly parallel to the grit bands, and therefore to the original lamination and stratification of the rock."[1] But the principle of *lit par lit* injection was not present to his mind; and the foliation in the granites of Donegal and Connemara[2] only served to connect them, in his view, with the sediments, as representing an extreme of metamorphism. Unpopular as this view has since become, the student finds it replaced by one presenting equal difficulties. As we have said, he is asked to regard the marginal streakiness and darkening of the granite as due to pressure acting on it.[3] The surfaces of

---

[1] *Student's Manual of Geology*, third edition, revised by Sir A. Geikie (1872), p. 229.

[2] *Ibid.* p. 145.

[3] Sollas, "Contributions to a Knowledge of the Granites of Leinster," *Sci. Trans. R. Dublin Soc.*, vol. xxix. (1890) pp. 496 and 501.

mica within the granite are " planes of flow or sliding surfaces,"
the primary arrangement of this mica having been due to viscous
flow near and parallel to a bounding surface. " On the application of pressure, shearing evidently commenced along lines
determined by the mica plates, and was accompanied by the
crushing of the quartz mosaics, while the felspar as a rule
yielded but little : the crushed material set flowing moved in
stream-lines round the immersed felspar, which remains as the
' eyes' in the 'flaser' structure."

The foregoing description by Prof. Sollas of a granite in
Co. Wicklow applies to most cases of foliated granites containing micaceous bands ; but the microscopic evidence of
crushing is by no means always so apparent. The comminu-
tion of the crystals of quartz, their flowing round the felspars,
the consequent "eye-structure," and the wavy banding of the
mica flakes, are all paralleled in cases of undoubted igneous
flow.[2] Indeed, as a molten rock becomes viscous and begins
to crystallise, it flows—as it probably did from the beginning—
under the influence of some earth-pressure far away behind it.
The same pressure that rears the dome of superincumbent
sediments, and permits of the entry of the molten magma,
forces that magma into every crevice long after crystallisation
has begun. Even the fracture of crystals and the production
of the well-known optical "strain-shadows" need not in all
cases be assigned to a period of crush subsequent to that of
consolidation.

The fact that igneous rocks move forward, and are operative
as true intrusive masses, after many of their constituents have
appeared as prominent crystals, is proved by many interesting
features in the zones of contact. Barrow[3] has described the
filtering out of oligoclase crystals from a granite, and their
accumulation in the main mass of the rock, because the
veins that had opened on the margin were too narrow
to admit them. The ultimate intrusive material in these
veins thus differs, through physical causes, from the granite
from which it is derived. This is, of course, only a striking

[1] Scrope long ago noticed this sliding action among the constituents of gneiss
(*Considerations on Volcanos*, 1825, p. 234).

[2] Cole, " Metamorphic Rocks in E. Tyrone, etc.," *Trans. R. Irish Acad.*,
vol. xxxi. (1900) p. 440.

[3] " On certain Gneisses with round-grained Oligoclase," *Geol. Mag.* 1892,
p. 65.

example of the intrusion of a magma from which crystals have already separated, such as that observed by Teall[1] in the glassy infillings of vesicles in certain lavas. The lava has almost solidified, gases have blown up hollows in it, and yet the residual magma is still fluid enough to run in and occupy the cavities. Another pleasing instance occurs on the coast of Co. Down,[2] where a fine-grained granite, already rich in crystals of quartz and orthoclase felspar, larger than those that developed in the groundwork of the rock, has invaded a dyke of basalt. Blocks of the basalt have been carried off by the granite, and have melted within the invader. In turn, they have reacted on and absorbed material from the granite magma, but have been unable to completely destroy the larger crystals that had already developed in it. In consequence, they now include red orthoclase and quartz, in a ground the constitution of which is not far different from that of the original basalt. Harker[3] has confirmed this observation during his researches in the Isle of Skye, where acid felspars and quartz grains have been transferred from a granite into fragments of a basalt into which it has intruded.

Salomon,[4] moreover, notes the transference of great crystals of biotite and triclinic felspar from an enveloping quartz-diorite to included fragments of a schist.

At Green Point, again, near Cape Town, there is an exposure of more than ordinary interest, since it was described by Charles Darwin,[5] the author of the term "foliation," in 1844. The shaly Malmesbury Beds, the oldest recognised rocks in Cape Colony, here come up almost vertically on the shore ; granite, with large crystals of orthoclase, occupies the country under Table Mountain to the north, and penetrates the shales at Green Point in a series of parallel sheets. Indeed, as Darwin wrote, films of altered clay slate are "isolated, as if floating, in the coarsely crystallised granite ; but, although completely detached, they all retain traces of the uniform north-west and south-east cleavage."

[1] "On the Amygdaloids of the Tynemouth Dyke," *Geol. Mag.*, 1889, p. 482.
[2] Cole, "On derived crystals in basaltic andesite, Co. Down," *Sci. Trans. R. Dublin Soc.*, vol. v. (1894), p. 239.
[3] "Igneous Rocks of Skye," *Mem. Geol. Survey of United Kingdom* (1904), p. 219.
[4] "Geologische Studien am Monte Aviolo," *Zeitsch. d. deutsch. geol. Gesell.*, Bd. xlii. (1890), pp. 476 and 493.
[5] *Geological Observations on Volcanic Islands*, Minerva ed. p. 264

Darwin goes on to point out how fissures parallel to the planes of cleavage would develop during the arching-up of the rock by a mass of molten granite, and "these would be filled with granite, so that wherever the fissures were close to each other, mere parting layers or wedges of the slate would depend into the granite. Should, therefore, the whole body of rock afterwards become worn down and denuded, the lower ends of these dependent masses or wedges of slate would be left quite isolated in the granite; yet they would retain their proper lines of cleavage, from having been united, whilst the granite was fluid, with a continuous covering of clay slate."

Here Darwin anticipates the principle of *lit par lit* injection, put forward by Lévy forty-three years later; but he does not foresee the connexion of this phenomenon with the structure of the adjacent gneiss. In accordance with the views then prevalent, the gneiss of Cape Town was regarded by him as belonging to the shale-series rather than to the granite, and as produced from the former by extreme metamorphism.

I have had the pleasure of noting on the spot, under the guidance of Prof. Andrew Young, a few further details of this section. The filmy shale-bands are so intimately penetrated by the granite that the large felspar crystals of the latter often lie in these bands as if they were fragments from some older rock. Composite gneiss of many types has been built up along this complex junction. Blocks of shale, completely detached, lie in the granite, and are in various stages of "granitisation" by the development of new minerals, or by the addition of material from the igneous rock.

Farther north, at Clifton, the granite is eminently gneissic, the dark bands coming up steeply on the face of a cliff cut for the road. Coming from this side towards Seapoint, any one familiar with Donegal or Wicklow, to take only these examples, would be prepared for meeting a mass of shales at no great distance. The gneiss of Clifton is no doubt of composite origin, and represents a large incompletely digested inclusion of the Malmesbury series.

I have been able to study in greater detail the very similar junction on Carbane, near Glenties, in Co. Donegal,[1] where crystals from the granite appear freely in the flakes of shale.

[1] "On Composite Gneisses in Boylagh, W. Donegal," *Proc. R. Irish Acad.*, vol. xxiv. Sect. B (1902), p. 213.

Pressure-changes are here traceable, but the main features are clearly original, and subsequent attempts at shearing have obscured the gneissic structure rather than intensified it. This is in accordance with the views of Bonney, already quoted.

So far, perhaps, a case has been stated, and it is well to return towards a more judicial attitude. We have from time to time noted the inclusion of dark blocks in our typical gneiss or granite, and have without question regarded them as remnants of some pre-existing rock through which the granite broke. But this is very far from representing the prevalent opinion of geologists. Every crystal or group of crystals in an igneous rock represents a separation of material, a segregation, whereby the surrounding magma is left the poorer by some constituents. From various causes, mere gravitation being one of them, at some time or other in the earth's history, and perhaps in every caldron of molten rock beneath our feet, a differentiation of material may have taken place, giving us a magma rich in silica on the one hand, and a magma rich in ferro-magnesian silicates and iron ores on the other.[1] Long before Brögger so brilliantly developed the differentiation-theory, in his great paper on the " Ganggefolge des Laurdalits " (1898), to account for the series of igneous rocks in the neighbourhood of Christiania, "segregation-veins" and "segregation-patches" had been freely spoken about, and it had been felt that mineral aggregates of very varied nature could separate out from a once homogeneous mass. J. A. Phillips,[2] in discussing the lumps of various kinds found in granite, referred the more crystalline and granitoid ones to segregation-patches, and the more angular and obviously sedimentary blocks to inclusions picked up during flow. Although the margin of a block might be very sharp, yet a crystal from the granite might be found running into it; and this was rather naturally accepted as evidence that the two types of rock consolidated from the same magma. We have already seen that this is by no means a necessary conclusion. The microscopic study of contact-products had, however, not progressed very far even five-and-twenty years ago, and the literature of true inclusions has been immensely enriched since

[1] An admirable discussion of this question, on which so much has been written, is to be found in the fifth chapter, "Die Differentiation der Magmen," of Doelter's work on *Petrogenesis* (1906), p. 71.

[2] *Quart. Journ. Geol. Soc. London*, 1880, p. 1, and 1882, p. 216.

then by the work of Lacroix, Doelter and Hussak, Salomon, and many others. Hawes,[1] however, had already published a remarkable study of the assimilation of detached blocks by a granite magma; and since that date the absorption theory, which seems to many workers easily justified by field observation, has found fair but by no means general support. Again and again the "segregation patches," which in some cases are mere black masses of mica, in other cases true diorites, in others composed of garnet, quartz, various felspars, pyroxene, and amphibole, are found to be marginal phenomena, and to lead one, by their increase in abundance in certain directions, towards the contact rocks into which the granite has intruded. Again and again, the streaking out of these patches is responsible for the local conversion of the granite into a banded gneiss. The gneiss of the valley of Boutadiol in the Pyrenees arises thus from the elongation of inclusions.[2] The Transvaal granite[3] contains "smaller or bigger masses of the Swaziland beds . . . forced off the main stock of the formation, and arranged in a parallel manner, all following the direction of the main mass." We thus observe in South Africa "long-drawn-out lenses of quartzite, chlorite, actinolite, and other schists, swimming, as it were, in a granitic magma." Observations of this kind are now numerous; yet the theory of "segregation-patches" by differentiation holds its own. Sollas[4] was one of the first to seriously question its general applicability, even in the cases brought forward by J. A. Phillips; but his suggestive criticism has borne but little fruit. Corstorphine and Hatch,[5] for instance, have quite recently referred the boulders of eclogite found in the diamond-pipes of South Africa to a concretionary origin, because their appearance is quite comparable to that of the basic patches so common in many granites."

Mention of these eclogite blocks, which are generally regarded as torn off from some deep-seated mass during the

[1] "The Albany Granite and its Contact Phenomena," *Amer. Journ. of Science*, vol. xxi. (1881), p. 31.

[2] Lacroix, "Le granite des Pyrénées," 2me mém. *Bull. Carte géol. de la France*, No. 71 (1900), pp. 21 and 25.

[3] F. W. Voit, "Gneiss Formation on the Limpopo," *Trans. Geol. Soc. S. Africa*, vol. viii. (1905) p. 142.

[4] *Trans. R. Irish Acad.*, vol. xxx. (1894), p. 502, etc.; see also Salomon, *Zeitsch. deutsch. geol. Gesell.* Bd. xlii. (1890).

[5] *Geology of South Africa* (1905), p. 298.

movement of the material that fills the volcanic diamond-pipes, reminds us that "eclogites"—*i.e.* crystalline rocks consisting mainly of garnet and pyroxene or garnet and amphibole—are among the commonest inclusions—or segregation-blocks, whichever we prefer—found in gneiss or granite.[1] In many cases they have been traced to their parent-rocks, which prove often to be limestones; and their occurrence thus gives useful aid to those who oppose the view of their origin by magmatic differentiation within the granite mass.

Among the best-known rocks of the eclogite class are the so-called "trap-granulites," or "pyroxene-granulites," of Saxony.[2] These are fine-grained masses, sometimes schistose, closely resembling the "amphibolites" and "pyroxenites" with garnet that are common as inclusions in our Irish granitoid gneiss. They occur in elongated lense-like forms of all sizes, sometimes by hundreds, within the pale Saxon "granulite." The latter rock, which has clearly metamorphosed the neighbouring schists, is an intrusive mass allied to granite. At one time it was held that the contact schists were very ancient, and that, by dynamic metamorphism, an already cooled granite had been forced in among them, the whole series becoming rolled out into a banded and gneissic complex. Now, however, geologists have been brought to ascribe the alteration of the schists to the igneous intrusion of the granulite,[3] and to regard the "trap granulites" as pieces picked off from the invaded rocks and partly digested in the normal granulite. Instead of being fundamental and Archæan, the gneissoid granulite becomes a granite with fluidal and contact structures, which intruded into strata as high in the series as the Carboniferous.[4]

W. A. Humphrey[5] has recently found that the "Urgneiss," or so-called fundamental formation, in Styria has affected, by contact metamorphism, conglomerates of Carboniferous age; and we thus meet another instance where a gneiss, once believed to be of high antiquity, becomes placed in the group of gneisses

---

[1] This matter is discussed in *Trans. R. Irish Acad.* vol. xxxi. p. 460. See also Doelter, *Petrogenesis* (1906), p. 177.

[2] See discussion in Lepsius, *Geologie von Deutschland*, 2ter Teil (1903), pp. 146, 169.

[3] *Ibid.* p. 171.

[4] M. H. Credner has presented a remarkable summary of the modern view in the *Compte rendu du Congrès geol. internat.* Vienna (1904), p. 116.

[5] *Jahrbuch d. k.k. geol. Reichsanstalt*, 1905, p. 363.

that have arisen at various dates by the intrusion of igneous rock into sediments or schists. Perhaps a reference should be given here to Lawson's [1] famous evidence of the absorption of Huronian rocks by the so-called Laurentian, or fundamental gneiss, at Rainy Lake. J. W. Gregory,[2] to quote an extreme case, has written of the gneisses of the Cottian Alps: "Paradoxical though it may appear, the evidence renders it most probable that the Waldensian gneiss, instead of being of Laurentian age, is really Pliocene, and, with the exception of the Saharian and recent alluvium and the glacial moraines, is the newest rock in the Cottians."

Harker,[3] again, has given us an example, admirably studied in the field, of the formation of gneiss in Tertiary times in the Isle of Rum. Before referring to this in detail, we should note that Callaway,[4] as far back as 1887, the year in which Lévy published his paper, "Sur l'origine des terrains cristallins primitifs," showed how gneiss has arisen in Co. Galway by the intrusion of granite into rocks rich in hornblende. The latter underwent partial melting and streaking out, and the resulting banded crystalline rock is thus of composite origin. In 1893,[5] again, Callaway, while dwelling strongly on the influence of pressure-metamorphism, cited a number of gneisses produced by injection of igneous rocks among the ancient masses of the Malvern Hills. A. Geikie and Teall,[6] in dealing with handsomely banded gneisses of igneous origin among the Tertiary rocks of Skye, have presumed that a separation of material, on the one hand more rich in silica and on the other more basic, took place in the plutonic caldron underground, and that the two types of magma moved forward together in a streaky viscid flow. But there are abundant cases where strong

---

[1] *Ann. Report Geol. Surv. Canada* for 1887, p. 133 F. Cf. Geikie on enclosures in the Hebridean gneiss of Loch Carron, *Ancient Volcanoes*, vol. i. p. 117.

[2] "The Waldensian Gneisses," *Quart. Journ. Geol. Soc. London*, vol. l. (1894), p. 232.

[3] "The Overthrust Torridonian Rocks of the Isle of Rum, and the Associated Gneisses," *ibid.*, vol. lix. (1903), pp. 189-215. Especially pp. 201-13.

[4] "Alleged Conversion of Crystalline Schists into Igneous Rocks in Co. Galway," *Quart. Journ. Geol. Soc. London*, vol. xliii. (1887), pp. 521, 523.

[5] *Ibid.* vol. xlix. p. 413.

[6] "On the banded structure of some Tertiary Gabbros in the Isle of Skye," *Quart. Journ. Geol. Soc.*, vol. l. (1894), p. 656. Also Geikie, *Ancient Volcanoes of Great Britain*, vol. ii. p. 342, etc.

banding has arisen by *lit par lit* injection of rocks already solid, but with well-marked foliation-planes which provided easy passage for the invader ; and it may be urged that, if we accept the differentiation-view in the case of Skye,[1] other cases may arise where a gneiss may result from the intrusion of one igneous rock into another of far older date.

The study of these relatively modern fluidal gneisses in Skye led the authors named to point out the importance of original flow in the structure of the pre-Cambrian gneisses of the west of Scotland ;[2] and Harker's work in Rum,[3] to which we now return, is an equally important contribution to the study of gneisses of all ages.

The gneisses of Rum, "with well-marked parallel banding and foliation, and frequent alternations of different lithological types, are perfectly characteristic gneisses in the ordinary descriptive sense of the word. Indeed, their appearance led Sir Archibald Geikie to assign them to an Archæan age." They are related to a highly disturbed belt of country, and are believed to represent material forced into place under considerable pressure. But their main masses do not show evidence of crushing, and their foliation is produced by igneous flow. A granitic magma, the Tertiary granite of the Inner Hebrides, has, in fact, flowed into pre-existing basic rocks. "Given a granitic magma enclosing débris of more basic rocks, an irregular distribution of the débris such as is seen where the xenoliths" (derived fragments) "are still traceable, reactions between the basic rocks and the acid magma of a kind familiar in many other districts, and that drawing out of the whole by flowing-movement which is proved by the banded structure, we have a complete explanation of the principal part of the Rum gneisses."[4] The "basic rocks appear to have been of the nature of gabbros, now transformed by metamorphism, and in some measure by interchange of material with the acid magma." Sollas's work at Carlingford[5] on the interaction of granite and gabbro, which was often regarded as a localised instance, or even as a misreading of the field-evidence, has been more than

[1] See Harker's discussion of these rocks, *Igneous Rocks of Skye*, pp. 90 and 119.

[2] See also Geikie, *Ancient Volcanoes*, vol. i., p. 116.

[3] See also Harker, "Tertiary Crust-Movements in the Inner Hebrides," *Trans. Edinburgh Geol. Soc.* vol. viii. (1904), p. 346.

[4] Harker, *Quart. Journ. Geol. Soc.*, 1903, p. 212.

[5] *Trans. R. Irish Acad.*, vol. xxx. (1893), p. 477.

justified by Harker's observations in another area within the British Isles. Geologists will remember that opposition to these views came from a reluctance to accept the large demands of Lévy, Lacroix, and Barrois in France, not to say of Hawes in America in 1881. Critics still urge the greater probability of the differentiation of molten magmas, as against the admixture of already differentiated materials, and see considerable difficulties in the way of producing composite rocks. F. D. Adams[1] in Canada would, it seems, refer banded gneisses to the metamorphism of sediments rather than accept as a general proposition the theory of interpenetration and partial absorption. Harker himself argues for an essential difference between true composite rocks and rocks of composition intermediate between two well-known petrographic types. The "intermediate rocks" are for him, as with Brögger, representatives of a special magma, and cannot be imitated by admixture. Yet the experimental researches of Doelter[2] point to the unlimited possibilities of admixture, provided that the temperature is sufficiently high in the caldrons underground. Doelter also points out the corroding effect of the gases associated with a molten and invading magma[3]: "In oberen Schichten scheint die Korrosionswirkung mitunter recht gering gewesen zu sein, das würde aber nicht hindern, dasz sie in tiefen Schichten bedeutend sein kann und dasz Magma sich langsam hinauffriszt, denn die Gase wirken wie eine Lötrohrflamme." The gases, moreover, that we know, in various combinations, among the emanations of volcanos, are potent influences in promoting the crystallisation underground of such minerals as quartz, mica, garnet, epidote, etc. May we not be closely approaching a time when we shall generally regard dynamic metamorphism, combined with movement, as the destroyer of distinct crystallisation in a schist, and contact-metamorphism, on a regional scale, as the main promoter of such crystallisation? Sometimes a sediment, foliated by pressure, in which all trace of original bedding has been lost, becomes invaded and heated, and new minerals arise along the foliation-planes; at other times a sediment is invaded

[1] "Some recent papers on the Influence of Granitic Intrusions upon the Development of Crystalline Schists," *Journ. of Geology*, vol. v. (1897), p. 293.

[2] These researches are admirably summed up, with those of other workers, in his recent book, *Petrogenesis* (1906). On admixture, see pp. 80, 118, etc.

[3] *Ibid.* p. 119.

in a little altered state, and the "mineralisation" takes place as an emphasis of the original stratification. Again, dynamic metamorphism may work its will on the complex series, and roll out indistinguishably the insidious igneous veins and the invaded sediments. But the vast majority of well foliated gneisses, and especially those with well differentiated bands, may now safely be ascribed to conditions of original flow. Most of them, moreover, will be found to represent the marginal phenomena of intrusive granite domes.

If absorption on a large scale of the surrounding material goes on we have a rational explanation, and practically the only one, of the relation of such intrusive masses to the sedimentary, metamorphic, or igneous rocks traversed by them. The mere uplift, by arching, of the rocks above them does not explain the phenomena of their *mise en place* as studied in the field. The features to be expected from plutonic absorption have elsewhere been discussed in connexion with the invasion of a mass of basic lavas and diorites by granite in the county of Londonderry;[1] and it was then pointed out that, by a process of internal differentiation, the absorbed basic materials might be carried away into the depths, the marginal granite remaining comparatively pure. Loewinson-Lessing[2] has laid stress on the influence of density in promoting such differentiation; and R. A. Daly[3] has developed the assimilation-theory with skill and, to say the least, unhesitating boldness. As our geological researches in the field do not often allow us to reach the depths to which absorbed matter may be supposed to plunge,[4] we have near the surface the spectacle of a pure igneous mass underlying that which it has invaded. Yet, even then, its viscosity during cooling very often enables it to support numerous blocks in what American writers have called the "shatter-zone," where the final attack on the superincumbent rocks remains recorded;[5] and movement in this region during the last stages of liquidity will result in the production of a streaky and apparently antique and "fundamental" gneiss.

[1] Cole, "Geology of Slieve Gallion," *Sci. Trans. Royal Dublin Soc.*, vol. vi. (1897), p. 242.

[2] *Compte rendu, Congrès geol. international*, 1899, p. 344.

[3] *Amer. Journ. of Science*, vols. xv. (1903), p. 269, and xvi. (1903), p. 107.

[4] Compare Doelter, *Petrogenesis*, p. 123.

[5] R. A. Daly, "Secondary Origin of Certain Granites," *Amer. Journ. Sci.* vol. xx. p. 208.

Is it this fact that led Humboldt to the observation, quoted by Scrope,[1] that the older—*i.e.* the lower—granites contain least mica? Fantastic as the statement may appear as a generalisation, I confess that I believe that almost all our granites with two micas, and the amphibole granites practically without exception, result from processes of admixture in the crust, and that the true subterranean granite magma, if such is to be defined, is near in composition to the rock known as aplite, and would crystallise out as a mixture of quartz and acid felspar, with far less mica than we see in the masses that are ordinarily exposed by denudation, Against this we must set Daly's view that the really pure igneous rocks are the basic ones, which move rapidly and thus gather little by the way, and that pure granites may be formed from the fusion of quartzose sediments, any basic material associated with them becoming drained off, by gravitational differentiation, into the depths.

Here it is indeed time to pause, for our hillside in Donegal has led us far afield. Let us come back from speculation to the sober facts on the surface of this pleasant land. The sunlight gleams on the micaceous rocks above Gartan Lough, and picks out each coarsely crystalline patch that juts forth, well foliated, from the decaying surface of the gneiss. This gneiss is full of muscovite, with a foliation-structure parallel to that of the inclusions of muscovite schist. The granitoid rock and the schist are indeed inextricably involved; the gneiss is composite beyond a question. This possibility must be faced in every area where "fundamental" masses, affected by serious crushing, are at present asserted to exist. If hundreds of such masses are then found wanting, on a revision of the evidence in the field, it will only increase the interest of those regions which have successfully withstood the test.

[1] *Considerations on Volcanos*, (1825), p. 233.

THE ARTIFICIAL PRODUCTION OF
NITRATE OF LIME

BY JOHN B. C. KERSHAW, F.I.C.

SINCE Sir William Crookes startled the scientific world, eight years ago at Bristol, by pointing out that the corn supply of the world was dependent upon the ample provision of nitrates to the soil, and that we were rapidly depleting our reserves of the only naturally occurring nitrate—namely, Chili saltpetre—scientists in all countries have been attempting to solve the problem of the extraction of nitrogen from the air in the form of nitrite or nitrate. At the meeting of the Bristol Association at Bristol, referred to above, Sir William Crookes pointed out that experiments made in earlier years, by Lord Rayleigh and others, had proved the possibility of burning the nitrogen of the air in the electric arc, for production of nitrites and nitrates; and basing his estimate upon the results of these experiments, Sir William Crookes stated that 14,000 Board of Trade units of electricity, employed in this way, would yield the equivalent of one ton of nitrate of soda.

Sir Joseph Wilson Swan, F.R.S., another noted scientist, in an address delivered before the Society of Chemical Industry in 1901, referred to the problem and its great importance in the following words:

"The manufacture of nitrate of soda in this way is still undeveloped industrially. The idea is nevertheless alluring, for surely chemical science and chemical industry can be put to no higher or fitter use than to help to increase the fertility of our fields. Those immense sources of motive power, the waterfalls of the world, which are now for the most part running to waste, require some great employment such as this, the production of nitric acid. If the quantity of nitric acid produced per k.w. hour can be increased yet a little further—and such a development appears far from hopeless—this immense application of electricity would become profitable. The subject is certainly worthy of the serious attention of chemists."

Chemists and engineers have followed Sir Joseph Swan's advice, and have been busy with the problem since the above words were written and published, five years ago. After many experiments, and the failure of an American company with works at Niagara Falls, the difficulties are believed to have been overcome, the successful scientists being Prof. Birkeland, of Christiania University, and S. A. M. Eyde, a Norwegian engineer.[1]

Air contains roughly 21 per cent. by volume of oxygen and 79 per cent. by volume of nitrogen. If an electric spark be passed through dry air, a small amount of nitrous oxide gas is formed by oxidation or " burning " of portions of the nitrogen. This experiment was performed in the laboratory by Cavendish over a hundred years ago. The problem that confronted the chemist was—How to enable this combination of the oxygen and the nitrogen to become sufficiently rapid, to render the process a profitable and practicable one for the manufacture of nitrates ? The earlier inventors used an enormous number of small sparking points for obtaining the electric discharge, and found even then that the yield of nitrous gases was too small to render the process of practical utility. The reaction was also found to be a reversible one, and the same electric spark energy which caused the combination of the two elements (oxygen and nitrogen) also brought about the disruption of the union. These and other difficulties caused the failure of the Atmospheric Products Company of Niagara Falls, which had been founded in 1902, with some flourish of trumpets on the part of our American cousins, to supply the world with nitrates by the Bradley and Lovejoy process. The problem was, in fact, more difficult than the American chemists and electricians had realised ; and, as already stated, it has been reserved for two Norwegian scientists to show the way in which it may be solved. Messrs. Birkeland and Eyde, in place of using a large number of small sparking arcs, make use of one large flame arc, produced by alternating current at 3,000 to 4,000 volts pressure, this flame being made to take the form of a disc, 4 to 5 ft. in diameter, by a special method, the details of which were discovered by

[1] Another method of extracting nitrogen from the air, by means of calcium carbide, has been patented by two German chemists, Messrs. Frank and Caro, and is now being developed commercially in Italy. This method will be described in a later article.

Prof. Birkeland by accident. Ordinary air is forced through this disc of roaring flame, and emerges charged with nitrous gases. This charged air is then quickly removed from the vicinity of the flame, and is subjected to chemical treatment in order to obtain the gases in the form of nitrite or nitrate of lime.

This process has been undergoing gradual industrial development in Norway since June 1903, and, after being tried upon scales of increasing magnitude, appears now to have reached a position of considerable industrial promise and importance. The first large installation of plant and machinery commenced work at Notodden in May of last year, with three furnaces of 700 h.-p. each, and the necessary absorbing towers and other plant. After the operation of the Birkeland and Eyde process at this factory had been witnessed by a number of independent experts, and tests had been made of its working efficiency during the summer months of 1905, a company was formed to promote the further industrial development of the process. A large number of German and Norwegian banks are lending their support to this new undertaking, and a capital of 7,000,000 kr. (£375,000) is being raised. The factory at Notodden has been taken over by the new company, and extension of the works, until it is in a position to utilise 30,000 h.-p., are contemplated.

The annual output of calcium nitrate by this company, when the works at Notodden have attained their full development, is estimated at 20,000 tons.

Ordinary nitrate of lime possesses the disadvantage, from the agriculturist's point of view, that it is very hygroscopic, and very readily soluble in water. This difficulty has been overcome, it is stated, by preparing the basic nitrate in its place; and experiments have shown that this salt, as a manure, is equal, if not superior, to natural nitrate of soda—the "Chili saltpetre" of commerce.

As regards cost, no very detailed estimates are yet available; but Dr. Otto Witt, who has himself investigated the process, states that the equivalent of half a ton of nitric acid (100 per cent.) can be produced per k.w. year, and that at Notodden, where the water power can be very cheaply developed, this amount of electricity costs only 16s. Since the value of the lime required for operating the process is not great, when compared with the value of the final product, it is expected that the new fertiliser will easily compete in price with Chili saltpetre for agricultural purposes.

In time, therefore, Sir Joseph Wilson Swan's prophecy, of the water power of the world being utilised to furnish the nitrate required for its fertilisation, may be realised; and the great problem of the continued supply of nitrogen to the soil may be solved, by tapping the practically inexhaustible reservoir of nitrogen contained in the atmosphere which surrounds our earth.

THE ECONOMICS OF UNIVERSITY EDUCATION

BEING AN ADDRESS DELIVERED TO THE STUDENTS OF UNIVERSITY COLLEGE, CARDIFF

By SIR ARTHUR RÜCKER, D.Sc., LL.D., F.R.S.

Principal of the University of London

THE political element in the science of political economy is apt to become so predominant that it is difficult to deal with economic problems in the judicial spirit which is essential to the student of science, but the subject which I have chosen for my address to-night, viz. the Economics of University Education, is, fortunately, still outside the range of controversial politics.

It was, moreover, treated at some length by the father of the science, Adam Smith, so that, in view of the importance attached to his opinion in other matters, it is worth while to inquire how far the circumstances have altered since his great work was given to the world, and how far his assertions and predictions have stood the test of time.

The principles which underlie the whole of Adam Smith's discussion of the subject are that the best teaching is that which most directly meets the desires of the student, or, if he is of tender years, the desires of his parents or guardians; and that the best way of securing such instruction is to make the teacher immediately dependent upon fees which the student pays. Adam Smith is thus, as a matter of theory, against endowments, against State support of education, and against free education; though, as will be shown hereafter, he admits that it is necessary to depart, to some extent, from the strict rigour of his principles.

A discussion based on his statements must be in part *à priori* and theoretical; but the fact that many years have elapsed since the *Wealth of Nations* was written makes it possible to appeal to experience, and to test abstract principles by the results of practice.

I propose, therefore, to take each of the propositions to which I have just referred, to prove, in each case, that I am not misrepresenting our author by asserting that he supports them, and to inquire how far they can be maintained in the light of the wider knowledge and longer experience of to-day.

It is difficult to discuss separately the questions as to whether the most useful teaching is that for which there is most popular demand, and as to how such teaching is best to be secured, as the two are very closely connected; but it may be pointed out in the first place that Adam Smith admits that the proposition that the best teaching is that which most directly meets the desires of the student, cannot apply to very young children. "Force and restraint," he says, "may, no doubt, be in some degree requisite, in order to oblige children, or very young boys, to attend to those parts of education which it is thought necessary for them to acquire during that early period of life."

He also admits that the principle under discussion will not apply to the elementary education of the children of the poorer classes.

"For a very small expense," he says, "the public can facilitate, can encourage, and can even impose upon almost the whole body of the people, the necessity of acquiring [the] most essential parts of education.

"The public can facilitate this acquisition by establishing in every parish or district a little school, where children may be taught, for a reward so moderate, that even a common labourer may afford it; the master being partly, but not wholly paid by the public; because if he was wholly, or principally paid by it, he would soon learn to neglect his business. . . .

"The public can encourage the acquisition of those most essential parts of education, by giving small premiums, and little badges of distinction, to the children of the common people who excel in them.

"The public can impose upon almost the whole body of the people the necessity of acquiring the most essential parts of education, by obliging every man to undergo an examination or probation in them, before he can obtain the freedom in any corporation, or be allowed to set up any trade, either in a village or town corporate."

This passage is remarkable for the wide difference between the means recommended to promote elementary education and those which are sanctioned by the present practice of most civilised nations. The principles that public money may be spent in providing elementary education, and that some form of compulsion may be used, are accepted; but the "little school" in each parish or district is nowadays represented by a building which must comply with the sanitary and educational ideals of experts. The fee, which was not to be "wholly paid by the public," is entirely provided by the community. The direct responsibility of the teacher to the parent, which the payment of fees by the latter was to secure, is replaced by responsibility to a public authority. The "small premiums and little badges of distinction," which were to encourage attendance, are represented by scholarships to Secondary Schools, and thence to the Universities. The indirect compulsion of requiring a minimum of attainment before the pupil is allowed to earn his living by setting up "any trade, either in a village or town corporate," is replaced by the knock of the School Board visitor, with the magistrate in the background.

I do not enlarge on these points, because my subject is University and not Elementary education; but it is impossible to avoid two conclusions. Firstly, that the public purse is providing for elementary education an enormously larger sum than Adam Smith contemplated, or would in all probability have approved; and, secondly, that, whether the bases of his arguments were valid or unsound, the conclusions he drew from them are not in accord with the practice or the public opinion of to-day.

The cases of the very young and the poor, important as they are and costly as the latter has proved to be, are, however, from Adam Smith's standpoint, only exceptions to the general rule which requires the direct dependence of the teacher, both financially and also in respect of the subjects taught and the manner of teaching them, on the student and his friends. He assumes throughout that, if this control is exercised, it is necessary and desirable that the student should pay the greater part of the cost of his own education. "The institutions for the education of youth," he says, "furnish a revenue sufficient for defraying

their own expense. The fee or honorary which the scholar pays
to the master, naturally constitutes a revenue of this kind."

Our author regarded all additions to this natural source of
income, that is, all endowments, as devices to enable the teacher
to defy public opinion. The teacher, being more or less independent of the approval of his students and their friends,
would be apt to neglect his duties. He would continue to
teach his subject as he had himself learned it and to ignore
modern developments. "Exploded and antiquated" systems of
science "can subsist nowhere but in those incorporated societies
for education, where prosperity and revenue are in a great
measure independent of their industry." The "salaries, too,"
of endowed teachers, "put the private teacher, who would
pretend to come into competition with them, in the same state
with a merchant who attempts to trade without a bounty,
in competition with those who trade with a considerable one."
"Thus the endowments of schools and colleges have in this
manner not only corrupted the diligence of public teachers,
but have rendered it almost impossible to have any good
private ones."

Finally, there were, in Adam Smith's time, "no public institutions for the education of women," and the instruction of
girls was in his opinion in a much healthier state than that of
men. They learned nothing "useless, absurd, or fantastical."
"They are taught," he says, "what their parents and guardians
judge it necessary or useful for them to learn, and they are
taught nothing else."

Nor has our author much faith in any system of control. If
the Governing Body consists wholly or in large part of the
Professors, "they are likely to make common cause, to be very
indulgent to one another, and every man to consent that his
neighbour may neglect his duty provided he himself is allowed
to neglect his own." If it consists of some external person such
as a bishop, governor, or minister of state, the most that he can
do is to compel the teacher to give lectures; he cannot compel
him to give them well.

Our author, then, held that the student should pay at all
events a part of the cost of his education, and the only case
in which part-payment is directly sanctioned is that of the
elementary education of the poor.

Secondly, he was of opinion that this was not only natural

and just in itself, but that it gave the pupils a desirable control over the subjects taught and the method of teaching them.

Thirdly, he held that the only other forms of control over the teacher which suggested themselves to him were inadequate.

It follows directly from the first of these points that those who desire University education should pay for it themselves. This opinion was expressed before the necessity for scientific and technical education was appreciated, and, indeed, before either science or technology had reached the point at which such education could be given on the scale to which we are now accustomed.

It is, however, certain that the fees or "honoraries" of the students cannot provide the laboratories and apparatus without which scientific education cannot be carried on. In the case of University Colleges, site, buildings, and initial equipment must be provided from other sources, and, even then, it is still necessary to supplement the highest fees which can safely be charged with donations and subscriptions from public and private funds. When all this is done, the fees are still too high for the pockets of students from the poorer classes, who have to be assisted by the remission of payment or by bursaries.

At the present time, therefore, it is simply not true that institutions for the higher education of youth can furnish from fees a revenue sufficient for defraying their own expense. The last report on University Colleges shows that at Cardiff the whole of your capital and nearly three-quarters of your general income are derived from extraneous sources. The students pay only one-fourth of what is spent on educating them.

The fact is that Smith did not, and could not be expected to, foresee the great cost of modern scientific and technological education; a form of education which is vital to the prosperity of the community as a whole, but which involves so great an immediate outlay on the part of the student that none but the comparatively wealthy could avail themselves of it. Nor had he fully grasped the importance which modern opinion rightly attaches to giving ability, wherever it is found, every chance of development. I believe I am right in saying that, according to the statisticians, the children of able parents are themselves able to an extent which is far above the average; but that the number of parents who are not specially endowed is so great that the total number of their progeny who display remarkable gifts is greater than the number of children in whom ability is inbred. The ideal to be aimed at is, of course, that the most should be made of all the ability of each generation, and, thus, the ability-catching net must be thrown far and wide and not confined to the best fishing grounds. This is the scientific justification, from the point of view of the community, and apart altogether from questions as to the happiness or success of individuals, which justifies large expense in helping ability which might otherwise never overcome the initial difficulties of want of money or of social position. Thus wider experience, wider knowledge, and, last but not least, the greater intensity of the struggle for existence have led to the general approval of methods of endowment and State and municipal support which our author condemned.

But, if the fact that this great economic revolution has taken place in the conditions under which University education is provided be admitted—and few of us would either deny or disapprove it—various consequences follow, which are fatal to many of Adam Smith's contentions.

He regarded it as desirable that the student or his friends should control education by the power of giving to or withholding from particular individuals or institutions the fees which they themselves paid, and that, in this way, the teacher should be made responsible to the student and his friends only.

It is unnecessary to enter into a lengthy *à priori* discussion of this principle; but it may be pointed out in passing that, if endowments led to the deplorable condition of the older Universities in the eighteenth century, they are not inconsistent with the enormously enhanced efficiency and activity which those great institutions display in the twentieth.

If free competition permitted the establishment of many excellent private schools, it is responsible for the evils of cramming. If the wretched condition of poor "Biler, the Charitable Grinder," was caused by endowments, free competition was responsible for Dr. Blimber and Mr. Whackford Squeers.

But, *à priori* arguments apart, and for reasons I have given, education is now, and probably always will be, largely supported by endowments and subventions, and the more important question is how this custom affects the relations of teacher and pupil. Teachers are no longer paid, or solely paid, by their pupils. They cannot, therefore, be solely responsible to them, but must, in large measure, be responsible to those who control the funds which make the education of the pupils possible.

As matters of practical educational politics, the first and second propositions on which Adam Smith's position is based are thus not, perhaps, disproved as questions of theory, but are, by almost universal consent, ignored as matters of practice. He disapproved of endowments for University education, whereas to-day efficient University education is only possible by the aid of endowments and subventions. He thought that the student should practically demand the education which he thought most suitable, whereas he must now, within certain limits, take what is thought best by those at whose expense he is in whole or in part educated.

And this brings us to the third question raised by our author. If the control of the student over the teacher is lessened or removed, by what is it to be replaced, or is education to be entirely dominated by the teachers?

Nowhere in this country has the latter ideal been accepted. In the older Universities large powers are entrusted to the graduates—that is, to past students, as members of Convocation. In the newer Universities the powers of Convocation are in general only advisory; and I frankly admit that I think this is the better alternative. Adam Smith, however, does not refer to Convocations or any form of control other than that of a bishop, governor, or minister of state. It is a curious proof of how great men are limited by their surroundings, that it should never have occurred to him that the principles of representative government could, with suitable modifications, be applied to educational institutions. Nearly all modern Universities in this country are governed by bodies on which laymen and experts are mingled. Provided the shares of power are properly distributed, this forms perhaps as good an arrangement as can at present be devised. It brings into close touch the educational expert and representatives of public opinion; it secures external supervision over the educational work without sacrificing that freedom of the teacher which is the essential basis of a University system.

Have "public endowments," said Adam Smith, "contributed in general to promote the end of their institution? Have they contributed to encourage the diligence and to improve the abilities of the teachers ? Have they directed the course of education towards objects more useful both to the individual and the public than those to which it would naturally have gone of its own accord ?"

The answer of the twentieth century to these inquiries is, that whatever may have been the correct historical reply, we believe that all these results can be obtained concurrently with the use of endowments by means of a system of governing public educational institutions which unites the representation of public and expert opinion by a method which he did not even discuss. It must, therefore, either not have occurred to him as a feasible plan, or he must have condemned it by means of the general proposition that authorities can provide teaching but cannot ensure its quality.

Taking, however, the system of government of modern Universities as an established fact, and admitting that Adam Smith's objections or possible objections have been overruled and his ideals abandoned, a series of interesting questions remains as to what principles should guide these relatively new University authorities in the exercise of their powers.

By their endowments and subventions they have in part relieved the teacher of the direct economic incitement to diligence. By their control over the courses of study which lead to a degree, they have in part deprived the student of freedom of choice as to what he shall study. How are the evils which might attend this relaxation and this restriction to be minimised ?

As regards the first, Adam Smith would, no doubt, have been in favour of paying the teacher as much as possible by a share of the fees, and as little as possible by endowment; but this leads to a difficulty with regard to courses of study for the degrees. The number of students who attend a professor's class must depend not only on his ability and energy, but on the place which his subject takes in the authorised curricula. It may be desirable that a great University should possess a Professor of Assyriology, but it would be absurd to blame the poor man if he could not make a decent living by his fees. On the contrary, he would be regarded as an ornament to the University if he could succeed in securing a small class of earnest post-graduate students. It would be equally absurd to attempt to support him by forcing students to study his subject as a necessary preliminary to a degree. Without, therefore, wholly condemning the system of paying partly by a share in the fees, which obtains in many Colleges, it is sufficient to say that it is open to the objection that in framing the curricula both the Professors and the Governing Body are apt to be influenced by considerations as to the effect of their regulations on the finances of the teachers, and not solely by educational advantages.

And this leads me to the general criticism that Adam Smith's arguments appear to be based almost entirely on the view that a University is a place where instruction is bought and sold, not a place where professor and student are linked together as leader and follower in a common search after knowledge.

For the mere purpose of securing the minimum amount of knowledge necessary for a particular purpose, such as entering a profession or passing an examination, it is, I think, possible to maintain that direct financial reward for imparting the precise information most likely to attain the object produces efficient teaching. It is a system of piece work which has some of the merits of the now discredited system of payment by results; but it has the inherent vice that it aims at imparting the minimum of necessary information. To teach anything not immediately required for the object in view is a breach of the implied contract to concentrate attention and effort only on what is likely to "pay." It is avowedly a system of instruction, and of instruction devoted only to an end which the student is to secure, or to fail to secure, in the immediate future.

A system of education is, however, based on more far-sighted views. In the course of his studies the pupil will receive a great deal of instruction which may, and indeed should, be calculated to be of direct use to him hereafter. But the object to be attained is not immediate utility, but that training of his intellectual powers which will not only give him knowledge, but also enable him to wield it as a weapon. The candidate for success has first to break down the barrier which bars entrance to his profession. But, once inside the gate, he has the harder task of fighting his way to the front among a crowd of competitors, and in that struggle not merely strength but dexterity is needed. It is one of the aims of education, as distinguished from instruction, to impart this intellectual dexterity, and experience shows that it is best attained, not by training the student to be skilful in deciding on the minimum effort required for each purpose, but by teaching him to be alert in detecting the importance of apparent trifles which others are likely to overlook. Thus, as research is largely concerned with the elucidation of the results of hitherto neglected facts, it is found that for many objects mental dexterity can best be fostered by turning the attention of the abler student from the known to the unknown, from information to investigation. This can only be done by men who are themselves investigators, and thus the modern view of the professor and student as partners in the search after knowledge is replacing Adam Smith's conception of the student as a customer who asks from one side of a lecture table to be supplied with so much knowledge of a particular kind, which ought to be duly served out to him by the affable gentleman who stands behind it.

On all the side issues which are connected with this change of view I have not time to dwell. I can only deal with one result of the modern view of a Professor's duties, and with the difficulties attending the greater rigidity introduced by stricter regulations as to courses of study.

As to the first, it is a curious commentary on Adam Smith's doctrines that the salaries of Professors of Science are certainly less than they otherwise would be, because the command of a laboratory is a condition favourable to the carrying out of those researches which are to them a source of interest and of reputation which they value as highly, or in many cases much more highly, than direct pecuniary reward.

" Great objects," says our author, " alone and unsupported by the necessity of application, have seldom been sufficient to occasion any considerable exertion. In England, success in the profession of the law leads to some very great objects of ambition : and yet how few men, born to easy fortunes, have ever in this country been eminent in that profession."

The reply to this remark is that very few men enter upon a task, the primary object of which is to gain a livelihood, unless the necessity of earning a living is imposed upon them. The conduct of other persons' business in the law courts, the detailed supervision of and care for other persons in illness, are tasks which, though often illuminated by the noblest self-devotion, are in general undertaken, not with any active hope of obtaining the highest rewards which the professions of Law or Medicine can bestow, but as more or less congenial means of earning a livelihood. It is therefore improbable that they will often be undertaken by persons who already possess means adequate to their desires.

The truth is that, so far from ignoring "great objects," the well-to-do man is attracted to work in which "great objects" form a larger and pecuniary reward a smaller, proportion of the inducement than that which leads to the adoption of one or other of the professions. He does not walk the hospitals, but he often devotes time, energy, and money to hospital management and finance. He does not administer the laws, but throws himself heart and soul into the task of making them. He does not go on circuit and plead his neighbour's cause, but as a diplomatist or Member of Parliament he devotes himself to the interests of his country.

In no country in the world is more labour given to the service of the city and the State by those who have little to get in return than in Great Britain. In no country has more of the best of the higher intellectual work been done by amateurs—that is, by persons who were not financially dependent on the results of their labours. To take a few examples only: Gibbon, Hallam, and Freeman were all men of independent means; Macaulay was not attracted to literature by any "necessity of application," though the decline of the family fortunes afterwards compelled him to earn his living; Ruskin was the son of wealthy parents; Boyle, as we know, was the "father of Chemistry and brother to the Earl of Cork"; Cavendish was a man of large fortune; Joule was a brewer; Darwin was not constrained to work by financial need; at the present moment a peer, *concensu omnium imperii capax*, is the President of the Royal Society. It is at least as true to say that the necessity of earning a living has made many a man, potentially great, mute and inglorious, as that "great objects, alone and unsupported by the necessity of application, have seldom been sufficient to occasion any considerable exertion."

And between these extremes there are many intermediate stages. Thus, it is fair to say that the "great object" of taking some share in adding to knowledge induces many a man to take a professorial chair at a smaller salary than he believes he could earn by other means. It may, of course, in some cases, be true that research occupies time and attention which should be given to students; but provided that Governing Bodies recognise, as they ought to recognise, that a reasonable amount of time should be reserved for research, the tasks of teaching and investigating are not antagonistic but supplementary. Given equal powers of exposition, the investigator brings to the lecture-room a freshness and first-hand knowledge which cannot be attained by the mere explanation and criticism of other men's labours. The students are proud of being associated with an institution which has not merely great traditions, or the hopes of a distinguished future, but which exercises a living influence on the solution of the problems which confront us to-day. They themselves feel that their studies are not merely reminiscences of what others have thought and done, but are giving them a glimpse of what they themselves may do. We, in this country, are too apt to forget that the indirect effects of a teacher's intellectual status cannot be measured by the number of lectures he gives or the number of exercises he corrects, and yet may be no small part of his influence and usefulness. If he attains distinction, the institution to which he is attached gains in reputation, the students in intellectual vitality. If it be true that "knowledge is power," the gift of adding to knowledge is yet more powerful, and, for the cultivation of this gift, time and opportunity should be allowed in every College of University rank.

As against Adam Smith, then, I should contend that "great objects," apart from direct financial gain, do exercise at the present moment a very remarkable influence in University education, and that the direct and indirect effects of this change are almost wholly beneficial. Economic pressure is not the only driving power by which men are urged to great exertion.

The other point on which I should like to add a few words is the influence of Universities in directing the studies of the students by means of authorised degree courses and curricula.

Of late the tendency has been to leave as much freedom of choice as possible to the student and his teacher. With this policy I am in hearty sympathy, though I must in all sadness say that it involves an enormous increase in the expense of examinations. But the most interesting side of the question is as to the method of dealing with students of technology.

It cannot be denied that many of them are chiefly anxious for instruction, and are comparatively indifferent to the finer influences of true education. "My son, sir," said a parent to a friend of mine, "is to enter the copper trade. I wish him to learn the chemistry of copper, and not to be delayed by studying oxygen, nitrogen, and other superfluous substances." This remarkable sentiment was uttered many years ago, and was, no doubt, an extreme case; but it raises the question whether Universities can properly deal with subjects which are studied chiefly from the utilitarian point of view.

One answer is to be found in Germany, where, in Berlin, the chief Technical High School has been endowed with the power of conferring degrees. Thus, two degree-giving bodies, in effect two Universities, exist side by side, the one devoted chiefly to pure literature and science, the other chiefly to technology.

The circumstances under which this arrangement has been contrived are so different to those which exist in our own country, that I will not arouse Germanophile susceptibilities by impugning it; but I should deeply regret if it were to form a precedent here.

Nothing would be more disastrous than this public division of subjects of study into the useful, and those which the public would immediately class as the useless. Nothing would more easily lead to the belief that public and private generosity should be reserved for the support of bread-winning studies; that only the future Watts and Stephensons should be encouraged, the Faradays and the Joules ignored.

Nothing would more tend to destroy that important part of University education which is given, not by the teachers, but by the students themselves. The mingling of lads whose subjects of study and whose aims and objects in life are different, is in itself an education. The young scholar or investigator, absorbed in antiquarian lore or the last scientific problem, would be all the better for rubbing shoulders with the future engineer or miner. His tendency to priggishness would be elbowed out of him. He would learn that some of the greatest ability in the world is devoted to practical objects. The student of technology would be an equally successful and a more cultivated man if he learned in youth to respect those who aimed at successes which they knew would not be rewarded with a shower of gold, but would none the less be worth winning.

To place these lads in rival institutions, to teach them that they were classes apart and distinct from each other, would be for each institution to abandon the proud title of " University"; and to introduce an artificial cleavage into the unity of knowledge.

But if the title of " University" is to retain its full meaning, it is necessary that full weight and adequate power should be given to those who represent the practical rather than the theoretical side of its work. Effete theories of education, even so-called academic principles, must, if necessary, be laid aside. The theorist and the practical man must confer together, must sit side by side on the same Governing Body. The education of the technical student must be such as the leaders of his industry consider a good preparation for his after career; together with such tests of general culture and a knowledge of scientific method as the best of these leaders would be the first to approve. The degree may, as is the case with medical degrees, be a sign of more ability and knowledge than the minimum that is necessary to secure entrance into a profession; but it must not only secure the sanction of the leaders of that profession, but be recognised by the student as something which it is not only honourable but useful to attain. Then, and then only, will the graduate achieve in industrial circles in this country the position which the " college man" has long held in America.

So far I have dealt exclusively with the education of the mind; but Adam Smith makes another exception to his general rule of the non-interference of the State with regard to the education of the body. After describing, in the passage I have already quoted, the means by which the State might assist, encourage, and even compel, the education of the poor, he adds :

" It was in this manner, by facilitating the acquisition of their military and gymnastic exercises, by encouraging it, and even by imposing upon the whole body of the people the necessity of learning those exercises, that the Greek and Roman Republics maintained the martial spirit of their respective citizens.

" That in the progress of improvement, the practice of military exercises, unless government takes proper pains to support it, goes gradually to decay, and, together with it, the martial spirit of the great body of the people, the example of modern Europe sufficiently demonstrates. But the security of every society must always depend, more or less, upon the martial spirit of the great body of the people."

Again—

" Even though the martial spirit of the people were of no use towards the defence of the society, yet, to prevent that sort of mental mutilation, deformity, and wretchedness, which cowardice necessarily involves in it, from spreading themselves through the great body of the people, would still deserve the most serious attention of government; in the same manner as it would deserve its most serious attention to prevent a leprosy, or any other loathsome and offensive disease, though neither mortal nor dangerous, from spreading itself among them; though, perhaps, no other public good might result from such attention, besides the prevention of so great a public evil."

For my part I think that the very fact that we have departed so far from Adam Smith's doctrines in other matters adds strength to the arguments which he urges with none the less force because the language in which they are expressed is somewhat old-fashioned.

In many ways the State, the municipality, and the pious founder have done for all of us much that Adam Smith would have disapproved. Some of us have held scholarships and fellowships. Some have, perhaps, had elementary education provided free, and the fact that the State will probably benefit because such persons will be more useful citizens, in no way diminishes the magnitude of the benefit they have received. National, municipal, and private funds have been showered on the members of every great College. The possibilities of a useful life have been extended for most of us by funds we could not of ourselves have amassed. We have shared in a portion of the wealth of past ages, and of that which other members of the great society to which we belong are accumulating to-day. Do we owe nothing in return? If we ignore, and I agree that we ought to ignore, Adam Smith's doctrine of "no educational endowments," can we treat with the same indifference his dogma that a man incapable of defending himself " evidently wants one of the most essential parts of the character of a man"? In my opinion each University student who receives from his country the skill to handle apparatus, or to handle a pen, should, in return, learn to handle a rifle to defend her in her hour of need.

And now to sum up. We have seen that the great economist's opinions as to the proper relations between teacher and pupil, between education and the State, have all or nearly all been abandoned by universal or nearly universal consent. He disapproved of free and compulsory education, and elementary education is both compulsory and free. He disapproved of endowments and State aid, but we live in an age of munificent founders and of grants to University Colleges. He approved of a form of education for women that is not the most popular type. He approved of direct economic relations between the teacher and student, whereas we interpose Governing Bodies to control both the one and the other. He did not believe that "great objects" had much influence apart from the necessity for earning a living, but there never was a time when so many teachers were also serious students who aimed at adding to knowledge. We have invented a new form of governing Universities which works well, though he ignored it. He called a man who was incapable of defending himself a "coward," and most of us take no pains not to deserve the epithet. Was he altogether wrong or altogether right, or were truth and error mixed in the *Wealth of Nations* as elsewhere? If I am to choose I should say that I think he was right in two points. I doubt if posterity will always believe that identical educations are necessarily the best for men and women; and I am sure that if disaster ever befalls this country posterity will say that it was because long immunity from imminent danger to our families and our homes had deadened in us the instinct of self-defence.

SOME RECENT DEVELOPMENTS OF THE ELECTROLYTIC DISSOCIATION THEORY

By GEORGE SENTER, Ph.D.

Lecturer on Chemistry, St. Mary's Hospital Medical School

	PAGE
INTRODUCTION .	381
OSMOTIC PRESSURE .	384
COMBINATION BETWEEN SOLVENT AND SOLUTE. THE HYDRATE	
THEORY	386
SOLVENTS OTHER THAN WATER	392
COMPLEX IONS .	394
SOME CRITICISMS OF THE THEORY.	396

INTRODUCTION

THE electrolytic dissociation theory, according to which the molecules of the dissolved substance in solutions which conduct the electric current are split up to a greater or less extent into ions—atoms or groups of atoms associated with electric charges—was proposed by the Swedish physicist Arrhenius in 1887. Thirty years before, Clausius had been led by purely physical considerations to the view that, in some cases, free ions are present in solutions of electrolytes; but he considered that the proportion was probably too small to allow of detection by chemical means. On the other hand, Arrhenius assumed that in solutions which have a high conductivity the greater proportion of the molecules of the solute are split up into ions, which alone are effective in conveying the current, that the degree of ionisation increases with dilution, and that in very dilute ($\frac{1}{1000}$ normal) solutions of such substances as sodium chloride and hydrochloric acid, ionisation is practically complete. According to this theory, the specific conductivity of a solution depends only on the number of ions present per unit volume and on their speed; and as the latter in dilute solution will not be greatly affected by further dilution, the friction being practically the same as in pure water, the coefficient of ionisation at a particular dilution—in other words, the proportion of the solute split up into ions—can be determined by comparison of

25

the molecular conductivity at this dilution with the limiting molecular conductivity when ionisation is complete.

Not long before, Van 't Hoff (1885) had brought forward his theory of solution, according to which a substance in dilute solution obeys the gas laws, osmotic pressure being taken as the analogue of gas pressure, and, further, the osmotic pressure of a dissolved substance is numerically equal to the ordinary pressure which it would exert if present in the gaseous form in the same volume as is occupied by the solution. Van 't Hoff found, however, that the osmotic pressure of salts, strong acids, and bases in aqueous solution (obtained indirectly from Raoult's freezing-point and boiling-point determinations) was greater than the calculated value—the solutions behaved as if more particles were present than were to be expected according to the amount of substance used—and it was found necessary to introduce a factor i, which represented the ratio of the observed and the calculated osmotic pressures.

The chemical world had not long to wait for an ingenious and fruitful hypothesis as to the significance of Van 't Hoff's factor i. Arrhenius suggested that the apparent increase in the number of particles was due to ionisation, and, in his classical paper in the *Zeitschrift für physikalische Chemie*,[1] gives a table in which the coefficient i is calculated from the results of conductivity and of osmotic pressure measurements for over ninety substances, including non-conductors, acids, bases, and salts. The agreement is in the great majority of cases extremely satisfactory.

It would seem to be a severe test of the theory to apply it to cases where the possible number of ions derived from a single molecule is large. Thus the normal sodium salt of hexabasic mellitic acid, $C_6(COONa)_6$, should dissociate according to the equation, $C_6(COONa)_6 = 6Na + C_6(COO)_6''''''$, and may therefore be expected to have an osmotic pressure in dilute solution about seven times as great as that of a solution of cane sugar of equivalent concentration. It has been shown indirectly, by freezing-point determinations, that this value is nearly reached in 0·0018 molar solution, so that the requirements of the theory are satisfactorily fulfilled.

The introduction of the electrolytic dissociation theory has been abundantly justified by the great number of diverse facts

[1] *Zeit. physik. chem.* 1887, i. 631.

which it correlates, and by the impetus which it has given to chemical investigation. One or two of the observations for which it offers a satisfactory interpretation may here be referred to.

(1) The properties of dilute solutions are generally additive in character, and can usually be represented as the sum of two factors, one pertaining to the positive, the other to the negative part of the solute molecule. This is true of the electrical conductivity (Kohlrausch, 1876), the specific gravity (Valson, 1871), the refractive index (Bender, 1890), the internal friction (Reyher, 1888), the magnetic rotation (Jahn, 1891), etc. The rotatory power of an optically active acid is, in dilute solution, independent of the base with which it is combined, and the same is true of the light-absorbing power of solutions containing coloured ions (Ostwald, 1892). All these observations at once become intelligible on the theory of the independent existence of ions.

(2) The phenomena of hydrolysis are readily accounted for on the assumption that water itself is slightly ionised. The amount of ionisation has been determined by four distinct methods, and the agreement is satisfactory. Further, the temperature coefficient of the conductivity of the purest obtainable water, as determined by Kohlrausch, agrees with that calculated from Van 't Hoff's equation representing the influence of temperature on the equilibrium.[1]

(3) The, at first sight, surprising fact that on neutralisation of equivalent amounts of different strong acids by strong bases in dilute solution the same amount of heat is set free, also finds a ready explanation on the theory. Since ionisation in dilute solution is practically complete, the process of neutralisation of potassium hydroxide by hydrochloric acid, for example, is represented by the equation, $K^{\cdot} + OH' + H^{\cdot} + Cl' = K^{\cdot} + Cl' + H_2O + x$ calories—in other words, the only reaction in this and other similar cases is the combination of $H^{\cdot}$ and OH' ions to form water, so that the amount of heat evolved in each case is necessarily equal for equivalent quantities.

No other theory so far proposed gives an adequate representation of the observations enumerated above. More or less plausible attempts[2] have been made to account for the abnormal

[1] Van 't Hoff, *Lectures on Physical Chemistry*, English edition, i. p. 157.

[2] Poynting, *Phil. Mag.* 1896, **42**, 298 ; Armstrong, *Encyc. Brit.* 10th edition, vol. 26, p. 741 ; Kahlenberg, *Journ. physical Chem.* 1906, **10**, 141.

osmotic pressures observed with salts, strong acids, and bases on the basis of the hydrate theory,[1] which we owe more particularly to Mendeleeff[2] and Pickering,[3] and which postulates that in aqueous solution the (undissociated) solute enters into chemical combination with the solvent; but this theory has proved inadequate to account *quantitatively* for the observed phenomena.

In recent years attention has been directed more particularly to direct measurements of osmotic pressure, to the question of combination between solvent and solute, to complex ions, and to the investigation of other solvents than water; each of these points will now be shortly considered.

Osmotic Pressure

As we have already seen, there is an intimate connection between the phenomenon of osmotic pressure and the electrolytic dissociation theory, since Arrhenius was able, on the basis of the latter, to account quantitatively for the abnormal osmotic pressures shown by solutions of electrolytes. As regards the experimental basis of his theory of solutions, Van 't Hoff chiefly relied upon the direct measurements of osmotic pressure made by Pfeffer, with the help of a semi-permeable membrane of copper ferrocyanide, as well as on the indirect measurements by the freezing-point and boiling-point methods carried out by Raoult. Within the last two or three years very careful measurements of the osmotic pressure of cane sugar solutions have been made by Morse and Frazer[4] in America with a semi-permeable membrane deposited by an electrical method; they find that "cane sugar in aqueous solution exerts an osmotic pressure equal to that which it would exert if it were gasified at the same temperature and the volume of the gas were reduced to that of the *solvent* in the pure state."

Van 't Hoff[5] was further able to prove thermodynamically that the osmotic pressure and gas pressure must have the same absolute value if the solution is sufficiently dilute. The principles underlying the thermodynamical proof have been put very clearly by Larmor.[6] According to his view, "the change of

---

[1] For a clear account of the hydrate theory, see Watts's *Dictionary of Chemistry*, Art. "Solution."

[2] *Principles of Chemistry*, i. 66, etc.   [3] *Watts's Dictionary*, loc. cit.

[4] *Amer. Chem. J.* 1905, **34**, 1.   [5] *Lectures*, ii. 33.

[6] *Encyc. Brit.* 10th ed. vol. 28, p. 170; *Phil. Trans.* 1897, **190** A, 205.

available energy on further dilution, with which alone we are concerned in the transformations of dilute solutions, depends only on the further separation of the particles . . . and so is a function only of the number of dissolved molecules per unit volume and of the temperature, and is, per molecule, entirely independent of their constitution and that of the medium "—the assumption being made that the particles are so far apart that their mutual influence is negligible. The change of available energy is thus brought into exact correlation with that which occurs in the expansion of a gas. In the proof there is no assumption as to the mechanism of the osmotic pressure; it may be due to molecular bombardments, to affinity between solvent and dissolved substance, to surface tension, or to some other cause. It has further been shown theoretically by Van 't Hoff[1] that the osmotic pressure is independent of the nature of the membrane, provided the latter is perfectly semi-permeable.[2]

Kahlenberg[3] has recently described a series of experiments on osmotic pressure which, in his opinion, invalidate Van 't Hoff's theory. He used a rubber membrane, and in his quantitative experiments determined the osmotic pressures exerted by lithium chloride and by sugar dissolved in pyridine. In both cases the equilibrium pressures were much smaller than those required by theory, and in equivalent concentration the sugar exerted the higher pressure. In the course of the work it was found necessary to stir the solutions inside the cells, and the author considers that, on account of this omission, the results of former observers are unreliable. It should be remembered, however, that the pressures observed by Pfeffer and by Morse[4] were equilibrium pressures, and remained constant for long periods.

Whetham,[5] on the basis of the thermodynamical proof referred to above, contends that the experimental results obtained by Kahlenberg have not the importance the latter claims for them.

Further, as the membrane was slightly permeable both for

---

[1] *Loc. cit.* p. 32.
[2] On the employment of ideal processes in thermodynamics, see Planck, *Annalen der Physik*, 1903, **10**, 436.
[3] *Journ. physical Chem.* 1906, **10**, 141.
[4] *Loc. cit.*
[5] *Nature*, 1906, **74**, 54, 295 ; Kahlenberg, *ibid.* 222.

lithium chloride and for sugar, an element of uncertainty is introduced in the measurements, and it seems desirable to reserve judgment on the matter, pending further investigation.

Combination between Solvent and Solute. The Hydrate Theory

The question as to whether solution is a chemical or physical process has long been one of the standing problems of chemistry. According to the hydrate theory, already referred to, the solute enters into chemical combination with the solvent, with formation of hydrates of varying complexity. The chief evidence adduced in favour of this theory was that the curves representing the variation of the physical properties of such solutions with dilution showed breaks at points corresponding with the composition of various hydrates, and in many cases it was possible to isolate the hydrates. With the development of the electrolytic dissociation theory, although the possibility of hydrate formation was not lost sight of,[1] most stress naturally came to be laid on the independent existence of ions, and by many the solvent was supposed to play a passive part, simply acting as a medium for dissociation. In recent years, however, more and more evidence has accumulated to show that the solvent plays an essential part in the phenomenon. The evidence in favour of the view that the ions, and in many cases the undissociated molecules, are associated with molecules of the solvent, will now be considered.[2]

(1) The first important contribution to this subject is due to Werner,[3] as a result of his investigation of complex compounds

[1] Ostwald, *Lehrbuch*, ii. 1, 801.

[2] It is important to realise clearly the distinction between the old hydrate theory of Pickering and the electrolytic dissociation theory, including the possibility of hydrated ions. On the basis of the former the molecules of the solute remain intact in solution and the abnormal osmotic pressures shown by solutions of electrolytes are regarded as being connected with hydrate formation ; on the other hand, the fundamental point of the electrolytic dissociation theory is the independent existence of the ions ; the increased osmotic pressure in solutions of electrolytes is considered as being due to an increase in the number of particles of solute owing to ionisation, and this number will not be affected if only one ion enters into a complex with solvent molecules. The question of hydration, though of great interest, is thus of secondary importance for the electrolytic dissociation theory.

[3] For a summary of Werner's work, by himself, see *Neuere Anschauungen auf dem Gebiete der Anorganischen Chemie*, Braunschweig, 1905.

containing ammonia and water. The following series of compounds containing trivalent cobalt were prepared, and their constitution determined by chemical and physical methods:

$$\left[Co(NH_3)_6\right]^{...}Cl_3''',\left[Co(NH_3)_5Cl\right]^{..}Cl_2'',\left[Co(NH_3)_4Cl_2\right]^{.}Cl',\left[Co(NH_3)_3Cl_3\right].$$

As the formulæ indicate, the first compound consists of a trivalent complex ion consisting of 6 NH_3 groups associated with one cobalt atom, and all the chlorine is ionised; in the second compound a chlorine atom has replaced a NH_3 group in the positively charged complex and the remaining two chlorine atoms are ionised. The chemical behaviour completely bears out the different character of the halogen atoms in the second compound; thus only two-thirds of the chlorine can be precipitated by silver nitrate.

Werner[1] further showed that the place of the basic ammonia in such compounds can be taken by neutral groups such as water molecules; thus the following series of cobalt compounds are known:

$$\left[Co(NH_3)_6\right]^{...}Cl_3''',\left[Co\frac{H_2O}{(NH_3)_5}\right]^{...}Cl_3''',\left[Co\frac{(H_2O)_2}{(NH_3)_4}\right]^{...}Cl_3''',\left[Co\frac{(H_2O)_3}{(NH_3)_3}\right]^{...}Cl_3'''$$

$$\ldots \text{ up to } \left[Co(H_2O)_6\right]^{...}Cl_3'''$$

In this case the positive ion remains trivalent, as the substituent is neutral. A further instructive example is the green hexahydrate of chromium chloride, which, according to Werner, has

the formula $\left[Cr\frac{(H_2O)_4}{Cl_2}\right]^{.}Cl' + 2H_2O$; and, in fact, the substance

in question can lose two molecules of water without any alteration in its characteristic properties.

Werner's results have been confirmed by many other investigators, so that if the presence of free ions in solution be granted, there can be no doubt that at least in some cases they are associated with solvent molecules.

Werner further expressed the view that the cause of electrolytic dissociation is to be found in the association of the metallic constituent of the salt with water molecules, and was thus able to bring the electrolytic dissociation theory into line with the old hydrate theory. He writes in 1893:[2] "The first condition for the electrolytic dissociation of a salt is the capacity of its

---

[1] *Loc. cit.* p. 132.   [2] *Zeit. anorg. Chem.* 1893, **3**, 294.

metallic constituent to unite with a definite number of water molecules with formation of a radical in which the water molecules are so arranged that direct combination between the metallic component and the acid residue is no longer possible." Although in this statement he does not refer to the capacity of the acid residue for combining with water molecules, several instances of this are given in his recent book.[1]

(2) Another argument in favour of the hydration of the ions is based on determinations of their relative migration velocities. The rates of migration of the halogen ions have the relative values: $Cl = 65\cdot9$, $Br = 66\cdot7$, $I = 66\cdot7$, and those of the alkali metals $Li = 35\cdot5$, $Na = 44\cdot4$, $K = 65\cdot3$. Thus, although of very different weights, the halogen ions migrate at practically equal rates, whilst the speed of the lithium ion is only about half that of the much heavier potassium ion. Bredig (1894)[2] and Euler (1897)[3] suggested as a possible explanation that the ions are associated with a large number of water molecules, so that their size is negligible in comparison with that of the molecular aggregate, but our most exact knowledge on this point is due to Kohlrausch.[4] He showed that the temperature coefficient of conductivity is a function of the mobility of the ions and, further, that the influence of temperature on the electrolytic resistance of dilute solutions of salts approximates to the temperature coefficient of the mechanical friction of water. Kohlrausch points out that these facts are most readily accounted for on the association hypothesis indicated above.

(3) Further evidence in favour of this view is to be found in the abnormal course of the freezing-point, boiling-point, and vapour pressure curves of many electrolytes with change of concentration—a phenomenon which has been investigated more particularly by Jones and his co-workers,[5] by Biltz,[6] and by Smits.[7] Jones and Biltz find that in many cases the molecular freezing-

[1] *Loc. cit.* p. 138.

[2] *Zeit. physikal. Chem.* 1894, **13**, 262.

[3] *Ann. der physik.* 1897, **64**, 273.

[4] *Sitzungsber. Preuss. Akad. Wiss.* 1902, **26**, 579; *Proc. Roy. Soc.* 1903, **71**, 338 ; Bousfield and Lowry, *Proc. Roy. Soc.* 1904, **74**, 280 ; Bousfield, *Proc. Roy. Soc.* 1905, **74**, 563 ; *Phil. Trans.* 1906, **206A**, 101.

[5] *Amer. Chem. Journ.* 1904, **31**, 356; 1905, **33**, 534 ; *Zeit. physikal. Chem.* 1906, **55**, 385.

[6] *Zeit. physikal. Chem.* 1902, **40**, 217 ; 1903, **43**, 41 ; 1906, **56**, 462; *Berichte*, 1904, **37**, 3036.

[7] *Zeit. physikal. Chem.* 1902, **39**, 385.

point depression—the depression produced by a gram-molecule of the salt × the volume of the solution in litres—gradually diminishes with increase of concentration, reaches a minimum in 0·1 to 0·5 normal solutions, and beyond this point gradually increases. The molecular elevation of the boiling-point shows the same anomaly, though not so distinctly, and the same is true of the molecular lowering of vapour tension.[1]

A possible explanation of these facts on the modified hydrate theory will be evident. With increase of concentration, the coefficient of ionisation will become less, and consequently the molecular depression will diminish. Further, owing to hydrate formation, the " free " solvent will be diminished, the effective concentration will be much greater than the apparent, and the molecular depression will be abnormally high—this effect will clearly be greatest in concentrated solutions. The molecular depression may thus be expected to be great in dilute solution, owing to practically complete ionisation, and in concentrated solution owing to hydrate formation, and to attain a minimum value at intermediate concentrations—a deduction which exactly corresponds with the experimental facts.

It is interesting to note that, according to Biltz,[2] potassium and cæsium nitrates do not show this anomaly; they behave as ideal binary electrolytes, showing a gradual increase of the molecular depression with increased dilution. It is well known that strong electrolytes do not obey Ostwald's dilution law, and it occurred to Biltz to test its validity for solutions of cæsium nitrate, as the deviations with ordinary electrolytes might possibly be due to hydrate formation. The coefficient of ionisation at different dilutions was determined by freezing-point measurements, and it was found that the results were in complete accordance with the law of mass action; the agreement is very much better than for any other strong electrolyte. It may be added that the coefficient of ionisation of this salt calculated from conductivity measurements did not agree with the values obtained by the cryoscopic method, and this speaks in favour of the view which has been expressed more particularly by Jahn,[3] that the conductivity method does not give accurate values for the degree of ionisation.

[1] Smits, *loc. cit.*

[2] *Loc. cit.*

[3] *Zeit. physikal. Chem.* 1900—1902, **33**, 545 ; **35**, 1 ; **37**, 349 ; **38**, 487, etc.

Jones [1] and his co-workers find by this method that, whilst the great majority of organic substances are not hydrated in solution, glycerine, cane sugar, and fructose are exceptions.

Biltz,[2] and also Jones,[3] have attempted to determine the "hydration" of electrolytes—the number of molecules of the solvent associated with a molecule of the solute—but the results are very uncertain, and must be regarded as of a preliminary nature. The values for potassium chloride, according to Biltz, vary from 25 to 19 with 0·1 to 0·5 normal solutions, whilst Jones finds for 3·2, 0·32, and 0·08 normal solutions of lithium chloride, the values 10, 18·5, and 53 respectively. Biltz finds that the number of associated water molecules increases with dilution, whilst Jones was at first of opinion that the hydration decreases on progressive dilution, but in his most recent paper has adopted the contrary view, as the numbers just quoted indicate. Bousfield,[4] from entirely different considerations, also arrives at the conclusion that the hydration increases progressively and continuously with increasing dilution.

The application of the law of mass action to this problem is a matter of considerable importance; it was first discussed by Nernst.[5] If one molecule of a salt S forms a hydrate with n molecules of water, according to the equation $S + nH_2O = S(H_2O)_n$, we obtain, on applying the law of mass action, $C_S C^n_{H_2O} = K C_{S(H_2O)_n}$, where C_S and C_{H_2O} denote the concentration of salt and water respectively and $C_{S(H_2O)_n}$ denotes the hydrate concentration. In dilute solution C_{H_2O} is much greater than the other two; it is therefore practically constant when the others are varied, and thus the amount of water in combination with the salt is nearly proportional to the salt concentration; in other words, the hydration is practically independent of the dilution. It can also be shown by thermo-dynamical considerations that the active mass of the water is proportional to its vapour pressure; and as, in dilute solution, the latter is only slightly affected by further dilution, a rapid increase of hydration on dilution, as postulated by Jones and

[1] *Loc. cit.*
[2] *Loc. cit.*
[3] *Zeit. physikal. Chem.* 1906, **55**, 385.
[4] *Proc. Roy. Soc.* 1906, **77**A, 377; *Zeit. physikal. Chem.* 1905, **53**, 257.
[5] *Theoretische Chemie*, 4th ed., p. 457.

by Bousfield, is very improbable. In very concentrated solutions, on the other hand, the complexity of the hydrates will probably increase up to a point with dilution.

(4) If ions are associated with water molecules, the latter must also be transported through the solution during electrolysis, and if the positive and negative ions are hydrated to a different extent, changes of concentration might be expected to occur at the electrodes, and could be detected by adding a neutral substance to the electrolyte. On Nernst's initiative, this suggestion has been tested by Garrard and Oppermann,[1] who submitted to electrolysis solutions of strong acids and bases with boric acid as the non-electrolyte. On the assumption that the hydrogen ion is not hydrated, the authors find that SO_4'', Cl', Br', and NO_3' ions are associated with 9, 5, 4, and 2·5 molecules of water respectively. Similar experiments have been made by Buckbock,[2] who used mannite and resorcin as non-electrolytes and hydrochloric acid as the electrolyte. Assuming that the hydrogen ion is associated with one molecule of the solvent, he finds that in very dilute solution four molecules of water are transported with the chlorine ion. As the experimental error in such observations is large, the results must be regarded as of a preliminary nature, but it is significant that the hydration found is by no means large, and, as we have already seen, this is *à priori* much more probable than the high values found by Jones and others.

(5) Additional confirmation of the presence of hydrates in solution is obtained from the results of solubility determinations in solutions of electrolytes.[3] In general, inactive gases are less soluble in solutions of electrolytes than in water, as would be expected if, owing to hydrate formation, only part of the solvent is free, and the solubility is most depressed by the salts which, from other considerations, are regarded as being most highly hydrated in solution.

Further, the fact that so many hydrates have been obtained in crystalline form may be considered as proving that there is such an affinity between solvent and solute as might conceivably lead to the formation of hydrates in solution. We may

---

[1] *Göttinger Nachrichten,* 1900, p. 86.
[2] *Zeit. physikal. Chem.* 1906, **55**, 563. Compare also Lobry de Bruyn, *Receuil. Trav. Chim.* 1904, **22**, 430. Morgan and Kanolt, *J. Am. Ch. Soc.* 1904, **26**, 1635.
[3] Results summarised by Baur, Ahrens' *Sammlung,* 1903, **8**, 466.

summarise this part of the subject by the remark that, whilst the evidence in favour of association between solvent and solute seems quite conclusive, no method is known of differentiating between the hydration of the undissociated molecules and the ions, and very little is known of the absolute amount of hydration.

SOLVENTS OTHER THAN WATER

In recent years a great amount of experimental material has been collected, more particularly by Walden and by Kahlenberg and their co-workers, with the object of elucidating the nature of non-aqueous solutions. The guiding idea in this work, from the point of view of the electrolytic dissociation theory, has been the suggestion of J. J. Thomson[1] and Nernst[2] that, since the attraction between contrary electric charges is inversely as the specific inductive capacity or dielectric constant of the medium, a solvent of high dielectric constant should have great ionising power. The results of numerous experiments show, as a matter of fact, that there is an undoubted parallelism, though not direct proportionality, between the two properties. Thus Centnerszwer[3] found that the conductivity of potassium iodide in liquid hydrocyanic acid, which has a higher dielectric constant than water, is about four times as great as in the latter solvent, whilst Jones[4] has shown by the boiling-point method that the same salt is rather more than half as much ionised in methyl alcohol as in water, the dielectric constants being in the ratio 32·5 to 81·0. There are, however, a good many exceptions to this rule. Dutoit and Aston[5] consider that the ionising power of a solvent is connected with its capacity for polymerisation, and other observers, more particularly Brühl,[6] maintain that the property in question depends on the unsaturated character of the solvent, which is almost, but not quite, the same thing as its capacity for polymerisation. Most solvents which are unsaturated have also a high dielectric constant.

[1] *Phil. Mag.* 1893 [v.], **36**, 320.
[2] *Zeit. physikal. Chem.* 1893, **11**, 220.
[3] *Ibid.* 1901, **39**, 217.
[4] *Ibid.* 1899, 31, 129.
[5] *Compt. Rend.*, 1897, **125**, 240.
[6] *Zeit. physikal. Chem.* 1898, **27**, 319; 1899, 30, 1.

It must be borne in mind, in considering this question, that the electrical conductivity method for determining the coefficient of ionisation is not available with many non-aqueous solvents, as it has not been found possible to estimate the limiting molecular conductivity.[1] The coefficient of ionisation cannot be estimated even approximately by comparison of the conductivity of a salt in aqueous solution with that in another solvent, as the conductivity depends on the mobility of the ions, as well as on their number, and the former property varies considerably in different solvents.

There are certain cases, as Kahlenberg[2] more particularly has pointed out, in which cryoscopic and ebullioscopic measurements indicate polymerisation of the solute, notwithstanding which the solution is a good conductor. Kahlenberg regards such observations as incompatible with the electrolytic dissociation theory, while its supporters maintain that both ions and polymerised molecules are present in the same solution. It is clear that the experimental observations could be accounted for in this way, as the decrease in the number of particles owing to polymerisation might more than counterbalance the increase owing to ionisation. It may be noted that solvents of high ionising power have also, in general, a high capacity for depolymerising associated molecules, and this probably explains why complications of the kind just mentioned are very rare in aqueous solution.

Another problem, to which attention has been devoted in recent years, is the conductivity of pure substances. The classical example in this connexion, to which reference has already been made (p. 383), is the conductivity of water. It is interesting to note that the conductivity of most pure liquids which have been so far investigated is of the same order as that of water. This is true of liquid ammonia, according to Franklin

---

[1] Since the above was written, the results of a comprehensive series of experiments on the electrical conductivity of halogen salts in twelve inorganic and organic solvents have been communicated by Dutoit (*Zeit. Electrochemie*, 1906, **12**, 642). Solutions up to $\frac{1}{50000}$—$\frac{1}{200000}$ molar have been investigated, and in all cases a limiting value for the molecular conductivity has been attained. Kohlrausch's law (p. 383) holds for the majority of solvents, but Ostwald's dilution law does not in general apply. In very dilute solution, the results are in excellent agreement with the electrolyte dissociation theory.

[2] *Journ. Physical Chem.* 1901, **5**, 339.

and Kraus,[1] and of liquid sulphur dioxide, according to Walden.[2] The latter investigator has also determined the conductivity of about forty carefully purified organic substances,[3] and in the majority of cases has obtained minimum values which vary from $1 - 5 \times 10^{-7}$ reciprocal ohms at $25°$, whilst Kohlrausch found $0·4 \times 10^{-7}$ for the purest water. A few examples may be given. The value for ethyl alcohol is 2×10^{-7}, for acetone $2·3 \times 10^{-7}$, methyl alcohol $1·45 \times 10^{-6}$. A few substances give exceptionally high values: thus for formic acid $1·5 \times 10^{-5}$ was found, and for acetamide 29×10^{-5}. When one takes into account the impossibility of obtaining absolutely pure substances, it is clear that the values cited can only be regarded as upper limits.

Many fused salts, as is well known, have a high conductivity, as have the filaments of Nernst lamps at high temperatures; the latter, however, are not pure substances. A reference to these interesting questions must here suffice.

COMPLEX IONS

Our knowledge of complex ions is due, in the first instance, to Hittorf. In the course of his classical experiments on the migration of the ions, he observed that in the electrolysis of solutions of potassium ferrocyanide the iron was transported, along with the cyanogen, to the anode. This can be most readily accounted for by the presence in solution of negatively charged ions containing iron and cyanogen. In the same way Hittorf observed that in the electrolysis of solutions of cadmium chloride, part of the cadmium moved towards the anode; in this case it must be assumed that cadmium forms one of the components of a negatively charged complex ion, perhaps $CdCl_4''$.

As an illustration of the methods employed in the investigation of complex ions, the experiments of Bodländer and Fittig, by which it was shown that solutions of silver chloride in ammonia contain the ion $Ag(NH_3)_2'$ may be cited. If this ion is present it must, according to the laws of chemical dynamics, be in equilibrium with its components according to the equation

[1] *Amer. Chem. Journ.* 1898, **20**, 520.
[2] *Zeit. physikal. Chem.* 1902, **39**, 513.
[3] *Ibid.* 1903, **46**, 103.

$Ag(NH_3)_2 \rightleftarrows Ag^{\cdot} + 2NH_3$, and by the law of mass action the expression $\dfrac{C_{Ag^{\cdot}} \times C^2_{NH_3}}{C_{Ag(NH_3)_2}}$ must be constant. This was tested by varying the concentrations of the ammonia and of the silver ions, and good constants were obtained, so that the validity of the original assumption is rendered probable. Complex ions containing ammonia, which have been investigated by Werner, Jörgensen, and others, have already been referred to (p. 387).

An interesting light is thrown upon the question of complex formation by the theory of electro-affinity of Abegg and Bodländer,[1] a short account of which will now be given. According to their views, complex compounds are formed by the combination of ions with neutral molecules ; they only differ in degree from double salts, in that the latter are more completely split up into their components in solution. The tendency shown by the different ions to enter into complex compounds is so much the greater the smaller their electro-affinity ; in other words, the ions which have the least affinity for electricity have the greatest tendency to enter into complexes. The electro-affinity is measured by the "decomposition potential" ("Zersetzungsspannung") of the ion ; it is the equal and opposite potential which must be applied to discharge it, and the order of the electro-affinity of the elements is the same as their order in the "tension series."[2] The alkali metals have the greatest affinity for positive electricity, while such metals as mercury, platinum, and gold have very little. We now find that, whilst the alkali metals occur almost exclusively as simple ions in solution—that is, show no tendency to complex formation—such metals as platinum and mercury enter so readily into complex ions that they are scarcely known in the free condition. This is just what is to be expected on Abegg and Bodländer's theory, so that the latter affords us a convenient method of representing the facts, and has been of great service in the experimental investigation of the question. It should, however, be mentioned that the decomposition potential is no exact measure of the electro-affinity, as is already evident from the fact that the former alters with dilution. The difficulty of obtaining an exact measure of the affinity between an ion

[1] *Zeit. Anorg. Chem.* 1899, **20**, 453.
[2] Van 't Hoff, *Lectures*, iii. 90.

and an electric charge is a considerable drawback to an extended use of the theory.

One further application of Abegg and Bodländer's theory may now be referred to. The electro-affinity of a complex ion is greater than that of its component simple ion ; in other words, the electro-affinity of an ion is strengthened by association with a neutral molecule. On this basis we can understand how weak ions, which are forced to dissociate to some extent when combined with oppositely charged strong ions, will strive to strengthen themselves by entering into combination with neutral molecules. As an example, we may consider acetic acid and ammonium acetate. As the hydrogen ion is comparatively weak the acid is only slightly ionised, but by association with the NH_3 group its electro-positive character is strengthened, so that the salt is strongly ionised.[1] We have also examples of the strengthening of the electro-negative character of cations by association with certain groups, as in the case of the hydrated platinic chloride investigated by Hittorf and Salkowski.[2] This hydrate has distinctly acid properties in aqueous solution, so that hydrogen ions and negatively charged $PtCl_4OH'$ or $PtCl_4O''$ ions must be present. In this case, according to the theory, the negative O'' or OH' ion has been strengthened by association with the neutral group $PtCl_4$. Reference has already been made to the occurrence in solutions of cadmium chloride of $CdCl_4''$ ions ; for such phenomena the term auto-complex formation has been suggested. Further contributions to this subject have been made by Steele[3] and others. Donnan and Bassett[4] have recently adduced evidence to show that solutions of cobalt chloride contain complex $CoCl_4''$ ions.

Considerations of space preclude reference to many other interesting applications of the electrolytic dissociation theory.

SOME CRITICISMS OF THE THEORY

It seems appropriate to conclude with a short account of the criticisms to which the theory has been subjected from time to time, and to consider, in the light of what has already

[1] Werner has arrived at the same conception of the nature of the NH4 group in another way. *Neuere Anschauungen*, p. 97.

[2] *Zeit. physikal. Chem.* 1899, **28**, 546.

[3] *Ibid.* 1902, **40**, 722.

[4] *Journ Chem. Soc.* 1902, **81**, 939.

been said, how the objections might be met. In this connection it will be most instructive to consider some of the points raised by Kahlenberg,[1] as this investigator has collected a wealth of excellent experimental material with the object of showing that the theory in its present form is untenable. Some of the difficulties of the theory are also enumerated by Armstrong in his article on "Chemistry" in the *Encyclopædia Britannica*.

One point very often made is that the law of mass action does not hold for solutions of strong electrolytes. In the same category is the observation that the values for the coefficient of ionisation of many electrolytes, as calculated from conductivity and from cryoscopic measurements, do not agree within the limits of experimental error. With reference to the latter point it must, in the first place, be borne in mind that cryoscopic and conductivity observations can only be expected to give the same results when ionisation is complete, and no complex ions are present. This is clear from the consideration that, according to the theory, the freezing-point depression depends upon the number of particles present, whilst the conductivity depends upon the number of ions, and if an erroneous assumption be made as to the nature of the ionisation—if, for example, formation of complexes takes place and is not allowed for—the results obtained by the two methods will be different, but that will not be a valid argument against the theory.

Various suggestions have been made to account for the anomalous results obtained with strong electrolytes. Jahn and Nernst[2] consider that the deviations from the law of mass action are due to the mutual influence of the ions and undissociated molecules, whilst other possible disturbing causes are the formation of complexes and interaction between the undissociated molecules or ions and the solvent. With reference to the last point we have already seen (p. 382) that cæsium nitrate, which on other grounds is believed to be only slightly, if at all, hydrated in solution, behaves in complete accordance with the law of mass action. The presence of complexes has been definitely proved in many cases, and as all these effects may play a part, the question becomes a very complicated one.

Cases where cryoscopic and ebullioscopic measurements

[1] Summary, *Trans. Faraday Soc.* 1905, **1**, 42.
[2] *Loc. cit.* p. 389. Compare Arrhenius, *Zeit. physikal. Chem.* 1901, **37**, 490.

indicate polymerisation whilst the solutions are fairly good conductors, have already been dealt with (p. 393).

A matter of great interest is Kahlenberg's experimental proof[1] that instantaneous chemical reactions are possible in solvents which have no appreciable conducting power. He has shown, for example, that perfectly dry hydrochloric acid gas causes an immediate precipitate of cupric chloride when passed into a solution of cupric oleate in dry benzene. On this point it may be observed that, while it is no essential part of the electrolytic dissociation theory that all chemical reactions are ionic, yet this view is undoubtedly held by some of its supporters, more particularly by those belonging to the school of Arrhenius. He himself says[2]: " We may assume that in the case of reactions which take place very slowly ions are present, though not in measurable quantity so far as our present methods are able to detect them "; and Euler[3] has attempted, though without conspicuous success, to prove the validity of this assumption. There is, on the other hand, a certain amount of evidence, besides that adduced by Kahlenberg,[4] to show that reactions can take place between undissociated molecules. Thus Shenstone[5] has found that perfectly dry chlorine acts on mercury; Donnan,[6] and also Slator,[7] are of opinion that the activity of the halogen alkyls cannot be ascribed to the presence of ions, and I have recently obtained results which seem to show that in the splitting up of monochloracetic acid by hot water according to the equation—

$$CH_2ClCOOH + H_2O = CH_2OHCOOH + HCl—$$

it is the undissociated acid molecule which is acted on. For these reasons it seems preferable to assume for the present that, although the ions show exceptional chemical activity, reactions between undissociated molecules also take place in certain cases.

With Kahlenberg's general conclusion,[8] that "whether a solution will conduct electrically or not depends on the specific character of both solvent and solute," it does not seem probable

[1] Journ. physical Chem. 1902, 6, 1.
[2] Electro-chemistry, English edition, p. 180.
[3] Zeit. physikal. Chem. 1901, 36, 405, 641 ; 1902, 40, 498 ; 1904, 47, 353.
[4] Loc. cit.
[5] Journ. Chem. Soc. 1897, 71, 471.
[6] Ibid. 1904, 85, 555.
[7] Ibid. 1904, 85, 1286.
[8] Journ. physical Chem. 1906, 10.

that many of the supporters of the electrolytic dissociation theory will be disposed to quarrel.

In conclusion, the question as to the motive for electrolytic dissociation—the mechanism of the process—may be referred to. We have seen that the ions show very great differences in their affinity for electricity, and it seems plausible to suppose that this is one of the factors conditioning ionisation. It is, however, impossible to carry this idea through quantitatively, as Abegg and Bodländer have pointed out,[1] since salts in which the ions have comparatively small electro-affinity are almost as highly dissociated as those in which the affinity is very much greater. These observers therefore suggest that the molecule first decomposes into neutral atoms, which then take up the electric charges, so that the degree of ionisation also depends on the strength of the bond between the individual atoms. This "separation into neutral atoms" seems quite as difficult to account for as the original question at issue, and as we know so little of chemical affinity, which may itself be electrical in nature, nothing definite can be said with regard to it. Werner[2] and Lowry[3] lay stress on another aspect of the question; they consider that ionisation is brought about by the affinity of the solvent for the components of the solute molecule. In this case also chemical affinity between the components of the molecule might be supposed to counteract the ionising tendency of the solution. This suggestion as to the part played by the solvent seems plausible, since, as we have already seen, the best ionising solvents are those having "subsidiary valencies,"[4] and it is natural to suppose that these are effective by entering into chemical combination with the solute. There would thus seem to be at least three factors of importance for ionisation : (1) the chemical affinity between solvent and solute ; (2) the chemical affinity between the components of the solute molecule ; (3) the affinity of the components of the solute molecule for electricity. As to the relative importance of these three factors, practically nothing can be said in the present state of our knowledge.

[1] *Loc. cit.* p. 463. Cp. also Abegg, *Zeit. anorg. Chem.* 1904, **39**, 330.
[2] *Zeit. anorg. Chem.* 1893, **3**, 294.
[3] *Trans. Faraday Society*, 1905, **1**, 197.
[4] Brühl, *loc. cit.*

THE DECADENCE OF AMMONITES

By FELIX OSWALD, D.Sc. (LOND.)

HARDLY any class of fossils is better known to the general public than that of the Ammonites; their beautiful spiral forms, often ornamented with spines and ridges, appeal to every lover of nature. Whilst the majority of Ammonites are rolled up to form spiral discs, it is well known that quite a number of more or less unrolled, turreted, and even straight species frequently occur in rocks of Cretaceous age, previously to the total extinction of the whole group; the abundance and variety of these abnormal forms is obvious even to the most casual visitor of the exposure of the Gault in the Warren, near Folkestone, where these fossils are so lavishly scattered on the seashore.

The development of more or less uncoiled species prior to annihilation is manifested, however, not merely in one family but in several; and it has occurred not only in the Cretaceous period, but in the Jurassic, and even so early as the Trias, when an important division of Ammonites became extinct. It would seem probable, therefore, that this phenomenon of uncoiling in different families may have had the same determining cause, and the present article is an attempt to trace graphically the parallel evolution of these forms and to hazard a speculation as to their origin.

The table on the opposite page of the range in time of Ammonites will show that the important sub-order of the *Discocampyli* appeared in the Permian, reached a great development in the Triassic, and died out completely at the end of that period.

One of the best known types of the *Discocampyli* is the genus *Ceratites* of the Muschelkalk (fig. 1); its comparatively simple lobes and saddles are in marked contrast to the increasing complexity which they display in the Jurassic and Cretaceous Ammonites. The degenerate families of the *Choristoceratidæ* and *Cochloceratidæ* not only show unrolled forms, but their lobes and saddles undergo a corresponding retrogressive simplicity by being reduced to six in the former and four in the latter family. The transitional stages in the process of unrolling can be traced from the normal spiral of *Polycyclus* (fig. 2) to the partly unrolled spiral of *Choristoceras* (fig. 3), the pulled-out turret-shell of *Cochloceras* (fig. 4), and finally to the straightened-out, rod-like *Rhabdoceras* (fig. 5).

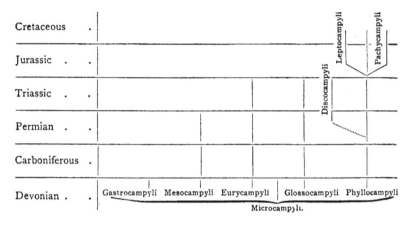

The *Phyllocampyli* was the only sub-order of Ammonites to survive through the Jurassic and Cretaceous periods. It has been subdivided into the *Leptocampyli* and *Pachycampyli*, both of which contain species showing the phenomenon of unrolling. In the *Leptocampyli* the transitional forms may be typically represented in gradation from the normal spiral of the Tithonian *Lytoceras* (fig. 6) to the partially unrolled *Macroscaphites* (fig. 7) and *Scaphites* (fig. 8), the straightened, hook-like *Anahamulina* (fig. 9), and *Ptychoceras* (fig. 10), in which the straight rod is bent upon itself, finally leading to the slightly curved or perfectly straight forms of *Baculites* (figs. 11 and 12). Even *Baculites*, in its infantile stage, starts as a closely coiled shell. The affinity of all these genera with each other is demonstrated by their similar suture lines, although these became much simplified in the straighter forms.

Whilst these anomalous forms were all developed in the Cretaceous period, there is a family of the *Pachycampyli*—the *Spiroceratidæ* (belonging to the group of *Morphoceratidæ*)—which displayed unrolled forms so early as the Callovian formation of the Jurassic; they are probably derived from the family of *Reineckidæ*, for their sutures are similar to those of *Reineckia* (fig. 13), but they have been reduced to the primitive formula of six lobes. *Œcoptychius* (fig. 14) already shows a tendency to being pulled out; in *Spiroceras* (fig. 15) the spiral has become quite an open one; and in *Baculina* (fig. 16) it has become completely straightened out into the shape of a needle.

Other groups of the *Pachycampyli* show a very similar parallelism. For instance, in the *Acanthoceratidæ* we can pass from the normal *Lyticoceras* (fig. 17) to the open and pulled out spiral of *Helicoceras* (fig. 18), and the more irregular, half-closed, half-open *Heteroceras* (fig. 19); finally we reach the open bow of *Toxoceras* (fig. 20), and the straightened double hook of *Torneutoceras* (*Hamites*) (fig. 21). It will be noticed that all these forms, as well as those of the next group, are of Cretaceous age.

The group of the *Cosmoceratidæ*, the members of which are so highly ornamented with ribs and tubercles, show even more aberrant variations of the spiral curve. Here again we can trace a transition from the normal *Douvilleiceras* (fig. 22) to the open spiral of *Crioceras* (fig. 23), and the hook-like *Anisoceras* (fig. 24) and *Ancyloceras* (fig. 25); the last-named may be compared with the more closely coiled *Macroscaphites* (fig. 7). The adult *Anisoceras* combines the open spiral of *Helicoceras* with the bow of *Toxoceras*, and in its old age terminates in a crook or retroversal bend. This group does not contain any straight forms, but the turreted spiral is well exemplified by *Turrilites* (fig. 26), which was already foreshadowed in the Trias by *Cochloceras* (fig. 4); and in *Bostrychoceras* (fig. 27) the turret has become in part an open spiral. *Emperoceras* shows a remarkable instability, varying according to age; the young remain in a Hamitean stage for a prolonged period, and then suddenly develop into a form like *Helicoceras* or even *Turrilites*.

A key to the cause of this unrolling of the spiral in these families of Ammonites, prior to their extinction, is perhaps furnished by the fact that the number of lobes in these aberrant species has undergone reduction, and reverts to the primitive number of six. The evolution of more or less uncoiled or straightened forms may, therefore, be due to a struggle, so to speak, to return to primitive simplicity both of outward form

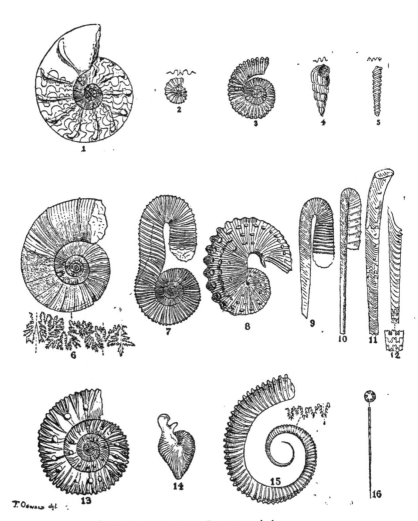

1. *Ceratites nodosus*, de Haan ; Trias (Muschelkalk).
2. *Polycyclus nasturtium*, Dittmar ; Trias (Keuper).
3. *Choristoceras Marshi*, Hauer ; Trias (Rhætic).
4. *Cochloceras Fischeri*, Hauer ; Trias (Keuper).
5. *Rhabdoceras Suessi*, Hauer ; Trias (Keuper).
6. *Lytoceras Liebigi*, Oppel ; Jurassic (Tithonian).
7. *Macroscaphites Ivani*, d'Orb. ; Cretaceous (Neocomian).
8. *Scaphites spiniger*, Schlüter ; Cretaceous (Senonian).
9. *Anahamulina subcylindrica*, d'Orb. ; Cretaceous (Neocomian).
10. *Ptychoceras Puzosianum*, d'Orb. ; Cretaceous (Barremian).
11. *Baculites incurvatus*, Dujardin ; Cretaceous (Turonian).
12. *Baculites anceps*, Lam. ; Cretaceous (Danian).
13. *Reineckia Brancoi*, Steinmann ; Jurassic (Callovian).
14. *Œcoptychius refractus*, de Haan ; Jurassic (Callovian).
15. *Spiroceras bifurcatum*, Quenst. ; Jurassic (Callovian).
16. *Baculina acuaria*, Quenst. ; Jurassic (Callovian).

      In some cases the suture-lines are given.

and internal structure. Now it is generally agreed that the Ammonites are derived, through such simple coiled forms as *Anarcestes* (fig. 29) and *Mimoceras*, from a primitive straight form such as the Devonian and Ordovician *Bactrites* (fig. 28),[1] which in turn may be connected with the straight forms (*Orthoceras*, fig. 30) of the Nautiloids by means of an intermediate type like the long pencil-shaped *Protobactrites*, which ranges from the Silurian to the Carboniferous.

All the available evidence points at present to the conclusion that *Bactrites* is an ancestral primitive genus, and is not descended from a coiled shell. It is therefore quite dissimilar in its origin from the equally straight *Baculites*, which (as already indicated) has been derived by retrograde development from a closely coiled spiral. As Hyatt[2] has concisely stated, "the straight *Bactrites*-like young of some forms of *Anarcestes*, the gyroceran young of others of the *Goniatitinæ* and the gyroceran adults and young of *Mimoceras*, indicate the derivation of the *Goniatitinæ* to have been from Silurian straight shells similar to *Bactrites*, if not directly from that genus itself." The close genetic relation between the coiled *Mimoceras* and the straight *Bactrites* was also prophetically foreshadowed by Hyatt,[3] when he described the species of *Mimoceras* as "separable from *Bactrites* in no essential characteristic except the presence of a permanent protoconch upon the apex. . . . Their characteristics and the protoconch ally them, however, even more closely with *Anarcestes*. . . . This evidence appears to need but one more link—the finding of a *Bactrites* with a globular protoconch." This missing link in the chain of evidence was eventually supplied by Branco,[4] who describes this protoconch (or initial chamber of the shell) as agreeing in shape with that of *Mimoceras compressum*, and differing altogether from that of the Nautiloids.

Among the Nautiloids a case of unrolling of a somewhat parallel nature to that of the Ammonites may be furnished

---

[1] Whilst this genus is not yet known to occur in the Silurian, it has been recorded from the Ordovician ; the earliest known species is Barrande's *Bactrites Sandbergeri* from his Etage D, in its two extreme horizons : viz., in d 1 (equivalent to the Arenig group) and in d 5 (equivalent to the upper part of the Bala group) ; but its presence has not hitherto been revealed in the intermediate stage.

[2] "Genesis of the Arietidæ," *Smithsonian Contrib. to Knowledge*, 1889, p. 1.

[3] *Proc. Boston Soc. Nat. Hist.* xxii. (1884), p. 309.

[4] "Ueber die Anfangskammer von Bactrites," *Zeitschr. deutsch. geol. Gesell.*, 1885, Bd. xxxvii. Heft 1, p. 1.

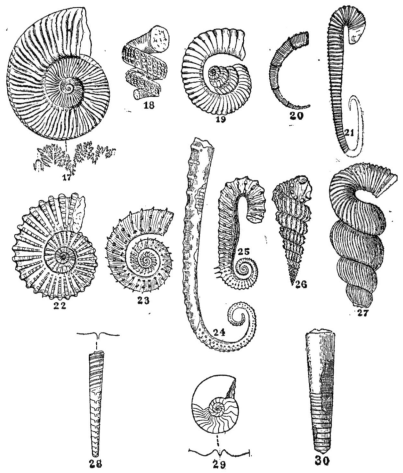

F. Oswald del.

17. *Lyticoceras Noricum*, Sowby ; Cretaceous (Neocomian).
18. *Helicoceras Robertianum*, d'Orb. ; Cretaceous (Neocomian).
19. *Heteroceras Emerici*, d'Orb. ; Cretaceous (Neocomian).
20. *Toxoceras annulare*, d'Orb. ; Cretaceous (Neocomian).
21. *Torneutoceras* (*Hamites*) *rotundatum*, Sowby ; Cretaceous (Gault).
22. *Douvilleiceras mammillare*, Schloth. ; Cretaceous (Gault).
23. *Crioceras Emerici*, Lév. ; Cretaceous (Neocomian).
24. *Anisoceras Saussureanus*, Pictet ; Cretaceous (Gault).
25. *Ancyloceras Matheronianum*, d'Orb. ; Cretaceous (Neocomian).
26. *Turrilites catenatus*, d'Orb. ; Cretaceous (Gault).
27. *Bostrychoceras polyplocum*, Roemer ; Cretaceous (Senonian).
28. *Bactrites elegans*, Sandb. ; Devonian.
29. *Anarcestes subnautilinus*, Schloth. ; Devonian.
30. *Orthoceras* (*Geisonoceras*) *timidum*, Barr. ; Silurian.

In some cases the suture-lines are given.

by the Palæozoic families of the *Lituitidæ* and *Trochoceratidæ*; here also it appears to be a mark of senility in the race. In the Ordovician and Silurian genus *Lituites* (in its present restricted sense) the shell is at first coiled in four whorls and then is unrolled in a very long straight portion, gradually widening in diameter, with an aperture partially contracted by lobes. The straight piece is bent slightly inwards near its commencement, just as in *Macroscaphites* (fig. 7). Noetling[1] considers that the development of *Lituites lituus* shows indications of the descent of this genus from *Nautilus*-like ancestors. The allied genus *Trochoceras*, which ranged from the Cambrian to the Devonian, now contains several well-known species formerly ascribed to *Lituites*—e.g. *T. cornu-arietis, T. arietinum, T. giganteum*, and *T. convolvans*; but the whorls rarely exceed two in number, and the unrolling has not proceeded so far as in *Lituites*; it only shows an incipient unrolling in the last whorl.

With regard to the Turrilite form of the unrolled Ammonites, Quenstedt, Oppel, and in later years Hyatt,[2] have shown that this form, for instance, often occurs transitorily in the young shells of several species of Ammonites, and becomes completely suppressed during the later period of growth, when the whorls develop normally in the same plane. The persistence of a Turrilite form in the adult is, therefore, clearly retrogressive and reversionary, and may be regarded as a more highly developed phase of a juvenile variation. Even all this bizarre variation of form did not avail to perpetuate the race of Ammonites, and not a single species survived the Cretaceous period. It must be remembered that the Ammonites had diverged in habit from the swimming Nautiloids and had adopted littoral habits, crawling along the bottom with their shells above them, and very rarely swimming. The presence of a rostrum, in fact, points to the disappearance of the hyponome, or swimming organ of *Nautilus*. The strikingly progressive complication of the suture-lines of the Ammonites was correlated with their usefulness in assisting the balance of the shell above the extended parts of the animal whilst crawling along the sea

---

[1] "Ueber *Lituites lituus*," *Zeitschr. deutsch. geol. Gesell.*, 1882, Bd. xxxiv. Heft 1, pp. 156–95, Taf. x. xi.

[2] "Embryology of Fossil Cephalopods," *Bull. Mus. Comparat. Zoology, Harvard Coll.* iii. No. 5, p. 72.

bottom. But this very complication would militate against the race retaining sufficient plasticity to vary effectively when conditions so adverse to them prevailed during and at the close of the Cretaceous. These conditions probably consisted in the main of a marked alteration in the distribution of land and sea, and a consequent radical change in currents—a change by which littoral organisms would be more easily affected adversely than free swimming creatures.

THE RUSTING OF IRON

By W. A. DAVIS

MUCH attention has been directed recently to the study of the rusting of iron. Simple as the investigation of this problem would at first sight appear, it is only within the past year that it has been given a definite solution. The importance of the question can hardly be overestimated. In rusting, all those properties of iron are destroyed which make it of pre-eminent value as a metal. Moreover, in view of its almost universal prevalence, the change is a very costly one. It has been stated that the London and North-Western Railway Company lose eighteen tons of metal daily from their rails owing to wear and rust—principally rust; this representing, at the present price of steel rails, a loss of nearly £40,000 per annum. In painting the Forth Bridge, where the greatest care is taken to prevent rusting, an expenditure of £2,100 per annum is incurred.[1] The increased use of iron and steel in modern structures makes it indispensable that an accurate knowledge should be obtained of the conditions under which the metal is converted, more or less rapidly, into a material which resembles the earthy ores from which it was originally extracted.

It has long been known that dry air does not cause iron to rust but that if the unprotected metal is left exposed to the weather, it rapidly becomes coated with the reddish-brown,

[1] I am indebted to the Secretary of the Forth Bridge Railway Company for kindly furnishing me with the following particulars. The whole of the bridge is painted every three years, one-third being painted per annum. Special care is taken to see that all parts affected with rust are properly cleaned and that the paint is laid on a clean, dry ground, each coat being allowed to dry thoroughly before the next coat is applied. The paint used consists chiefly of oxide of iron, red lead and boiled linseed oil. Twenty-eight painters are continuously employed throughout the year, their wages amounting to £1,700. The cost of the mixed paint ready for use is about £20 per ton ; about twenty tons are used annually. The part of the bridge most affected by corrosion is that from high water to a height of about twenty-five feet; such parts have to be dealt with more frequently than others.

crumbling material known as rust. The condition most favourable to rusting appears from every-day experience to be alternate or simultaneous exposure to air and water, as in the case of the surface of a railway line on which the rain is drying after a shower.

During the greater part of the nineteenth century rusting was regarded as a simple process of oxidation. At first no attempt was made to explain the part played by water in the change, although the presence of water was recognised as essential. The experiments of Crace Calvert in 1871 indicated that the presence of carbon dioxide was an important factor in rusting. Professor Crum Brown in 1888 definitely formulated the theory that iron changes into rust under the combined action of the oxygen and carbonic acid of the air in presence of *liquid* water. The carbonate (or bicarbonate) of iron formed as the initial product was supposed to be converted gradually by the action of atmospheric oxygen into a ferric hydroxide or rust. The successive changes were summarised by Crum Brown in the following equations :

$$4\,(Fe + H_2O + CO_2) = 4\,FeCO_3 + 4H_2 \dots \dots (1)$$
$$4\,FeCO_3 + 6\,H_2O + O_2 = 2\,Fe_2(OH)_6 + 4\,CO_2 \dots (2)$$

Liquid water is essential in the process and acts by producing a solution of carbonic acid. Once the action is initiated it becomes continuous, because as much carbon dioxide is regenerated by the process represented in the second equation as is required in the first. This theory of rusting, which may be called the "*carbonic acid theory*," was generally accepted and its correctness not called into question for several years.

In the year 1898, however, in a lecture delivered at the Royal Artillery Institution, Woolwich, Prof. Wyndham R. Dunstan brought forward reasons for considering that the change was essentially of a different character; the full account of Prof. Dunstan's experiments, made in conjunction with Drs. Jowett and Goulding, on which this view was based, was not, however, published until 1905. Prof. Dunstan considered that the earlier experiments of Calvert were in some respects erroneous. It was contended that when the utmost care is taken to exclude carbon dioxide, iron will still rust in pure oxygen in contact with water. The older theory that rusting was due to carbon dioxide was considered "quite untenable since it has been

shown that rusting can take place in absence of carbonic acid."
Previous to this it had been considered that the action which
certain materials, such as lime and the alkalis, are known to
exercise in preventing the formation of rust, was due to their
power, as alkalis, of absorbing carbon dioxide; but a number
of other substances were cited by Prof. Dunstan as sharing with
alkalis the power of inhibiting rusting. These were sodium
nitrite, potassium ferrocyanide, chromic acid, potassium chromate
and potassium dichromate. A new theory of rusting was
suggested which was based on the fact that hydrogen peroxide
was known to be formed in certain previously studied cases of
oxidation; this had been established for certain physiological
processes by Hoppe Seyler and for the oxidation of certain
metals, such as zinc, by Schönbein and Moritz Traube. Prof.
Dunstan considered that the rusting of iron took place as
represented by the equations—

$$Fe + O_2 + H_2O = FeO + H_2O_2 \ldots (3)$$
$$2\,FeO + H_2O_2 = Fe_2O_2(OH)_2 \ldots (4)$$

According to this view the action of pure oxygen and water
on iron is to produce ferrous oxide and hydrogen peroxide,
the latter then interacting with the ferrous oxide, forming
hydrated ferric oxide or rust. This view, originally put forward
in 1898, was also advocated two years later by Prof. Dunstan
in a Report on the Atmospheric Corrosion of Steel Rails
presented to the Steel Rails Committee of the Board of Trade
(1900). It was considered to give a simple explanation of the
action of such substances as alkalis and oxidising agents in
preventing the conversion of iron into rust, this property being
attributed to the power possessed by these substances of decomposing hydrogen peroxide at the moment of its formation. It
was adduced as evidence in favour of this view that, if a plate of
highly polished steel is immersed in a strongly alkaline solution of the peroxide, decomposition of the latter occurs very
rapidly, bubbles of oxygen being liberated on the surface of
the steel; yet no rusting is seen to occur. The composition
of several samples of ordinary iron rust was found to be
approximately that corresponding with the formula $Fe_2O_2(OH)_2$;
it was pointed out as "a curious coincidence, if nothing more,
that this is equivalent to—

$$2\,FeO + H_2O_2."$$

The hydrogen peroxide theory is, however, at the outset open to objection. First, the formation of hydrogen peroxide during the rusting of iron has never been directly proved, in spite of the fact that hydrogen peroxide is a substance easily detected when present in even the most minute quantity and that, according to equations 3 and 4, it might be expected to accumulate to some extent during the process of rusting. Secondly, there is no reason why the destruction of hydrogen peroxide, the *product* of the action represented by equation 3, should prevent the action of oxygen and water on metallic iron from continuing; it might, indeed, be thought that it would favour such an action.

It must here be emphasised that Prof. Dunstan considered that, in view of his results, the older theory which attributed rusting to carbonic acid was practically to be abandoned altogether: "So far as ordinary rusting is concerned undoubtedly it must be, because carbonic acid gas is not necessary for rusting to occur." And again: "This hypothesis has been proved by the results of the present investigation to be quite untenable, since it has been shown that rusting can take place in the absence of carbonic acid." An alternative method of explaining rusting had, indeed, appeared necessary when all attempts to arrest the change by excluding carbon dioxide had failed. The validity of the hydrogen peroxide theory, therefore, depends on the question whether pure water and pure oxygen are together sufficient of themselves to cause iron to rust. It therefore becomes of importance to consider the method by which the air and water were purified in Prof. Dunstan's experiments, so as to ensure the absence of traces of carbon dioxide.

The apparatus used is illustrated in fig. 1.[1] The iron, the surface of which had been thoroughly cleaned and polished, was contained in the tube A, which was attached at *c* by the only rubber joint in the whole apparatus. Hydrogen, which had been passed through several wash bottles and towers containing caustic potash, was driven from B through the distilling flask C, which contained water, lime and potassium permanganate, so as to fill the apparatus completely, the clamp *c* being closed. A quantity of water amounting to about 100 c.c. was then distilled in a current of hydrogen from the flask C into

---

[1] I am indebted to the Chemical Society for permission to use the blocks illustrating this article.

the bottle D. The whole apparatus being still full of hydrogen, the taps b and d were closed, c was opened and the portion of the apparatus above tap b was exhausted by means of the air pump. The taps c and a were now closed, b was opened and water was distilled into the T-piece above tap a until the level of the distilled water reached tap c. On opening the tap c the water, saturated with pure hydrogen, entered the tube A and completely covered the piece of iron. The tap b was now closed and the portion of the apparatus above it filled with pure oxygen admitted through tap d. "The utmost care was taken that this oxygen was completely free from carbon dioxide by passing it through three wash bottles containing

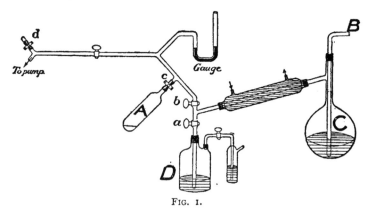

Fig. 1.

aqueous potassium hydroxide and afterwards through a tower of solid potassium hydroxide." It was found that as long as the iron was left in contact with the water saturated with hydrogen and in an atmosphere of the same gas, no action occurred. Directly oxygen was admitted, however. "action immediately commenced" and in a short time rust became visible.

Sufficient as the precautions adopted might appear for the purpose of excluding carbon dioxide, their inadequacy has been clearly demonstrated by the results obtained by Dr. G. T. Moody and published in the *Transactions* of the Chemical Society for April 1906. These results established beyond doubt that when very special precautions are taken to eliminate carbon dioxide, iron may be left in contact with purified oxygen and water for many weeks without undergoing change. In discussing the foregoing results, Dr. Moody points out that

"Dunstan and his co-workers do not appear to have been aware of the extreme difficulty of completely excluding carbonic acid. That they, in their most important experiment, did not take adequate precautions to ensure the absence of carbonic acid is clearly shown, for in their description of it they make the significant admission that on allowing oxygen to pass into the vessel containing water and iron 'action immediately commenced, a substance of a green colour being produced, which rapidly changed to the red colour characteristic of rust.' No more conclusive evidence that carbonic acid was present in the materials could be afforded than the formation of this green colour, which invariably accompanies the early stage of attack of carbonic acid on iron in presence of air or oxygen."

After several failures to exclude the last traces of carbon dioxide, Dr. Moody succeeded in devising an apparatus by means of which it was found possible to leave iron, oxygen and water in contact during long periods without even a speck of rust appearing on the surface. The apparatus finally adopted is shown in fig 2. The side tube of the distilling flask A passes through the long condenser case B and forms a bend at C in which the piece of purified iron is placed; the other end of the

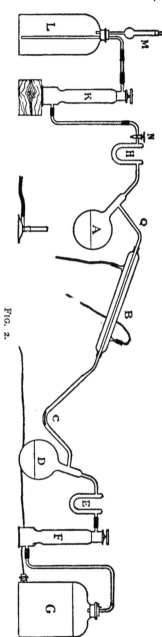

FIG. 2.

bend is connected to the flask D which acts as a receiver. D is connected in series with the large aspirator G by means of a U-tube E, tightly packed with soda lime, and a tower of caustic potash F. The U-tube H, also containing soda lime, and the caustic potash tower K are connected with the air reservoir L containing sticks of moistened potash. When the aspirator G is working, air enters the apparatus through a minute orifice at the upper part of the tube M, which is filled with soda lime. From the stopcock N to the side tube joining E and F all parts of the apparatus are fused together, so that when reasonable precautions are taken, leakage of carbon dioxide from the air into bend C is entirely prevented. The other parts of the apparatus are securely joined together by pressure india-rubber tubing wired in place and having its surface covered with vaseline.

In an actual experiment a small cylinder of nearly pure polished iron about 40 mm. long and 2 mm. in diameter was placed in a bent piece of dry glass tube containing sufficient of a 1 per cent. solution of chromic acid just to cover the iron. A slow current of air was drawn through the apparatus for about three weeks so as to remove all traces of carbon dioxide. The dilute chromic acid solution served to keep the iron clean and free from rust while the carbon dioxide was being removed. Water was then distilled from the flask A, which contained a 1 per cent. solution of barium hydroxide, until all the chromic acid was washed into the receiver D. During the distillation the cock N was closed and, with the object of preventing carbon dioxide from finding its way into the apparatus through a sudden inrush of air, care was taken to cool the apparatus very ʃslowly after distillation. Air was then drawn slowly through the apparatus during about six weeks.

In the earlier experiments it was found that after this period the iron remained perfectly bright with the exception of those parts which rested on the glass, where a slight discoloration was visible. As interaction between the glass and iron appeared to take place, precautions were taken to avoid contact between these substances. The ends of the small iron cylinders were therefore covered with small blobs of purified paraffin wax: in this way the metal was caused to lie in the bend of the tube without touching the glass, the position being such that as each bubble of air passed the bend, the upper surface of the iron was exposed to the oxygen whilst the lower surface remained immersed in water. In one experiment, after distillation an amount of air was passed during five weeks which thirty times exceeded the quantity required completely to convert the iron into ferric oxide: not even one speck of rust appeared. To show that the treatment with chromic acid had not in any way modified the tendency of iron to rust in presence of carbon dioxide, in a similar experiment in which iron had been exposed to air and water during three weeks, the glass tube was cut at the point Q immediately above the distilling flask. Air containing carbonic acid, cleaned by previous passage through a tower containing pumice stone moistened with water, was then passed slowly through the apparatus. After six hours the surface of the iron was distinctly tarnished; after seventy-two hours the whole of the surface of the metal was corroded and a considerable quantity of red rust had collected in the bend of the tube.

These experiments show beyond any doubt that purified oxygen and purified water are insufficient to cause the rusting of iron. The view that iron, pure oxygen and water interact directly to form ferrous oxide and hydrogen peroxide is thus incorrect, so that the theory of rusting based on this assumption becomes untenable. The peculiar behaviour of substances which are known to arrest the rusting of iron cannot be attributed to their power of decomposing hydrogen peroxide but must be assigned to another cause. This question has been investigated by Dr. Moody with especial care. The action of alkalis, such as lime or caustic soda, is at once explained by their power of destroying carbon dioxide. The influence of salts of weak acids such as sodium nitrite and potassium ferrocyanide is also due to their power of absorbing carbon dioxide, the volatile acids, nitrous and hydrocyanic acid, being liberated. In presence of sodium nitrite, ferrous oxide is not oxidised to the ferric state. Weak solutions of chromic acid, which rapidly decompose hydrogen peroxide, do not prevent iron from rusting rapidly, whilst potassium iodide, which also decomposes hydrogen peroxide, not only does not inhibit rusting but actually accelerates it.

It has also been made clear that during the formation of rust under natural conditions a large proportion of ferrous carbonate is produced. The composition of rust in the course of formation is, indeed, altogether out of harmony with the hydrogen peroxide theory, since this theory postulates that double the quantity of hydrogen peroxide necessary to oxidise the whole of the ferrous oxide to ferric oxide is produced by the action of water and oxygen on iron (see equations 3 and 4). The following analysis I. shows the composition of rust freshly formed in an iron tank the metal of which was alternately and in rapid succession exposed to the action of air and water. Analysis II. gives the composition of the same rust after eight days' exposure in the form of powder to ordinary air; compared with I., it shows the readiness with which ferrous carbonate and ferrous oxide undergo oxidation in ordinary air and accounts for the low percentages of ferrous iron found in most samples of rust by previous observers.

			I.	II.
Percentage of iron as ferric oxide .	.	. .	55·73	80·27
„ „ ferrous oxide	.	.	32·86	14·11
„ ferrous carbonate.	.	.	11·40	5·62

It would therefore appear to be finally proved that in the process of rusting, the production of hydrogen peroxide plays no part; the influence of carbonic acid in ordinary rusting is shown by the large proportion of ferrous carbonate in newly formed rust. But carbonic acid is not the only acid which may promote the atmospheric corrosion or rusting of iron. In some cases sulphurous acid or nitric acid may exert a similar influence. Notably is this so in the case of the iron roofs and girders of several of our railway termini; the iron is here constantly exposed to the action of steam, condensed on the cold surface of the metal and charged with carbonic acid and sulphurous acid produced by the combustion of the coal of the locomotives. The conditions for rusting in such a case are nearly ideal and very special precautions become necessary to ensure protection of the metal from corrosion. The rapidity of rusting in such conditions is shown by the fact that it has been found necessary to specially support, by masonry, girders at Blackfriars Station (District Railway) which have undergone severe corrosion.

Primarily the rusting of iron is the result of acid attack and the conditions for rusting to occur must be the same as those known to be determinative of chemical action in general : namely, the possibility of the existence of an electric circuit. The interaction of iron with water and oxygen appears to be impossible in the absence of an electrolyte, just as the union of hydrogen and oxygen has been shown by recent experiments to be also impossible in absence of impurities. In the case of iron the presence of a trace of acid, by rendering the water an electrolyte, fulfils the conditions requisite for action to occur. In the case of ordinary atmospheric corrosion the acid is usually carbonic acid.

The misapprehension or misconception of this position has given rise to some discussion on the subject in the columns of *Nature*. Thus, it has been suggested that whilst carbon dioxide, oxygen and water are essential for the rusting of *pure* iron, the last two alone may be sufficient to cause the rusting of impure forms of the metal. But rusting in such cases appears to be due to the production of acids owing to the oxidation of impurities in the iron, these acids playing the same part as carbonic acid in the rusting of pure iron.

The remarkable sensitiveness of iron to attack by water containing even small amounts of carbonic acid may be illustrated by a simple experiment. Distilled water which has been shaken with or left in contact with ordinary air is poured on a perfectly clean, polished plate of iron. After forty seconds, when the metal is seen to be perfectly bright, the water is run into a porcelain basin containing a drop of a dilute solution of potassium ferricyanide ; a marked blue coloration shows the presence of dissolved iron. Using rain water, the amount of iron dissolved in the ferrous state (as ferrous bicarbonate) is still greater and may be detected after thirty seconds' contact. Distilled water which has been well boiled and thus freed from carbonic acid, does not, however, dissolve a trace of iron. On the other hand, when iron borings are left in contact with water saturated with carbonic acid, hydrogen is steadily evolved from the surface of the metal and the latter passes into solution. The influence of carbonic acid in promoting rusting can be well realised from the following comparative data, which show the rate of absorption by iron of oxygen from ordinary air and from air almost entirely freed from carbonic acid.

	Percentage of total oxygen in 100 c.c. of air absorbed by 10 grams of iron.	
	Ordinary air and distilled water.	Air and water nearly freed from carbonic acid.
After 6 hours' exposure . . .	5·7	None.
„ 24 „ „ . . .	29·1	None.
„ 72 „ „ . . .	61·3	0·9
„ 168 „ „ . . .	94·3	3·8

These experiments show clearly that when the amount of carbonic acid is very small the absorption of oxygen takes place with extreme slowness. In ordinary air absorption takes place fairly rapidly, beginning almost immediately and continuing until nearly the whole of the oxygen is removed.

The rusting of steel as distinguished from iron requires consideration as presenting certain peculiar features. It was pointed out in the Report of the Committee on Steel Rails already referred to, that the condition and mode of distribution of the constituents of steel, not merely their nature and quantity, should be taken into account in judging of the suitability of the metal for the manufacture of rails. The character of a steel depends not only on its chemical composition but much more on the thermal treatment to which it has been subjected. Steel may appear, when examined microscopically, to be nearly homogenous or it may appear more or less heterogeneous; the iron may be present in two forms, a- and β-ferrite, and also in the "carbide" forms—"pearlite" or "cementite." The micro-structure appears largely to determine the behaviour on rusting. In steel, the iron (ferrite) is attacked during atmospheric corrosion much more rapidly than the carbide of iron. In certain cases, particularly when manganese is present, the iron (ferrite) is found to form large distinct veins and laminæ; these appear to be more readily attacked than the rest of the mass during atmospheric corrosion, forming superficial furrows. It is plainly of importance that a micro-structure as resistant as possible to rusting should be assured in cases where resistance to rusting is one of the principal requirements to be met. The influence of the addition of certain elements such as nickel, in diminishing the liability of steel to rust, has recently been recognised; and certain varieties of steel containing nickel are now known which are almost entirely resistant to atmospheric corrosion. Much remains to be learned in this direction.

To sum up: the cause of rusting is the action of liquid water containing traces of acid on iron in presence of atmospheric oxygen. Certain kinds of iron are more susceptible than others to attack, in virtue of their different degrees of purity. To prevent rusting it is necessary primarily to exclude every trace of acid ; and, as this is generally impracticable, the alternative is to prevent contact of the iron with water and the atmosphere by means of some protective coating such as paint. Whether, in the case of steel, the internal structure can be so modified by a suitable and inexpensive treatment that the metal shall be nearly rustless, is a problem that still remains open and urgently needs investigation.

SOME ASPECTS OF
"DOUBLE FERTILISATION" IN PLANTS

By ETHEL N. THOMAS, B.Sc.

University College, London

THE last few years have seen several interesting and important advances in botanical science, and although some of these were initiated at the close of the last century, one is yet tempted to think of them as twentieth-century discoveries. The application and elucidation of them certainly belongs to the present century, and they may at least be looked upon as new foundation stones for the erection of twentieth-century superstructures.

Among these, though by no means the most important, is the discovery of the obscure process known as "double fertilisation."

The phenomenon known as double fertilisation—perhaps rather unfortunately so called—was first described by a Russian observer, Nawaschin,[1] to a Russian audience in August of 1898 at Kief. It was brought further into prominence the next year by the publications of Prof. Guignard,[2] who also had discovered, in the same plant, the process described by Nawaschin.

Lilium Martagon had long been the classical plant for the study of fertilisation; nevertheless these workers found that it amply repaid a 'Revision der Befruchtungsvorgänge.'

The processes of approximation of the male and female elements, which culminate in the fusion of the male and female nuclei, may be said to be initiated when, by the agency of wind or insect, pollen grains are deposited upon the receptive stigma, where they germinate with the production of a tube.

When the pollen tube sets out upon its long and complicated journey in search of the ovum deeply seated in the tissue of the ovule, it bears within it two male nuclei, or sperms, one of which

[1] "Resultate eines Revision der Befruchtungsvorgänge bei Lilium Martagon." *Bull. de l'Acad. imp. des Sciences de St. Pétersb.* ix. 1899.
[2] "Sur les anthérozoïdes et la double copulation sexuelle chez les végétaux angiospermes." *Comptes rendus Acad. d. Sci.*, Paris, 1899.

420

effects fertilisation by fusion with the ovum ; but what is the fate of the other ?

If a botanist had been asked this question previous to the publications of Professors Nawaschin and Guignard, he would have said that the second male nucleus merely withered away after its successful fellow had fulfilled its destiny. But the papers of '98 and '99 conclusively demonstrated, by independent researches, that in *Lilium Martagon* the second sperm also fuses with two neighbouring nuclei of the ovum, thus forming a triple nuclear fusion. As the result of the fusion of one male nucleus with the ovum, an embryo is produced; as the result of the fusion of the other male nucleus with two nuclei, the endosperm upon which the embryo feeds is produced.

These new results were received at first with some caution by botanists, not surprising when we consider that the history of fertilisation and the development of embryo and endosperm in *Lilium Martagon* and many other flowering plants had been described repeatedly. The previous failures to observe the triple fusion were probably due partly to a slight difference in time between the fertilisation of the ovum and the fusion with the endosperm-producing nuclei, and partly to the frequent occurrence of fusion of *two* of the three nuclei of the triple fusion before the arrival of the third.

The production of endosperm from the fusion of two nuclei of the embryo sac had always been looked upon as an interesting and rather puzzling phenomenon, for the explanation of which many diverse hypotheses had been put forward. The intrusion of a *male* nucleus into this fusion seemed to some infinitely to increase the incomprehensibility of the process, and many took refuge in the view that fusion with endosperm-producing nuclei was largely accidental, and without meaning, and might equally well take place with one of the other nuclei present.

For the next few years, therefore, the statement given above as to the fate of the second male nucleus would have received a qualifying addition, to the effect that in some cases it apparently persisted and fused with the endosperm-producing nuclei, or possibly with another sister cell of the ovum. The latter fusion, though often referred to as a possibility, has, I think, never been demonstrated.

During the last few years, however, so numerous and widespread have been the species in which the triple endosperm-producing fusion has been described that probably no botanist would hesitate now to speak of this as the normal function of the second male nucleus.

The probable universality of the process, then, being admitted, let us inquire into its possible history and significance.

The ovum and the other nuclei in question are enclosed within a membranous bag set in the heart of the ovule, and known as the embryo sac. The embryo sac is undoubtedly a retained and enclosed spore, never set free from the plant. Its contents are therefore homologous with the products of germination of a fern spore, that is, to the little green prothallus together with the reproductive organs that it bears. Now there are other plants, of the same general rank as ferns, which differ from them in that their spores produce prothallia of two kinds, bearing one or other sexual organ. The embryo sac of flowering plants thus more strictly corresponds with a spore producing female prothallia.

In the Gymnosperms, which form a group roughly intermediate in complexity between the ferns and flowering plants, the spore (= embryo sac) destined to produce female reproductive cells is retained by the plant as in the Angiosperms, but the ovum is produced in an organ which, although much reduced, is plainly comparable to the archegonia of fern prothalli. Thus the reproductive tissue is quite plainly marked off from the vegetative.

In the Angiosperms, on the other hand, reduction has taken place to such an extent that the vegetative and reproductive tissues within the spore—together represented by only eight nuclei—show no obvious differentiation, and would-be theorists must fall back upon the guides of "origin" and "destiny" so much used by zoologists.

Origin shows that one of the nuclei with which the second male nucleus fuses is sister nucleus to the ovum—*i.e.* produced by the same nuclear division—and therefore presumably a potential ovum; *destiny* shows that it behaves as an ovum in fusing with a male nucleus, and were it not for the presence of another nucleus—probably vegetative in character—making up the much talked of triple fusion, we should have here a second act of fertilisation within the embryo sac. Should we also have a second embryo, and is the presence of the intrusive vegetative nucleus causally related to the degenerate production of a mere food tissue? It has been suggested by Miss Sargant[1] that it is, and that we have in the endosperm an embryo withheld from its true development by the mass of vegetative nuclear matter in its composition. If this view be accepted, then we may suppose the Angiosperms to be derived from forms which habitually produced two embryos in each embryo sac. In this connection it must be remembered that the embryo in Gymnosperms survives at the expense of several of its brethren.

On the other hand, there have been those who look upon the endosperm derived from the triple fusion as part of the vegetative contents of the spore, the production of which has been delayed until after the formation and fertilisation of the female organ. This would be as though a fern spore should form archegonia at the beginning of germination, and after these had been fertilised, produce the main bulk of the vegetative prothallus for the support of the young embryo. Prof. Strasburger,[2] who is the chief exponent of this view, looks upon the fusion of the male and female nuclei to form the supposed belated prothallial tissue, as purely secondary, and more or less accidental.

It will be remembered that one of the tests of fertilisation is the carrying on of paternal qualities, and that this is effected by "double fertilisation" has been brought out very strikingly by hybrid experiments.

The very remarkable transmission of endosperm characters belonging to the paternal side of crossed races of maize had long been a puzzle to plant breeders. Very obvious and unmistakable differences exist in the endosperms of the varieties of maize. While starch is the form of food storage in many, in some it appears as sugar. Another constant and differentiating feature is found in the colour of the endosperm which, showing through the thin covering layers, colours the whole seed. When two such widely differing races are crossed, the colour and chemical nature of the endosperm of the variety from which the pollen was obtained may appear in the ripening seed of the female plant,[3] and this was always quoted as an extreme instance

[1] "Recent Work on the Results of Fertilisation in Angiosperms." *Annals of Botany*, vol. xiv., Dec. 1900.

[2] "Einige Bemerkungen zur Frage nach der 'doppelten Befruchtung' bei den Angiospermen." *Bot. Zeit.* ii. 1900.

[3] H. de Vries, "Sur la fécondation hybride de l'albumen." *Comptes rendus de l'Acad. d. Sci.*, Paris, 1899.

of the far-reaching effect of the act of fertilisation upon the maternal tissues.

Now that " double fertilisation " has been proved to occur in these crosses,[1] the matter receives some measure of explanation, for, whatever the view entertained as to the morphological value of the tissue thus produced, there can be no doubt that the characters of the hybrid endosperm referable to the race of the male parent are directly transmitted by the male nucleus entering into its composition, and not due to some mystic influence spreading through the parent, from the act of fertilisation producing the embryo. In this sense at least, the endosperm is a new generation whose characters are of double origin. Whether we accord the title of true fertilisation to the triple fusion or not, this much is certain, that we have true gametes— with the reduced number of chromosomes—degraded to help in the development of the victorious pair. Such a fate is not without analogy in the animal kingdom, for it is said that in the Sauropsida many sperms may enter the egg, one of which fuses with its nucleus, while the remainder render service to the fertilised egg by working up the food materials of the yolk. In some of the lower animals also, supernumerary sperms which have entered by chance become absorbed by the fertilised egg, together with the polar bodies—potential female gametes. The extra sperms have even been known to conjugate with the polar bodies, a process very similar to the fusion of the second male nucleus with the sister cell of the ovum in flowering plants. Their development ceases, however, after a few divisions.

I have spoken of the theory which regards the endosperm as the equivalent of a once thriving embryo, debased by the introduction of a third vegetative nucleus at the time of fertilisation; and also of the view which holds endosperm formation to be a fresh start of the retarded prothallial tissues, inaugurated by chance nuclear fusions. A third possibility suggests itself, which has the somewhat dubious merit of partly reconciling the other two, but at the same time is not entirely without foundation.

In the archegonia of all ferns, the last division which gives rise to the ovum cuts off from it a nucleus known as the ventral canal nucleus. This nucleus has been identified in nearly all

[1] Guignard, "La double fécondation dans le mais." *Journal de Botanique*, xv. 37-50, 1901.

Gymnosperms, and is probably always formed. It persists in many instances until after the fertilisation of the ovum, and has been known to divide a few times.[1] This is very suggestive, considered in connection with the fact that in some Gymnosperms *both* the sperm nuclei enter the same archegonium. The second disappointed male nucleus has also been seen to divide.[2] It seems not unlikely, therefore, that under some circumstances these gametes,[3] which would certainly have an attraction for one another, should fuse and make an abortive attempt at embryo formation, analogous to the abortive fusion of sperm and polar body in animals quoted above.

If this is granted as a possibility, it does not require a great stretch of imagination to suppose that the decadent organism might readily be stimulated to new vegetative growth by the fusion with it of a vegetative nucleus. We should then see in the endosperm of Angiosperms, not the deflection of an embryo from its proper development, but rather the elevation of supernumerary gametes—the female of which we may suppose to have been shelved for some time—to a new but subordinate existence.

It is just conceivable that this tissue may have arisen in some such way. I have assumed that the sister cell of the ovum concerned in the triple fusion may be regarded as the homologue of the ventral canal cell of Gymnosperms, although this is largely conjectural. The third nucleus entering into the fusion is one of four nuclei arising in the base of the sac and generally supposed to represent the vegetative tissue of the prothallus. Although it has been suggested that it is this nucleus which gives the "vegetative" turn to the development of the fusion nucleus, it must be borne in mind that there is very little to distinguish the early stages of endosperm formation from the early divisions of the embryo in some Gymnosperms, notably in the cycads, where the production and differentiation of cell tissues take place late. So that I would regard the vegetative

---

[1] Coker, "Notes on the Gametophyte and Embryo of Podocarpus." *Bot. Gaz.* 1902.

[2] Arnoldi, "Beitrage zur Morphologie der Gymnospermen. III. Embryogenie von Cephalotaxus Fortunei." *Flora*, 1900.—Miss Ferguson, "The Development of the Egg and Fertilisation of Pinus Strobus." *Annals of Botany*, 1901.—Miss Robertson, "Studies in the Morphology of Torreya Californica." *New Phylologist*, 1904.

[3] Blackman and also Chamberlain regard the ventral canal nucleus as a potential female gamete.

nucleus as giving to the effete gametes a new impetus to development, rather than determining the nature of that development. That this may be necessary is shown by the case of cyclops mentioned above, in which the product of fusion of a supernumerary sperm and polar body aborts after a few divisions.

Whatever the true history of this obscure process may turn out to be, direct advantage would seem to accrue to the embryo from the supply of a food substance of the same double origin as itself.[1] This is brought out very well in the hybrid maize races alluded to before, where the dissimilarities are obvious. It is very significant to find that whatever the nature of the endosperm produced by the cross, the embryo bears the same endosperm character.[2] We cannot doubt, therefore, that the power to benefit by a particular kind of endosperm is inevitably correlated with its production.

The unravelling of the history of the phylogenetic evolution of the process of endosperm formation should prove one of the most interesting developments in botany, and if accomplished will go far to solve the problem of the origin of Angiosperms.

[1] See "Recent Work on the Results of Fertilisation in Angiosperms." *Annals of Botany*, xiv. 1900.
[2] Bateson, "Reports to the Evolution Committee."

RECENT WORK
ON PROTEIN-HYDROLYSIS

By J. REYNOLDS GREEN, Sc.D., F.R.S.

Professor of Botany to the Pharmaceutical Society of Great Britain

THE decomposition of proteids under the influence of the various enzymes which attack them has been for many years the object of much research. The early workers obtained from such digestions a great number of bodies whose relationships were very inadequately interpreted, and whose identities were the subject of much disputation. It was not till the time of Kühne that any course of decomposition was suggested that appeared to possess any great probability.

Kühne's hypothesis was known as the "cleavage" theory of proteolysis, from the fact that he supposed that the proteid attacked—a globulin or an albumin—was at once separated into two approximately equal varieties, which were possessed of distinctly different properties. One of these he termed *hemi-albumose,* the other *anti-albumose.* The subsequent decompositions of these two albumoses he held to go on side by side, but independently, so that among the products of digestion two distinct groups of substances appeared, a hemi- and an anti-group. The pepsin of the stomach converted the two albumoses into *hemi-peptone* and *anti-peptone* respectively, and the two peptones, existing in the same mixture, and not being readily capable of separation, formed *amphopeptone.* The action of the trypsin of the pancreas converted the hemi-peptone into a number of crystalline bodies, chiefly consisting of amino-acids, such as leucin and tyrosin, with a certain number of amido-bodies such as asparagin.

This view of Kühne's, though evidently only provisional, was accepted by physiologists for many years, and the various discoveries which were made by different workers in both animal and vegetable physiology were fitted into it with, however, gradually increasing difficulty.

Kühne himself and one of his pupils, Chittenden, disturbed it seriously by ascertaining that his hemi-albumose, comprising half the original globulin on his hypothesis, was a mixture of at least three albumoses, with another substance of very indigestible character in small proportion. These three albumoses, known as proto-, hetero-, and deutero-albumoses were shown to be quite different from each other in many respects, and not to stand upon the same level as each other, as the theory demanded that they should. The formation of the albumoses was successive and not simultaneous. When proto- and hetero-albumoses were submitted to the action of pepsin they formed deutero-albumose before, or coincidently with, peptone.

When the three albumoses were further examined, neither was found to be a pure hemi-product. Nor was anti-albumose much like hemi-albumose in its reaction ; indeed, it soon became doubtful whether it could properly be called an albumose at all. There was little or no evidence of its occurring during digestion, as it should do in considerable amount if the cleavage hypothesis were correct.

When the further stage of cleavage by trypsin was scrutinised, the albumoses of the hemi-group were not found to behave as the theory required. Both proto- and hetero-albumose were found to give rise to a certain amount of anti-peptone and to amino-acids, but in the case of the proto-albumose the latter were the chief products, while in that of the hetero-albumose the anti-peptone was in excess.

Kühne's hemi-peptone has not been proved to occur during digestion—its existence, in fact, is somewhat hypothetical. No hemi-compound seems to occur under the action of trypsin, so that if such a body is formed it is at once split up into the amino-acids.

Besides these critical objections raised against Kühne's hypothetical scheme of proteoclastic decomposition, a more theoretical one may be advanced, based upon the probability that the zymolytic breaking up of large molecules would be comparable in different cases. In the case of the action of diastase on starch we have undoubtedly a gradual decomposition of the complex molecule of the substrate, marked by the successive appearances of members of the amyloïn or malto-dextrin groups of substances, closely resembling each other and showing gradually increasing differences from the original starch. Analogy points to a course of decomposition of proteids, similar but more complex, as the substrate itself is more complex than starch.

This line of reasoning has met with considerable support from the investigations of many physiologists since Kühne's hypothesis was first critically examined.

Kutscher, applying more modern chemical methods to the investigation of Kühne's anti-peptone, ascertained that it is not a homogeneous substance at all, but a mixture of bodies which are mainly basic in character. He used as a precipitant a hot saturated solution of phosphotungstic acid, which separated them from the amino-acids, leucin, tyrosin, etc., which were present with them in the digestion.

These basic bodies, now for the first time found as zymolytic products of proteid, included lysin and lysatin, two bodies which had already been discovered and prepared by Drechsel, and two others which have been named argenin and histidin. The amounts of the several basic substances varied in different cases, but the argenin was generally prominent among them.

Lysin, as described by Drechsel, is a dextro-rotatory body, apparently having the constitution of diamido-caproic acid, being therefore related to leucin. It loses its power of affecting polarised light when heated with baryta water to ·150° C. Lysatin appears to be homologous with either creatin or creatinin.

Both lysin and lysatin form crystalline salts with various metals, particularly platinum and silver.

These basic bodies show certain connections with urea which carry our knowledge of the relation between the intake and output of nitrogen somewhat further, pointing to a direct enzymic formation of urea from proteid, as well as the more indirect method of its origination in the tissues of the body. Kossell and Dakin obtained an enzyme from the liver of the dog which is capable of splitting up argenin into ornithin and urea. According to Thompson the latter only is formed in the body by the enzyme, to which the name *argenase* has been given.

The close relations existing between lysatin and urea were pointed out by Drechsel, who by heating the former, purified by crystallation of its silver salt, with excess of baryta water, obtained one gramme of urea nitrate from ten grammes of the silver compound.

Besides these basic bodies the number of amino- and amio-acids known to result from the zymolysis of proteid has been increased, aspartic and glutaminic acids and asparagin having been identified.

The occurrence of a body giving a purple colour with a few drops of chlorine or bromine water was many years ago described by Nencki, and for a long time its composition remained undetermined. It has been called protein-chromogen by some writers, but has more recently been termed trypto-phane. Through the investigations of Hopkins and Cole it has now been isolated and its constitution determined to be skatol-amido-acetic acid. On decomposition it yields considerable amounts of indol and skatol. Though the latter have long been known to occur among the faecal products of digestion, they have been considered to arise from putrefactive products of digestion in the intestines. Hopkins and Cole have shown that they are to be attributed, like the others mentioned, to the action of the proteoclastic enzymes. In the course of their researches they found cystin to occur side by side with tyrosin and tryptophane.

Other substances have been found to result from the decomposition of proteids by hot acids and hot alkalis which have been very completely investigated by various workers, and which are beginning to find places in the scheme of proteid zymolysis.

Perhaps the most interesting and important of these are those discovered by Fischer and Abderhalden about three years ago, as they throw a light upon the relation between peptone and the crystalline bodies accompanying or succeeding it in intestinal digestion. On decomposing with hot acids casein, edestin, egg-albumin, fibrin, the globulin of serum, and certain other proteids, they found among the products a considerable quantity of a-pyrrolidin-carboxylic acid, which accordingly they were inclined to consider, like the ordinary amino-acids, as a primary decomposition-product of the initial proteid. When, however, they acted on these bodies with a strong pancreatic enzyme, and continued the action for some long periods, as long sometimes as several months, no trace of a-pyrrolidin-carboxylic acid could be found. The solution was found, however, to contain a new product, a derivative of peptone, which, when acted upon by hydrochloric acid, readily yielded the acid in considerable quantity.

This new substance, which the authors call a polypeptide, is easily precipitated by phosphotungstic acid, by which means it can be separated from the amino-acids. By repeated solution and reprecipitation by the same reagent it can be purified to a considerable extent. It then gives no biuret reaction, or only a very faint one. This shows, of course, that it is not the ordinary peptone. On its hydrolysis with hot hydrochloric acid it gives rise to a-pyrrolidin-carboxylic acid and phenylalanin, as well as to certain quantities of the ordinary amino-acids, leucin, glutaminic and aspartic acids, etc. It is uncertain whether lysin and the other basic bodies are formed also.

When casein was subjected to the successive action of pepsin and " pancreatin" (presumably trypsin), the polypeptide was found to occur, but side by side with it were small quantities of a-pyrrolidin-carboxylic acid and phenylalanin. The polypeptide was found to be slightly alkaline, to give a reddish-purple colour with a very little dilute cupric sulphate, and a precipitate with tannin, with platinic chloride, and with alcohol.

These results appear very far-reaching. In the first place they point to peptone not splitting primarily into the amino-acids, as taught by Kühne, but rather yielding a polypeptide of definite constitution intermediate between them, adding thus another member to the classes of proteids.

Then they indicate a very intimate association between pepsin and trypsin in the working of the body. By their successive actions the formation of members of the phenylalanin group is effected, though by the action of trypsin alone the process seems to be arrested at the polypeptide stage.

It is clear that further investigation is called for here, for the action of pepsin precedes that of trypsin, while in the formation of the phenylalanins by the successive action of trypsin and acid, the last stage is effected by the latter and is apparently due to it.

It is interesting to note with reference to the occurrence of the polypeptide in the course of the zymolysis, that according to the same writers several of the polypeptides which they have succeeded in synthesising are capable of digestion by pancreatic juice obtained from a fistula, and rendered active by the addition of 5 per cent. of the secretion of the intestine.

The opening years of the present century were marked by another discovery in the field of zymolysis, which has given rise to a startling modification of our views of the mechanism of proteid decomposition, and which threatens to revolutionise our ideas of the relationship between the different proteases. This discovery was made by Otto Cohnheim in the course of some investigations on the fate of peptone, which though very evident in the contents of the alimentary canal during digestion cannot be detected in the blood, even in the vessels going from the intestines. Apparently peptone is not absorbed as such by the blood, and its fate has given rise to much speculation. In the years 1880 and 1881 the view was put forward by many physiologists that its disappearance is due to changes effected by the intestinal mucous membrane, causing its reconversion with a more stable form of proteid, possibly one of the proteids of the blood plasma. Hofmeister held that such a conversion was effected by the leucocytes, while Heidenhein and Shore thought the action was due rather to the cells of the intestinal epithelium in the region of its absorption. The general trend of opinion was that these cells rather than the leucocytes were responsible for its disappearance.

In 1890 Neumeister, while not denying the possibility of a reconstruction of proteid from the peptone, put forward the view that a further splitting was possible.

To this problem in 1901 Cohnheim directed his attention, trying at the outset to discover the reconstructed proteid in the epithelium of the intestine, but he failed to obtain any evidence of its occurrence there.

In view of the importance of the question a summary of his experiments may be given. Digesting lean meat by pepsin in the presence of oxalic acid he obtained the peptone with which he began his work. After removal of the oxalic acid by a salt of calcium he found his peptone to be practically free from albumoses. Using this preparation he repeated Neumeister's experiments, and found that about one-third of the small intestine of a cat altered 0·6 gm. of peptone in two hours so completely that after removal of all coagulable proteids the digested fluid no longer gave a biuret reaction. The method he adopted to remove these proteids consisted in boiling the mixture of peptone solution and intestinal epithelium with sodic chloride and acetic acid. The coagulum thus obtained was filtered off, and the filtrate examined. It gave no biuret reaction, but yielded a crystalline precipitate when a solution of phosphotungstic acid was added to it. The idea that the peptone had been converted into coagulable proteid and so removed was negatived by the discovery that the filtrate from the coagulum contained nitrogen equal in amount to that of the original peptone used.

In subsequent experiments Cohnheim used an extract of the epithelium instead of the mucous membrane itself. He ground up with sand the lining of the intestine of various animals killed during active digestion, extracted the mass with alkaline normal saline solution, and pressed out the extract in an iron tincture-press. On standing for some time a good deal of coagulable proteid was deposited, which was filtered off, and the filtrate employed in the same way as originally he had used the intestinal epithelium itself. Such extracts were found to possess the power of causing the disappearance of peptone.

The experiments clearly pointed to the existence in the intestinal mucous membrane of a new proteoclastic enzyme, differing from those known up to the time of the experiments in having no action on the coagulable proteids, such as albumins and globulins, nor upon fibrin, but possessing the power of decomposing proteoses and peptones, and forming from them crystalline bodies which do not give the biuret reaction, and which are thrown out of solution by the presence of phosphotungstic acid.

It is evident that these observations throw some light upon the uncertainty which has hitherto existed as to the digestive value of the *succus entericus*. Very conflicting statements have been made as to its power of acting on coagulable proteids in liquids of different reactions. As the experiments with the juice were mainly made on coagulable proteids and not on peptone alone, the results could only be discordant.

Cohnheim isolated the enzyme from his extracts by gradually adding ammonium sulphate and separating fractionally. He gave it the name *erepsin*.

The enzyme was thrown out of solution when the extract contained 60 per cent. of ammonium sulphate. In later preparations of it Cohnheim mixed two volumes of the extract with three volumes of saturated solution of the salt, suspended the resulting precipitate in water, and subjected it to dialysis, freeing it in this way from much proteid matter which was associated with it.

The enzyme when thus prepared was found to work in an alkaline medium, but not in an acid one.

Erepsin, though acting energetically upon peptone, was found capable of decomposing both primary and secondary proteoses also, though the rate of decomposition varied considerably in different cases.

Cohnheim found that fibrin, coagulated proteids from serum, vitellin from the seeds of the pumpkin, beef, the proteids of the intestinal wall, globin, and the Bence-Jones proteid all resisted the action of the new enzyme. Among the products of digestion he identified, besides the amino-acids occurring in the alimentary canal, the basic bodies argenin, histidin, and lysin.

Casein appears to be easily and rapidly digested. The position of casein among the proteids is, however, such that this need cause no surprise, and its behaviour does not contradict the view already stated that erepsin acts not on the coagulable proteids proper, but on the products of their digestion.

Cohnheim's discovery does not stand alone. As we shall see later, investigations made independently on the vegetable proteases by Vines led to the anticipation and subsequent verification of the existence of a peptoclastic enzyme in plants. Vines's work was carried out in large measure before Cohnheim's results were announced, and his experiments were strikingly confirmatory of the latter.

The discovery of a new enzyme dependent for its activity in the body upon the previous action of other proteases has led to a reinvestigation of the digestive juices. Erepsin seems to share the properties of trypsin, but its discovery at once suggests the question of the real nature of the latter enzyme. It seems at any rate possible that what we have hitherto called trypsin may be a mixture of two enzymes, one peptonising coagulable proteids, the other attacking the peptone when formed. The same question has long disturbed the minds of workers on the hydrolysis of starch, some holding the view that a particular enzyme converts it to dextrin, and a second hydrolyses the dextrin to sugar. There has, however, been so far advanced no satisfactory proof of the existence of dextrinase. The occurrence in the same secretion of two enzymes separable with greatest difficulty, if at all, is exemplified by the simultaneous presence of trypsin and rennin in pancreatic juice.

Very careful researches have been made by Vernon on the secretions of the pancreas and small intestine during recent years. At the outset he ascertained that while an extract of pancreas possessed the power of splitting up fibrin, and of forming amino-acids from peptone, these two powers did not run in the least degree on parallel lines. Fresh pancreatic extracts in which the tryptic enzyme was present only in the condition of its zymogen, nevertheless possessed the property of decomposing Witte's peptone with great energy, and they did not gain greater power in this direction after the zymogen had been converted into trypsin. The trypsin in the zymogen condition being inert, it necessarily follows that a special peptoclastic enzyme was present in the extract.

The coincident occurrence of the two enzymes in the pancreas is supported by the further observation that when extracts of this organ are kept for some time under aseptic conditions, the powers of fibrin digestion and of peptone decomposition do not disappear with the same relative rapidity, but show considerable variation in this respect. If the extract is prepared with glycerine, the tryptic effect is maintained longer than the ereptic; if the solvent is dilute alcohol, the reverse is the case.

Vernon made also a series of experiments on the facility with which the enzymes could be destroyed by the action of alkalis. He had shown previously that trypsin, when exposed to the action of sodium carbonate in as low a percentage as 0·4, underwent destruction with considerable rapidity, which was greatly increased as the percentage of the alkali became higher. Erepsin also shows this susceptibility to change under similar exposure. Comparing the two enzymes, he found that when kept at 38° C. in the presence of 0·4 per cent. of sodium carbonate, trypsin is at first more rapidly destroyed than erepsin, and that later, after nine or ten hours, the reverse is the case; the erepsin being decomposed the faster.

Another difference between the two was observed in their behaviour with alcohol, pancreatic erepsin being the more readily precipitated.

These reactions completely negative the idea that the whole of the proteoclastic changes due to pancreatic juice are the work of one enzyme only. There are clearly two present, though so far the actual division of labour between them does not seem clearly established. Vernon's experiments leave open the question of the exact rôle of trypsin, *i.e.* whether it is pepto-clastic, or whether the whole of the peptone-splitting is due to the erepsin.

Besides its existence in the digestive juices of the pancreas and the small intestine, erepsin has been ascertained to be more or less abundant in almost all the tissues of the body, and not to be restricted to those of warm-blooded animals. Much of our knowledge on this head, again, is due to Vernon, who has made very exhaustive researches in this direction. It seems, as a tissue enzyme, to be most abundant in the kidney; then, in order of quantity, in the intestine, pancreas, spleen, liver, and muscles. Whether its prominence in the kidney is connected with the profound proteoclastic action which produces urea may be a subject for further research.

The distribution of erepsin has not been exhaustively ascertained at present. Cohnheim holds it, as does Vernon, to be very widespread in the animal kingdom. From a study of the products of digestion in the Octopoda, he has been led to believe in its existence in their alimentary canal. Suspending the latter, under appropriate precautions, in a bath of the blood of the animal, kept oxygenated by leading a stream of the gas through it, he found that the canal maintained its spontaneous movements, and absorption of the foodstuffs it contained went on. After some time crystalline derivatives of proteid were found in the blood-bath, but nothing giving a biuret reaction. The blood circulated through the alimentary canal contained neither.

Vernon also has discovered erepsin in various groups below the Mammalia, tracing it in almost all the tissues of the pigeon, the frog, the eel, and the lobster, and in the organ of Bojanus and the muscles of the Anodon. The amount obtainable was, however, relatively small in the tissues of the cold-blooded animals, with the exception of the pancreas of the frog, which was singularly rich.

Researches made upon the proteoclastic enzymes of plants during the period under consideration have been very fertile in results, and have led to the identification of enzymes strikingly similar to if not identical with those now shown to exist in the animal body.

It has been known for many years that a great number of proteases exist in plants, though how far they are specifically distinct from each other cannot well be affirmed, on account of the differences in the plants in which they have been found and of certain differences in the conditions of the experiments that have been carried out with them. Most of them act intracellularly, only those of the so-called insectivorous plants being formed out of glandular tissues. The best known of them are papaïn, from *Carica Papaya*; bromelin, from *Ananas sativa* (the pineapple); nepenthin from Nepenthes (the pitcher-plant); and the trypsins from germinating seeds. When they are compared with the proteases of animal origin they have always appeared to resemble trypsin on the whole rather than pepsin, as, though sometimes working in acid solutions, they have been found to produce amino-acids as well as peptone. Indeed no definite evidence pointing clearly to the existence of a pepsin or pepsin-like enzyme in the vegetable kingdom has been forthcoming.

The enzymes alluded to had up to the beginning of the century been found in only a limited number of plants, and even in those they were not proved to be widely distributed. The opinion was held, however, by many physiologists that the vegetable organism needed for its ordinary metabolic processes a distribution of proteases as widespead as that of diastases, starch and proteids playing similar rôles in its nutrition. The difficulty of finding these looked-for proteases was considerable, as no rapid method of detecting their activity had been hit upon. The idea that they should attack such substances as fibrin or coagulated albumin was always underlying the method of research.

The recognition of the existence of tryptophane as a mark of the progress of so-called tryptic digestion was seized upon by Vines, and in his hands became extremely useful as an indication of the existence of proteasic enzymes. With its assistance he succeeded in showing that the latter are very widely distributed, finding them in a very great variety of plants taken from all groups of the vegetable kingdom, and from almost all parts of the plants containing them.

Many plants, however, gave no evidence of the possession

of such enzymes by decomposing fibrin. Reflecting on this anomalous behaviour, Vines was led to substitute in his experiments proteids more closely allied to those existing in plants, such as the various forms of proteose. Using these he found evidence of proteoclastic action in cases in which the plant had failed to act on fibrin. This result seemed surprising; the action attributed to trypsins, to which these enzymes were held to belong, was clearly divisible into two separate parts, and enzymes were contained in many plants which could only begin in what had been thought to be the middle of the digestion and carry it thence to completion.

From that time onwards Vines directed his attention to the question whether the so-called trypsin was a single enzyme, working in two separate stages, or whether it was not to be regarded as a mixture of a fibrinoclastic with a peptoclastic enzyme.

It was at this point in his researches that Vines says he became acquainted with the then recently published researches of Cohnheim, to which attention is called above, which naturally strengthened him in adopting the latter hypothesis of the constitution of the vegetable proteases. His conclusions are thus stated : " It would appear, therefore, that plants form two distinct kinds of proteases, the one a trypsin, the other an erepsin ; and so far as the facts go, they indicate that the former is generally associated with depositories of proteid nutriment, such as seeds, fruits, bulbs, laticiferous tissue, etc. ; the latter with ordinary foliage-leaves, stems and roots. But further research is required in order to definitely establish this distinction."

In the pursuance of this idea Vines turned his attention to papaïn as the enzyme which has been, perhaps, more frequently the subject of research than any other. The results that have been obtained with it by different workers have been somewhat conflicting, both as to the conditions of its action and the bodies to which it gives rise. This may be due to the fact that so far the experiments with it have been made with commercial preparations and not with fresh extracts of the actual fruit. The results can scarcely be concordant where there is no uniformity of material. On looking over the earlier work of Martin, Sharp, Halliburton, and others, the chief differences of opinion are found to relate to the question of the appearance of peptone. Martin found that he obtained it from fibrin, but not from globulin, and only traces from the proteases of the plant. Halliburton found peptone when he used animal proteids, but not when he employed vegetable. Other writers found sometimes traces of peptone and sometimes none.

Other differences of opinion may be found on the question of the most suitable reaction for the process of papaïn digestion. Wurtz and others, particularly Martin, found a neutral medium like that of fresh papaïn-juice was most advantageous, Davis says that a trace of acidity or of alkalinity causes acceleration, while Emmerling finds the best medium for its digestion of fibrin is a slightly alkaline one.

To test the matter more exhaustively Vines used three distinct samples of commercial papaïn, obtained from Messrs. Christy, Finckler, and Merck respectively, worked with digesting fluids of different reactions, and used various antiseptics, carefully controlling all the experiments. As he expected, he found a good deal of difference in the preparations on all the points under investigation.

The experiments not only threw a certain amount of light on the conflicting results of the earlier observers, to a certain extent explaining them, but they were found to have a very direct bearing on the question already stated as to the presence of one enzyme or two in the papaw. Christy's papaïn readily digested or dissolved fibrin in the presence of fluoride of sodium, but it could not split up peptone. Finckler's papaïn did the same in the presence of hydrocyanic acid; Merck's preparation resembled Finckler's when used in a weak solution. Christy's papaïn when digested with peptone in presence of toluol in an acid medium gave a strong tryptophane reaction, but had hardly any action on fibrin.

Altogether the experiments lent support to the theory that commercial papaïn contains a pepsin and an erepsin.

Vines found in other experiments that nepenthin, contained in the liquid in the pitcher of Nepenthes, converts fibrin into peptone with great rapidity, but only very slowly breaks up peptone when set at the outset to work upon it. The course of the experiments show, that the fibrin digestion is so much more active than peptolysis as to suggest the presence of pepsin and erepsin, the former in much the larger quantity.

In a very considerable number of cases in which he examined

the proteases that can be prepared from leaves, he found that
the ereptasic power was the only one present, coagulable
proteids remaining unaltered.

In the course of some very important researches published in
1902, and carried out in the well-known Carlsberg laboratory,
Weis came to the conclusion that in the course of the germination of barley two proteases are involved in the proteoclastic
processes. Their separate activities could be demonstrated by
the action of stannous chloride, or of prussic acid. The first
stage led to the formation of proteases, the second to the
appearance of crystalline, non-proteid compounds. Weis
referred these activities to a peptase and a tryptase respectively.
He claims to have been able to distinguish clearly between the
parts played by the two enzymes, the peptase being very
rapidly brought into action, and very soon stopped ; the tryptase
commencing more slowly and continuing long after the cessation
of the activity of the peptase. He says that true peptone only
appeared in very small quantity. Among the products of the
activity of the tryptase he found amino-acids, hexone bases
such as lysin, etc., and free ammonia.

In ungerminated barley Weis detected a very feeble peptasic
power, but no tryptasic fermentation occurred. The latter
appeared suddenly on the fourth day of germination.

In the light of the work published since 1902 Weis's results
show nothing which is opposed to the view that his "tryptase"
is really ereptase, as Vines contends.

In pursuance of the view already stated, Vines carried out
later a series of researches which had for their starting point the
different degrees in which, with his mixed extracts from various
sources, the fibrin-digesting and the peptolytic powers they
possessed were affected by modifications in the conditions
under which the digestions were carried on. As these led to
important conclusions it is well that they should be stated in
some detail.

Attention may first be called to the work of Mendel and
Underhill on the action of papaïn. Using a 1 per cent. solution
of sodium fluoride as an antiseptic they found this salt affected
prejudicially the digestive activity of the enzyme in so far as to
retard or altogether inhibit the decomposition of peptone, while
its formation from fibrin was not materially checked. Vines,
from using other antiseptics and obtaining contrary results, was

able to attribute the inhibition of the peptoclastic power to the presence of the salt, and to postulate that sodium fluoride can inhibit the action of erepsin but not that of pepsin.

This means of discrimination did not, however, go further than papaïn, which was thus shown to be probably a mixture. Vines adopted, consequently, another method, which was the study of various solvents to extract the enzymes from the same material. He was at once successful with yeast and with the mushroom. With these materials he found that a rapidly prepared watery extract could not digest fibrin, but could split up Witte peptone. Using a salt solution extract prepared equally rapidly, fibrin was digested in twenty-four hours, and Witte peptone was equally easily decomposed. Evidently the first enzyme, provisionally regarded as a pepsin, is not soluble in distilled water, but dissolves in salt solution, while the second, an erepsin, is soluble in both.

Further power of differentiation between them was afforded by varying the reaction of the digesting liquid. In his early work on the same two fungi, Vines found that peptolysis and fibrin digestion were affected in much the same manner, but not to the same degree, by the reaction, whether acid, alkaline, or neutral. He subsequently applied this method to bromelin, nepenthin, and the proteases obtained from germinating barley and from the bulb of the hyacinth. Certain precautions were found necessary; the acid and alkali used had to be added in due proportion to the concentration of the enzyme in the digesting liquid. The acid employed was hydrochloric, and the alkali carbonate of sodium. The antiseptic used was not always the same, but was always that which previous experience had shown to affect the decompositions as little as possible.

In the case of papaïn it was found that fibrin digestion was not materially affected by difference of reaction, taking place equally well in presence of 0·5 per cent. or 1 per cent. of sodium carbonate, and of 0·2 per cent. or 0·3 per cent. of hydrochloric acid, hydrocyanic acid being the antiseptic; on the other hand, the splitting of the peptone was diminished in alkaline liquids. In neutral solution the fibrin was digested more readily than the peptone. The experiments were in all cases made in pairs, the fibrin and the peptone being subjected to the papaïn in separate vessels simultaneously and side by side.

Vines concludes that papaïn contains a fibrin-digesting

enzyme having a wide range of action, limited in one direction by 0·5 per cent. of hydrochloric acid, and in the other by a greater amount than 1·5 per cent. of sodic carbonate, and a peptone-splitting enzyme of narrower range, not extending beyond 1·5 per cent. of the alkali.

With bromelin, peptolysis but not fibrin-digestion was completely inhibited by alkalinity.

With yeast the results were not quite the same—the range for peptolysis was greater than that for digestion of fibrin ; in the former case proteoclastic activity was shown to extend from about 3 per cent. of sodic carbonate to 0·5 per cent. of hydrochloric acid, in the latter from about 1 per cent. of the alkali to 0·1 per cent. of the acid. Yeast thus gives results which are the converse of those yielded by the papaw and the pineapple. Slightly different effects were observed in the case of the mushroom ; peptolysis was retarded by acid but promoted by alkali, while fibrin digestion was arrested by deviation from the normal reaction, whether acid or alkaline.

Vines confirmed the observations of Weis as to the existence of two proteases in malt, or germinating barley, using, however, proteids of animal origin in his experiments, while Weis employed chiefly glutin and legumin. He found further that the pepsin is more readily affected than the erepsin by dilute alkalis; the inhibition of the former was complete in the presence of 1 per cent. sodic carbonate, but the erepsin acted, though slowly, for twenty-four hours.

The extract of the hyacinth bulb can digest both fibrin and peptone in an alkaline medium, though the natural reaction of the juice is faintly acid. Fibrin digestion is more readily affected than peptolysis by increased acidity or alkalinity.

Experiments on nepenthin showed that the liquid in the pitcher of Nepenthes peptonises fibrin much more rapidly than it decomposes Witte-peptone ; fibrin-digestion is inhibited by alkalinity, but promoted by acidity ; peptone-decomposition, which is slow in any case, is much retarded by alkalinity, but is not inhibited by the small percentage (0·15 per cent. of sodic carbonate) that arrested the digestion of fibrin.

Summarising his results, Vines states that peptolysis takes place within a range extending from distinct alkalinity to a degree of acidity beyond the natural reaction of the plant from which the particular enzyme is extracted ; fibrin-digestion is much less uniform, showing such differences that it is possible to arrange the individual cases into two groups, thus:

(a) Those in which it is limited to an acid reaction; yeast, mushroom, malt, Nepenthes.

(b) Those in which it also occurs with an alkaline re-action; papaïn, bromelin, hyacinth bulb.

The work of Vines on the existence of erepsin in plants does not stand alone, for Delezenne and Mouton have discovered it in certain fungi belonging, like the mushroom, to the group of the Basidiomycetes.

We find thus that there is a considerable body of evidence pointing to the conclusion that trypsin as a distinct enzyme has no existence in plants, but that what has hitherto been described under that name is really a mixture of a pepsin and an erepsin. This does not imply an identity between such pepsin and the similarly named enzyme of the animal stomach. Indeed, we must admit that the names pepsin, erepsin, etc., apply, not to individual enzymes, but rather to groups of them, consisting of several members which show certain differences among themselves, differences which affect only the conditions of their actions, and not the ultimate products of their activity.

As this indicates a considerable probability in the case of the vegetable proteases, there arises at once a question as to the nature of the trypsin of the animal organism. Is this also a mixture of some member of the pepsin group with an erepsin? On this point it would be premature to make a definite pronouncement, but there are certain facts apparently having a bearing on the point which claim attention.

It has been put on record by several observers, some of them writing many years ago, that under certain conditions the pepsin of the stomach seems to have the power of producing amino-acids. Lubavin in 1871 digested proteids with pepsin for considerably longer than a week, and stated that leucin and tyrosin occurred among the products he obtained. Winternitz in 1899 found the tryptophane reaction given after an extract of the stomach of the pig had acted on fibrin for seven hours. These experiments were noteworthy as showing the result of *prolonged* peptic digestion. They have been amply confirmed by the researches of several physiologists in 1901 and succeeding years. Laurow, Salaskin, Zunz, Langstein, Malfatti, Glaessner, all speak of the formation by either the tissue of the stomach or by some gastric extract, of some of the products hitherto attributed to trypsin.   Malfatti has drawn attention to the frequent appearance of tryptophane among the products of a peptic digestion, and has sought to separate pepsin into two enzymes of the two classes we have under consideration. Failing to do so, he has felt drawn to a view long ago stated by Hoppe-Seyler, that pepsin has peptoclastic powers as well as trypsin, an opinion, however, which is opposed by the fact that a very intense peptic digestion is possible without any appearance of tryptophane.

Glaessner claims to have detected an enzyme existing in the stomach side by side with pepsin, which possesses proteoclastic power in a neutral or faintly alkaline medium, as well as in an acid one.   He has named it *pseudopepsin*, and says it is yielded mainly by the pyloric mucous membrane during autodigestion.   He adds that it produces tryptophane in the course of its activity, and thereby he distinguishes it from pepsin.

Glaessner's conclusions are disputed by Klug, who failed to find any evidence of the existence of pseudopepsin.

Other authors, however, have obtained results which confirm Glaessner to some extent, inasmuch as they find tryptophane can be produced by the action of gastric preparations.   Pekelharing describes some experiments made upon a juice extracted by Buchner's pressure method from the pyloric mucous membrane, which, after being freed from ordinary pepsin, was found to have proteoclastic powers in neutral and faintly alkaline media, and to form tryptophane. It was, however, very feeble. Pekelharing found it in the stomach of the pig and in the gastric juice of the dog.   He adds, however, that unaltered gastric juice working in an acid medium will give rise to tryptophane after several days' action, and says that his researches lead him to the opinion that this is a property of all pepsin, even when purified by means of acetate of lead and oxalic acid, provided that the solution is sufficiently concentrated, and that too much hydrochloric acid is not present.   With not very strong pepsin solutions the acid must not exceed 0·2 or even 0·1 per cent.

Admitting that pseudopepsin may exist in small quantity side by side with pepsin in the stomach of the dog, Pekelharing is not inclined to admit that it is a secretion of the gastric glands, but suggests that it is to be regarded as an autolytic enzyme in the tissue of the organ, and that it is set free only on the death of the cells.

Vines also has published some experiments on the peptoclastic powers of pepsin. He used the commercial preparation known as " pepsin, pure scales," as well as a glycerin-extract of the stomach of a pig, and digested fibrin by its aid in 0·2 per cent. solution of hydrochloric acid, employing sometimes no antiseptic, sometimes hydrocyanic acid. He obtained a tryptophane reaction after two days, which became very strong after a further 24 hours.

When we consider that so many observers have thus obtained evidence of the occurrence of tryptophane when extracts of the stomach have been employed, and when we remember the wide distribution of erepsin in the tissues as shown by Vernon, it does not seem impossible that in these experiments the observers may have been working with mixtures of pepsin and erepsin in various proportions. It is rather singular that Vernon says nothing in his paper on the occurrence of erepsin in the stomach walls, though he has made a very complete examination of most of the other viscera.

If this view be correct, there seems nothing impossible in the suggestion that the trypsin of the pancreatic juice may also prove to be a mixture of two enzymes ; one of the pepsin class, though differing from gastric pepsin in working in alkaline media, and the other an erepsin, similar to if not identical with Cohnheim's enzyme. This point will doubtless receive investigation in the near future.

The autolytic digestion of the tissues after death has passed out of the domain of ordinary bacterial putrefaction, and attention may be called to certain agents which play a prominent part in it, and are members of the group of proteoclastic enzymes. The limits of this article forbid more than a passing reference to them ; but no doubt they will throw a light on some features of proteolysis.

An enzyme of this kind was first suggested by Salkowski in 1890, when he observed that when certain viscera, especially the liver, were crushed and kept at blood heat for a considerable time in presence of antiseptics, the tissue became changed in so far that it yielded an increased amount of nitrogenous substances soluble in water. Several of his pupils have studied these decompositions subsequently, and have found these nitrogenous substances to be similar to those produced from proteids by trypsin.

The liver enzyme was precipitated from the digested liver substance by Jacoby, who also prepared a similar body from the thymus.

Several investigators have shown that other proteoclastic enzymes can be prepared from the spleen and from the kidney. Attention may be drawn to the work of Hedin and Rowland, of Leathes, and of Hedin, on the former viscus, and to that of Hedin and Rowland and of Dakin on the latter.

The spleen appears to furnish two distinct proteases, which Hedin has named lieno-a and lieno-β protease respectively. The former acts chiefly in an alkaline medium, the latter preferring an acid one. Both enzymes appear to be capable of combination with nuclein substances, in which condition they become insoluble in dilute acetic acid. In the absence of the nuclein substances acetic acid dissolves them. They could be separated from each other to a certain extent by fractional treatment with sulphate of ammonium. Hedin attributes to them both a power of action like trypsin, but he does not discuss their actual products. According to ⌐Leathes the β-protease carries the decomposition to the same extent as trypsin.

Hedin holds that these enzymes exist in the leucocytes of the spleen, rather than in the actual tissue of the viscus.

Dakin shows that the enzyme of the kidney is of a very powerful type. It acts, unlike trypsin, in an acid medium ; but it produces a very thorough decomposition, the products of its activity including ammonia, alanine, and aminoisovaleric acid, leucin, and pyrrolidin-carboxylic acid, phenylalanin, tyrosin, lysin, histidin, cystin, hypoxanthin, and certain indol derivatives, but not argenin or aspartic acid.

The number of proteoclastic enzymes which have within recent years been found to exist, and the very varying conditions under which their activity is manifested, must lead the way to a new classification of them. Among the group two classes at least, possibly three, can be clearly recognised, each composed of several members which show certain small differences of working. Similar small differences appear as to their relative stabilities, or the readiness with which they are destroyed by various unfavourable conditions. There is no good reason for classifying them according to the reaction of the medium in which they act; for we have seen that in the cases of many of them there is considerable latitude in this particular.

There is at present but little evidence bearing upon the identity of animal and vegetable proteases, but it seems unlikely that differences in this respect will help us much in classification.

BIBLIOGRAPHY

COHNHEIM, *Zeit. physiol. Chem.* 33, 451 ; 35, 134, 396 ; 36, 13.

DAKIN, *Journal of Physiology*, 30, 84.

DELEZENNE and MOUTON, *Comptes Rend.* 136, 633.

EMMERLING, *Ber. d. deut. Chem. Ges.* 35, 695.

FISCHER and ABDERHALDEN, *Zeit. physiol. Chem.* 39, 81 ; 85, 215.

—— *Sitzungsber. k. Akad. Wiss.* Berlin, 1905, p. 290.

GLAESSNER, *Beitr.* 2 *Chem. Physiol. u. Pathol.* B. S.

HEDIN, *Journal of Physiology*, 30, 155.

HEDIN and ROWLAND, *Zeit. physiol. Chem.* 32, 341, 531.

HOPKINS and COLE, *Journal of Physiology*, 27, 418.

JACOBY, *Beitr.* 2 *Chem. Physiol. u. Pathol.* 1 (1902) 147.

KLUG, *Pflüger's Archid.* 92, 281.

KOSSELL and DAKIN, *Zeit. f. physiol. Chem.* 41, 321 ; 42, 181.

LEATHES, *Journal of Physiology*, 28, 360.

LUBAVIN, *Hoppe-Seyler's Med. Chem. Untersuch.* 1871, p. 463.

MALFATTI, *Zeit. physiol. Chem.* 31, 43.

MENDEL and UNDERHILL, *Trans. Connecticut Acad.* 11, 13.

MENLIN, *Journal of Physiology*, 5, 230 ; 6, 360.

SALKOWSKI, *Zeit. f. Klin. Med.* 1890 Suppl.

THOMPSON, *Journal of Physiology*, 32, 137.

VERNON, *Journal of Physiology*, 30, 330 ; 32, 33.

VINES, *Annals of Botany*, 15, 563 ; 16, 1 ; 17, 237, 597 ; 18, 289 ; 19, 149, 171 ; 20, 113.

WEIS, *Comptes. Rend. des travaux du Lab. de Carlsberg*, 5, 133.

A YEAR'S PROGRESS IN VERTEBRATE PALÆONTOLOGY

BY R. LYDEKKER.

To record all the work that has been done in any branch of science during one particular year till some months after its completion is manifestly an impossible task ; and the following notes accordingly relate only to such books and memoirs on vertebrate palæontology published during the last twelve months or so which have come under the writer's notice. At the time when these notes were written the year had still a couple of months to run, and some of the papers published during the latter part of 1905, or copies of which did not reach this country till about that time, are therefore included in the survey. As already stated, the review must of necessity be an imperfect one ; and among the papers which have been perused only such as are of a certain amount of interest and importance receive mention.

Although the past twelve months, so far as the writer is aware, have not produced any startling discovery in the past history of vertebrates, yet a good tale of work has been accomplished, and much light thrown in several instances on obscure points.

The most important work issued during the period under review is the British Museum *Catalogue of the Fossil Vertebrata of the Fayum, Egypt*, by Dr. C. W. Andrews. Since, however, preliminary notices of many of the discoveries recorded therein—such as the ancestry of the elephants—were published considerably earlier, while the extinct vertebrates of Egypt will form the subject of another article in SCIENCE PROGRESS, very brief mention of the volume will suffice on the present occasion. The importance of the Egyptian discoveries consists not only in the revelation of the pedigree of the elephant and the addition of several totally new types to the list of mammals, but in the evidence as to the existence of a relationship between two mammalian groups previously regarded by most naturalists as widely sundered. This linking-up of the proboscideans with the sea-cows is indeed most important, while the additional evidence in favour of a connection between the zeuglodonts and the creodont Carnivora is scarcely less so. The discoveries, as the title of the volume indicates, are, however, by no means restricted to the mammalian class; and the identification in the Egyptian Eocene of a giant tortoise related to the modern species of the Mascarene Islands, of a southern type of freshwater pleurodiran tortoise, of a gigantic snake, and, above all, of an apparently ostrich-like bird, is of the very highest importance, alike from a phylogenetic and from a distributional point of view.

As regards the geographical distribution of animals in past times, the Egyptian discoveries will indeed render it necessary to recast many of the prevalent theories. On this occasion it must suffice to mention that Dr. Andrews is in favour of the view that Africa was connected by land with South America during the late Cretaceous, and possibly even in the early Tertiary period. It is added that if the existence of this land-bridge be admitted, the Tertiary carnivores of Patagonia, generally known as sparassodonts, may be the descendants of the Eocene creodonts of Africa.

The suggestion conveyed in the last sentence at once brings us face to face with a problem which looms large in the year's work. Dr. Andrews, if he does not actually class them as creodonts, is evidently of opinion that the Patagonian sparassodonts are very closely allied to that group, from which, in his opinion, they are derived.

On the other hand, Mr. J. W. Sinclair, of Princeton University, in an elaborate and sumptuously illustrated memoir—forming part 3 of the fourth volume of *Reports of the Princeton University Expeditions to Patagonia*, 1896–9 (Princeton, 1906)—definitely classes these debatable carnivores as marsupials. Indeed, he even goes so far as to place some of them (such as *Prothylacinus* and *Cladosictis*) in the same family as the existing Tasmanian thylacine. To discuss in detail the arguments in favour of this view is obviously out of the question in the present article; but it may be mentioned that the sparassodonts lack the palatal vacuities and likewise the epipubic bones (vestigial in the thylacine) of modern marsupials, while, according to the observations of Dr. F. Ameghino, they have also a fuller series of successional teeth.

These features, coupled with the marked general resemblance of the sparassodonts to the creodonts, might well make us pause before giving adherence to the views of Mr. Sinclair. But, probably while his memoir was in the press, a new factor has been introduced into the question. In a paper read before the London Zoological Society in January last, Mr. C. S. Tomes announced that microscopic examination of sections of the teeth of creodonts and sparassodonts revealed the fact that in internal structure the enamel of these is of the type characteristic of modern Carnivora, and quite different from that of marsupials. He concludes by the statement that, while in one feature the enamel of the teeth of the Carnivora may possibly indicate remote marsupial affinity, yet in this respect creodonts (and sparassodonts) carry us no farther than their descendants, the recent members of the order.

If these conclusions are well founded, the case for the marsupial nature of the Patagonian sparassodonts will, at any rate, have to be reconsidered.

The reference to Mr. Tomes's paper has introduced the question of the affinities of the true creodonts (altogether apart from whether sparassodonts are members of that group or marsupials); and to this subject Mr. W. D. Matthew, of the American Museum, has made a contribution in a paper published during the year in the *Proceedings* of the United States National Museum (No. 1449), under the title of " The Osteology of *Sinopa*, a Creodont Mammal of the Middle Eocene." After referring to Dr. Wortman's view that creodonts and modern carnivores are divergent branches from a Cretaceous marsupial stock, the author expresses himself as follows: "I think it safe to say that if we set aside superficial and adaptive characters, and rest principally upon deep-seated resemblances, such as are found in the characters of the base of the skull, the dental and dorso-lumbar formulæ, etc., we find every known creodont very much nearer to the modern Carnivora than to the modern marsupials. On the other hand, the little that is known of Cretaceous marsupials bears distinctly the marsupial stamp in every detail, and does not show any approach to the early placentals."

It is thus evident that the creodont-sparassodont-marsupial question is one that is very far from approaching a settlement;

almost the only point on which most authorities are agreed being that creodonts include the ancestors of the modern Carnivora. Much depends on the interpretation of the dental formula of the carnivorous marsupials—that is to say, whether it corresponds (as the writer believes it does) with that of creodonts, or whether it is altogether different. That the creodonts are a very primitive group (possibly the direct descendants of anomodont reptiles) seems to the writer almost certain. That they are also in some way related to the sparassodonts likewise appears to him most probable. On the other hand, it is difficult to believe that there is not some affinity between sparassodonts and marsupials, although the writer cannot now (as he once did) definitely support the view that the two should be referred to the same division.

To revert for a moment to Mr. Sinclair's memoir, it should be added that it contains much valuable information with regard to the undoubted marsupials of the Patagonian Tertiaries.

Another American memoir of considerable interest, of which copies reached this country in November, 1905, is one by Mr. Earl Douglass on Tertiary Mammals from Montana, published in the *Memoirs* of the Carnegie Museum, Pittsburgh. In addition to the remains of a number of creodonts, more especially of the genus *Ictops*, a small skull is described under the name of *Xenotherium unicum* and provisionally referred to the Mono-tremata. If trustworthy, such an identification would be of the highest interest; but the question is whether the evidence is sufficient even to justify the provisional reference. In both groups of existing monotremes the skull is of an extremely specialised type, and in the adult at any rate edentulous; and this being so, it seems almost essential to know the nature of the shoulder-girdle before referring a toothed species (unless the teeth recall the deciduous dentition of *Ornithorhynchus*) to the order.

Continuing our brief survey of American work, the next on the list is a paper by Messrs. Matthew and Gidley on the fossil *Equidæ* of the Dakota Miocene. The paper is, however, merely a preliminary one, dealing with the description of new specimens; but the results obtained during their study are deemed of such importance as to justify a future memoir in which all the species of Miocene three-toed horses will undergo revision.

The preliminary paper is published in the *Bulletin* of the American Museum (xxii. art. 8).

Of very considerable interest, on distributional grounds, is the description by the second of the two naturalists last mentioned (in the *Proceedings* of the *U.S. Mus.* No. 1447) of the skull of a ruminant nearly allied to the musk-ox, obtained during irrigation works at Zuni, New Mexico. For this ruminant the new generic name *Liops* is proposed.

Despite the abundance of their skulls and teeth, complete skeletons of the Creodonts or "ruminating hogs" of the North American Oligoceni are very rare; and it is therefore satisfactory to learn that a skeleton of the smallest species, *Merycoidodon gracilis*—an animal the size of an ordinary fox—has been set up in the American Museum. (See C. W. Gilmore, *Proc. U.S. Nat. Mus.* xxxi. p. 513.)

Not much importance can be attached to Mr. J. W. Gidley's notice of the occurrence of remains of an extinct racoon in a Californian cave-deposit of Pleistocene age (*Proc. U.S. Nat. Mus.* xxix. pp. 553, 554); but if he is right in identifying a carnivore allied to *Amphicyon* in the same deposit, we have a discovery of very considerable interest, considering that in Europe the genus is of middle Tertiary age.

Remains of fossil seals from the Calvert Miocene of Maryland and Oregon have formed the subject of two papers during the year. In the first of these (*Proc. U.S. Nat. Mus.* No. 1475), Mr. F. W. True describes a humerus from Maryland, which he regards as representing a new genus, *Leptophoca*. To the same genus he refers a fossil seal from Bessarabia, described in 1860 by Nordmann. The subject of the second paper (*University of Oregon Bulletin*, iii. suppl. No. 3) is a seal-skull from Oregon, which is regarded by its describer, Mr. T. Condon, as representing a genus (*Desmatophoca*) with characters intermediate between those of the *Phocidæ* and the *Otariidæ*. It is remarkable that no reference is made to a paper published a year previously (*Smithsonian Miscel. Collect.* xlvi.), in which Mr. True described the skull of a seal from the same State and formation, referred to the *Otariidæ* under the name of *Pantoleon*. So far as the writer can see, there appears no reason why these skulls should not belong to different sexes of one and the same species. They are the oldest remains of sea-lions at present known.

Another pinniped—this time a walrus—has been named on the evidence of one-half of a lower jaw derived from a Miocene Tertiary littoral deposit at Yorktown, Virginia. It has been described by Messrs. Berry and Gregory in the *American Journal of Science* for June last as representing a new genus and species, under the name of *Prorosmarus alleni.*

Before leaving this portion of the subject it may be added that, in the paper already cited, Mr. Condon expresses the belief that seals have originated from terrestrial (creodont ?) Carnivora. This, it may be observed, is in direct opposition to the view of Dr. Andrews, who (in the volume cited above) states that certain presumably aquatic creodonts, with otter-like limb-bones, may have been the progenitors of the modern Pinnipedia (or, at all events, one section of that group).

A paper by Mr. O. A. Peterson, published in the *Memoirs* of the Carnegie Museum, Pittsburgh (ii. art. 8), on new Suilline remains from the Miocene of Nebraska, is devoted to the description of the osteology of a new peccary of the genus *Thinohyus.* A second, by Mr. F. B. Loomis, on Eocene Primates, published in the *American Journal of Science*, is likewise merely descriptive.

Two memoirs on the mammals of the Santa Cruz beds, Patagonia, have already received mention. Besides these, two others, by Prof. Albert Gaudry, on the same subject have been published in the first volume of the new Paris journal *Annales de Paléontologie.* In the first of these the author discusses the attitudes of many of these Patagonian ungulates, as deduced from the articular surfaces and proportions of their limb-bones ; contrasting them at the same time with European types. In the second memoir, on the other hand, the fossil Patagonian fauna is discussed as a whole in connection with its bearing on the theory of a great Antarctic continent.

As regards the Pleistocene mammals of South America, the only memoir that has come under the writer's notice is one by Mr. Max Rautenberg, of Breslau, on the skeleton of a ground-sloth from the Arroyo Pergamino, Argentina. It is described in vol. liii. of the Stuttgart *Palæontographica* as a new species of that group of mylodons which is often separated from the typical genus on account of peculiarities in the dentition and the shortness of the nasal bones, as *Pseudolestodon.* The name *P. hexaspondylus*, by which it is designated, may perhaps refer to the intimate connection between the first two neck-vertebræ.

Reverting to the Old World, M. Marcellin Boule, in the first volume of the *Annales de Paléontologie* (Paris), has instituted a careful comparison between the osteology of the cave-lion and that of its existing representative. More important is his statement that the smaller *Felis arvernensis* of the Upper Pliocene agrees in the characters of the lower jaw and dentition with the lion rather than with the tiger, thereby suggesting that it may have been the ancestor of the former.

A point of some interest has been raised by Dr. E. Lönnberg, of Upsala (*Arkiv. Zool.* Stockholm, iii. art. 14), with regard to the systematic position of the giant extinct Irish deer, or so-called Irish elk, *Cervus giganteus*. Hitherto this ruminant has been regarded as a near relative of the fallow deer. Dr. Lönnberg, on the other hand, would have us believe that it is a kind of aberrant reindeer ; this opinion being based in some degree on the fusion of the vomer with the adjacent bones of the skull.

In a notice of this paper in the *Field* newspaper, the present writer has, however, pointed out that the front cannon-bones of the Irish deer appear to indicate that the lateral metacarpals were of the "plesiometacarpalian" type characteristic of the fallow-deer and its relatives. If this be correct, there will be strong evidence of the correctness of the old view.

In connection with Ireland, it may here be mentioned that Mr. R. J. Ussher, in the *Irish Naturalist* for November, has given a preliminary account of the excavation of the contents of the ancient "hyæna-dens" in the Mammoth Cave, near Doneraile, County Cork. Although all the remains at present identified appear referable to the ordinary cave-species, this is the first record of the occurrence of the cave-race of the African spotted hyæna in Ireland. In another serial (*Proc. Roy. Irish Acad.* xxvi. No. 1, 1906), Dr. R. F. Sharff has described and figured jaws from the Newhall Caves, county Clare, which he regards as referable to the Egyptian wild cat (*Felis ocreata* or *maniculata*), although it is doubtful if the evidence is sufficient.

Passing to Cyprus, we find that Miss Dorothy Bate (*Geol. Mag.* decade 5, iii. pp. 241-5) has given an account of the skeleton of the pigmy hippopotamus (*Hippopotamus minutus*) obtained by herself from a cavern in that island, and now mounted in the British Museum (Natural History). The features distinguishing the species from its ally, the living pigmy hippopotamus of Liberia (*H. liberiensis*), are indicated in this communication.

Going still farther south, special interest attaches to a note by Dr. R. Beck, in the serial last mentioned (pp. 49-50), on a fragment of the tooth of a mastodon from Barkly West, South Africa. Assuming it to be authentic, this specimen affords the first evidence of the occurrence of this group of Proboscidea in Africa south of the Sahara, and has an important bearing on distribution.

Two papers—in addition to the one by Mr. Tomes already cited—on the subject of teeth claim brief notice. In the first of the twain, published in the *Proceedings* of the Washington Academy of Sciences (viii. pp. 91-106), Mr. J. W. Gidley confirms the opinion of certain other observers that the cusps which go to form "tritubercular" molars are by no means homologous with one another in different genera of fossil and recent mammals. Nevertheless, despite the fact that the "protocone" may not be the original primary cusp, the author advocates the retention of Prof. H. F. Osborn's names for the cusps of the tritubercular molar. In this we venture to think he is wrong. The names are cumbrous and difficult to assign to their respective positions ; and if they do not represent homologous elements, the sooner they are consigned to oblivion, and replaced by the older terms, the better.

The second of the two papers on dentition is one by Dr. O. Abel, of Vienna, on the milk-molars of recent and fossil sea-cows (Sirenia), published in the second volume of the *Neues Jahrbuch für Mineralogie*, etc., for 1906. The author has been enabled to demonstrate that the sea-cows, although now monophyodont, were originally diphyodont, and that the reduction of the dentition commenced with the posterior premolars.

Finally, in the *Comptes Rendus* of the Paris Academy (cxlii. p. 610) Mr. C. Depéret directs attention to the important influence which ancient migrations have had on the evolution of mammals.

Turning to work on fossil reptiles, a brief reference will suffice to Dr. W. J. Holland's memoir on the dinosaurian *Diplodocus* (*Mem. Carnegie Mus.* Pittsburgh, ii. art. 6), in which special reference is made to the model of the skeleton recently set up in the Natural History Branch of the British Museum.

There is also a discussion with regard to the true position of a bone which has been regarded as the clavicle of this dinosaur.

Of much wider scope is the description by Dr. F. von Huene, published in *Geologische und Palæontologische Abhandlungen,* Jena (xii. art. 2), of dinosaurian remains from the Trias of extra-European countries. The genera described include the South African *Euscelesaurus, Massospondylus,* and *Thecodontosaurus* (typically European), and the North American *Anchisaurus, Cælophysis,* and *Ammosaurus,* of the Connecticut Valley. Of these the first five are included in the Theropoda, or carnivorous group—the first in the family *Plateosauridæ,* the next three in the *Thecodontosauridæ,* and the fifth in the *Cæluridæ.* The sixth genus, on the other hand, is referred to the Orthopoda, in which it probably represents the family *Nanosauridæ.*

In vol. xxii. art. 2 of the *Bulletin* of the American Museum of Natural History, Prof. H. F. Osborn has given a complete description and restoration of the skeleton of the recently discovered carnivorous dinosaur *Tyrannosaurus* from the Upper Cretaceous of North America. This reptile stood about 16 ft. to the crown of the head, and there is a possibility that it may have carried armour. The most remarkable feature in its osteology is the presence of a series of abdominal ribs comparable to those of the tuatera (*Sphenodon*), such structures having hitherto been unrecorded among either dinosaurs or crocodiles. The author states, however, that they exist in the allied genus *Allosaurus,* and suggests that they may also be represented in the herbivorous sauropodous dinosaurs, in which group they have been regarded as referable to the shoulder-girdle.

From the Lower Jurassic of Victoria, Australia, Dr. A. Smith Woodward (*Ann. Mag. Nat. Hist.* ser. 7, vol. xviii. pp. 1-3) has recorded the claw of a dinosaur apparently nearly allied to *Megalosaurus.* The identification is of considerable importance as strengthening the evidence as to Australia having been in connection with continental land-masses during the epoch in question.

In connection with the last paragraph, it may be noted that in the *Beitr. Pal. Oester.-Ung.* Franz Baron Nopcsa elaborates a previous account of the greater part of the skeleton of a large carnivorous dinosaur from the Oxford Clay of Oxford in the collection of Mr. J. Parker, of that city. In place of referring this specimen to *Megalosaurus*, Baron Nopcsa considers that it indicates a genus apart, which he identifies with *Streptospondylus*, typified by a few vertebræ and limb-bones in the Paris Museum from the Kimeridgian of Havre. Among other peculiarities, the Oxford dinosaur is stated to differ from *Megalosaurus* in possessing four (in place of three) hind-toes. It may, however, be mentioned that Phillips, in his description of the typical species of the last-named genus, stated that he was uncertain whether there might not have been a fourth hind-toe.

Before leaving the subject of dinosaurs, it may be mentioned that American palæontologists have recently attempted to estimate the "live-weights" of *Diplodocus* and *Brontosaurus*. Although the attempt might seem hopeless, their method of going to work has been so thorough that there appears to be considerable probability of the estimates presenting a fair approximation to the reality. The plan adopted was to make a model of the entire reptile, as deduced from a study of the skeleton, on a scale of one-sixteenth the natural size. The cubic contents of such a model multiplied by the cube of 16 would indicate the probable amount of water displaced by the reptile when in the flesh. To arrive at this result, one of the miniature models of *Brontosaurus* (the length of whose skeleton is 66½ ft.) was cut into six pieces of convenient size for purposes of manipulation; and the equivalent water-displacement of each of these fragments determined. From this the water-displacement of a model of the natural size was calculated by means of the above-mentioned formula, which gave as a result the displacement of 34¼ tons by the entire animal. Since, however, *Brontosaurus* is believed to have walked along the bottom of lakes in search of food to depths which caused its whole body to be submerged, it is probable that the reptile in life was slightly heavier than water, and to allow for this an addition of about 10 per cent. was made to the calculated weight, thus bringing the estimate to a total of 38 tons. Vast as is this weight, it is, however, only about two-thirds of the estimated weight of the heaviest whales, which is presumed to be not less than 60 tons. It may be added that the weight of the African elephant " Jumbo " was only 6½ tons.

The Mesozoic crocodiles with amphicoclous vertebræ of North America receive treatment in the American *Journal of*

Geology (xiv. pp. 1-17) by Mr. S. W. Williston, who is of opinion that the remains from Wyoming described some time ago by Dr. W. J. Holland are rightly referred to the European *Goniopholis* or a nearly allied genus. Other remains from the same state the author describes under the new generic designation of *Cœlosuchus*.

The next memoir for notice is one by Mr. J. H. McGregor, published in the *Memoirs* of the American Museum (x. art. 2) on the Phytosauria (or Belodonts) of the Trias. These reptiles were classed by Prof. Huxley with the Crocodilia; but since his time the opinion has been steadily gaining ground as to their right to form a separate ordinal group. According to our author, it is safe to conclude that phytosaurs, crocodiles, and carnivorous dinosaurs form three divergent branches, or "phylæ," from a common Permian, or early Triassic stock, nearly allied to the tuatera (*Sphenodon*) with a thecodont dentition and two-headed ribs. Mr. McGregor is also of opinion that phytosaurs represent the nearest known relatives of the ichthyosaurs, or fish-lizards—although this is a point which may require further investigation before it can be definitely accepted. The following classification of the group is proposed:

PARASUCHIA.—A. AËTOSAURIA.—*Aëtosaurus.*

B. PHYTOSAURIA.—1. *Phytosaurus.*
2. *Mystriosuchus.*
3. *Rhytidodon.*
4. *Stagonolepis.*
5. *Parasuchus.*
6. *Episcoposaurus.*
7. *Rileya.*
8. *Palæosphenia.*

Of these, Nos. 1 and 2 are continental, and Nos. 4 and 7 British, No. 7 being based on two vertebræ and a humerus from the Keuper of Bristol, named and described in 1902 by Dr. F. von Huene. Of the rest, Nos. 3, 6, and 8 are North American, while No. 5 alone is Indian. The type of No. 2 is *Belodon planirostris.*

The list is, however, incomplete, for some time ago Dr. R. Broom gave a preliminary notice of certain reptilian remains from the Karoo series of the Aliwal North district, South Africa, which he regarded as representing a large phytosaur, under the designation of *Erythrosuchus*. Of these remains a fuller account has been given by their original describer during the period under review in the *Annals* of the South African Museum (v. pp. 187 *et seq.*), where the same view of their affinity is maintained.

Dr. Broom, in the *Proceedings* of the Zoological Society of London (1906, pp. 591 *et seq.*) has also given an account of another Karoo reptile—*Howesia*—from the Aliwal North district. In his original notice Dr. Broom thought that *Howesia* might be included in the Rhynchocephalia ; but he is now of opinion that, together with *Gnathodon*, it is best regarded as representing another group, the Gnathodontia. This group, to which ordinal rank is assigned, is considered to be related in some degree to the rhynchocephalians, but more nearly to the phytosaurs.

Taking next the ichthyosaurs, we find that Dr. Smith Woodward (*Geol. Mag.* Decade 5, iii. pp. 443-4) has described two specimens containing in their interior the skeletons of fœtuses. One of these, from the Lias of Somerset, is of special interest as having been described so long ago as 1849 by Dr. J. Chaning Pearce. This, however, is not all relating to this group, for Mr. C. W. Gilmore, who published a memoir on the osteology of the American fish-lizard commonly known as *Baptanodon* in the *Memoirs* of the Carnegie Museum, Pittsburgh, in 1905, has communicated a further notice of the osteology of the genus in a later issue of that serial (ii. art. 9, 1906), in which he describes a new species. In referring to the plates in the sclerotic of the eye, this author states that their mode of articulation admits of a certain amount of contraction and expansion in the bony ring. Mr. Gilmore still maintains the distinctness of *Baptanodon* from the European *Ophthalmosaurus*, although admitting the occurrence of a representative of the latter in American strata.

Next to that on the Phytosauria, perhaps the most important memoir on fossil reptiles issued during the period under review is one by Mr. B. Brown on the Upper Cretaceous genus *Champsosaurus*, published in the *Memoirs* of the American Museum (ix. part 1, December 1905). The genus has often been placed in the Rhynchocephalia, as the representative of a special suborder—Choristodera. In the author's opinion that group is, however, considered worthy of ordinal rank. "*Champsosaurus*," he writes, "cannot be considered ancestral to the Rhyncho-cephalia proper, because it is already a long-nosed type derived from a short-nosed form. It has lost the notochord. The ptery-goids are highly specialised, compressed, and extended backwards and forwards, completely obscuring the basisphenoid, while the ethmoid is developed in front of the prevomers. Although similar in many characters to *Sphenodon*, this similarity emphasises rather the very persistent primitive features of *Sphenodon*." Why these admittedly specialised characters of *Champsosaurus* necessarily entail ordinary separation from the Rhynchocephalia is perhaps not very clear. There is a marked tendency among some writers to an unnecessary multiplication of reptilian orders.

As regards chelonians, the most important memoir is one by Mr. G. R. Wieland, forming article 7 of the second volume of the *Memoirs* of the Carnegie Museum, on the osteology of the huge Niobrara Cretaceous turtle, *Protostega gigas*. In discussing the affinities of this and allied turtles, the author concludes that *Protostega* should be referred to the family *Chelonidæ*; and like-wise that the family itself (instead of being widely sundered therefrom) should be placed next to the *Dermochelyidæ*, as re-presented by the existing leathery turtle (*Dermochelys*) and the extinct *Eosphargis* and *Psephophorus*. Another paper on fossil chelonians is one by Mr. O. P. Hay (*Bull. Amer. Mus.* vol. xxii. art. 3), in which the new genus *Xenochelys* is proposed for a tortoise from the Oligocene of Dakota referred to the family *Dermatemydidæ*, while Cope's *Emys septaria* of the Bridger Eocene is raised to separate generic rank under the title of *Echmatemys*, and part of a shell from the Pliocene of Peace Creek is made the type of the new species *Terrapene putnami*. In a third paper on chelonians, Mr. E. S. Riggs (Field Columbian Museum, Geological Series, ii. No. 7) describes a fossil tortoise from the Laramie Cretaceous of Montana, under the name *Basilemys sinuosus*; the generic determination being provisional.

The year's literature on fossil snakes (in addition to the full description of *Gigantophis* in Dr. Andrews's volume) appears to comprise only a single memoir in the *Beitr. Pal. Oester.-Ung.*, in which Dr. W. Janesch describes a specimen of *Archæophis proavus*, a species from the Eocene of Monte Bolca, first named in 1849. As the result of his investigations, the author concludes that this snake, which is specially characterised by its sharp muzzle, was aquatic and probably marine in habits, and that it is so distinct from all living groups as to be entitled to rank as the representative of a separate family—the *Archæophidæ*.

On labyrinthodonts (or stegocephalians) little work seems to have been done during the year. Very important, however, is the identification by Dr. Smith Woodward of an archegosaurian from beds in the Vihi Valley, Kashmir, apparently underlying marine Permian strata. The labyrinthodont was accompanied by remains of certain fishes and plants—the latter of the Gondwana type; and the whole series is described by Messrs. Woodward and Seward in the *Palæontologia Indica* (new series, ii. art. 2). The labyrinthodont is referred to the European *Archegosaurus*, no reference being made to *Gondwanosaurus* of the Bijori beds in Central India, which was probably contemporaneous, or thereabouts. The fishes are referred to the Lower Permian genus *Amblypterus.* The importance of the paper is connected with stratigraphical rather than with purely palæontological considerations.

In the preceding passage we have passed imperceptibly from amphibians to fishes, which now claim the remaining space. Perhaps the most important paper in this group is one by Dr. O. Abel, on fossil flying-fishes (*Fossile Flugfishe*), published in the *Jahrbuch* of the Austrian Geological Survey (1906, lvi. part 1). At the present day there are two distinct types of marine flying-fishes—namely, the flying-gurnards and the flying-herrings, the latter being what may be called the typical flying-fishes; while there is also the fresh-water African *Pantodon.* It is quite evident that each of these has acquired its powers of flight independently of the other; and similarly, Dr. Abel shows that in past geological times several kinds of fishes, totally distinct from the modern types, possessed long pectoral fins, which were intended, in all probability, to enable their owners to skim the surface of the water in flying-fish fashion. The earlier of these fishes—*Thoracopterus* and *Gigantopterus*—occur in strata belonging to the period of the Trias, or New Red Sandstone, and, like their non-flying contemporaries, had their bodies encased in an armour of quadrangular enamel-covered scales. The so-called flying-fishes of the Chalk (*Chirothrix*) are regarded, however, by the author, despite their long pectoral fins, as deep-sea forms.

As a memoir of high morphological value, special mention

30

may be made of Dr. E. Henning's "Gyrodus und die Organisation der Pycnodonten," published in the Stuttgart *Palæontographica* (liii. pp. 137 *et seq.*). The author claims that he is the first to give a full and detailed account of all the individual skeletal elements of these fishes, more especially those of the skull. In many respects, particularly in dental characters, the pycnodonts are evidently related to the lepidodonts (*Semionotidæ*). Both, indeed, seem to be derived from the same stock ; the pycnodonts being, however, specialised for a particular mode of life, which has resulted in the loss of many primitive characters. *Gyrodus, Mesturus, Microdon,* and *Stemmatodus* form the more primitive group, connected by *Palæobalistum,* especially in cranial characters, with the Eocene *Pycnodus.* These fishes probably fed on sedentary or slow-moving invertebrates ; and form an analogous type to the modern wolf-fish (*Anarrhichas*). From the deep form of the body in many, such as *Mesodon,* they were probably slow swimmers ; but the somewhat elongated shape of *Pycnodus* suggests more rapid movement.

Dr. Smith Woodward's descriptive work includes a paper on Carboniferous fishes from the Mansfield district of Victoria (*Mem. Nat. Mus. Melbourne,* No. 1). The remains were discovered in 1888, and a notice of them published by the late Sir F. McCoy in the following year. Of the six generic types recognised, one is too imperfectly known for its affinities to be exactly defined ; four others, *Acanthodes, Ctenodus, Strepsodus,* and *Elonichthys,* occur in the Permian and Carboniferous of Europe and the Carboniferous of North America ; but the sixth, *Gyracanthides,* although related to a northern Carboniferous type, is peculiar and of exceptional interest. It appears, indeed, to be an acanthodian referable either to the *Diplacanthidæ* or a kindred family group, but of a highly specialised nature, the specialisation displaying itself in the enlargement of the pectoral fins, the reduction and forward displacement of the pelvics, and the absence or modification of the intermediate spines. A restored figure of this remarkable shark is given.

In the paper referred to above, in connection with a dinosaurian claw, the same palæontologist records *Ceratodus* from the Jurassic of Victoria, thus confirming the evidence of the extension of the Gondwana fauna to Australia. From the Lias of Lyme Regis he describes (*Quart. Journ. Geol. Soc.* lxii. pp. 1-4) a new species of the chimæroid *Myriacanthus*; and to the

Proceedings of the Geologists' Association (xix. pts. 7 and 8) he contributes a thoughtful paper on the study of fossil fishes in general.

Although morphological rather than palæontological, a paper by Mr. E. S. Goodrich, of Oxford, on the mode of development and origin of the fins of fishes (*Quart. Journ. Micr. Sci.* June 1906), has such an important bearing on palæontology that a reference to it in the present article seems imperative. It is shown that the mode of development of the dorsal fins is essentially the same as that of the paired fins, both arising as longitudinal folds, into which grow buds from the myotomes, these being subsequently affected by concentration and fusion. The observations of the author practically give the death-blow to the theory that the paired fins of fishes (and consequently the limbs of vertebrates generally) are derived from modified gill-arches, for that theory gives no explanation of this remarkable structural resemblance of the paired to the median fins. On such a theory the resemblance is inexplicable, whereas on the lateral (and median) fold theory such a resemblance is not only easy of explanation, but is precisely what might be expected to occur.

Those interested in fossil otoliths of fishes should refer to a paper by Mr. G. F. Bassoli on bones of this nature from the Pliocene and Miocene strata of Emilla, Italy, published in the *Rivista Ital. Pal.* xii. pp. 36 *et seq.*

The three remaining papers deal with the classification and phylogeny of fishes. Of the first of the triad, by Mr. L. Dollo (*Bull. Soc. Belge Géol. et Pal.* xx.), the mere quotation of the title, "Sur quelques points d'Ethologie Paléontologique relatifs aux Poissons," must suffice. The second paper is one by Mr. C. R. Eastman on the Dipnoan affinities of Arthrodires, published in the *American Journal of Science* for 1906 (xxi. pp. 131 *et seq.*). The object of this communication seems to be to support the views of those who (like Dr. Smith Woodward) regard the Palæozoic Arthrodira (as typified by *Coccosteus*) as an ordinal group of the Dipnoi, or Dipneusti (lung-fishes). "Arthrodires and Ctenodipterines [the Palæozoic *Ctenodontidæ* and *Dipteridæ*]," writes the author, "may be regarded as specialised off-shoots which diverged in different directions from primitive Dipnoan ancestors ; and the more generalised descendants of these latter have alone survived till the present day. . . . The recognition of arthrodires as an order of Dipneusti precludes their association with ostracophores [*Cephalaspis*, etc.] in any sense."

The third paper of the triad, and the last on the whole list, is one on the classification of Selachian fishes communicated by Mr. C. T. Regan, of the British Museum, to the *Proceedings* of the Zoological Society (1906, pp. 722-58). It deals, of course, with both recent and fossil forms; and perhaps its title is a little misleading, since the author includes in the Selachii (= Elasmobranchii) not only the sharks and rays, but the chimæras and their extinct relatives (Holocephali). It is this marked departure from the classification adopted in the British Museum *Catalogue of Fossil Fishes* which renders it necessary to refer to the paper in the present article. Which of the two views will ultimately prevail remains to be seen, although the present writer has a very strong opinion of his own on the subject. A startling suggestion in Mr. Regan's paper is one to the effect that the "Selachii" will eventually have to be separated from the class Pisces as typified by the Teleostomi. To this the reply is, "Cui bono?"

In conclusion, the writer must again crave the indulgence of his readers for such imperfections as may exist in a record which from the very nature of the case must be incomplete.

NOTE ADDED

As this proof was going to press an important memoir (unfortunately in Danish) by Mr. Herluf Winge, on extinct and living Ungulates of Lagoa Santa, Brazil, was received. It is published in *E. Museo Lundi* Copenhagen, and deals not only with the subject forming the title, but likewise with the classification of the Ungulata generally.

THE BEHAVIOUR OF OVER-STRAINED
MATERIALS

By A. O. RANKINE, B.Sc.

Assistant in the Department of Physics, University College, London.

IN 1835 Weber[1] discovered a phenomenon which threw an entirely new light on the question of the elasticity of bodies. He found that when a silk thread was stretched by a weight, the immediate stretch which resulted was not the only effect observable, but that, if the weight were allowed to remain, the length continued slowly to increase for many hours. On the removal of the weight, although an immediate shortening took place, the thread did not at once regain its original length, but continued to contract perceptibly for twenty days longer. This effect, which Weber called " Elastische Nachwirkung," has been the subject of continued investigation from that time to this, and much important information has been gathered; but, curiously enough, very little reference to it can be found in the standard treatises on the properties of matter. This lack is particularly noticeable in English text-books. They contain complete records of the known properties of solids subjected to stresses less than the elastic limit, and of the viscous properties of fluids, but little is said about the behaviour of solid bodies in a state of over-strain, when they appear to partake both of the properties of a truly elastic solid and of those of a viscous liquid. Perhaps the results of investigations in this field are regarded as too indefinite, or possibly even somewhat inconsistent. Neverthe-less, they are of sufficient interest and importance to justify consideration; for an effect which, since its discovery in 1835, has been shown to exist in very many other substances besides silk (so that, indeed, it may be regarded as a property common to all solid bodies), must be capable of throwing much light on the internal structure of matter.

[1] W. Weber, *Pogg. Ann.* 34 (1835); 54 (1841).

It will be well, at the beginning, to describe in detail the behaviour of solid bodies when subjected to stress—that is to say, applied force. Let us suppose, for the sake of exactness, a body in the form of a wire held fixed at one end, and to which a longitudinal stress may be applied by loading the other end with weights. When the load is attached, there results an immediate deformation in the form of an extension of the wire, which, if the load is sufficiently small, does not increase with time, even though the weight is allowed to remain on the wire. The removal of the stress effects a contraction of the wire, which at once regains its original length. A repetition of this procedure produces identical results. Now let the wire be loaded with a greater weight. The increase in length produced will again be immediate and afterwards constant, provided that this weight also does not exceed a certain limit. The extension will, however, be greater than in the first case in the direct proportion of the weights, but the contraction subsequent to the release from stress will again be instantaneous and equal to the extension. Results similar in all respects to these are obtained as greater and greater loads are experimented with, until, when the stress has reached and exceeds a certain limiting value depending on the material and dimensions of the wire, other effects appear, the investigation of which forms the subject of this article. The stress on the wire is now said to have reached the "elastic limit." The wire no longer behaves as a perfectly elastic body, *i.e.* it does not now immediately regain its original length upon the removal of the load. The deformation or "strain," as it is called, resulting from the applied stress, does not remain equal to its initial value, but increases with the progress of time. It is no longer a function of the weight only, but also depends on the time during which the latter has been applied. If the load is allowed to remain on the wire, the gradual increase of length continues for many hours, and, when the load is at length removed, the immediate contraction is followed by a slow but long-continued recovery towards the original length.

An exactly similar series of observations may be made on a wire clamped at one end and twisted at the other. Provided the torsion-couple does not exceed the elastic limit for the particular case, it produces torsion of constant value, and the recovery is complete on release. But so soon as the couple exceeds that limit, the initial twist is followed by an increase with time, and the subsequent initial recovery by a gradual untwisting.

The general nature of the effect of a constant stress which exceeds the elastic limit is shown in fig. 1, where the strain or deformation produced at any time is given by the ordinate to the curve. The time is reckoned from the moment when the stress is applied, and the length O A represents the initial immediate strain. This is followed by the gradual increase of deformation denoted by the part of the curve A C. At the time corresponding to the point B the stress is removed, with the

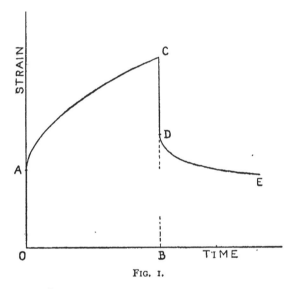

FIG. 1.

result that an immediate recovery of magnitude C D occurs, and afterwards a slow diminution of strain shown by the curve D E. The chief method of research on this question of over-strain has been to observe the changes in deformation taking place in wires of various substances when stretched by a constant force or twisted by a constant couple, particularly the gradual recovery from strain subsequent to the cessation of the stress indicated by the curve D E. Another means of investigation has, however, also been much used. Let us again suppose a wire being gradually stretched by a weight, but let this weight be in such a form that it can be removed a little at a time. When the wire attains a certain length, remove a small portion of the stretching weight. This will cause a small shortening of the wire, which, even though in a state of over-strain, still possesses certain elastic properties. It will be seen that, provided this diminution in stress does not reduce the total stress to a value below the elastic limit, this decrease in length will only be temporary, and that soon the wire will again be stretching. As soon as it reaches the same length as before, remove another small amount of the weight, and wait until the wire is again of the length decided upon above. Similar removals of weight will produce similar results, until at last the remaining weight is equal to the elastic limit, when no further tendency to increase in length will be observable in the wire. By this procedure the wire has been kept approximately at constant length. It is obvious that by removing the stress continuously at a suitable rate, instead of in finite portions after certain intervals, an absolute constancy of length could be maintained, and the necessary rate of removal of stress recorded.

The two chief methods of investigation are therefore:

(1) To observe the changes in length and torsion in wires subjected to longitudinal and torsional stresses respectively, and the recovery following the removal of the stress.

(2) To observe the diminution of stress in wires kept at constant length or constant torsion.

The aim of these experiments has been, in the first case, to discover the law connecting strain with time, both while the stress is applied and after it has been removed. In the second case it has been to find what function of the time the stress is when the strain is unaltered. But before recording the results of these attempts, attention may be called to certain general features of the elastic after-effects. Referring again to fig. 1, it will be noticed that the immediate extension represented by o A is shown as being equal to the immediate recovery indicated by c D. This is actually the case even when the stress is applied for a very long time. That is to say, wherever the point B is chosen, c D remains equal to o A. The same statement is, however, not true of the after-effect represented by D E. The shape of the curve D E is not independent of the time of duration of the stress. N. A. Hesehus[1] has shown that if the stress is applied only momentarily, the after-effect is not perceptible, and that it increases for stresses

[1] N. A. Hesehus, *Journal de la Soc. Phys. Chem. russe* (1882).

of increasing duration. If the point B approaches O, the curve D E becomes less steep, until, ultimately, when B is practically coincident with O, the vertical line C D represents the whole of the recovery observable; or, provided the stress is only momentary, the wire under test immediately regains its original form, and behaves like a perfectly elastic body. Again, it is obvious that some kind of connection must exist between the curves A C and D E. It has been shown that the immediate deformation O A corresponds and is equal to the immediate recovery C D, and the question arises, " Is the curve D E identical in shape with A C, and will they coincide when superposed; or, in other words, is the recovery from strain the exact counterpart of the whole deformation effected by the application of the stress?" If the answer be in the affirmative, it is clear that the curve D E will cut the time axis at a point at double the distance O B from O, which would mean that the strain will be zero after a time reckoned from the moment of release equal to the time during which the wire was under stress. Athough there appear to be certain metal wires for which this is true, there are also many other materials for which it is not the case, and therefore a general rule cannot be stated. In Weber's account of his experiments on the stretching of silk fibres, he records that a fibre subjected to the action of a weight for thirty-six hours was still visibly shortening twenty days subsequent to release. The gradual after-recovery was, in this case, therefore, very much less rapid than the increase of strain following the initial elastic strain. Similar results have also recently been found by Prof. Trouton and the writer[1] in the case of lead wires under longitudinal stress. On the other hand, Mr. Phillips[2] has found that for some metal wires, including copper and gold, the recovery corresponds exactly to the strain, recovery is complete after a fixed finite time, and there is no permanent deformation.

In the light of this knowledge, it seems that we may in general divide the strain occurring in over-stressed wires into three parts, one of which may, in certain cases, be absent. There is the initial elastic deformation O A (fig. 1), which has its counterpart in the immediate recovery C D. Besides this, there is a secondary elastic strain included in A C, which

[1] Prof. Trouton and A. O. Rankine, *Phil. Mag.* Oct. 1904.
[2] P. Phillips, *Phil. Mag.* April 1905.

accounts for the recovery represented by D E; and finally there is a viscous flow, also included in A C, from which there can be no recovery. If this third element be absent, as it apparently is in certain cases, then A C and D E are identical in shape, and complete recovery occurs after a finite time. But if part of the strain is viscous in nature, then we cannot expect the recovery to correspond exactly to the deformation, nor will the original shape be ever regained. In some materials this viscous flow forms a very large part of the strain; indeed, in the cases of pitch[1] and lead the curve A C (fig. 1) after a short time becomes practically a straight line, indicating a uniform flow similar to that exhibited by liquids under constant stress, and the recovery following release is comparatively small.

Of all those who have conducted research on this question of " Elastische Nachwirkung," probably Kohlrausch[2] has been responsible for most of the information acquired. In 1863 he began an investigation of the effect of torsional forces on glass, indiarubber, and various metal wires. He has shown that the after-effect—*i.e.* that portion of the recovery represented by D E in fig. 1—is nearly proportional to the initial deformation; or, in other words, the rapidity with which a twisted wire untwists bears a constant ratio to the initial angle of torsion. He also found that the after-effect was greatly increased by rise of temperature. An over-strained wire may be caused to recover more quickly by raising its temperature. But perhaps the most suggestive effect discovered by Kohlrausch is the result of the following experiment carried out by him:

A wire, clamped at one end, was twisted at the other end, and, after being held for some time, released. The gradual recovery followed the immediate one; but before it was complete the wire was again twisted through a small angle, this time towards the original equilibrium position, *i.e.* in the direction in which motion was at the moment taking place. What now happened is of great interest. The wire, instead of continuing to untwist towards the original unstrained position, first turned in the direction opposite to that of the second twist applied, or in the direction of initial twisting, came to rest, and then continued its first unwinding course.

Apparently it is possible to superpose two of these after-

---

[1] Prof. Trouton and E. S. Andrews, *Phil. Mag.* April 1904.

[2] F. Kohlrausch, *Pogg. Ann.* 119 (1863); 128 (1866); 158 (1876); 160 (1877).

effects, and the possible number of superpositions is not limited to two. We may twist a wire first in one direction and then in another, back again in the first, reverse again, and so on. We shall thus superpose a large number of strains, the recovery from which is in each case a function of the time, and we shall find that the wire, when at length released, will commence to recover from the final strain applied, then the next previous, and so on, following in the reverse order, and in the opposite direction, all the series of strains. When this fact is borne in mind— that the behaviour of a wire at any time depends to a greater or less extent on all the deformations it has previously experienced—it is surely not to be wondered at that some apparent inconsistencies arise in experiments on elastic after-effects. It is not enough that the wires under test should be of the same material. Consistency of results cannot be expected unless, also, they have exactly the same past history. It is evident, too, what great care must be taken to ensure this being the case. This behaviour of over-strained wires is, in many respects, analogous to that of iron and steel under the action of magnetising forces. Elastic hysteresis is exhibited, and hysteresis curves similar to those in the magnetic case have been obtained by Cantone[1] in 1893 and 1898, and Weinhold in 1899.

Yet another effect in connection with the slow recovery after strain has been noticed by Maxwell.[2] It has been shown that a body, even when in a state of over-strain, retains certain elastic properties. Hence it is possible for it to execute vibrations. Suppose that a wire clamped at one end is caused to execute small oscillations at the other end. If the amplitude is not too large, the wire will continue to vibrate about the unstrained position. The position of equilibrium is that unstrained position. But if the wire is over-strained by turning it through a large angle in either direction, then, although vibrations may still persist, the point of equilibrium about which they occur will no longer be the original unstrained position. Neither will it be itself stationary, but will continue to move slowly towards its initial position, gradually undoing the effects of the over-strain. Maxwell has shown that this recovery is more rapid when vibrations exist than when they

---

[1] M. Cantone, *Rendic. R. Acc. Lincei*, 2, 1893 ; L. Weinhold, *Zur Elastizität der Metalle, Diss.* (Leipzig), Chemnitz, 1899.

[2] J. C. Maxwell, *Enc. Brit.* vol. vi. p. 313.

are absent, or that decrease of strain is facilitated by super-posed motion in the wire.

The fact that the magnitude of the deformation in a body stressed beyond the elastic limit depends upon the time should be of importance to those who require to know what will be the strain under given conditions; yet it seems at present to be little enough appreciated. One cannot expect to be able to predict the value of any physical quantity unless all the conditions which may affect its value are clearly specified. Yet it is common to find diagrams purporting to give the magnitude of the strain in a body subjected to stress, even when the latter exceeds the elastic limit, without any reference being made to that other factor upon which, as we have seen, the strain largely depends—viz. the time of application of the stress. Since the strain is a function of the time in the region beyond the elastic limit, it is clear that no definite meaning can be attached to a curve representing stress against strain unless the duration of the stress is also recorded. In the case of stresses less than the elastic limit, there is, of course, no. necessity to specify the time of application. The initial immediate deformation is the only one which occurs, does not increase with time, and has a per-fectly definite value for every value of the stress. The strain is a function of the stress only, and the curve mentioned above, in so far as it refers to the region below the elastic limit, can be rightly interpreted without reference to the question of time at all. But when this limit is passed (under which circumstances the strain may have, as time goes on, practically any value for any one particular stress), it is absolutely necessary that the duration of the stress, as well as the stress itself, should be known before the magnitude of the strain can have any useful meaning. A series of curves showing strain against stress, and also specifying how long after the application of the stress the determination of the resulting strain was made, would certainly be useful. The curves shown in fig. 2 are not experimental curves, but probably indicate roughly what might actually be determined. In this diagram strain is plotted vertically and stress horizontally. Let us suppose it to represent the increases in length in a wire brought about by applying various forces, and further, that a unit of length horizontally represents unit force, and unit length vertically unit increase in length in the wire If O P is the elastic limit, then if a force O M less than

O P be applied, the resulting increase in length, M m, will have a certain fixed value depending solely on the magnitude of O M. The line O R, then, represents completely the relation between stress and strain, however long the wire may have been stretched. Now consider what will happen when the wire is stretched by a force greater than the elastic limit O P, say O S. The wire will exhibit an immediate increase in length, followed by a further slow increase with time—hence the strain has different values according to the duration of the stress. It would be possible to determine the increase in length after a

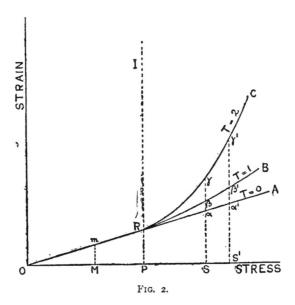

FIG. 2.

series of intervals—for example, every minute. The strain after one minute might be S β, after two minutes S γ, and so on. If the force were only applied momentarily, a strain, S a, less than any of these, would be obtained. Similarly, a second series of points a', β', γ', etc., could be obtained by observing the increase in length produced by a still greater force O S' after the same minute intervals. A whole series of such observations could be made, thus obtaining several sets of points, a a' a'', etc., β β' β'', etc., and γ γ' γ'', etc. By joining corresponding points the curves R A, R B, R C would be obtained. The curve R A will be very nearly that which would represent the variations of strain with stress in the ideal case when the stress is only applied for an infinitely short space of time. It has been pointed out earlier in this article that under these circumstances the recovery from strain after release is at once complete. The wire behaves just as it would do under similar conditions for stresses less than the elastic limit. It seems reasonable to suppose, therefore, that the curve R A will simply be a prolongation of the straight line O R, which represents strain against stress in the truly elastic region. The curve R B will give the value of the strain for any particular value of the stress, at a time one minute after the application of the latter, and in all cases this value will be greater than that corresponding to a stress of momentary duration. Similarly the curve R C indicates the various values of the strain after two-minute intervals. A diagram founded on experiment, in order to be complete, would contain many more curves than those indicated, and would probably consist of curved lines, as there is no adequate reason for supposing that they would be straight. The vertical straight line R I represents a limit beyond which these curves cannot be. If a force were applied to the end of a wire and continued to act for an infinite time, the strain could not become more than infinitely large, and would probably not do so. If we denote by T the time during which the stress is in action, then the curve for $T = o$ is R A, and that for $T = \infty$ a curve somewhere to the right of R I; and between these two all the other curves, representing various values of T, must lie. A diagram consisting of a series of curves such as those just described would certainly be very interesting and instructive, and doubtless also very useful.

Several attempts have been made to develop a theory which will produce results consistent with those which have been observed, but none has been uniformly successful. The difficulties in the way of such an attempt are certainly very great. The facts observed cannot be accounted for by internal friction in the ordinary sense of the term. The bodies which exhibit this elastic after-effect certainly appear to possess elastic and viscous properties; but these properties are not independent, and cannot be treated as such. Besides this, there is the fact brought to light by Kohlrausch's experiment described above, that a body is capable of responding to more than one over-strain at one and the same time: in other words, that the strain in a body is not only dependent on the stress at the moment in question, but also on all the previous stresses to which it has been subjected, and that in a way which is very difficult to understand. The fundamental principles upon which the development of a theory may be based become, under the circumstances, insufficient, and the theorist is usually reduced to the necessity of determining experimentally the form of some function or other, in order to make progress. The result is that the theory, although perhaps to some extent useful, is not a true physical explanation of the observed phenomena. It has been usual among experimentalists on this subject of elastic after-effect to attempt to fit empirical equations to the curves obtained, both in the case of the increase in strain in a body subjected to a constant stress and in that of the recovery from strain after release; and, further, to those curves representing the value of the stress necessary to maintain the strain constant as time goes on. The early work of Weber and Kohlrausch was mainly upon the slow recovery after over-strain. They found that the equation

$$\log x = A - at^m$$

was capable of representing the after-effect in all the bodies investigated (which included silk, indiarubber, and several metal wires), where x denotes that deformation which persists in a wire or fibre after the torsional forces have ceased to act, and t is the time that has elapsed since release. In many cases the simple formula

$$x = \frac{C}{t^a},$$

or, better—

$$x = \frac{C}{(b + t)^a}$$

sufficed. In these formulæ, everything except x and t is constant. In Weber's very first results he used the value $a = 1$, but later found a value less than unity to be more suitable. O. E. Meyer,[1] in 1874, made the first attempt to develop a theory of "Elastische Nachwirkung," and was successful in obtaining for the after-effect a formula of the same form as that made use of by Weber and Kohlrausch for representing their observations, but did not extend the theory to the case of the variation of stress with time necessary to keep the strain in the body in question constant. This latter case would be expected to be more simple than that of actual recovery from strain, for, since no change in outward

[1] O. E. Meyer, *Pogg. Ann.* 151 (1874); 154 (1875); *Wied. Ann* 4 (1878).

form occurs, all that can be happening is an internal rearrangement of the particles of which the body is composed—a fact which has caused the method to be much used in later research.

The next theory of importance was that published by Boltzmann[1] in 1878. The total after-effect was taken as being due to a whole series of after-effects resulting from applications of stress at all times previous. He assumed, and afterwards verified by experiment, that if a body were acted upon by a stress for a very short space of time τ, then the resulting after-effect, after any time t, was proportional to $\frac{\tau}{t}$, and also to the magnitude of the stress. The total after-effect in a body subjected to stress for a considerable time will be made up of all the elementary after-effects due to the application of stress during the series of short times into which the total time may be divided. Further, he assumed that these short durations of stress produce in the aggregate the sum of the effects they would have produced individually. This principle has been called the "Principle of Superposition." Kohlrausch has shown, however, in the case of a silver wire, that this principle is somewhat removed from the truth, and that the calculated after-effect, on this assumption, is always greater than that actually observed. That is to say, the small durations of stress produce less effect when existing together than the sum of their individual effects. Notwithstanding this fact, however, Boltzmann's final results are in close accordance with observation. The chief results are the following :

1. That if a wire is suddenly twisted, then the couple necessary to maintain it at this initial torsion may be represented by—

$$D = c\left\{a - b \log\left(\frac{t}{\rho}\right)\right\},$$

where D is the couple and t the time reckoned from the moment of twisting ; a, c, b and ρ being constants. The restriction is imposed that the formula holds neither for very small nor very large values of t. The four constants shown can in reality be reduced to three, for $\log \rho$ may be added to the constant a, giving it a new value.

2. That if a wire is suddenly acted upon by a constant couple

[1] L Boltzmann, *Wiener Sitzungsberichte*, October 8, 1874 ; *Wied. Ann.* 5 (1878).

D′, then the value of the torsion after an elapse of time t is given by—

$$\theta = \frac{D'}{a}\left\{ 1 + \frac{b}{a} \log\left(\frac{t}{\rho}\right) \right\},$$

a, b, and ρ, as before, being constants.

When the wire has been for a long time under the action of this couple, and is suddenly released, then if θ is the angle reckoned from the position occupied by the wire at the moment of release, its value at any subsequent time is given by the same formula, *i.e.* the recovery corresponds in every respect to the strain originally produced. Here again, however, the restriction must be imposed that the formula is inadequate for very small values of t. It will be seen that if t is put equal to zero, then the corresponding value of θ is infinity in the negative sense, which is, of course, inadmissible.

Considerable experimental evidence is forthcoming in support of these results of Boltzmann's. Had all subsequent workers attempted to apply his theory to their results, it would probably have been even greater. In making this application to the case of variation of stress at constant strain, Kohlrausch found that the couple D necessary to keep the torsion constant in a suddenly twisted glass fibre could be represented with great accuracy (except for very small values of t) by the formula

$$D = a - b \log t,$$

which will be seen to be of precisely the same form as that given by Boltzmann's result for that case. The formula proved also to be valid for very large values of t, and Kohlrausch considered the small variations which did occur as being sufficiently accounted for by possible changes in temperature, which, as before pointed out, have considerable effect on the phenomenon. Recently, Prof. Trouton and the writer found that the stress-time curve at constant strain for lead wires, both for stretching and torsion, could be represented with considerable accuracy by the equation

$$S = a - K \log (pt + 1).$$

The introduction of the third constant in this empirical formula was an attempt to make it valid for $t = 0$, but for values of t not too small a formula precisely of the Boltzmann type sufficed. The formula made use of by Mr. Phillips for stretched indiarubber also supports that of Boltzmann, and the writer has found that it is also valid in the case of tough jellies when maintained at constant torsion. With the second of Boltzmann's formulæ—that which claims to give the value of the torsion at any time under the action of a constant couple, and the recovery after release—the experimental evidence is not so completely in agreement. In some substances the recovery curve is not of exactly the same form as that representing the increase of deformation resulting from a constant stress. Thus, in the case of the stretching and recovery of lead wires, although the recovery may be approximately represented by an equation of the type

$$\theta = a + b \log t,$$

the stretching cannot be so represented. This increase in length is partly due to a purely viscous flow from which there can be no recovery, and hence cannot correspond exactly to the latter. On the other hand, there are substances in which no viscous flow occurs, and for these the deformation and recovery curves are identical in shape, and can be fitted to the same equation. Mr. Phillips has observed that for indiarubber and certain metal wires, including gold and copper, this is the case, and that the curves of stretching and recovery are capable of being denoted by the equation

$$L = a + b \log t.$$

There is, however, a further exception. The same observer has found that this formula is not valid in the cases of iron and steel wires. But it is not surprising that certain exceptions should occur in common with other physical laws.

There are several points of interest in connection with the first law, viz. that the stress necessary to maintain constant the strain in a suddenly deformed body may be represented by

$$D = a - b \log t.$$

It is apparent that if this formula holds good until t has the value given by

$$\log t = \frac{a}{b},$$

D, the stress, has then a zero value, i.e. within a finite time the stress in the body has entirely disappeared. Or again, suppose that the body is perfectly elastic up to a certain limit, and let the limiting value of stress be D_i. Then when t has the value given by

$$\log t = \frac{a - D_i}{b},$$

the stress in the body is equal to the elastic limit, and, there being now no tendency towards increasing deformation, no further decrease in stress is necessary to maintain constant strain. In this case the stress would become reduced to the elastic limit also within a finite space of time. It is obvious that the formula must cease to be valid as soon as this point is

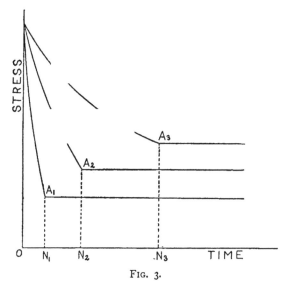

Fig. 3.

reached, for the stress will neither take a negative value nor a value less than the elastic limit. Possibly it does not hold good even up to that point, for one would expect that the stress would not be reduced to its final value until after an infinite time—*i.e.* the curve representing stress against time would approach asymptotically to that value. Some light is thrown on this question by the curves obtained by the writer[1] for jellies of various concentrations. They are shown in fig. 3, and represent the variation in the couple occurring with progress of time when the torsion is constant. It will be seen that the curves, instead of being smooth throughout, show breaks at

[1] A. O. Rankine, *Phil. Mag.*, April 1906.

A_1, A_2, and A_3, and that the parts following these points are horizontal, indicating a constant couple. Apparently $A_1 N_1$, $A_2 N_2$, etc., are the respective elastic limits for the specimens. No further decrease of couple occurs at constant torsion, and it seems that in these cases the limits are reached after the finite times ON_1, ON_2, and ON_3. These curves were obtained by an automatic process, and the time scale used was not sufficiently large to justify fitting equations of the Boltzmann type; but they at least appear to point to the fact that the stress may reach its final value after a finite time, instead of after an infinite time as might be readily supposed.

This account of the effects of over-straining materials is by no means exhaustive. It is only intended to call attention to a subject which has been somewhat neglected. More detailed accounts of individual work can, of course, be found in the actual papers to which reference is given, and if this abstract succeeds in creating sufficient interest to cause them to be read, it will have served its purpose. It would not be well to conclude, however, without mentioning certain effects resembling those of over-strain. It is known that analogies exist between the phenomenon of elastic after-effect and many other physical phenomena, and considerable research has been undertaken with the object of establishing possible connections between them; but much yet remains to be done, for, although the various effects have been shown to possess many similarities, the work has not yet proceeded sufficiently far to definitely establish, or otherwise, a general law. Hesehus, in his paper on " Elastische Nachwirkung " published in 1882, seems to have expected that some relation would be shown to exist, particularly in the cases of residual charge on electrical condensers, magnetic hysteresis, and the decay of phosphorescence. It is well known that when a condenser has been charged, the discharge is not at once complete upon connecting the coatings, but that a further charge develops on them if they are once more insulated. Or, if connection is permanently established, a small current continues to flow for some time after the initial discharge. This secondary discharge is in many respects analogous to the slow recovery which follows the immediate recovery in an over-strained body when suddenly relieved from external stress. Again, it has already been pointed out that it is possible to obtain curves of elastic hysteresis in solid bodies subjected to stresses greater than the elastic limit, thus presenting similarities to the well-known phenomenon of magnetic hysteresis. Prof. Ewing, whose theory of magnetism has been so successful, imagines the resemblance between the two phenomena to be so close that they may be accounted for by very similar molecular theories. In his presidential address to Section G of the British Association[1] this year, he conceives the molecules in an elastic body as being endowed with certain polarities whose mutual actions cause the molecules to arrange themselves, in an unstrained body, in a very stable order. Small stresses effect relative movement between them, but they are not sufficiently disturbed to destroy the stability of their configuration, and they return to their previous position when the stress is removed. But if the body is subjected to a stress great enough to produce too great relative motion, the arrangement of the molecules is no longer stable, and slipping occurs. (This slipping has actually been observed with the aid of a micro-scope.) This represents the stage at which the elastic limit is exceeded, and the deformation is gradually increasing. When the stress is removed, the molecules do not regain their original configuration, but arrange themselves in the stable order which happens to be nearest their present position. This movement corresponds to the immediate recovery on release, which is, however, not complete, and the gradual recovery which follows may be regarded as due to slow movements of the molecules into more stable positions.

A further physical phenomenon which is analogous is that of phosphorescence. A body which possesses this property, after being exposed to the action of light, continues for some time to emit light on its own account when placed in darkness. The intensity of this light falls off with the lapse of time. Mr. Whiteside, at the British Association this year, reported some experiments, the object of which was to examine this decay of intensity. He finds that if I denote the intensity at any time t after the phosphorescent body has been placed in darkness, its value can be denoted by the equation

$$I = \frac{1}{a + bt}.$$

It is worthy of note, in this connection, that phosphorescence

[1] Prof. J. A. Ewing, *Phil. Mag.*, September 1906.

is a property which has been detected only in solid bodies, *i.e.* bodies which possess elasticity other than bulk elasticity, and hence are capable of exhibiting elastic after-effects. It seems reasonable to expect, therefore, that something more than a mere analogy will be found between the two properties, and that a real connection exists.

Finally, it has been recently found by N. A. Hesehus[1] that the changes of resistance in a selenium cell resemble the variations in strain referred to in detail above. A selenium cell has different resistances according to the intensity of the light to which it is exposed, but the change of resistance is not immediately complete when the light intensity is altered ; in other words, the resistance is a function both of the intensity and the time. Thus, when the cell, after being exposed to light, is put in complete darkness, its resistance increases very rapidly at first, but more and more slowly as time elapses, and a considerable period of time must intervene before the resistance reaches its final constant value.

No doubt further instances could be enumerated—instances in which the value of one physical quantity is dependent upon the value of a second quantity and also upon time—all of which make wider and wider the field for research on elastic over-strain in its more extended sense.

[1] N. A. Hesehus, *Journal de la Soc. Phys. Chem. russe,* 1906.

THE PRINCIPLES OF SEED-TESTING

BY T. JOHNSON, D.Sc.

Professor of Botany in the Royal College of Science, Dublin

WHILE botany, like other subjects of study, will always have its followers who pursue it for its own sake, or because of the increase of knowledge of Nature its successful prosecution brings, there are others who through taste or other cause devote themselves to botany from the applied or economic point of view. The magnificent work done by the English botanists in the nineteenth century was due, in large measure, as Sir J. Hooker himself once told me, to the demands of people at home and abroad for information as to the nature and uses of the plants forming the flora of our newly acquired colonies. It was, on the other hand, the absence of colonies before 1880 which helped to produce the splendid results in anatomical and physiological botany for which Germany became famous.

There are two branches of economic botany, the foundations of which have been laid practically within the last fifty years— that bearing on plant diseases or plant-pathology, and that of seed-testing.

It may not be out of place here for me, as a former student and teacher of the Royal College of Science, London, to call attention to what appears to a botanist a peculiar lack of appreciation of the modern developments of botany in the hesitation shown as to the future of the Biological Division, in the Report of the Committee on that College's conversion into a more strictly technical or applied College of Science. If the Committee were familiar with the beneficial work carried out in Germany by Frank, Hartig, Von Tubeuf, Aderhold, etc., in the Biological Institutes of Berlin, Munich, and Hamburg, and in France by Prillieux and Delacroix, etc., there would surely be no inclination to suppress the Biological Division, but rather to develop it along economic lines. Even if there were no ordinary students this division could make an ample return for all expenditure on it by additions to our knowledge of benefit to agriculture, horticulture, and other branches of economic botany.

It was in the year 1869 that seed-testing, as now understood, started. Nobbe of Tharand was asked to examine a grass-mixture, and found the sample was not true to description, a remark which applied to many other samples he then obtained from various parts of Germany.

Although the credit of starting the first Seed-testing Station must be given to Nobbe, measures had been taken as long ago as 1816 in Switzerland to suppress fraud in the seed trade. Thus an inspector had the right of entry into a seed shop or warehouse for inspection of the seeds on sale, punishment following detection of fraud. In England in 1869 the Adulteration of Seeds Act was passed, making it penal to kill or dye seeds. The Royal Horticultural Society of England did much to expose the corruption which had crept into the seed trade. In its second report (*Farmer's Magazine*, February, 1869), the Royal Horticultural Society Committee says: ". . . Everything is thus thrown upon the honesty of the dealer. He fixes the prices, regulates the quality, and the purchaser is kept in the dark, and has no check upon either. This is a temptation beyond what the average frailty of human nature ought in fairness to be exposed. . . One of the chief functions of the association [of wholesale seedsmen] is . . . the regulation of prices . . . and the determination as to what kinds of seeds should have their average lowered and to what extent it should be done." With honourable exceptions trade catalogues offered in addition to "nett" or pure seed "trio" seed, *i.e.* seed killed for admixture purposes ! The Act of 1869 made the admixture of killed seed an offence, but did not provide machinery for the detection of the offence, as is now the case for artificial manures and feeding stuffs under the Fertilisers and Feeding Stuffs Act of 1893.

In 1900 the English Board of Agriculture appointed a Committee to inquire into the conditions under which agricultural seeds are at present sold and to report whether any further measures can with advantage be taken to secure the maintenance of adequate standards of purity and germinating power. This Departmental Committee recommended the establishment of one central Seed-testing Station under Government auspices, with the fees so fixed as to encourage seedsmen to sell subject to re-testing by the purchaser, should he desire it. So far this

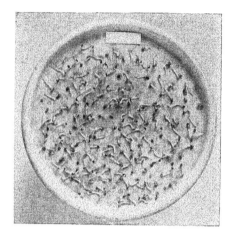

White Clover. High Germination.

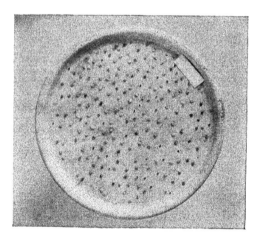

White Clover. Low Germination.

Face p. 485.]

recommendation has remained a dead letter in Great Britain. A Government Station was already in existence in Ireland when the Committee was appointed, and so far some 6,000 samples of seed have been tested by it.

The revelations of fraud and ignorance published in 1875 by Nobbe in his *Handbuch der Samenkunde* led to vigorous action, and Seed-testing Stations were started in nearly every country in the world, mostly under Government control. At the present time there are some 150.

The object of each Station is to assist the seedsmen and the farmers in securing the best and purest seed for agricultural or other purposes. Evidence abounds to prove that where in any trade there is a demand for inferior goods the supply will be forthcoming, and that where there is ignorance on the buyer's side the seller will, in too many cases, take advantage of it and try to profit by it. These statements are not universally applicable in the seed trade. They hold true, however, to such an extent as to necessitate the existence of Seed-testing Stations to determine for all parties concerned the nature of the goods under sale.

The accompanying illustrations are made from samples of seed on sale in Ireland this year, and tested in this Station. It is not necessary to label the good and bad kinds. While in appearance and price there is little to distinguish the two kinds of seeds from one another the test shows an enormous difference.

As many of my readers are probably not familiar with the qualities which characterise a good seed, or with the procedure followed in ascertaining these points, I propose in the following lines to describe briefly the more important points. In examining a sample of seed one's attention is directed to—

1. THE GENUINENESS

By this is meant that the seed is really what it is described as being. A farmer who orders meadow fescue and is supplied with the much cheaper perennial rye, is defrauded. The two seeds are very similar in appearance, and lend themselves to deception. A farmer who orders turnip seed and receives the cheaper rape, loses money in the purchase, and further gets a crop of green tops instead of fleshy roots.

Charlock, the yellow mustard-like pest in potato, corn, and other fields, has a seed very like turnip or rape, and often, apart from taste, needs the microscope for identification. At one time charlock was largely mixed with turnip or rape, and often, to hide its presence in the seed, was killed before mixing, to prevent its appearance in the field, on the principle, as was well stated, that "dead men tell no tales." Owing to the tendency of turnips and swede to bolt, the general similarity in appearance in the seeds of turnip, swede, rape, and charlock (all members of the genus Brassica), and to the consequent danger of fraud, cases occur every year, in my own experience, in which the Seed-testing Station is called in to decide as between buyer and seller. The clovers at one time were far from being genuine. Nobbe quotes a letter from a Continental firm offering to a seedsman quartz stones so agreeing in size and colour with red clover or white clover that the ordinary farmer could not distinguish them. English red clover has a high value in the trade, but every season is not favourable to the general maturing of its seeds. The supply falls short of the demand, and other less hardy red clovers are liable to be substituted. In such cases the genuineness is generally ascertainable by examination of the seed impurities.

Several seeds in falsely called English red clover are foreign, and never occur in England. By the examination of such impurities Stebler of Zurich has, during the past twenty years, collected most valuable information as to the country of origin of seeds, and his results are now in course of publication. Advantage of the knowledge of this means of identification is sometimes taken by the grower. Thus Russian flax seed, which has a reputation for strong growth, has a common impurity called "false flax" (*Camelina sativa*). The grower of flax in another country, knowing this, sometimes introduces a little false flax seed into his flax seed to give the impression that the seed is genuine Russian-grown. The meadow grasses (*Poa* sp.) differ greatly in value. The seeds of the different species possess definite botanical characters, but these characters can be so far removed by cleaning-machinery that the cheaper, more easily obtained seed can be substituted for the dearer, better kind. The ordinary buyer could not be expected to know the difference. Must he then continue to be, because of his unavoidable ignorance, the possible victim of fraud?

2. Purity

By practical purity is meant that the seed is not only true to name, but that it contains nothing else in measurable quantity. The two chief sources of impurity are inert matter, such as stones, particles of soil, broken seeds, stalks, chaff— all dead weight—and the seeds of other plants, chiefly weeds. Great improvement has taken place in this branch of the seed trade. The better houses have often elaborate machinery by which the impurities are removed. Unfortunately, however, ·these very impurities find a ready market, and the supply *within the trade* is not equal to the demand!

Many of the weed seeds are highly objectionable or injurious. The dodder is very generally present in some clovers, and may do a great deal of harm as a parasitic pest. The dodder seed is in size and colour so like a particle of soil as to be indistinguishable to the farmer. Fortunately for Ireland, the dodder does not thrive well, and there are few cases on record of injury to crops caused by it. I had one case before me in 1900 of the destruction of a flax field by the flax dodder, and the correspondent mentioned cases of great damage in some fields in earlier years.

In England the dodder has, I understand, done much harm locally, from year to year, but does not often ripen its seeds. In some of the Continental and American clovers, however, ripe dodder seeds are plentiful. In England an ounce of dodder in a ton of clover is the limit of impurity considered permissible. On the Continent the limit varies—from absolute freedom to from five to ten dodder seeds in one kilogramme of clover. Most foreign clovers contain dodder, and may require several sievings before becoming free from it. At the recent International Conference in Hamburg it was decided to invite from all over the world information as to the prevalence of dodder in the flora, etc., of each country.

The procedure followed in testing the purity of a seed is quite simple. A definite weight, varying according to the kind of seed, is taken from an average sample of the bulk. The impurities of the two kinds already mentioned are separated out, weighed, and expressed as a percentage. A difficulty arises in some cases where the seeds are blind or deaf, *i.e.* have all the external characters of the true seed, but lack the kernel. This is especially the case in such a grass as the meadow foxtail. In some cases this seed is pure chaff. When the seed is examined by transmitted light the presence or absence of the kernel is observable. In the great majority of Stations these blind seeds are treated as impurities, but where seed is bought by the bushel (or by volume) it would be more equitable to include them in the germination test.

3. GERMINATION

The seed from which the impurities have been removed is taken, and from it a definite number, varying according to the seed under inquiry—200 to 400 generally—is separated without selection, and placed under suitable conditions of temperature, air, and moisture for germination. At the end of a certain time, from a few days in the case of clovers and flax, to 28 or even 35 days in the case of some of the Poas, all the seeds capable of germinating will have sprouted, and can have been counted. In this way the percentage of germination is obtained. This varies from seed to seed, but should not vary much for the same kind, even from season to season, when the seed is ripe and fresh. Excellent work was done by Mr. W. Carruthers, F.R.S., the Consulting Botanist to the Royal Agricultural Society, when that Society was led to issue a schedule of agricultural seeds indicating the minimum percentage of germination of each kind which the seedsman should be asked to guarantee to the member when purchasing.

The riper and more perfect the seed, the more quickly and uniformly will it germinate. This is called the Germinating Energy, and is usually expressed in an Interim Report. Old and not well "filled" seeds have a low and irregular germinating energy. Some seedsmen seem to act on the view that a seed is never too old to be sold, and save seed from one year to another year on the ground that next season, owing to adverse weather, there may be a shortage of supply. Such prevision and provision are supported by the statement that, in some cases, two-year old seeds give better plants. In the majority of cases, one-year old seed is the best to sow. From two years on, seeds lose their vitality by degrees.

4. TRUE VALUE

It is obvious that one gets an imperfect idea of the quality of a seed by considering the purity and germination apart from one another. A seed may be pure, but of low germinating power through age, "heating," want of ripeness, etc. Another seed may be very impure, but what there is of it, true to name, may germinate well. Either report alone would be misleading. I had one sample of Timothy grass to test, intended for experimental work. It was pure and looked good seed, but germinated only 10 per cent. It was in consequence useless. Hence it is usual to combine the purity and germination percentage to get the True Value of the seed. This is expressed by the following formula, where—

$$P = \text{percentage of purity,}$$
$$G = \text{percentage of germination, and}$$
$$T V = \text{True Value,} \quad \frac{P \times G}{100} = T V.$$

Thus a sample of perennial rye showing 90 per cent. purity and 80 per cent. germination has a true value of 72—*i.e.* every 100 lb. contains only 72 lb. really good seed.

In the Irish Station the percentages of purity and germination are stated separately.

Many factors affect the germination of a given seed. We are not yet in a position to say what is the best or optimum condition in each respect for the production of the highest degree of germination in the laboratory of each particular kind of seed. In nature the seed is still less placed under the optimum conditions in every respect; it is in consequence usual to assume that the percentage of germination in nature will be 5 per cent. less than under the more or less controllable artificial conditions. Further, 5 to 8 per cent. is allowed as a margin of error in cases of dispute in estimating percentage of germination in the Seed Station. (Advantage is sometimes taken of this by the trade to add on to a Station's report 5 to 8 per cent.) The poorer and more uneven the seed the greater will be the difference in two tests of it even under apparently identical conditions. One is apt to lose sight of the fact that the object of testing the seed should be, not to obtain the highest possible degree of germination under the most perfect artificial conditions, but to secure a good working idea of the germinating

power of the seed when placed under field conditions. Let
us now consider in some detail the procedure followed in
ascertaining the germinating power of seed :—

(1) *Germinating Bed.*—Excepting in the case of some of the
larger seeds, they are placed, without previous soaking, directly
in the seed-bed, without being in contact with one another.
The seed-bed varies. In some cases strong folded blotting-
paper is used, in others porous clay dishes of varying thickness,
in others sand, and in yet others (especially cereals) ordinary
soil in flower-pots or saucers in a greenhouse. I shall not stop
to mention the practical details to be observed in keeping these
media pure. In this Station the same seed is tested in two or
more of these media and the average result taken, this being
further checked by the simultaneous germination of a "control"
seed of known germinating power. I have found this additional
precaution very useful in cases of dispute between buyer and
seller, and recommend it for general adoption.

(2) *Temperature.*—Whatever the nature of the bed may be, it
is essential that its temperature should be under control and
should remain constant. In most cases the temperature recommended is 20° C. Here our results have been more satisfactory
when the thermometer of the incubator was 24° C. Certain seeds
(Poa, Dactylis, Beta, etc.) are found to germinate better if they
are also exposed each day to a temperature of 30° C. for six hours,
in imitation of the diurnal rise of temperature in nature.

(3) *Moisture and Seed-bed.*—The process of dehydration
through which a seed passes in ripening has its counterpart
when germination is taking place. A dry seed will not germinate, no matter how favourable the other conditions may be.
If, on the other hand, a seed is left water-logged it will not
germinate, but in a few days die and rot away. The seed-beds
used in testing are of a porous nature, in contact by partial
immersion or otherwise with fresh water, so that the necessary
moisture reaches the seed, generally by capillarity. For most
seeds the amount of moisture so obtained is sufficient. This
holds true for the seed-bed, whether of porous clay, sand,
asbestos, or blotting-paper, and for most seeds.

(4) *Air.*—Fresh air is usually provided for in the incubators
by ventilators. Seeds are particularly sensitive to injurious
chemical substances, whether in the seed-bed or in the gas often
used in heating the incubator.

(5) *Light.*—Most seeds seem indifferent to sunlight, *i.e.* the seeds germinate equally well in darkness or light. It is a popular saying that a seed in nature germinates best when covered by a layer of soil equal in depth to the diameter of the seed. Exposure to direct sunlight brings with it difficulties in regulating the temperature of the incubator. In a few cases—*e.g.* the Poas—exposure to light has been shown to be distinctly beneficial, expediting the germination of the seed by one to three or four days, or increasing its percentage of germination.

Where the seed is of good average quality—*i.e.* such as a seedsman should sell and a farmer sow—the foregoing conditions can be kept so generally constant that the test will be quite reliable, and may be safely taken as an indication for sowing purposes of the quality of the seed.

Objections Raised against Seed-testing

(*a*) *Unreliability.*—Occasionally objectors to seed-testing quote cases—

> (1) In which several Stations have given widely different reports on the same seed.
>
> (2) In which the same seed tested by the same person under different methods has given different results.

The particulars as to the quality of the seed, the number of seeds tested, the use or not of a "control" seed, the temperature of the incubator, the degree of familiarity with the incubator, and the experience of the tester are all omitted.

The difficulty with the Poas is the usual case quoted. They are delicate seeds, but not beyond reliable testing, and, further, they form an infinitesimal part of the seed trade. A few years ago an Irish landowner sent independently to a seed firm, strongly opposed to seed-testing, and to me, samples of "tussock" grass seed. The fact that the firm identified this seed of a valuable fodder grass *Poa flabellata*, or *Dactylis cæspitosa*, as a worthless grass, *Aira cæspitosa*, cannot be quoted as proof of their general unreliability as seedsmen. The Committee was further told that only skilled hands could prepare a uniform mixture of seeds of grasses and clovers. The mixture made in the seed warehouse rearranges itself in the journey to the farm, and needs remixing before being sown. The purchase of mixtures is being, for obvious reasons, more and more discouraged by the farmers' advisers.

Again, some seedsmen object to a Seed Station on the ground that the Station might give to their rivals, perhaps by unfair use of the Station's report, a reputation which has taken them years to secure. The seedsmen in Germany were at first strongly opposed to the establishment of Stations. Now, however, they see the advantage of a properly conducted one, and in such an important centre as Hamburg they use it freely in their dealings with one another, and with their suppliers and customers.

One seedsman admitted before the Committee that the results of testing by a Continental expert and by his own expert uniformly came very near one another. He knew, he said, when he was buying worthless seed, and supposed the farmers in the West of Ireland went on buying the "blowings" of grass seeds because it paid them to sow them!

(b) *Pedigree.*—It is impossible by mere observation or by an ordinary germinating test to determine the pedigree of a seed. As the degree of fixation of the characteristics of an agricultural variety is often of an indeterminate character, the pedigree question is surrounded with natural difficulty, and often complicated by trade interests.

Confidence between buyer and seller, without which no trade could exist, must be largely the determining factor in accepting the statements as to the pedigree of a seed, especially where field experiments are not carried out. The originator of an agricultural or a trade variety would not be so foolish as to supply seed not in keeping with his description, and direct dealing with him should be guarantee enough, pending the field result.

Advocates of seed-testing have never contended that the pedigree of a seed can be tested in the incubator. One well-known firm brought before the Committee samples, in some cases intentionally mixed, of known seeds with very different pedigrees but so similar in appearance that no one, the firm asserted, could distinguish them from one another. Apart from the fact that there is a microscopical means of distinguishing rape, swede, cabbage from one another, it appears to have been overlooked that the common law could deal adequately with the cases. If a seedsman who submitted oats, germinating 78 per cent. and gathered in a rather wet autumn, to sulphur burning to preserve them was, to my knowledge, fined about £150, it is easy, if the law is reliable, to foretell the fate of a seedsman supplying, *e.g.*, mangel mixed with 25 per cent. wild beet, one of the samples submitted. A Seed Station has not, as its main function, the detection or prevention of fraud.

(*c*) *Time-Limit.*—Some firms state that quite half their business has to do with seed they never handle at all. Such speculative trade has its risks to run, and the speculating firm which no doubt receives a guarantee when buying the (foreign) supplies should be prepared to have the seeds it deals in submitted by the British farmer to test.

It has been argued by some that general seed-testing would paralyse the seed trade if customers waited for the results of the testing before sowing. Such delay is quite unnecessary. The same machinery for taking samples under the Fertilisers and Feeding Stuffs Act could be utilised in preventing the postponement of the sowing of seeds. It is only the samples taken from the bulk that are required for action under the Act.

It is true that seeds do not ripen equally well each year. Machinery exists for the removal of the unripe seeds, and there are seedsmen who guarantee year by year the same percentage of germination. The buyer should at any rate have the opportunity of knowing the germination percentage of the seed he is buying.

There could be a mutual agreement as to a unit of repayment or of compensation in cases where the test showed the seeds to be under or above the guaranteed standard. Such a mutual arrangement has been found workable. The common law could still deal with cases where there was evidence of deliberate fraud. It has come to my knowledge that in England, where certain landlords agree to pay half the cost of the seeds supplied to their tenants, these in some cases arrange for low-priced inferior seed to be actually supplied to them, while the landlord pays for superior seeds and so the whole bill. The tenant does not seem to realise that he more than pays for the difference later on, in poorer crops.

To meet the time difficulty attempts to ascertain the vitality or viability of a seed at once have been made by the use of chemical reagents. Further, Dr. Waller has shown in a very interesting manner that seeds which are alive give an electrical discharge which he calls a "blaze" current, when a current from an induction coil is sent through the seed. He has known only one case in which the seed failed to give the "blaze" current, and subsequently germinated. He has had no case of a seed giving a blaze current without subsequent germination. Further, by his method an idea of the degree of vitality of the seed is obtainable. I am hoping to be able, with Dr. Waller's help, to make practical use of his method in seed-testing.

Fungi on Seeds

One needs very little experience in seed-testing to realise to what an extent seeds harbour fungi. Many of these fungi are saprophytic, and their abundance is one sign that the seed is old and dead. Others are, however, parasitic. They exist as a hibernating mycelium in the substance of the seed and its coats or in the form of spores attached or clinging to the seed coats or concealed within them, and sprout under the conditions favourable to the germination of the healthy seeds. Several years ago I made an examination of mangel balls showing the pycnidia of *Phoma betæ*, a fungus which does an immense amount of harm in Ireland. As Appel has recently suggested in such cases as *Phoma betæ* and *Helminthosporium gramineum*, seed-testing might aid considerably in preventing unsuitable seed from being sown. In this connection it may be mentioned that Rolfs found the centrifugal apparatus helpful in detecting the spores of *Fusarium lini* on flax seeds.

Cost of Seed-testing

Apparatus.—The Station should have several incubators, a greenhouse if possible, a plot of ground, dissecting and compound microscopes, a reliable set of named seeds, certain books, at least two rooms—one for the staff, and one for the incubators. £100 would suffice to equip the Station with the necessary apparatus.

Staff.—The director should be a trained botanist, and should have at least one scientific assistant who would be responsible to the director for the accuracy of all the work.

The counting of seeds, washing of dishes, etc., take up a great deal of time, and for these purposes two or three smart boys or girls are necessary. In this Station, with 1,500 samples a year to be tested, the directorship is part of my duty as Professor of Botany in the Royal College of Science. One witness before the Seed Committee thought a central Station would need a Director receiving £1,000 a year, with the necessary staff and equipment.

Cost of Test.—The cost of a test varies very much according to the character of the seed and the extent of the test, being as a rule from 3s. upwards.

In this Station the seedsman pays 2s. per sample for reports on purity and germination; the farmer, however, only pays 3d. Our work is confined to the Irish seed trade, and all fees go to H.M.'s. Treasury.

It appears that no fewer than 850 samples are sent from the United Kingdom abroad each year to be tested, naturally at considerable expense and unavoidable delay. A well-conducted Station for Great Britain under Government control would be highly beneficial to British agriculture. There is evidence from time to time that the existence of this Station has vastly improved the quality, *e.g.,* of the flax seed sown in Ireland. A seedsman selling good seed has nothing to lose, but much to gain, by having his seeds inspected and certificated.

In the rubbish too often palmed off on the ignorant farmer, the Station proves when called in (and most of our testing is done for the farmers in Ireland) a necessary detective.

The Seed Committee makes what seems to me a fundamental error when it says that "the price at which these seeds (certain uncleaned seeds of inferior quality sold in Ireland) are sold not unfrequently corresponds fairly accurately to their value." The seed may be regarded as the farmer's raw material on which he expends his money, time, brain, land, men, and farm appliances. If the seed is impure and of low germination, he will have to spend not less, but rather more, of some of these in working up his raw material, and the resulting harvest will bear no comparison with that derivable from good seed. The disparity between the two results will bear no comparison with the difference in the original outlay in money on the two kinds of seeds. If the farmer knows by test that he is buying seed germinating only 75 per cent. instead of 95 per cent. he can increase the quantity sown, and so save loss in one direction. The Station would give him this information.

In the same way the cost to the State of a central Seed Station is trifling compared with the benefit to British agriculture its creation would mean.

THE CHEMISTRY OF INDIARUBBER

By SAMUEL S. PICKLES, M.Sc.

Scientific and Technical Department, Imperial Institute.

To-DAY the elastic substance indiarubber, or caoutchouc, as it it sometimes called, holds a position in the front rank of natural products. In a century it has advanced from being almost a botanical curiosity of little value, step by step at first, and latterly by leaps and bounds, until at the present time it must be classed amongst the most valuable of our raw materials of commerce.

On account of its peculiar nature and many characteristic properties, the position of indiarubber is unique. Its applications, at first few and limited, have gradually been extended, until now there is scarcely an industry in which it does not play some part. So wide and varied are its adaptations that it comes somewhat as a surprise to us if we recount the thousand-and-one articles, household and textile, in the construction of which it is employed—waterproof clothing, goloshes, rubber mats, erasing materials, teeth settings, hose pipes, vulcanite articles, elastic threads, scientific apparatus, electrical insulators, waterproof sheetings, and the tyres of vehicles are only a few of the numerous instances where rubber has found commercial application.

That much patient experimental inquiry, ingenuity, and technical skill have been concentrated on these applications is obvious, but the remarkable fact is that, in the manipulation of this substance, in its conversion into so many different products, the chemical reactions involved have never been clearly understood. It is only during the last few years that we have been able to arrive at even an approximately correct conception of the chemical nature of indiarubber itself.

Botanically, caoutchouc is one of the constituents of the milky juice or "latex" of certain trees and shrubs belonging chiefly to the natural orders *Euphorbiaceæ, Moraceæ, Artocarpaceæ* and *Apocynaceæ*. The caoutchouc is suspended in the latex in the form of minute transparent globules averaging about $\frac{1}{12500}$ in. in diameter, much in the same way as the oily particles are suspended in milk. On coagulation of the latex, which can be brought about by heating or by treating with certain chemicals, a clot or coagulum is formed, which consists chiefly of caoutchouc, always mixed, however, with varying quantities of resinous and albuminous impurities.

Chemically, caoutchouc is a hydrocarbon of the empirical formula C_5H_8. Its molecule is, however, much larger than such a formula would indicate, and in the absence of accurate data we are compelled to make use of the rather vague representation, $(C_5H_8)_n$ or $(C_{10}H_{16})_n$. It is an unsaturated hydrocarbon, containing one double bond or unsaturated linking for every complex C_5H_8. When pure, caoutchouc is an almost colourless elastic substance, transparent in thin sheets, and it has a specific gravity varying, in different samples, between ·91 and ·93·

It is insoluble in many of the usual solvents, but in chloroform, benzene, and several other organic liquids it dissolves in a rather remarkable manner. Up to a certain point it swells up and seems to absorb the solvent, then it goes gradually into solution ; but even moderately dilute solutions are quite gelatinous in character. Indiarubber thus behaves as a colloid, and it is principally this fact that makes its investigation a matter of such great difficulty. The known methods for ascertaining the molecular weights of compounds are useless in the case of caoutchouc. The difficulty of the problem is also increased by the fact that it is impossible to distil indiarubber, or, in fact, to heat it to much above 100° C., without effecting a decomposition of the molecule. Further, the substance is a hydrocarbon, and the absence of reactive radicles such as hydroxyl or carboxyl groups precludes the possibility of obtaining a number of derivatives which might otherwise have been useful for purposes of characterisation.

On heating caoutchouc to temperatures above 100° C. it begins to get viscous and sticky, and on cooling it does not again recover its elastic properties. If the heating is continued in a distilling flask, the whole mass becomes liquid, and eventually a very pungent-smelling distillate collects in the receiver. That the process is not one of simple distillation, but rather of decomposition, is apparent when the distillate is examined, for although the distillation does not commence until a temperature of about 160° C. has been attained in the flask, yet some portions of the distillate are found to have boiling points below 40° C. The indiarubber molecule has evidently suffered disruption, and, in the process, smaller molecules have been produced. Now, although it does not follow that in the decomposition products one is likely to find exactly the same arrangement of carbon atoms as in the parent substance, yet one might reasonably expect that an examination of the constitution of these products would yield some clue to the probable constitution of the indiarubber molecule. Therefore it is not surprising to find that, baffled probably by the complexity of the caoutchouc molecule itself, chemists have, in the past, chiefly turned their attention to a study of these products of destructive distillation. It would take too long to refer, in chronological order, to the whole of these investigations; for such an account the reader is referred to the author's report on this subject in the British Association Reports, York, 1906. It will be sufficient to consider the constitution of the compounds which have from time to time been isolated from the mixture which is obtained on distillation.

ISOPRENE, C_5H_8, BOILING POINT 37°—38° C., SPECIFIC GRAVITY ·682

Most of the earlier investigators observed the presence of a low-boiling portion in the distillate, but the separation of a single definite substance was not at first effected. W. H. Barnard[1] in 1833, Beale and Enderby[2] in 1834, John Dalton[3] (1834), Dr. Gregory[4] (1835 or '36), and Liebig[5] (1835) all obtained fractions boiling between 32° and 77° C. A. F. C. Himly[6] in 1835 described an oil b.p. 33°—44° C., and A. Bouchardat[7] (1837) was able to separate two fractions of the distillate, both of which were stated to boil below 20° C.

To Greville Williams[8] (1860), however, belongs the credit of

---

[1] Dr. Ure's *Dictionary of Arts*, etc., 1853, p. 358.
[2] *L'Institut*, 1834, p. 290.
[3] *London and Edin. Phil. Mag. and Journ. of Science*, 1836, p. 479.
[4] *Ibid.* p. 322.
[5] *Annalen der Chim. u. Pharm.* 1835.
[6] *Dissertation*, Gottingen, 1835.
[7] *Journ. de Pharm.* vol. xxiii. 1837.
[8] *Proc. Royal Society*, 1860, p. 517.

having isolated as a definite separate compound a volatile hydrocarbon, boiling, after careful rectification, between 37° and 38° C. To this he gave the name " isoprene." He observed that its vapour density was 2·44 times that of air, and gave its specific gravity as ·6823. Analysis showed it to have the formula C_5H_8. On standing in the air isoprene rapidly absorbed oxygen and became ozonised; the ozonised compound thus formed was found to be a strong oxidising agent. Williams also observed that isoprene readily absorbed bromine.

An interesting and important stage in the history of rubber chemistry was reached in 1879, when M. G. Bouchardat[1] took up the investigation of isoprene.

Polymerisation of Isoprene into Dipentene and into Caoutchouc

On heating isoprene in a sealed tube to a temperature of 280°—290° C. for ten hours, Bouchardat found that there resulted a viscous, sticky mass, which on distillation was shown to contain (1) some unchanged isoprene, (2) an inactive hydrocarbon, boiling at 170°—185° C. with an odour like lemons—a substance now known as dipentene—and (3) other high boiling hydrocarbons. The tendency to ready polymerisation by heat on the part of isoprene was thus demonstrated. The same investigator afterwards observed that polymerisation could also be effected by other means, viz. by the action of a cold saturated solution of hydrochloric acid upon isoprene. In this case, however, the polymerised product was found to be elastic, and in other respects possessed the properties of ordinary indiarubber.

These observations of Bouchardat were in 1882 confirmed by Prof. W. A. Tilden,[2] and similar results were obtained with isoprene derived from other sources, such as the decomposition of turpentine oil by heat. Tilden also showed that this polymerisation could be brought about by the action of other reagents than hydrochloric acid—for example, by nitrosyl chloride. A further important communication on this subject was made by Prof. Tilden in 1892[3] to the effect that on standing for a long time in bottles, isoprene is converted spontaneously into indiarubber. The following is an abstract from Prof. Tilden's

[1] *Comptes Rendus*, lxxix. p. 1117 (1879).
[2] *Chem. News*, 1882, pp. 120, 121 (vol. xlvi.).
[3] *Ibid.* 1892, p. 265.

paper read before the Philosophical Society of Birmingham in May 1892:

Specimens of isoprene were made from several terpenes in the course of my work on those compounds, and some of them I have preserved. I was surprised a few weeks ago at finding the contents of the bottles containing isoprene from turpentine entirely changed in appearance. In place of a limpid colourless liquid, the bottles contained several large masses of a solid of a yellowish colour. Upon examination this turned out to be india-rubber. The change of isoprene by spontaneous polymerisation has not, to my knowledge, been observed before. I can only account for it by the hypothesis that a small quantity of acetic or formic acid had been produced by the oxidising action of the air, and that the presence of this compound had been the means of transforming the rest. The liquid was acid to test-paper, and yielded a small portion of unchanged isoprene.

The artificial rubber, like natural rubber, appears to consist of two substances, one of which is more soluble in benzene or in carbon disulphide than the other. A solution of artificial rubber in benzene leaves on evaporation a residue which agrees in all characters with a similar preparation from Para rubber. The artificial rubber unites with sulphur in the same way as ordinary rubber, forming a tough elastic compound.

Previous to this time Wallach[1] had recorded that, after allowing isoprene to remain in a sealed tube for a long time in the light, and subsequently adding alcohol to the liquid, a tough indiarubber-like mass separated out, which, on standing in the air for some time resinified. C. O. Weber[2] in 1894 also confirmed Tilden's observation with regard to the spontaneous polymerisation of isoprene. After allowing 300 grammes of isoprene to stand in a bottle for nine months, he found that it was converted into a treacly viscous mass, which on treatment with methyl alcohol yielded 211 grammes of a spongy substance identical with indiarubber.

It thus appears that isoprene, besides being one of the decomposition products of caoutchouc, is also a possible, if not probable, source of caoutchouc in the plant organism.

Constitution of Isoprene

Greville Williams showed that isoprene contained carbon and hydrogen in the proportion C_5H_8, and that it readily combined with bromine, a fact which indicated its unsaturated character.

[1] *Ann. der Chem.* 1887 (vol. 238), p. 88.
[2] *Journ. Soc. Chem. Industry*, 1894, p. 11.

M. G. Bouchardat studied the action of the halogen acids upon isoprene, and found that two classes of compounds could be obtained by the addition of *one*, or *two*, molecules of the halogen acid, *e.g.* (1) $C_5H_8 \cdot HCl$ and (2) $C_5H_8 \cdot 2HCl$, thus indicating the presence of two unsaturated linkings or double bonds in the C_5H_8 molecule. Further proof of this was provided by Prof. Tilden[1] in 1882, who prepared the additive compound $C_5H_8Br_4$, isoprene tetrabromide. From a consideration of these derivatives as well as of the fact that on oxidation isoprene yields carbonic, formic and acetic acids, Tilden concluded that isoprene was β-methyl-crotonylene—

$$\begin{array}{c} CH_3 \\ \diagdown \\ \diagup \\ CH_2 \end{array} C - CH = CH_2,$$

and the subsequent researches of Gadziatzky, Mokiewsky, and Ipatieff supported this view. The synthesis of isoprene in 1897 by Euler[2] from β-methyl-pyrollidine settled beyond doubt the constitution of isoprene, and proved the correctness of the above formula. The question of how a compound possessing such a formula can pass in one case into dipentene ($C_{10}H_{16}$), and in another case into caoutchouc, will be considered at a later stage.

The low-boiling portion of the distillate obtained from rubber constitutes only a very small fraction of the total. At the first distillation, the thermometer shows a gradual rise with no very distinctive stationary points. On redistillation, however, it is found that after the low boiling portion, above referred to, has passed over, the temperature rises rapidly to about 140° C., and between 140° and 190° C. about one-third of the total distils over. The major portion of this is found to boil between 170° and 180° C.

DIPENTENE (CAOUTCHINE, DI-ISOPRENE, OR CINENE), $C_{10}H_{16}$, BOILING POINT (171·5°), (175°—176°), 180°—181°)

A. F. C. Himly, in the paper before mentioned (Göttingen, 1835), separated a fraction which, after distillation and purification, had a boiling point 171·5° C. This was a neutral oil with an orange-like odour, and of sp. gr. ·8423 at 16° C. It was a hydrocarbon, and analysis and vapour density determination showed it to have the formula $C_{10}H_{16}$. Himly gave to this liquid the name "caoutchine."

[1] *Chem. News*, 1882, p. 120 (vol. xlvi.).
[2] *Berichte*, 30 (1897), p. 1989-91.

Later, Greville Williams[1] isolated a similar fraction (b.p. 170°—180°) from the distillation products of both indiarubber and gutta-percha. On rectification this yielded a liquid b.p. 170°—173° C., identical with the caoutchine obtained by Himly, and which had a vapour density double that of isoprene. Williams observed that it belonged to a group of substances isomeric with turpentine oil, having the formula $C_{10}H_{16}$. Afterwards, in 1879, M. G. Bouchardat, by heating isoprene in a closed tube to 280° C. for ten hours, also obtained a hydrocarbon, of which he found the boiling point to be 174·6° C. This also had an odour of lemons, and its formula was $C_{10}H_{16}$. The name di-isoprene was given to this hydrocarbon, and Bouchardat suggested that it was identical with caoutchine and also with the inactive hydrocarbon of turpentine oil, because he was able to obtain the same derivatives, *e.g.* terpene hydrate, from all three. The same hydrocarbon was again obtained by Bouchardat when he distilled the elastic compound which he had obtained by the polymerisation of isoprene.

The identity of these hydrocarbons with one another, and with the terpene hydrocarbon cinene (now known as dipentene), was afterwards confirmed by Wallach.[2] It is thus seen that in some way or other dipentene is intimately connected with the rubber hydrocarbon. Its constitution, therefore, is a matter of considerable importance from this point of view. Fortunately, as in the case of isoprene, the constitution of dipentene has been thoroughly worked out.

Constitution of Dipentene

Dipentene belongs to that class of hydrocarbons known as terpenes. Most of the members of this class may be regarded as derivates of the hydrocarbon cymene, methyl-isopropyl-benzene—

$$CH_3 \cdot \langle \bigcirc \rangle \cdot CH {<}^{CH_3}_{CH_3}$$

Dipentene is one of the most commonly occurring terpenes; it is found free in nature, and may also be easily obtained from certain other natural products such as terpineol and pinene. Tilden[3] in 1879 showed that it was closely related to the alcohol,

[1] *Proc. Royal Soc.* 1860, p. 517.
[2] *Ann. der Chem.* 1884 (225), p. 311 ; 1885 (227), pp. 292-6.
[3] *Ber.* 1879 (12), p. 848.

terpineol, $C_{10}H_{18}O$, and later Wallach[1] showed that by spitting off the elements of water from this alcohol, dipentene was readily obtained. Wagner[2] first proposed the formula

$$CH_3 \cdot C \begin{array}{c} \diagup CH - CH_2 \diagdown \\ \diagdown CH_2 - CH_2 \diagup \end{array} CH - C \begin{array}{c} \diagup CH_3 \\ \diagdown CH_2 \end{array}$$

for dipentene as the result of an exhaustive consideration of its properties and of its relationship to terpineol, pinene, and sobrerol.

The proof of the correctness of this formula was supplied by Prof. W. H. Perkin, jun.,[3] in 1904, who, by accomplishing its complete synthesis from δ-keto-hexahydrobenzoic acid, removed all doubt as to its constitution.

The question now arises, How does isoprene pass into a compound having the above constitution? Ipatieff[4] suggested a probable transition by the condensation of two molecules of isoprene in the following manner :

2 molecules isoprene   →   1 molecule dipentene.

The relationship of dipentene to the rubber molecule will be considered later.

OTHER HYDROCARBONS OF THE FORMULA $C_{10}H_{16}$

From the indiarubber distillation products there have been also isolated several other hydrocarbons having the same empirical formula as dipentene, viz. $C_{10}H_{16}$. On carefully re-distilling the dipentene fraction—*i.e.* that portion boiling between

[1] *Annalen*, 1885 (230), p. 258.
[2] *Ber.* 27, p. 1636.
[3] *Journ. Chem. Soc.* 1904, p. 654.
[4] *Journ. f. Prakt. Chem.* 55 (1897).

150° and 200° C.—Prof. Harries[1] separated two other fractions, one of which boiled at 147°—150° C., sp. gr. ·8286, and contained no dipentene. For this hydrocarbon Harris revived the name di-isoprene and he considered it to have the formula—

$$\begin{array}{c} CH_3 \\ \diagdown \\ C - CH_2 - CH_2 - CH = C - CH = CH_2. \\ \diagup | \\ CH_2 CH_3 \end{array}$$

The other new fraction obtained had a boiling point 168°—169° C., sp. gr. ·8309, and this also contained no dipentene. On treatment with bromine a deep violet colour was produced. Harries concluded that this was a new terpene, and in a subsequent paper he advanced the view that instead of containing a six-carbon ring like dipentene, the hydrocarbon possibly contained a closed ring consisting of eight carbon atoms.

HEVÉÈNE

A large portion of the distillate boils above 200° C., but only one definite fraction in this region seems to have been separated and investigated. Alexander Bouchardat[2] in 1837 isolated a neutral yellow oil boiling at 252° C., to which he gave the name hevéène. Its specific gravity was ·921 and it combined directly with the halogens, but the combination was generally accompanied with evolution of the corresponding halogen acid. The constitution of this hydrocarbon has never been worked out, and very little is known regarding it. It is described in some places as a sesquiterpene of the formula $C_{15}H_{24}$, and in others as a diterpene—$C_{20}H_{32}$. The analytical results obtained by Bouchardat did not quite agree either amongst themselves or with either of the above formulæ. The boiling point would appear to indicate a sesquiterpene.

Four years ago Fischer and Harries[3] made experiments on the distillation of indiarubber in a vacuum. They found that when the distillation is carried out under a pressure of ·25 mm. only a very a small quantity of isoprene or dipentene is produced, the distillate consisting chiefly of high-boiling hydrocarbons.

We must thus assume that the low-boiling products are

[1] *Ber.* 35 (1902), pp. 3260, 3266.
[2] *Journ. de Pharm.* vol. xxiii. (1837).
[3] *Ber.* 35 (1902), pp. 2162-3.

the results of disruption caused by the somewhat drastic conditions to which the rubber molecules are subjected in the process of distillation at atmospheric pressure.

From a comparison of the above distillation products we can also form the following conclusions:

(1) That the caoutchouc is in some way connected with the terpenes, and any constitutional formula advanced should be capable of explaining the transition of rubber into such compounds as isoprene and dipentene.

(2) That in the indiarubber molecule we have at least five carbon atoms arranged in the order—

$$CH_3 - C - C - C$$
$$|$$
$$C$$

since this arrangement is found in all the known distillation products.

Derivatives of Caoutchouc

Probably the first chemical derivatives of indiarubber obtained were the vulcanisation compounds. Goodyear[1] in the year 1839 seems to have been the first to apply the process of heating rubber with sulphur. He found that after such treatment the indiarubber lost its stickiness and maintained its elasticity between wider ranges of temperature. For many years, however, the chemistry of vulcanisation was but imperfectly understood; the sulphur was generally said to be "absorbed," whilst some workers regarded the process as one of substitution, the sulphur replacing the hydrogen atoms of the indiarubber "resin."[2] C. O. Weber[3] pointed out that this hypothesis was improbable as only a comparatively small quantity of sulphuretted hydrogen is liberated during the vulcanising process. He further demonstrated that the process is much more probably one of addition than of substitution, the sulphur attaching itself at the double bonds or ethylenic linkings in the molecule.

Using sulphur chloride as a vulcanising agent, he prepared an interesting series of compounds, the composition of which supports the "additive" theory. Thus a completely vulcanised

---

[1] *Chemistry of Indiarubber* (Weber), 1902, p. 41.

[2] Thorpe's *Dictionary of Applied Chemistry*, vol. ii. p. 312.

[3] *Chemistry of Indiarubber* (Weber), 1902, p. 47; *Journ. Soc. Chem. Ind.* 1894 p. 11.

product was obtained, the analysis of which agreed with the formula $C_{10}H_{16}S_2Cl_2$. Other compounds were also prepared containing a much lower percentage of sulphur, and these seem to throw some light on the question of the size of the rubber molecule. For example a homogeneous product was obtained containing only 4·8 per cent. of sulphur, and to explain its constitution it is necessary to assume that the caoutchouc molecule is at least $C_{50}H_{80}$. Weber eventually concluded that the formula $C_{60}H_{96}$ was within the range of possibility, and this would yield a vulcanisation compound of the type $(C_{60}H_{96} \, SCl)$. A large number of sulphur compounds are possibly formed between these two limits, containing varying proportions of sulphur, and Weber regarded these as corresponding to the different degrees of vulcanisation which can be obtained with rubber and sulphur or sulphur chloride.

The formula $C_{10}H_{16}S_2Cl_2$ for the fully vulcanised compound, is in accordance with the view that the indiarubber molecule contains one double bond for every C_5H_8 complex. Additional evidence is provided by a consideration of the brom-derivative of caoutchouc, especially if, in the preparation, care has been taken to prevent substitution. Thus if bromine is added to a cooled solution of indiarubber in chloroform until the brown colour of the bromine just remains, and the solution is then poured into alcohol, a parchment or rag-like substance is precipitated which on analysis is found to have the formula $(C_5H_8Br_2)_n$. It was first prepared by Gladstone and Hibbert[1] in 1888 and afterwards examined by Weber,[2] who also prepared and examined the hydrochloride $(C_5H_8 \cdot HCl)_n$.

From a consideration of the physical properties of caoutchouc, Gladstone and Hibbert, on the other hand, came to the conclusion that for every $C_{10}H_{16}$ there were three double bonds. Some halogen derivatives were also prepared by them which also seemed to support this conclusion, but the great bulk of evidence is in opposition to this view. At the same time the same two workers endeavoured to arrive at the molecular weight of caoutchouc by the application of the Raoult method, but the depression of freezing point observed was so small, that if the method holds good in this case, the molecule must be represented as at least $C_{1000}H_{1600}$.

[1] *Journ. Chem. Soc.* 1888, p. 680.
[2] *Chemistry of Indiarubber* (1902), p. 31.

The action of strong nitric acid upon caoutchouc has from time to time been observed, and various results have been recorded. Recently R. Ditmar[1] has gone into the question systematically, and has been fortunate in obtaining a definite compound by this treatment. The compound produced was shown to have the formula $C_{10}H_{12}N_2O_6$, and to be a dinitro-mono-carboxylic acid, probably dinitrohydrocumic acid.

The treatment with nitric acid is, however, somewhat severe, and it is possible that the formation of a six-carbon-ringed body of this nature may be due to a secondary reaction.

Action of Gases on Caoutchouc

(a) *Nitrous Fumes.*—On passing nitrous fumes, obtained by the action of nitric acid upon arsenious oxide, through light petroleum in which rubber was suspended, Harries[2] obtained a colloidal mass which rapidly changed into a yellow amorphous substance. It was found, by analysis and molecular weight determination, to have the formula $C_{40}H_{62}O_{24}N_{10}$.

By varying the conditions other nitrosites were also obtained,[3] one of which (nitrosite " C ") is a yellow compound having the formula $C_{20}H_{30}N_6O_{14}$. An interesting point about this nitrosite is that it can also be obtained from the product of polymerisation of the hydrocarbon myrcene, which is found occurring naturally in Bay oil. As myrcene is an open-chained hydrocarbon, Harries, at first, considered this fact as favourable to the idea that rubber was also an open-chained unsaturated hydrocarbon.

(b) *Nitrogen Dioxide.*—Weber[4] found that on passing dry nitrogen dioxide gas, prepared by heating lead nitrate, through a solution of Para rubber in benzene, a coherent mass separated out, which, after washing and purification, was obtained as a straw-coloured powder. This, on analysis, gave figures corresponding to the formula $C_{10}H_{16}N_2O_4$. Other investigators[5] have, however, been unable to isolate a compound of this composition. The substances generally obtained more nearly agree with the formula $C_{10}H_{15}N_3O_7$ or $C_{20}H_{30}N_6O_{14}$, the same as that of Harries' nitrosite " C."

[1] *Ber.* (35), 1902, pp. 1401-2 ; *Monatshefte* (25), p. 464.
[2] *Ber.* (34), 1901, p. 2991.
[3] *Ber.* (35), p. 3256.
[4] *Ber.* (35), p. 1947.
[5] Harries, *Ber.* (38), pp. 87-90 ; Alexander, *Ber.* (38), pp. 181-4.

(c) *Ozone.*—The study of the action of ozone on solutions of caoutchouc by Prof. Harries of Kiel[1] has been accompanied by most fruitful results, so much so that our direct knowledge of the constitution of indiarubber may be said to date from Prof. Harries' first paper on this subject in 1904.

When Para rubber was dissolved in chloroform, and a stream of ozonised oxygen passed through the solution for several hours, caoutchouc ozonide was formed. On distilling off the solvent, a colourless syrup was obtained which, on drying in a vacuum, solidified to a hard glass, and this on analysis proved to have the formula $C_{10}H_{16}O_6$. Two molecules of ozone had evidently attached themselves to the $C_{10}H_{16}$ group. On decomposition of the above ozonide with water, hydrogen peroxide was formed, and the liquid remaining had the properties of a ketone or aldehyde. When, however, the mixture with water was heated for a long time, complete solution took place, the hydrogen dioxide disappeared, and levulinic acid was afterwards found to be present in the aqueous solution. The levulinic acid was obviously the oxidation product of levulinic aldehyde—

$$CH_3 \cdot CO \cdot CH_2 \cdot CH_2 \cdot CHO + H_2O_2 \rightarrow CH_3 \cdot CO \cdot CH_2 \cdot CH_2 \cdot COOH + H_2O$$
　　LeVulinic aldehyde.　　　　　　　　　　　　　LeVulinic acid.

and as in the process of oxidation of unsaturated hydrocarbons the oxygen is always found at the point where a double bond has previously been, Harries first proposed the following arrangement of carbon atoms in the indiarubber molecule—

$$\begin{array}{c} C \\ \diagdown \\ \diagup \\ C \end{array} C = CH - CH_2 - CH_2 - C = CH - C$$
$$CH_3$$

This would give on oxidation—

$$OCH - CH_2 - CH_2 - CO - CH_3$$

i.e. levulinic aldehyde.

In the aqueous solution, besides levulinic acid, there was also found a crystalline substance, m.p. 195° C., which Harries discovered to be another derivative of levulinic aldehyde, namely, levulinic aldehyde peroxide.

$$CH_3 - C - CH_2 - CH_2 - CH$$
$$O = O = O = O$$

This is formed along with levulinic aldehyde when the

[1] *Ber.* (36), 1903, p. 1923; (37), 1904, pp. 839, 2708; (38), 1905, p. 1195.

ozonide $C_{10}H_{16}O_6$ is decomposed with water. This decomposition can be expressed—

$$C_{10}H_{16}O_6 + H_2O = CH_3 - CO - CH_2 - CH_2 - CHO + CH_3 - C - CH_2 - CH_2 - CH$$

Ozonide	Levulinic aldehyde	O = O = O = O
		Levulinic aldehyde peroxide.

Ozonide   =   *Levulinic aldehyde*   +   *Levulinic aldehyde peroxide.*

On boiling with water the peroxide decomposes into levulinic aldehyde and hydrogen peroxide, H_2O_2. These were the only products formed by the decomposition of the ozonide, and, indirectly, of the caoutchouc molecule. The arrangement of carbon atoms and double bonds is identical in both the aldehyde and the aldehyde peroxide, viz.

$$CH_3 - C - CH_2 - CH_2 - CH$$

so it is necessary to have a formula for caoutchouc which contains multiples of this complex. Moreover, it must contain a closed, not an open, carbon chain, and, as the molecular weight of the ozonide was found to agree with the formula $C_{10}H_{16}O_6$, Harries concluded that the closed chain consisted of two of the above groups, forming an eight-carbon ring, thus:

$$CH_3 - C - CH_2 - CH_2 - CH$$
$$CH - CH_2 - CH_2 - C - CH_3$$

1 : 5 Dimethyl-cyclo-octadiene Δ 1 · 5

Accepting this formula for the chemical molecule of caoutchouc, the ozonide would be represented—

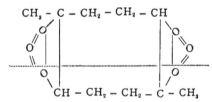

Splitting up as indicated by the dotted line, this would give one molecule of levulinic aldehyde peroxide, and one molecule of levulinic aldehyde—

$$CH_3 - C - CH_2 - CH_2 - CH$$
1 molecule
$$O = O = O = O$$
1 molecule $CHO - CH_2 - CH_2 - CO - CH_3$ +

the scheme thus providing a reasonable explanation of the observed facts.

33

It will be noticed that the above formula is spoken of as representing the "chemical" molecule of caoutchouc. The real molecule of Para rubber, or, as Harries terms it, the "physical" molecule, must still be expressed $(C_{10}H_{16})_x$, and be regarded as a sort of polymer of the chemical molecule. The polymerisation must, however, take place through very simple or loose additions, as the polymer is easily resolvable under the influence of such agents as ozone into single $C_{10}H_{16}$ "chemical" molecules.

Harries also indicates how a molecule, represented by the 1 : 5 dimethyl-cyclo-octadiene formula, might easily break down into isoprene, or become converted into dipentene or di-isoprene, the known products of the distillation of rubber:

(1)
$$CH_3 - C - CH_2 - CH_2 - CH \qquad CH_3 - C - CH_2 - CH_2 - CH$$
$$CH - CH_2 - CH \cdot H - C - CH_3 \qquad CH_2$$
$$\rightarrow$$
$$CH_2 = CH - C - CH_3$$
Di-isoprene.

(2)
$$CH_3 - C - CH_2 - CH_2 - CH \qquad CH_3$$
$$CH - CH_2 - CH_2 - C - CH_3 \quad CH_2 \rangle C - CH = CH_2$$
$$\rightarrow$$
Isoprene (2 molecules).

$$\downarrow$$

$$CH_3 \qquad CH_2 - CH$$
$$CH_2 \rangle C - CH \qquad \langle C - CH_3$$
$$CH_2 - CH_2$$
Dipentene.

This formula for the chemical molecule of Para rubber is also in agreement with the observed facts (1) that for every complex $C_{10}H_{16}$ there are two unsaturated linkings; (2) caoutchouc is an optically inactive hydrocarbon. It will be seen that the formula proposed does not contain an asymmetric carbon atom.

It might reasonably be expected that some of the depolymerised rubber molecules having the above structure would be found in the distillation products, and Harries considers it probable that such molecules are contained in the fraction boiling between 160° and 170° C., especially as this fraction gives an ozonide similar to the caoutchouc ozonide.

Referring to the formation of rubber in the plant, isoprene has been suggested as a possible starting point in the process of polymerisation. Prof. Harries considers that this is not the case, but that another hydrocarbon, pentadiene,

$$CH_3 - C - CH_2 - CH_2 - CH$$

is produced in the plant, probably by reduction of the sugars (principally pentoses), and that, in the nascent state, the above complexes condense, forming ultimately the rubber molecule $(C_{10}H_{16})_x$. In support of this hypothesis he instances the formation of the furans from sugar (Fischer and Laycock), and the fact that a-methyl-furan is obtained from beechwood tar, probably by decomposition of the celluloses (Harries).

a-Methyl-furan can easily be made to yield levulinic aldehyde, and this, by reduction, may give the above pentadiene.

For the present we are compelled to leave here the chemistry of indiarubber. It is probable that the last word regarding the constitution is not yet spoken, for there are several points which are not satisfactorily explained by Harries' formula. The fact that although several octacarbocyclic compounds are known, none have been found to polymerise into elastic substances ; the observation that, on distillation in a vacuum, only high-boiling hydrocarbons are formed, most of which, judging from their boiling points, have formulæ at least $C_{20}H_{32}$; and the unsatisfactory and indefinite assumption of the mode of polymerisation, render further investigation necessary.

MARCEL NENCKI,[1] 1847—1901

By S. B. SCHRYVER, D.Sc., Ph.D.,

Lecturer on Physiological Chemistry to University College, London.

By the death, in 1901, of Marcel Nencki, at the early age of fifty-four, and in the full height of his scientific activity, physiological chemistry was deprived of one of its most ardent workers and successful pioneers. The appearance of his collected works at the end of 1904 affords us an opportunity of appraising his contributions, and determining his influence on the development of the branch of science to which he devoted his activities.

The subject of this memoir was the son of a Polish landed proprietor, and was born in 1847, in Kalisz, in Poland. He received his school education in his native land, but removed, in consequence of political disturbances, to Crakow, in Austrian Poland, in 1864. He shortly afterwards removed to Jena, where he applied himself to the study of classical philology and philosophy. He soon, however, manifested his bent towards experimental science, and in 1867 he commenced the study of medicine at Berlin University.

The speculations of Liebig on the chemical processes in living organisms were at that time exercising considerable influence on scientific thought, and Nencki, at a very early stage of his career, decided to devote himself to the study of biological chemistry. He had the foresight to perceive the importance of a mastery of chemical technique, and for this purpose he enrolled himself as a student in the chemical laboratory of the Berlin "Gewerbeakademie," then under the direction of Baeyer. He found here as fellow-students Graebe and Liebermann, and the work in this laboratory at the time cannot have failed to exert a stimulating action on any student with a strong inclination to scientific work. He graduated in 1870, at Berlin, and shortly afterwards (1871) was appointed

[1] *Marceli Nencki opera omnia*, Braunschweig: F. Vieweg und Sohn, 1904.

assistant in the Pathological Institute in Bern. Here he was destined to remain for twenty years, and it was here that the most considerable part of his life's work was carried out. From such colleagues as Naunyn, Langhans, and Kocher he received every encouragement and stimulation, and he rapidly obtained the academic promotion he deserved. In 1877 a chair of physiological chemistry was created for him. In 1891 he was offered the direction of a laboratory at the newly created Oldenburg Institute of experimental medicine at St. Petersburg, and, attracted by the increased facilities in a splendidly equipped laboratory, he reluctantly agreed to leave Bern. Here he remained till his premature death from cancer in 1901. His association with Pawlow at St. Petersburg was perhaps the most memorable incident of this later stage of his career.

To summarise Nencki's work is no easy task, for his versatility, his skill as a chemist and his biological instinct, attracted him to many fields of research, and his collected works fill two large volumes, each containing more than eight hundred pages. They comprise works on pure organic chemistry, on physiological chemistry and pharmacology, and on bacteriology, the latter subject treated from both the chemical and morphological standpoints. It will be most convenient to consider first his researches on pure chemistry, and the morphological aspects of bacteriology, and to reserve the consideration of his physiological chemical work to the last—for it is to his contributions to this branch of science that he owes his principal title to fame.

Organic Chemistry, etc.

His first series of researches on organic chemistry were devoted to the preparation of various new derivatives of Thiurea, with the object of obtaining synthetically uric acid. In this object he was not successful, although he obtained many new derivatives of interest. He also prepared a large series of new derivatives from the sulphocyanides. Closely allied with urea and uric acid derivatives are those of guanidine, and from this latter body, by condensation with fatty acids, Nencki produced a series of crystalline products known as the guanamines, of which the crystals have well-defined forms. These guanamines Nencki often prepared subsequently for the purpose of identifying the various fatty acids obtained by him in the study of the putrefaction products from proteins. Allied to this subject were his investigations of the various guanidine condensation products, such as the melamines.

For the purpose of his pharmacological investigations he had occasion to synthesise various ketone derivatives of phenol, and he showed that these bodies could be obtained by the condensation of fatty acids with phenols in the presence of zinc chloride. Various substances of importance for the manufacture of aniline dyes were thereby obtained.

In connection with this work Nencki applied the use of ferric chloride as a condensing reagent in place of the previously used aluminium chloride; by the action of acidyl chlorides on various aromatic derivatives in the presence of ferric chloride he succeeded in obtaining a series of ketone derivatives.

Although Nencki published a large number of papers on organic chemistry, the latter form but a subordinate portion of his work. A few of the more important results are indicated in the above lines.

The morphological studies on bacteria were devoted to the investigation of the bacterial flora of the digestive tract, and also to the various organisms connected with infectious diseases. His official position in Switzerland led him to investigate the various bacteria connected with diseases of the udder of goats ; from St. Petersburg he was sent by his Government to the Caucasus to investigate Rinderpest, and he undertook various researches with the object of preparing an immunising serum. He was also officially entrusted with investigations in connection with cholera and diphtheria. Closely allied with these researches were his investigations on the efficacy of antiseptics, to which reference will be made again later.

Physiological Chemistry.

By far the most important part of Nencki's work was devoted to the chemical elucidations of biological problems, and it is proposed to consider his influence on this branch of science in some detail.

As already mentioned, the influence of Liebig when he commenced his scientific career was paramount in this branch of science, which up till then had attracted but few workers.

At that time there was no journal devoted exclusively to

this subject. Occasional papers appeared in the *Berichte* of the German Chemical Society, which had only been established shortly before, and in Voit and Pettenkoffer's *Zeitschrift der Biologie* (1865). In the early seventies appeared Pflüger's *Archiv*, and also Schmiedeberg's *Archiv*, both of which contained their quota of biochemical work. In 1877 Hoppe-Seyler's *Zeitschrift für Physiologische Chemie* made its first appearance. This remained the sole journal devoted exclusively to physiological chemistry till the appearance, in 1901, of Hofmeister's *Beiträge*. This year (1906), however, no less than three new journals have made their appearance—viz., the German *biochemische Zeitschrift*, edited by Neuberg, and two journals in the English language, the American *Journal of Biochemistry* and the English *Biochemical Journal*. If scientific progress can be measured by the increase of bulk in literature, we have made astonishing strides in the study of biological chemistry in the last few years.

It remains for us now to determine Nencki's own part in this development.

His researches in this branch may be conveniently divided into the following series :—

I. Those connected with the oxidation or destruction of various substances when introduced into the animal body.

II. Those connected with the chemical dynamics of putrefaction and fermentation.

III. Those connected with the study of the formation of urea in the animal body.

IV. Those relating to the chemistry of animal pigments.

I. Destruction and Oxidation in the Animal Body.

For a thesis on the oxidation of aromatic derivatives in the animal body Nencki received his doctor's diploma. His investigations on this subject may appropriately, therefore, be considered first.

The most important results in this series refer to the behaviour of benzene. The isolation of phenylsulphuric acid in the urine after injection of benzene was due to Baumann. It remained for Nencki to follow out the process of the oxidation of benzene to phenol quantitatively, and to devise a process for the study of physiological oxidation. In this work he was joined, amongst others, by Madame Sieber, with whose collaboration he afterwards produced so many valuable works. Nencki and Sieber showed that when benzene was administered to an animal the proportion of phenol that could be detected in the urine after hydrolysis with acid to destroy the phenyl sulphuric acid, was a constant for any particular animal. This constant was, furthermore, independent of the nutrition of the animal; the ratio of the phenol to the benzol administered was constant for a given animal, whether that animal were well fed or in a fasting condition. Certain toxic bodies are known, however, to diminish the oxidative processes of the body; amongst such poisons may be mentioned arsenic, phosphorus, antimony, etc. Nencki and Sieber showed that if benzene be ingested after administration of such poisons, the proportion of phenol excreted was much diminished. They made use of this fact for devising a process for measuring physiological oxidation in the body in health and disease, and applied it in the investigation of oxidation in cases of leukæmia and of diabetes mellitus. In the former case they found a general depression of oxidation; in the latter case, however, they found that oxidation was normal, and they concluded therefore that the excretion of sugar must be due to other causes than the depression of oxidative functions.

Amongst other substances of which the behaviour in the animal body was investigated may be mentioned the various aromatic hydrocarbons with side chains, the various phenols, acetophenones and allied ketones, and various esters.

The substituted benzene derivatives were found to undergo oxidation when administered to an animal. Toluene was oxidised to benzoic acid and excreted chiefly in the form of hippuric acid, the conjugated glycine derivatives. In hydrocarbons containing more than one side chain, it was found that one and one only of the side chains was oxidised, and substituted benzoic acids were excreted in a conjugated form, either as glycuronates or as substituted hippurates. Phenols were oxidised to bodies containing a larger number of phenolic hydroxyl groups, whereas benzoic acids and substituted benzoic acids were not further oxidised, but excreted either as glycuronates, or in some other conjugated form. Naphthalene was oxidised to naphthol. From these results Nencki makes important pharmacological generalisations. He shows that those bodies which undergo oxidisation, and thereby deprive the tissues of oxygen, are generally strongly toxic, whereas bodies which do not undergo oxidation, such as benzoic acid, are comparatively harmless.

Another important investigation of Nencki concerns the fate of the various esters in the alimentary tract. He showed that the pancreatic juice was capable of hydrolysing various aromatic esters, amongst others phenyl salicylate. This body in the small intestine undergoes scission into phenol and salicylic acid, two powerful antiseptics which become available only in the intestine. Through this discovery practical pharmacy was enriched by the addition of an important drug, phenyl salicylate, which has found a large application under its commercial name of salol.

When we summarise these researches on the fate of various bodies after ingestion, we find that pharmacology is indebted to Nencki for various important discoveries, and he must be ranked along with Schmiedeberg, Baumann, Jaffé, and others amongst those pioneers who applied accurate chemical work to pharmacological study.

II. RESEARCHES ON PUTREFACTION, ETC.

Nencki's work on the chemical action of micro-organisms must rank as his most important, not only on account of the biological facts discovered, but on account of the influence they have exerted on the general development of protein chemistry. Not only did Nencki investigate the conditions under which micro-organisms live and multiply; he isolated the degradation products of the various media on which they grow, and thereby determined their action both on proteins and carbohydrates; he investigated the influence of oxygen on this action, and improved the technique of the study of anærobiosis, and finally he attempted to imitate their chemical action by purely chemical methods.

It is of interest to study the evolution of his work on this subject. The first stimulus was given by the publication of the researches of Schunck and Jaffé. The former had already shown that indican, which is contained in normal urine, yields on hydrolysis with acids, indigo blue. The latter showed that the injection of indol, which had been recently discovered by Baeyer, caused a considerable increase in the amount of indican excreted.   These results stimulated Nencki to a further investigation of the subject, with the object of determining the origin of the indican excreted in the urine.   He had noticed the statement in some published work of Corvisart and of Kühne, that an odour resembling naphthylamine (or indol?) was produced, when gelatine was digested with pancreas.   What simpler hypothesis, then, than that indican was produced from indol, which itself was a product of the pancreatic digestion of proteins?

These few observations serve, then, as the starting-point of some of the most important of Nencki's works.   It was very soon noticed, however, that the peculiar and disagreeable odours were produced only when the digestion took place in the absence of antiseptics, and it was soon recognised that the production of bodies causing these odours was due to the presence of micro-organisms.   Nencki and his pupils then proceeded to isolate the various products that are produced when proteins of different origin are allowed to putrefy.   The earlier investigations were generally made by allowing pancreas to act on proteins in the absence of antiseptics.   During the course of these researches, others, such as Pasteur and Koch, were employed in the elaboration of the methods for obtaining pure cultures.   Nencki's later works on putrefaction were undertaken, therefore, with pure cultures, and the products of putrefaction under certain specified conditions were accurately studied.   His first important discovery consisted in the actual isolation of indol from amongst the products of protein degradation.   By allowing various proteins to putrefy in the presence of pancreas and then distilling the products, he found, in addition to fatty acids, a body, the properties of which had already been described by Bayer, and which he isolated in the form of the nitroso-derivatives, and also in the free state.   Nencki by this means definitely isolated indol as a product of protein degradation, although it was not recognised at first that it was not an ordinary pancreatic digestion product, but due to the intervention of bacteria. These results stimulated Nencki to the investigation of other products obtained under the same conditions.   Elaborate researches were made on the gases evolved, the fatty acids produced, and the crystalline degradation products.   It was soon recognised that the earliest stage in the protein degradation was a hydrolysis, and Nencki was one of the first to note that the chemical differences in the proteins were due to the fact that they are built up of different groups, for the amount of hydrolysis products, such as glycocoll, leucine, and tyrosine, vary greatly in proteins of different origin. These hydrolysis products, then, in the presence of bacteria undergo further degradation; scission of ammonia takes place, and various volatile fatty acids are formed, such as acetic, butyric, and valerianic acids. By the direct action of putrefactive bacteria on the hydrolysis products, some of the lower degradation products can be obtained. Thus, by treating leucine with putrefying ox-pancreas, products were isolated the formation of which could be accounted for by the following equations :—

$$C_6H_{13}NO_2 + O_2 = C_5H_9O_2NH_4 + CO_2,$$

ammonium valerianate being thus formed, the acid of which on further oxidation,

$$C_5H_{10}O_2 + O_2 = C_6H_8O_2 + CO_2 + H_2O,$$

yields butyric acid.

The discovery of the difference in the chemical action of the secreted enzymes and that of bacteria (at that time known as unorganised and organised ferments) induced Nencki to suspect that the real seat of formation of indol was in the large intestine, and that its formation was due, not to the ordinary secreted ferments such as pancreatin, acting on the proteins, but rather to the action of the micro-organisms which are always found in this organ. It seemed therefore probable that the products of putrefaction would be found in the fæces, and a chemical examination of these excreta was therefore undertaken. This work was entrusted to Brieger, who succeeded in isolating for the first time the important putrefaction product skatol. The empirical chemical relationship of this body to indol was soon recognised.

In the course of time various other products of putrefaction were isolated, amongst which may be mentioned phenylethyl-amine, skatolacetic acid, and mercaptan.

The question then arose—To what aromatic hydrolysis products of proteins are these various putrefaction products due?—for indol, skatol, etc., cannot be obtained by direct hydrolysis—they are the products of the action of bacteria on such bodies.

Salkowski, who with his pupils had also been working

on isolation of products of putrefaction, had succeeded in the meantime in obtaining other bodies, such as phenylpropionic acid, phenylacetic acid, parahydroxyphenylpropionic acid, the corresponding acetic acid, etc.; and Nencki concluded that these various bodies must be derived from three different aromatic bodies obtained by the hydrolysis of proteins. These three parent substances must be phenylaminopropionic acid (phenyl alanine), which ¦had been obtained by Schulze as a protein hydrolysis product, tyrosine, and another body at that time unknown, and having the empirical formula of a skatolaminoacetic acid. He represented the formation of these various putrefaction products from their present substances by the following series of equations :—

Decomposition products of phenyl aniline.

$$C_6H_5 \cdot CH_2 \cdot CH \cdot (NH_2)CO_2H + H = C_6H_5 \cdot CH_2 \cdot CH_2 \cdot CO_2H + NH_3$$
Phenyl propionic acid

$$C_6H_5 \cdot CH_2 \cdot CH_2 \cdot CO_2H + O_3 \quad = C_6H_5 \cdot CH_2 \cdot CO_2H + CO_2 + H_2O$$
Phenyl acetic acid

$$C_6H_5 \cdot CH_2 \cdot CO_2H + O_3 \quad = C_6H_5CO_2H + CO_2 + H_2O$$
Benzoic acid

and also

$$C_6H_5 \cdot CH_2 \cdot CH(NH_2) \cdot CO_2H \quad = C_6H_5 \cdot CH_2 \cdot CH_2 \cdot NH_2 + CO_2.$$

Decomposition products of tyrosine.

$$C_6H_4{\Big\langle}{}^{OH}_{CH_2 \cdot CH(NH_2) \cdot CO_2H + H_2} = C_6H_4{\Big\langle}{}^{OH}_{CH_2 \cdot CH_2 - CO_2H + NH_3}$$
Parahydroxyphenyl propionic acid.

$$C_6H_2{\Big\langle}{}^{OH}_{CH_2 \cdot CH_2 \cdot CO_2H + O_3} = C_6H_4{\Big\langle}{}^{OH}_{CH_2 \cdot COOH} + CO_2 + H_2O$$
Parahydroxyphenyl acetic acid.

$$C_6H_4{\Big\langle}{}^{OH}_{CH_2 \cdot COOH} = C_6H_4{\Big\langle}{}^{OH}_{CH_3} + CO_2$$
Paracresol.

$$C_6H_4{\Big\langle}{}^{OH}_{CH_3} + O_3 = C_6H_4{\Big\langle}{}^{OH}_{CO_2H} + CO_2 + H_2O$$
Parahydroxybenzoic acid.

$$C_6H_4{\Big\langle}{}^{OH}_{CO_2H} = C_6H_5 \cdot OH + CO_2.$$
Phenol.

From skatolaminoacetic acid.

$$C_6H_4\left\langle{{C(CH_3)}\atop{NH}}\right\rangle C \cdot CH(NH_2) \cdot CO_2H + H_2 = C_6H_4\left\langle{{C(CH_3)}\atop{NH}}\right\rangle C \cdot CH_2 \cdot CO_2H + NH_3$$

Skatol acetic acid.

$$C_6H_4\left\langle{{C(CH_3)}\atop{NH}}\right\rangle C \cdot CH_2 \cdot CO_2H + O_3 = C_6H_4\left\langle{{C(CH_3)}\atop{NH}}\right\rangle C \cdot CO_2H + CO_2 + H_2O$$

$$C_6H_4\left\langle{{C(CH_3)}\atop{NH}}\right\rangle C \cdot CO_2H = C_6H_4\left\langle{{C(CH_3)}\atop{NH}}\right\rangle CH + CO_2$$

Skatol.

$$C_6H_4\left\langle{{C(CH_3)}\atop{NH}}\right\rangle CH + O_3 = C_6H_4\left\langle{{CH}\atop{NH}}\right\rangle CH + CO_2 + H_2O.$$

Indol

At the time this scheme of reactions was published (1889) all the above degradation products (or bodies with the same empirical formulæ) had been obtained by either Salkowski or Nencki and their pupils, with the exception of the skatolamino-acetic acid. A body of this empirical formula was finally isolated by Hopkins and Coles in 1900, who thereby proved the correctness of Nencki's speculations as to its probable exist-ence as a protein hydrolysis product. It is the substance which gives the so-called tryptophane reaction, although it has not, as Ellinger has recently shown, the exact constitutional formula assigned to it by Nencki, for the body it yields on oxidation is not skatol carbonic acid, but indol acetic acid—*i.e.* it contains no methyl group.

The important bearing of Nencki's work on the chemical dynamics of putrefaction on the general development of protein chemistry will be readily perceived from the above short summary.

Nor was the biological aspect of the question neglected by Nencki. It was important to determine what part bacterial putrefaction played in the general economy of putrefaction. Do bacteria play any necessary rôle in the alimentary tract of the higher organisms, in causing the degradation of ingested proteins? A unique opportunity for investigating this question was offered by a case of fistula of the small intestine under

Kocher in the Bern hospital. An investigation was made of the bacteria found in the fistula and also of the excreted products. None of the ordinary protein putrefaction products, such as indol and skatol, were discovered, and furthermore, it was shown that there was always excess of acid present. The conditions were not favourable to the development of bacteria (as had been shown by a previous research in Nencki's laboratory by Madame Sieber), and Nencki with his co-workers on this research (Macfadyen and Madame Sieber) came to the conclusion that bacteria were not necessary for the utilisation of proteins in the alimentary tract. Nuttall and Thierfelder, by direct experiment, subsequently arrived at the same conclusion.

Another series of important investigations was undertaken, with the object of determining the relation of aerobic to anaerobic growth. It was shown that under anaerobic conditions the nutrient medium serves as a source of oxygen supply; under these conditions, furthermore, oxidation does not proceed so far. The results obtained were in accordance with what might be expected from the equations already given to represent the various processes.

Nencki also investigated the action of various bacteria on carbohydrates. He succeeded in demonstrating the important fact that, whereas some species of bacteria will produce an inactive lactic acid, others produce an active form. The relation of these stereochemical forms to living bodies remains to-day one of the great mysteries of biochemical action. Not less interesting were Nencki's researches on the preparation of products from proteins and carbohydrates of products obtained originally by means of micro-organisms by chemical means. He showed, for example, that skatol and indol could be obtained from proteins by fusion with potash, and that hydrogen, which is also often a product of putrefactive action was evolved at the same time.

Special interest attaches at the present time to Nencki's studies on the decomposition of sugars and other bodies in the presence of oxygen at body temperature. He showed that glucose, for example, in the presence of alkaline hydroxides at the temperature of 40° yields relatively large quantities of lactic acid. This body is probably a normal degradation product of the sugars, for it is excreted in the urine under conditions which cause a depression of the normal oxidative functions of the body. In recent years considerable progress has been made in the study of the chemical mechanism of sugar destruction, and successful attempts have been made to produce those products which had hitherto been obtained only by the agency of living organism. Mention may be made here of Buchner's researches on cell-free fermentation, and Duclaux's researches on the production of alcohol from sugar in the presence of sunlight. Lastly, in a very recent number of the *Zeitschrift für physikalische Chemie*, Schade has demonstrated that dextrose in the presence of alkaline carbonates decomposes into acetaldehyde and formic acid:

$$C_6H_{12}O_6 = 2(CH_3 \cdot CHO + HCOOH).$$

Furthermore, formic acid in the presence of rhodium, acting as a catalyst, decomposes into hydrogen and carbonic acid gas, and the former body acting in nascent condition on the acetaldehyde, reduces it to alcohol. In this way the production of alcohol by fermentation is explained. On the other hand, in presence of caustic alkalis, the main product is lactic acid, and this body is not the result of an oxidation of sugar, but of hydrolysis. This confirms Nencki's results, who showed that lactic acid could also be formed from sugar in the absence of oxygen. The production of different bodies from sugar by action of the carbonates and hydroxides is, according to Schade, analogous to the production of ketones and acids from substituted ethylacetoacetates—either one class of body or the other predominates, according to the concentration of the different ions present.

Nencki attempted to apply his results on sugar destruction to the elucidation of the pathogenesis of diabetes mellitus. It was considered possible that the sugar elimination was a consequence of the insufficient quantities of alkalis in the body. It was shown, however, that administration of bodies such as sodium citrate and sodium lactate, which are readily converted by the organism into the carbonates, causes no diminution of the sugar output in diabetes. Nencki then sought, but without success, to isolate sugar-destroying enzymes from different organs of the body; and, if we except the results of Stocklasa and the somewhat doubtful results of Lépine and Cohnheim, but little progress has been made in this direction since 1882, the year in which Nencki published his researches on this subject.

The oxidation of uric acid and proteins was also investigated, but the results were not very definite. Uric acid could be oxidised only as far as uroxanic acid—

$$C_5H_4N_4O_3 + O + 2H_2O = C_5H_8N_4O_6.$$

Since that time much progress has been made in the study of oxidation by means of enzymes, and the isolation of oxidases of specific character from different organs. It is only necessary to mention here the work of Bach and Chodat on the plant oxidases, the work of Martin Jacoby on the different oxidases of the liver, and the large series of researches on oxidases which destroy the bodies of the purine group—with which work are associated, amongst others, the names of Partridge and Jones, Schittenhelm, and Burian. (See recent volumes of Hoppe-Seyler's *Zeitschrift*.)

We see, from the above summary, that Nencki's researches dealing with putrefaction and fermentative action covered a large field. On the morphological side they were concerned with isolation of new species of bacteria and the preparation of pure cultures. On the chemical side they were concerned with the isolation of the various degradation products of proteins and sugars, and with attempts to reproduce by purely chemical methods reactions hitherto associated only with vital activity. The results obtained have exerted a great influence in many directions—in none more than in the furtherance of our knowledge of protein chemistry.

III. The Formation of Urea in the Animal Body.

In his first published work, carried out under the direction of Schultzen, Nencki showed that if glycocoll and other amino-acids be administered to an animal kept in nitrogenous equilibrium, there is an increased output of nitrogen in the form of urea. The administration of acid amides does not, however, cause a like increase in urea formation. This indicated that the ordinary products of tryptic digestion, such as glycocol, leucine, etc., rapidly undergo decomposition with the elimination of nitrogen, which is excreted in the form of urea. What is the mechanism of this urea formation, and what organs are concerned with the chemical processes by means of which it is brought about? These were the questions which Nencki set about solving many years after his first experiments with Schultzen on the subject, and when in St. Petersburg he had the advantage of the co-operation of Pawlow with his great surgical skill.

Nencki had already shown that putrefactive organisms cause the elimination of ammonia from the products of tryptic digestion, but he had also demonstrated, in conjunction with Madame Sieber and Macfadyen, that such organisms were not necessary for the normal assimilative processes. It remained, therefore, to seek for a locality where, supposing ammonia to be a precursor of urea in the animal body, the elimination of this body from the products of tryptic digestion could take place. Furthermore, in the event of the formation of ammonia, and ammonium carbonate (or carbamate), it was necessary to determine which organ was responsible for the conversion into urea. At the time the experiments dealing with these points were commenced, it was known from the researches of Schröder and his co-workers that the liver was the principal, if not the only, seat of urea formation. This fact had been determined by the perfusion of blood containing various nitrogenous bodies through the surviving liver and the demonstration of the increase of urea after the passage through this organ. Nencki, Pawlow and their assistants then endeavoured to determine what would happen supposing the liver were thrown out of the normal circulation. For this purpose they joined the portal vein with the vena cava inferior (Eck's fistula), and showed that the animal could be maintained in moderate health under these conditions provided that it did not receive a heavy protein diet. If, however, it was given a large meat diet, it rapidly developed toxic symptoms, and showed an increased ammonia output in the urine. These same toxic symptoms could be produced in a normal animal by the *intravenous* injection of ammonium carbonate, but *not* by the introduction of this body directly into the stomach. In the dog with the Eck fistula, however, the toxic symptoms were produced by the direct introduction of the carbamate into the stomach. These results indicated, therefore, that ammonium carbamate is a precursor of and is converted by the liver into urea; the toxic product of protein degradation is thereby converted into a harmless body.

34

The next part of the work was devoted to the determination of the seat of ammonia formation. In conjunction with Zaleski, a method was devised for the determination of minute quantities of ammonia in the blood and tissues, and by means of this method a series of estimations of ammonia in organs and blood from different parts of the body was made. It was found that the blood in the portal and mesenteric veins contained larger amounts of ammonia than other parts of the vascular system. Furthermore, large amounts of ammonia were obtained from the mucous membrane of the digestive tract. Finally, the tissues of well-fed animals yield more ammonia than those of fasting animals. These results leave little doubt that it is in the tissues of the alimentary tract that the elimination of ammonia from the products of protein digestion takes place. Passing from this locality to the liver in the form of ammonium carbonate or carbamate, they are then converted from the toxic form into the non-toxic urea, which is then rapidly eliminated in the urine.

It was shown that after Eck's fistula atrophy of the liver takes place. We now know that unless this organ is supplied with excess of alkali, the autolytic enzyme comes into play and brings about hydrolysis. We do not yet know why, however, the animal only survives the total extirpation of this organ but a very few hours, as was demonstrated by Nencki and Pawlow in the case of dogs, and by Minkowski in the case of geese.

The results on urea formation are of considerable interest from the clinical standpoint. They indicate the importance of full urine analyses for the purpose of diagnosis, especially in diseases of the liver. Much remains to be done in this direction, but it is hoped that the simplification of urine analysis by the methods of Mörner, Folin, and others may stimulate further work in this direction.

IV. Researches on the Animal Pigments.

These investigations were carried out principally in conjunction with Zaleski and Madame Sieber. More than fifty years ago, Teichmann had observed that when a drop of blood is mixed with glacial acetic acid and allowed to evaporate at a gentle heat over a small flame, crystals separate, which have since that time been known as Teichmann's crystals. Nencki and Madame Sieber were the first, in 1884, to prepare the crystals on a large scale and to determine their composition by chemical analysis. The subject was afterwards investigated by other observers, amongst whom may be mentioned Cloetta, Mörner, Schalféjew, and Rosenfeld. Various methods were employed and various formulæ assigned to the crystalline products obtained. The final credit, however, for bringing order into the chaos created by so many contradictory results must be assigned to Küster, who showed that the products obtained by the different observers were identical, only in different states of purity. He showed that the crystalline "hæmines" were not all entirely soluble in chloroform containing quinine or pyridine; if such a solution, after filtering off from the insoluble residue, be thrown into acetic acid containing hydrochloric acid, a crystalline compound separates out, to which the composition represented by the formula $C_{34}H_{33}O_4N_4ClFe$ must be assigned. Nencki and Zaleski prepared various derivatives from hæmatin. The most important of these was hæmatoporphyrin, which can be obtained by the action of hydrobromic acid. This in its free state is a halogen-free and iron-free body, and its preparation from bromine can be represented by the following equation:

$$C_{34}H_{33}O_4N_4ClFe + 2HBr + 2H_2O = C_{34}H_{38}O_6N_4 + FeBr_2 + HCl.$$

An additional stimulus was given to the investigation of the blood pigments by some researches which were being carried out by Schunck and Marchlewski on chlorophyll. From this plant pigment these observers obtained a series of derivatives, the formation of which can be schematically represented as follows:

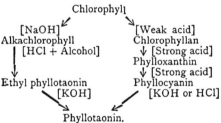

Chlorophyll

[NaOH]   [Weak acid]
Alkachlorophyll   Chlorophyllan
  [HCl + Alcohol]   ↓ [Strong acid]
                      Phylloxanthin
                      ↓ [Strong acid]
Ethyl phyllotaonin   Phyllocyanin
    [KOH]   [KOH or HCl]

Phyllotaonin.

Now phyllotaonin was found by Schunck and Marchlewski to yield, on treatment with alcoholic potash, a body having a spectrum very similar to hæmatoporphyrin. To it was assigned

the name phylloporphyrin. The chemical relationship of the
two bodies could be expressed by the two following formulæ :

$$\text{Hæmatoporphyrin} = C_{34}H_{38}O_6N_4$$
$$\text{Phylloporphyrin} = C_{34}H_{38}O_2N_4$$

The relationship of the empirical formulæ between the blood-
pigment and chlorophyll demonstrated by Schunck and
Marchlewski induced Nencki and Zaleski to attempt to convert
by reduction the hæmatoporphyrin, derived from the former
body, into the phylloporphyrin derived from the latter. They
succeeded, however, in obtaining another product — viz.
$C_{34}H_{38}O_4N_2$, which stands intermediate between the two—to
which they gave the name mesoporphyrin. In addition to this
a volatile substance, hæmopyrrol, was obtained, which readily
oxidises to the pigment urobilin, which is found in urine after
blood extravasations.

Although phylloporphyrin was not directly obtained from
hæmatoporphyrin, there seemed very little doubt of their
intimate chemical relationship. From this stage Nencki and
Marchlewski continued their work conjointly, and they subsequently showed that phyllocyanin (see above), on reduction,
also yields hæmopyrrol.

This body has probably the following formula :

i.e. it is methyl-propyl-pyrrol.

As the result of their researches, and those of Küster, who
had in the meantime been investigating the oxidation products
of hæmatin, Nencki and Zaleski endeavoured to assign a
formula to hæmatin. The constitution of this body, however,
is still far from being established with certainty, and it is very
improbable that the formula assigned to it is correct. It is
hardly necessary to reproduce it here.

The constitution of chlorophyll has recently been the subject
of investigations by Willstätter. Some preliminary results
have been already made public, which are of great interest, and
lead to the expectation that much may be accomplished in the
immediate future.[1]

[1] See recent number of Liebig's *Annalen.*

Researches on other pigments were also carried out in Nencki's laboratory, such as those on urobilin, the melanines, and hair pigments. The results are not of the same general interest as those relating to the blood-pigments.

Before concluding this review, it is necessary to mention other researches on physiological chemistry, which do not fall under any of the four headings mentioned above. Amongst these are the investigations on the chemical nature of pepsin, carried out principally by Madame Schumow-Simanowski, and those on the excretion of hydrochloric acid and other haloid acids by the mucous membrane of the stomach. It was shown that if chlorides be partially substituted by bromides in the salts of the foods ingested, hydrobromic acid is excreted by the gastric mucous membrane, together with hydrochloric acid. The latter acid cannot, however, be replaced by hydriodic acid.

There are also other researches of interest which might be mentioned, but sufficient has already been discussed to indicate the great scope of Nencki's work.

It remains now to attempt to assign to Nencki his position as an investigator. There is no epoch-making discovery of far-reaching influence which is associated with his name, and for this reason he cannot be placed in the front rank of scientific workers. Furthermore, he owed much, when generalising his results, to the labours of his contemporaries working in the same fields. Valuable as were the conclusions that he could draw from his own investigations on putrefaction, he owed much to the labours of Salkowski, whose results supplemented his own. Similar remarks may apply to his pharmacological investigations, and the relation of his work on this subject to that of Baumann and others. In his conclusions on the chemistry of hæmatin, he owed much to the results obtained by Küster.

If, however, we cannot assign to Nencki a position in the front rank, his skill as an organic chemist, his great grasp of biological problems, his versatility and industry, and, above all, the success which attended his work, will entitle him to a highly honourable place amongst the pioneers of physiological chemistry.

RECENT ADVANCE IN THE STUDY OF FUNGI

By A. LORRAIN SMITH, F.L.S.

In the early days of botanical study fungi were considered to be of very little importance, either for good or evil. The mushroom and one or two others of the larger kinds, as affording toothsome dishes for the table, were allowed to be of some value; but, as a class, they were either objectionable or negligible, mysterious in origin, and useful only to poet and artist to heighten the effect of some picture of gloom and decay:

> And agarics and fungi, with mildew and mould
> Started like mist from the wet ground cold;
> Pale, fleshy, as if the decaying dead
> With a spirit of growth had been animated!

At the present time fungi have come to their own, and no branch of botany now receives more attention. The *rôle* that the mycelium alone plays in the soil, in disintegrating the yearly accumulations of plant-wreckage, proves the once-despised mould to be indispensable to further plant-life. Though fungi do not, like nitro-bacteria, directly enrich the soil by collecting nitrogen from the air, recent researches have shown that they furnish carbohydrate material in the form of mannite glycogen, etc., to these bacteria. Since Prof. Frank's discovery of mycorhiza, the fungus that lives in symbiotic relationship with the roots of plants, and aids or supplements the absorptive function of the root-hairs, many experiments have been made, and much has been written for and against Frank's conclusions, but the balance of proof seems largely to uphold the advantage of the symbiosis to the higher plant. Gustave Kunze has shown quite recently that the fungal filaments excrete acids—mainly oxalic acid—*to a much greater extent than do the roots of the higher plants.*[1] They bring into solution the salts necessary for plant-growth,

[1] See Hall, *Science Progress*, (1906) i. 51, etc., "Solvent Action of Roots upon Soil Particles."

and for this reason alone mycorhizæ and fungal hyphæ generally are of enormous importance in plant economy.

Fungi are overwhelming by mere force of numbers : over
50,000 forms have already been recorded and described ; and,
as much territory is still unexplored fungologically, the lists of
new genera and species still to be added to the world's flora
assume alarming proportions. Even in our own well-searched
Islands new species are found year by year. The pursuit of
this branch of the subject knows, as yet, no pause. All this
collecting and describing, this spade-work of the science, must
however be done before a true and lasting knowledge can be
attained. Close on the heels of the collectors and systematists
follow the more critical workers, who examine and compare,
and on the data supplied build up a coherent system.

Thus E. S. Salmon, in recent years, has monographed the
Erysipheæ on morphological lines. He has found that many of
the species recorded, though growing on different hosts, are not
really autonomous ; he has given them their true position as
synonyms of species already described, and has thus largely
reduced the number of species in that family. Another order
of work, carrying much the same results, has been undertaken
by Von Höhnel, who publishes, now and again, notes on
various species, pointing out synonyms and deleting redundant
descriptions.

One of the most notable of recent works on fungi is Klebahn's
Heterœcious Uredineæ, an exhaustive treatise on the rusts that
grow at different stages on different hosts. His work has
formed a starting point for further effort, and has given a
great impetus to students of this group. All advance on this
line is by continual inoculation experiments from one host-plant
to another. The work is slow and often disappointing, but
successful results are constantly being published, and the
number of species, in which the life-cycle has been followed
throughout, is increasing rapidly.

Klebahn has also taken up as a study the life-histories
of Ascomycetes. They have frequently—if not always—both
pycnidial and conidial forms in addition to the perfect fruit,
the apothecium or perithecium with the asci and spores. He
has added several instances of relationships to those already
known ; but here again the work is slow, and results are
uncertain. Tracing out these life-histories, and so linking up

the different forms assumed by one fungus, leads to simplification, and the burden of nomenclature should be lightened for the fungologist of the future, even though new species continue to crop up.

The ease with which fungi can be cultivated in the laboratory has made them favourite objects for physiological and cytological investigation. Nuclear division takes place by karyokinesis, as in other groups of plants, though the phenomena of division vary considerably for the different forms and stages of development. In many cases the nuclei are so small that the details cannot be followed.

The occurrence of sexuality has long been known in the filamentous fungi, both in the oomycetes and the zygomycetes, though in some of the species the sexual organs are now functionless. A. H. Trow has studied *Achlya*, and has traced the sexual process in that genus of Saprolegniaceæ. Ruhland, a German botanist, has recently described the variations that take place within the genus *Albugo*, which tend to prove its close relationship to Peronosporeæ. A. F. Blakeslee reports most interesting work on the Mucorini. The rare and uncertain occurrence of the sexual zygospores in that Order has been explained to be due to this or that unfavourable condition of the substratum, and various recipes have been recommended by fungologists for their successful cultivation. Blakeslee has discovered, and proved beyond a doubt, that plants of most species of the Mucorini are unisexual, and zygospores are only formed when two different strains, termed by him (+) and (−), are grown together. Zygospores are then formed at the points of contact of the two plants.

The general question of sexuality in the higher groups was accounted by many as settled when that lifelong investigator Brefeld announced "that no fertilisation process occurred in any group except in the Phycomycetes." Improved methods of fixing, staining, etc., have made it worth while to attack the subject again, and R. A. Harper led the way by his work on a minute ascomycetous fungus, *Sphærotheca castagnei*. He found that both antheridium and ascogonium were present, and that fusion undoubtedly took place between the nuclei of these organs. This discovery has stimulated others to work in the same field: some, like M. Dangeard, to a fierce denial of Harper's interpretations; others to verify and support his findings, or to extend our knowledge to closely related forms. The constantly accumulating mass of evidence must, by this time, have convinced the most sceptical as to the increased range of sexuality in the ascomycetes. It has been proved by Harper to be present in the genera *Sphœrotheca*, *Erysiphe*, and *Pyronema*. Claussen has worked it out in *Boudiera*; he did not, indeed, see the actual opening between the antheridium and ascogonium, but the nuclei in the female cell doubled in numbers as the male cell became quite empty, and the nuclei thus congregated fused in pairs before the ascogenous hyphæ began to grow out from the ascogonium. In some forms the female cell alone is present. In one such, *Humaria granulata*, V. H. Blackman and H. Fraser have found that the large number of nuclei in the ascogonium (upwards of a thousand) fuse in pairs. There is here no question of a male organ, but the writers interpret the fusion as a very reduced form of conjugation. Gustav Ramlow describes the ascogonium in *Thelebolus stercoreus*, a minute ascomycete with only one ascus. There is no fusion between the ascogonium and any other cell, and no fusion between nuclei until the ascus is formed. It would almost seem as if the ascomycetes would need to be examined species by species, as no two appear to behave exactly alike.

A great deal of attention has been directed to the development of the ascus, a branch of investigation which Harper also started afresh. The most recent paper on this point is by H. Faull, who examined and compared a large number of species. He finds that the asci bud out from the penultimate cell of the ascogenous hyphæ only in some cases. In a few of the species examined they arose from the terminal cell, in others they grew apparently from any of the cells. In every case definitely determined by him, and by previous workers, there were two nuclei present in the young ascus which fused, and, after a resting stage, divided to form the ascospores. Harper held that the astral rays which radiate from the poles of the nucleus coalesced and bent round the nucleus to form the spore-membrane. Faull sees no evidence for this: the spore wall, he contends, is formed from the cytoplasm of the ascus independently of these rays. This view would certainly show less of dissimilarity between the spore of the ascus and the spore of the various sporangia of the lower groups, and so Faull comes to the general conclusion that the phenomena of spore-formation, as observed by him in the ascus, are not incompatible with the view that homologises the ascus with the oomycetous sporangium. He thinks the ascomycetes may possibly have arisen from some phycomycetous group.

Remarkably interesting work has been done on fertilisation in Uredineæ by V. H. Blackman. The cells of the hyphæ, as well as the spores in this Order, are binucleate from the mature æcidium stage until the teleutospore is formed, when the two nuclei fuse together. *Phragmidium violaceum*, with its simple æcidium, was selected for examination to determine at what stage in the life-history of the uredine the binucleate condition arose. It was found that the first hyphæ, produced after infection by the teleutospores, had uninucleate cells, and that the cells of the young æcidium were also uninucleate, but at the base of the latter there were present a row of rather large cells termed by Blackman the female cells. These cells become binucleate by the nucleus of a neighbouring cell migrating through the cell wall. The two conjugate nuclei do not fuse— they divide, and a daughter nucleus of each passes on to form the series of binucleate æcidiospores, with the subsequent binucleate mycelium and spores of uredo and teleutospore stages. Fusion between the paired nuclei finally takes place in the maturing teleutospore. Blackman looks on the first association of the two nuclei as a reduced form of fertilisation that came into play as the primary male organs—the spermogonia and spermatia—became functionless. He recognises in this life-history an alternation of generations : the gametophyte with single nuclei, represented by the mature teleutospore, sporidia, spermogonia, and mycelium, succeeded by the sporophyte which includes all the binucleate stages. Christman, an American investigator, found in another species of the same genus that two of the larger æcidial cells themselves fused, their nuclei remaining distinct. There is evidently no hard and fast rule of procedure, but the end—the binucleate cell—is always achieved before spore formation begins. In forms such as *Puccinia malvacearum*, where no æcidium is formed, the binucleate condition arises just before the formation of the teleutospores ; in other similar species nuclear migration takes place at a very early stage after teleutospore infection.

The Uredineæ have a bad record as disfiguring and destroying rusts. They are all parasites, and form a well-marked group, closely allied to the Basidiomycetes. A recent mono-graph by P. and A. Sydow gives a complete account of the genera and species, arranged in the order of the host-plants. D. M'Alpine has since published an account of Australian rusts. He notes the curious fact that only four of the indigenous species are heterœcious; all the others are autœcious, and complete their life-history on one host.

Plant diseases caused by parasitic fungi form the subject of much study and research. We have not recently suffered in this country from any such widespread and devastating epidemics as those caused by potato disease or by hop mildew; but all cultivators know, to their cost, of the prevalence of one form or another of fungal parasites that injure their crops. Stem and root, leaves and fruit, all are liable to disease from a large variety of fungi, and need to be guarded against attack. A host of workers is engaged on this branch of plant pathology, and new facts are being brought to light day by day. The life-histories of these inimical fungi are being worked out, so that a remedy can be applied at the precise time when it will have most effect. A very valuable discovery was made by Eriksson in working on rusts. He found that the morphological species may include a number of biological or physiological species—*i.e.* that there is some peculiarity in the parasite that cannot be distinguished by the microscope, but which yet effectually prevents it from changing its habitat. In many cases the rust that grows on one form or variety of the host-plant will infect no other. This has a most important bearing on the spread of rust disease, and it is thus possible to avoid the cultivation of the varieties most subject to attack, and to select those that are proved to be immune to the prevalent disease. Marshall Ward followed on the same lines, experimenting with a rust that grows on brome grass, and he proved the narrow range of choice of host in the parasite; he proved, however, the existence of what he called "bridgeing species." A plant, otherwise immune to the rust experimented with, could be infected by the rust if it had previously been grown on some intermediate or "bridgeing" species.

E. S. Salmon has demonstrated biological species also among the *Erysipheæ*, the mildew of hops, vines, peas, etc. A species of *Erysiphe* may grow on a large number of host-plants, and morphologically be always the same, but it cannot easily be transferred from one host to another. By wounding or weakening the host, he found he could induce the growth of an alien form, but with ordinary conditions one host is adhered to.

It is not always possible to trace outbreaks of disease to the source of infection. Spores are very light bodies, and a high wind might transport them to great distances. There is, however, no doubt that they are conveyed from one country to another along with their appropriate host. Quite recently a disease of gooseberries was introduced into Europe from America ; it is reported from Austria, Denmark, and Russia, and has lately appeared in Ireland. Mr. Salmon has been insisting on the duty of measures being taken by Government to prevent the importation of infected seeds or seedlings from foreign countries. It would be well to protect ourselves from such diseases, if it were at all possible, by destroying or disinfecting suspected material. We have seen in the past how an insignificant mould may become a national calamity; and the old adage always holds good that " prevention is better than cure."

List of Papers.

BLACKMAN, V. H. On the Fertilization, Alternation of Generations, and General Cytology of the Uredineæ, *Ann. of Bot.* xviii. 1904, p. 323.

BLACKMAN, V. H., and FRASER, H. C. J. Further Studies on the Sexuality of the Uredineæ, *Ann. of Bot.* xx. 1906, p. 35.

—— —— On the Sexuality and Development of the Ascocarp of Humaria granulata, Quél., *Proc. Roy. Soc. B.* vol. lxxvii. 1906, p. 354.

BLAKESLEE, A. F. Sexual Reproduction in the Mucorineæ, *Proc. Am. Acad.* xl. 1904, p. 205.

CHRISTMAN, A. H. Sexual Reproduction in the Rusts, *Botan. Gazette,* xxxix. 1905, p. 267.

CLAUSSEN, P. Zur Entwickelungsgeschichte der Ascomyceten, Bondiera, *Bot. Zeit.* lxiii. 1905, p. 1, 3 pls.

DANGEARD, P. A. Recherches sur le Developpement du Périthèce chez les Ascomycetes, *Le Botaniste,* 9ᵉ série, 1904, p. 59.

ERIKSSON, J. Neue,Untersuchungen über die Spezialisierung, Verbreitung, und Herkunft des Schwarzrostes (*Puccinia graminis,* Pers.), *Jahrb. f. wiss. Bot.* xxix. 1896, p. 499.

—— Weitere Beobachtungen über die Spezialisierung des Getreideschwarzrostes, *Zeitschr. f. Pflanzenkr.* vii. 1897, p. 198.

FAULL, J. HORACE. Development of Ascus and Spore Formation in Ascomycetes, *Proc. Boston Soc. Nat. Hist.* xxxii. 1905, p. 77.

HARPER, R. A. Die Entwicklung des Peritheciums bei Sphærotheca Castagnei, *Ber. d. deutsch. bot. Ges.* xiii. 1895, p. 475.

—— Sexual Reproduction in Pyronema confluens, *Ann. of Bot.* xiv. 1900, p. 321.

HÖHNEL, FRANZ VON. Mykologische Fragmente, *Ann. Mycol.* iii. 1905, p. 402, etc.

KLEBAHN, H. *Die wirtswechselnden Rostpilze* (Berlin: Gebrüder Borntraeger, 1904).

—— Zusammenhänge von Ascomyceten mit Fungis imperfectis, *Centralbl. Bakt.* xv. 1905, p. 336.

KUNZE, G. Ueber Saureausscheidung bei Wurzeln und Pilzhyphen und ihre Bedeutung, *Jahrb. wiss. Bot.* xlii. 1906, pp. 307-93.

MCALPINE, D. *The Rusts of Australia* (Melbourne : R. S. Brain, 1906).

RAMLOW, GUSTAV. Zur Entwickelungsgeschichte von Thelebolus stercoreus Tode, *Bot. Zeit.* lxiv. 1906, pp. 85-99, 1 pl.

RUSHLAND, W. Studien über die Befruchtung von Albugo Lepigoni und einiger Peconosporeen, *Jahrb. wiss. Bot.* xxxix. 1904, p. 135.

SALMON, E. S. On Specialization of Parasitism in the Erysiphaceæ, *Ann. Mycol.* iii. 1905, p. 172.

—— A Monograph of the Erysiphaceæ, *Mem. Torr. Bot. Club,* xxix. 1900, p. 224.

SYDOW, P. u. A. *Monographia Uredinarum* (Leipzig : Gebrüder Borntraeger, 1904).

TROW, A. H. On Fertilization in the Saprolegnieæ, *Ann. of Bot.* xviii. 1904, pp. 541-69, 3 pls.

WARD, H. MARSHALL. The Bromes and their Rust Fungus (*Puccinia dispersa*), *Ann. of Bot.* xv. 1901, p. 560.

—— On the Relations between Nest and Parasite in the Bromes and their Brown Rust (*Puccinia dispersa*), *Ann. of Bot.* xxi. 1902, p. 233.

—— Further Observations on the Brown Rust of the Bromes (*Puccinia dispersa*) and its Adaptive Parasitism, *Ann. Mycol.* i. 1903, p. 132.

THE CEYLON RUBBER EXHIBITION, AND RUBBER CULTIVATION IN THE EAST.

By J. C. WILLIS, Sc.D.,

Director of the Royal Botanic Gardens, Ceylon.

An Exhibition of Rubber has lately been held in the Royal Botanic Gardens, Peradeniya, Ceylon, and marks a distinct stage in the development of this great new industry. As the cultivation of rubber is, at the present time, probably the most profitable agricultural pursuit in the world, and is entirely the creation of the scientific departments kept up (by their Governments) in the tropical colonies of Ceylon and the Straits Settlements aided by the central gardens of Kew, it may be useful to review at this time the way in which science has assisted it. Not only was the establishment of the industry due to the forethought of men of science, but it has been developed in a very scientific spirit, and is being advanced mainly by the labours of scientific men.

For its inception we must go back more than thirty years, when, inspired by Sir Clements Markham, expeditions were sent by the Indian Government to the river Amazon, to collect and bring back the seeds of the South American rubber plants, which looked as if they would one day cease to supply the large quantity of rubber consumed in the world. The expeditions, headed by Mr. Wickham and Mr. Cross, successfully brought to Kew seeds of the Para rubber (*Hevea brasiliensis*) and other species. It was intended that these should be sent to India, but it was evident that at that time there was no Botanic Garden in India with a suitable climate, and the bulk of the plants were in consequence sent, in charge of a special gardener, to Ceylon, where they were established in the branch garden at Henaratgoda, near Colombo, which was opened by Dr. Thwaites for their reception. A few were also planted at Peradeniya, but

this has proved on the whole rather too cold for them.   Others
were sent to Singapore and elsewhere.

For many years the Hevea did not seed, and meanwhile
much Ceãra rubber (*Manihot Glaziovii*) was planted, but
the returns from this were but poor, and the cultivation
was given up in most districts, the more so as the same
land would do for tea, which was then spreading rapidly in
Ceylon.

Meanwhile, however, the Botanic Gardens, under the late
Dr. Trimen, were not idle, and a large plantation was set
out, about 1886, from the seed of the forty-six original trees.
In 1888, when the latter were twelve years old, Dr. Trimen
commenced to tap the largest of them, making V-shaped incisions
with a hammer and chisel.   In that year this tree yielded 1 lb.
13 oz. and it was tapped every second year thereafter till 1896.
In the nine years that thus elapsed it yielded an average of
1½ lb. a year.   This rubber, being simply dried in coco-nut
shells, was very dirty.

The next stage was in 1897, when the writer, perceiving
the weak point of the previous experiment, tapped a whole
plantation of younger trees, to get an average yield, and
found that about 120 lb. an acre might be annually expected
from trees twelve years old.   At 2s. a pound this was estimated to give a profit of 27 per cent., and caused a rush to
plant Para rubber, the only check being the still prevailing
scarcity of the seed.

At the same time the writer made a discovery which has
proved of the greatest importance to the rubber industry, but
which is still without a scientific explanation, viz. the " wound-
response."   If a given area of the bark be tapped a second time
within a week or ten days, it will yield a larger amount of
latex than at the first tapping, and another tapping may
produce an even larger amount.   We may illustrate this by
some actual figures.   Four Hevea trees were tapped at
Peradeniya in 1899 at intervals of about a week, and the
yields of latex, in cubic centimetres, were 61, 105, 220, 208,
255, 290, 276, 253, 264, 275, 255, 262, 328, 449.   The experiment
had then to be stopped, owing to the departure of the
operator for Europe.   This is a very remarkable fact, and at
once showed the reason for the bad reports which had so
often been made of this tree—the experimenters had only

tapped once. At the same time it practically quadrupled the yield to be expected.

At about this same time the Director of the Botanic Gardens in Singapore commenced to push rubber cultivation, and that with considerable success, as the planters in the Malay Peninsula were in very low water owing to the low prices of coffee, and had not, like the Ceylon planters, stable and prosperous industries such as tea, cardamoms, coco-nuts, cacao, and so forth. The result has been that Malaya has rather more older trees at the present time than Ceylon, though it has not so large an area planted.

During the following year, 1898, Mr. John Parkin was in Ceylon as Scientific Assistant at the Royal Botanic Gardens, and worked out the question of the wound-response in detail (the figures above given are some of his). He came to the conclusion, which has not as yet been generally accepted, that the latex, in Hevea at any rate, is mainly of the nature of a water-store, the plant drawing upon it in dry weather, when the latex is found to be much thicker, and richer in caoutchouc.

In addition to this, Mr. Parkin worked out in detail the method of making " biscuits" which has been universally followed ever since in Ceylon and Malaya (for the sheet is simply a larger biscuit). Biffen had shown that the essential constituents of the smoke used to clot the rubber in South America were acetic acid and creosote, and Parkin applied this to the coagulation of the latex. Collecting it in tins containing a little water (to prevent immediate coagulation) he then mixed with it the calculated amount of acetic acid and a little creosote, clotting the latex into the form of the vessel in which the reaction was performed. It was then rolled, and hung up to dry. This process has been used ever since in Ceylon, and, as the milk is filtered before coagulation, a pure biscuit is obtained, getting on the open market a very high price.

On the strength of these results of 1897-9, many people began to plant rubber, but the rapid extension of the cultivation was checked till about 1902 by the scarcity of seed. After that time, a vast quantity of seed began to be also available from private sources, and extension of cultivation has gone on rapidly, so that now there are about 105,000 acres opened in rubber in Ceylon, about 60,000 in the Malay Peninsula, and perhaps 13,000 in South India, besides possibly about another 20,000 in other countries, such as Samoa, Seychelles, Java, Cochin-China, tropical Africa, etc. Already, then, one of the largest planting industries in the tropics has been created.

Not only so, but, thanks to the work of Parkin and others, this has proved to be one of the most, if not actually the most, profitable industries ever established.   Many estates in Ceylon have been lately harvesting rubber at the rate of 200 lb. an acre, at a cost of about 1s. a lb., and selling it for 5s. to 6s.   Even at the lower figure, there is a profit of £40 an acre per annum.

The early crude method of tapping in V's was soon given up, and two or three systems are now in vogue, especially the "herring-bone" and the "half-spiral." In the one a narrow vertical channel is cut from a height of 6 ft. down to the ground, and lateral channels made from it at angles of about 45°. In the other, spiral grooves are cut half-way round the tree at angles of 30° to 45°.   The wound-response is obtained in these by paring the edges of the "spiral" or "herring-bones," and the various knives that have been designed for the purpose take off very thin parings.

During the last two years the planting public of Ceylon and Malaya has been mainly interested in rubber planting, which has been the one absorbing topic of conversation, and the shares in the many companies started have already risen, when there was any rubber yield in fairly near prospect, to three to eight times their par value; and they are firmly held even at these figures, showing what practical men, who as yet have been the principal investors, think of their prospects.   This being so, the attention of the whole rubber world was naturally focussed upon the Rubber Exhibition held from September 13 to 27, 1906, in the Royal Botanic Gardens at Peradeniya, in Ceylon. Large buildings, covering over half an acre, were mainly filled with raw, and to some extent manufactured, rubber, tools for tapping, machinery for treatment, and other objects of interest.

It would be foreign to the purpose of this paper to enter into any description of the exhibits; but it may not be amiss to point out one or two of the directions in which the exhibition demonstrated that science must still be called to the aid of the

35

rubber planter. He needs the aid of specialists in entomology and mycology almost every month to attack the diseases that threaten his trees, and of the chemist for their scientific manuring; but there are also many directions in which new discoveries have to be made before we can look upon the rubber industry as beyond the fear of competition from the wild materials collected in South America, Africa, etc. Thus we want to find out, as soon as may be, why the cultivated rubber is less elastic than the wild South American; it does not immediately go back to its original form on being stretched. We want to find out the very best way of tapping to get the maximum yield with the smallest consumption of bark. We want to find out how best to coagulate rubber into marketable form, and many other things.

To those who look forward more than a few months, one of the most interesting exhibits in the whole show was the samples of coloured and sulphurised rubbers exhibited by Mr. Kelway Bamber, Government chemist in Ceylon. At present the colouring, vulcanising, and mixing reagents are added to the rubber after coagulation. The raw rubber is macerated in powerful engines, and thoroughly mixed with the various substances, and then worked up into the finished articles, and heated at the end, when vulcanisation takes place. It seems waste of time, energy, and money to do this mixing in the dry rubber, when it can be much more easily and cheaply done in the milk. The milk is filtered, in the Bamber process, and then mixed with the vulcanising and colouring agents, which, of course, become absolutely intermixed with a little stirring. The vulcanising agents used are such as give free milk of sulphur on the addition of an acid, and acid is added to the mixture to such an extent as to free the sulphur and coagulate the rubber, which is thus formed with sulphur in contact with almost every molecule. This rubber can then be treated in the ordinary way, and made up into rubber goods, which, when heated at the finish, vulcanise. In the same way colours, fibres, mixing substances of every description (provided that they can be wetted) can be added to the milk, and mix homogeneously with it. While requiring as yet much elaboration in detail, this method holds the germs of the future treatment of rubber, and there is probably a great future before it.

RECENT LITERATURE UPON THE RUBBER INDUSTRY, ETC.

WILLIS, J. C., Rubber Cultivation in Ceylon, *Circs. Roy. Bot. Gdns.* i. 4, 1898.

PARKIN, J., Caoutchouc or India-rubber, *Ibid.* i. 12, 1899.

WRIGHT, HERBERT, *Para Rubber.* Colombo: A. M. & J. Ferguson, 2nd ed. 1906.

WILLIS, J. C., BAMBER, M. K., and DENHAM, E. B., *Handbook of the Ceylon Rubber Exhibition* (containing the series of lectures given there by specialists, the judges' reports, descriptions of machinery, etc., and altogether forming an indispensable book for any one interested in rubber). Colombo: Government Printing Office. London: Dulau & Co. and Wyman & Sons. 7s. 6d. net.

THE REFORM OF THE MEDICAL CURRICULUM :

A PROBLEM IN TECHNICAL EDUCATION

By HENRY E. ARMSTRONG,

Professor of Chemistry in the Central Technical College, South Kensington, London ; Member of the Mosely Educational Commission in the United States of America.

In addressing the students of the Faculty of Medicine in University College, London, in 1870, Huxley protested against medical students being required to devote their energies to the acquisition of any knowledge which might not be absolutely needed in their subsequent career. "Any one," he said, "who adds to medical education one iota or tittle beyond what is absolutely necessary is guilty of a very grave offence." Thirty-five years are gone and we remain much as we were—there has been no little "grave offence" committed in the interval.

I am spurred to these remarks by the knowledge of the fact that, at the instance of the Faculty of Medicine, a resolution was moved at a recent meeting of the Senate of the University of London to the effect—"That the Examination in Organic Chemistry be part of the Preliminary Scientific Examination at the end of the first year and that the Syllabuses of both Organic and Inorganic Chemistry be so modified as to admit of the work being done in the first year of the Curriculum." This is said to be proof that medical men are enemies of science. Are they ? I opine not—and perhaps my right to hold views on such a matter is not altogether questionable. As a student, I heard Ludwig's course of lectures on Physiology at Leipzig. I was during twelve years a teacher in St. Bartholomew's Hospital Medical School of the special class in Chemistry for students proceeding to the London degree. Through my life-long friend, Dr. Horace Brown, I was brought into contact with fermentation problems in the early seventies and have always continued to take the deepest interest in them. Of late years it has been my good fortune to see my eldest son—as well as a number of my own pupils—engage in the study of the action of enzymes; in fact, I educated him to that end, feeling that the physiological field is that which offers to the investigator most real plums and nuts sweet to eat when cracked—the field in which work of the highest public importance is to be done, especially in connexion with the many intricate problems which the study of food presents. Another son, who has passed recently through the first three years of the medical course at Cambridge, has brought me into close touch with the latest developments of the curriculum of preliminary medical studies. I have some idea, therefore, both of the possibilities and the requirements. Moreover, I am not thinking of the subject for the first time: as far back as 1885, in my address at Aberdeen as President of the Chemical Section of the British Association, I gave utterance to the following opinion, based on the experience which I had had in a medical school:—

" We may also hope that it will be possible ere long to teach Chemistry properly to medical students. Seeing that the practice of medical men largely consists in pouring chemicals into that delicately organised vessel the human body; that the chemical changes which thereupon take place or which normally and abnormally occur in it are certainly not more simple than those which take place in ordinary inert vessels in our laboratories, the necessity for the medical man to have a knowledge of chemistry—and that no slight one—would appear to ordinary minds to stand to reason: and that such is not generally acknowledged to be the case can only be accounted for by the fact that they have never yet been taught *chemistry*; that the apology for chemistry which has been forced upon them has been found to be of next to no value. No proof is required that the student has ever performed a single quantitative exercise: and I have no hesitation in saying that the examinations in so-called practical chemistry, even at the London University, are beneath contempt: after more than a dozen years' experience as a teacher under the system, I can affirm that the knowledge gained is of no permanent value and the educational discipline *nil*. Here the reform must be effected by the examining boards: it is for them to insist upon a satisfactory preliminary training and they must so order their demands as to enforce a proper system of practical teaching; and if chemistry is to be of real service to medical men more time must be devoted to its study. Physiological chemistry is taught nowhere in our country, either at the Universities or at any of our great medical schools."

Not a few important changes have been made in the interval, so that many of the complaints possible twenty years ago no longer hold. Simple quantitative work in chemistry has now taken the place of much of the worthless test-tubing which formerly monopolised attention. But Physiology has profited most. Huxley, in 1880, complained of the lack of special classes—of the unreality, of the bookishness, of the knowledge of the taught. Now the subject is taught everywhere practically. Here and there, also, courses of so-called Physiological Chemistry have been instituted. Meanwhile the subjects themselves are grown to an extraordinary extent and a new science— Bacteriology—has sprung into existence. Consequently, the educational burden thrown upon the shoulders of the unhappy student of medicine is now far heavier than it formerly was, so that the effect of any improvements made in the teaching is more than counterbalanced by the increased demands.

If not at the parting of the ways, we are within measurable distance thereof. It is obvious that the medical man stands more in need than he ever did of *exact training* in chemistry— that medicine is very largely a branch of applied chemistry. " The chemist, the physiologist, the pathologist (the latter nowadays almost synonymous with bacteriologist) will be the physicians of the future and it is to their efforts that men may look for new and greater victories over the terrible power of disease "—are the concluding words in the article on *Science in Medicine* published in the preceding number of this journal. In the same number, in his article *On the Physical Basis of Life*, Mr. W. B. Hardy, F.R.S., speaks of the greatest advance since Huxley wrote on this subject forty years ago as being in the domain of chemical physiology. Dr. Bayliss, F.R.S., in the article on *Enzymes* deals with the highest problems in chemistry. Again, Prof. Halliburton, F.R.S., in his Report on Physiological Chemistry, in the volume of Reports on the Progress of Chemistry during 1904, published by the Chemical Society, comparing the condition of the subject with what it was some years ago, writes :

" If even a superficial survey of modern literature is taken, one is at once struck with the great preponderance of papers and books which have a chemical bearing. Chemistry is coming to be recognised more and more as one of the foundations of physiology and a mainstay of the art of medicine."

Indeed, the physiologists see quite clearly that lack of chemical technique is at present the greatest hindrance to progress from which they suffer. The problems they have to solve are the most difficult problems in chemistry. Let there be no misunderstanding therefore: medical men require a far deeper, more intimate knowledge of chemistry than that they possess at present. From this point of view, the resolution moved in the London Senate was a mistaken one and it was properly defeated: there must be no shortening of the time devoted to the study of the subject. But behind the resolution was the just feeling that the present course is a totally unfit preparation for medical practice—this is the feeling which we have to meet and provide for.

I saw that such was the case even when I first began to teach and adopted a somewhat unconventional course in consequence—but it was scarcely possible until recently to formulate a course which satisfied most requirements: our knowledge was too vague; the directions in which attention could be turned with advantage were not clear; in fact, the issues were scarcely before us.

Many things must now be done without further loss of time to secure the needed reforms.

Huxley's highly developed 'homoceatic' sense led him to assert that it was a mere affair of mechanical arrangement to provide the remedy for the shortcomings in medical education the existence of which he deplored—or, as he put it, "for that imperfection of our theoretical knowledge which keeps down the ability of England in medical matters" (he should have said, in all matters).

He urged that the theoretical branches of the profession— "the Institutes of Medicine," he called them—should be taught in not more than three central institutions. We are still talking of taking this step but doing very little towards making it an accomplished fact. The next thing to be done, Huxley said—he should have said to be done at the same time—was to go back to primary education and to insist upon the teaching of the elements of the physical sciences in all schools : to insist on the elements of chemistry, the elements of botany, the elements of physiology being taught in our ordinary and common schools, so that there shall be some preparation for the discipline of medical colleges. Teaching of the theory or Institutes of Medicine might then be confined, he argued, to physics as applied to physiology—to chemistry as applied to physiology—to physiology itself and to anatomy. Afterwards, when thoroughly grounded in these matters, the student might go to any hospital he pleased for the purpose of studying the practical branches of his profession. No better programme could be laid down now.

What has London done to see it carried out? Whereas formerly some knowledge of science was asked for from all candidates at matriculation, in the reconstituted University the subject is made alternative with Latin—truly a substitution of chalk for cheese. And snippet science is now required : the broad treatment thought of by Huxley, which was gradually coming into being in the schools, having been disallowed by the new regulations—to the great detriment of the schools. The University has taken many of the schools under its ægis and is seeking to influence them by inspection—but the inspectors are mostly amateurs ; no attempt is being made to utilise the services of those amongst us who really have paid attention to such matters, who could give useful practical advice to the schools 'as well as make them directly acquainted with the requirements of the colleges. Those concerned in the work generally and who determine what is done are no doubt actuated by the very best of motives ; like so many well-meaning people, however, they simply do not understand the problems with which they are nominally undertaking to grapple. Moreover, no attempt whatever has been made to " accredit" the schools to the colleges, in the manner followed in the Middle Western Universities of the United States : a plan full of promise, for it is said that in the States where this plan has been adopted the whole educational system has been improved and strengthened ; that the University is looked up to as a counsellor and friend of the schools, the University teachers learning much by continued intercourse with their scholastic colleagues and *vice versa*.[1]

[1] Cp. Prof. T. Gregory Foster's report in the Report of the Mosely Educational Commission.

It is much to be regretted that the scheme put forward by the Consultative Committee, a couple of years ago, for a school-leaving examination conducted in and partly by the schools and with direct reference to the work of the school—the one sane and safe way of testing their educational efficiency—was still-born : the Board appears to have lacked the courage to act upon the recommendation of the Committee and Heads of Schools openly declare that they are afraid of asking to be trusted ; it is only too clear that they are disinclined to accept the responsibility or take the trouble which the execution of such a scheme would entail. Unhappy England : everybody seems to be anxious to be helped by somebody else, in these latter days ; and when this is not the case—especially in the Civil Service—all excepting a very few highly placed persons are denied the opportunity of showing that they can act as responsible, discriminating individuals ; at most, the opportunity comes in later years when the desire and willingness to accept responsibility is lost through complete atrophy of the centres of independence.

Most unfortunately, University examinations are in the highest degree academic—medical students, as a rule, simply cram for them, knowing perfectly well that they will in a short time inevitably forget almost everything that they are forced to learn for the examination. The disciplinary value of the work they are called upon to do in chemistry, at all events, in most cases, is very slight, as it in no way brings them into touch with the problems of their profession or leads them to understand, in the slightest degree, the nature of the materials and the processes with which they have to deal in the living organism.

I despair when I think how little we have done since Huxley spoke out—when I think how much we might have done had we only all learnt to work harmoniously together—had we but been willing to co-operate. Enough proposals have been made. It would almost seem, however, that instead of being welcomed, any one is regarded as the common enemy who ventures even to suggest that improvement is desirable and possible and to point out the way in which change could be made with advantage. It is our peculiarly English way of progressing, I suppose : every reform has to be effected at the point of the bayonet.

However, the time cannot be far off when we shall be forced to use bayonets in reforming the medical curriculum, if change do not come quietly without compulsion. Medical men are becoming so increasingly important as guardians of the health of the community—it is clear that they can do so much when they are really well trained—that the public will not be content to stand by and see their education neglected, more particularly in its preliminary stages, as it is at present.

If the schools will but teach their pupils to read and to write English—to use books with profit as sources of information and inspiration and to express themselves properly—there should be little difficulty in training them when they become students at college, provided always that attention be paid also to the development of the faculties on the practical side: the use of fingers and of eyes *must* be encouraged, the art *must* be acquired of experimenting with a definite, logical purpose; powers of observation *must* be cultivated. When proper habits are developed in reading, writing and simple experimenting, all else will follow of itself—if we can only learn to build a sound foundation, the superstructure will not be difficult to erect. At present there is no proper foundation whereon to build: the aims and objects of the schools are so unreal—so out of touch with the requirements of life, their courses are so barren of information and interest, so little attention is paid in them to the manner of knowing and to what is known. Surely a University should help us to do these things—what is the use of our being banded together if we do not work together? Moreover, the problems with which we are confronted in teaching are of infinite complexity and difficulty—we need, therefore, to recognise our individual limitations and to help and support one another.

If we are to introduce the reforms which are needed into the medical curriculum, we must take a comprehensive view of the subject and seek to improve our system in all its stages.

The present Academic Registrar of the London University has himself directed attention to the almost dishonest way in which we conduct ordinary examinations—to the manner in which credit is given for nibbles instead of exacting honest answers. Not even a fishing club could be managed on such principles—the whole fish must be caught before the fisherman can score. Of course, on the other side of the picture, we have to consider whether the questions set deserve honest answers or can receive them even. But there cannot be a doubt, if simple practical questions were set with all due care and substantially correct answers were required, entrance examinations would be revolutionised and rendered an effective means of directing education : now they neither direct it nor do they ensure the supply of properly trained students to the colleges. The game is mainly one of Crammer *v.* Examiner, in which the former usually wins, the poor deluded examinee being the victim.

If the Chinese can resolve to enter upon so vast an interference with social custom as is involved in proscribing the use of opium, surely we can prescribe saner methods of examination. At present, most examinations serve merely to pick out the more studious, who are not necessarily the more able for service in the workaday world—in fact, there is not the least doubt that by examinations we are gradually selecting out a special kind of person, of a literary type, for all professions for which University study is a preparation ; professions in which alertness of mind and practical common sense are requirements are bound to suffer in consequence. Armchair study cannot develop these attributes : materials must be handled and processes and actions studied if such qualities are to be educed or cultivated to perfection.

The Indian Civil Service examination is a case in point : for its purpose, probably it is one of the most irrational tests that could well be devised ; some day perhaps we shall see that this is the case and admit that it is almost as barbarous a mode of testing those whose ability is to be gauged as is the method adopted by engineers in testing the strength of materials—that of smashing the pieces tested, so injuring them that they cannot afterwards be used. The moral injury done to candidates by forcing them just to cram is not considered ; the mental injury which must accrue is never thought of.

In higher examinations the standard is set at such a pitch that passing becomes impossible except to those who have the faculty of cramming—of acquiring knowledge of a mass of facts, of arranging these in orderly fashion and then reproducing them at word of command. It is a more than significant fact that women and Oriental students are taking the higher places in such examinations, which exactly suit the acquisitive type of mind. But it cannot be seriously held that this type of mind is that which we desire to select out of all others for the service of the Empire. It has been clearly demonstrated, even in recent times, that an attitude of philosophic doubt is not conducive to sane and serious government.

We need practically trained, useful and alert men, with fully developed powers of insight—not walking dictionaries: yet men who know how to use dictionaries when required. It is more than strange that we should continue to encourage what we know to be waste of brain-power by forcing our students to learn dictionaries off by heart and exact no proof that they really know how to use them.

There is also urgent need of reform in the case of the scholarship examinations at the Universities: in these again the standard set is preposterously high and altogether academic. The subversive influence of the examinations on the work of the schools is only too well known—yet neither the Universities nor the schools make the least effort to modify them.

It is not so long ago that we were influenced almost entirely by Oxford and Cambridge: now Universities are springing up everywhere; and County Councils are offering scholarships *ad libitum*; consequently, examinations are dominating education in all its branches: the outlook is very serious if, as I have contended, undue encouragement is being given to an unpractical type of scholar.

Whilst the Chinese—led by Japanese if not by Western example—are engaged in reforming their ways, we seem to be bent more and more on aping the methods they are in process of abandoning. We alone seem to be walking with eyes shut into the bottomless pit of dependence. In London, we seem to be thinking of nothing—talking of nothing—but examinations; we absolutely cringe before them, worship them. The last thing to occur to us is to talk of our work—to consider our methods of working. The fact is, we dare not: we know them to be so antiquated—so perfectly Erewhonian in style. In my early days, we took up subjects because they interested us or because it was represented to us that they would prove to be of value—not because we were compelled to undergo an examination in them. Thus, at Leipzig, although professedly a student of chemistry, I not only attended Ludwig's course but also the botany course; and a number of us persuaded one of the Professors to give us a short special course of anatomical demonstrations—on Sunday mornings, as he had no other spare time! Now there seems to be no honesty of purpose in us—every subject is forced on by an examination; each specialist in turn is allowed to come forward and insist that salvation is only possible if his subject be included in the compulsory curriculum. The correspondence early in December in the *Times* on the omission of Geography from the Higher Civil Service examinations is of this complexion—the underlying assumption seems to be that, unless it be examined on, the subject must necessarily fall out. If the art of reading were taught and proper books were written—not the pemmican type of manual now provided—Geography should be a subject for armchair study and quiet inquiry.

Let us admit what we know to be the case: that for all practical purposes the present type of matriculation examination is entirely unsatisfactory—not one which influences the schools in practical directions but simply a literary, Chinese form of examination; and that this is true of most other examinations. Having purged our consciences in this manner, we can begin afresh. "Fifty years ago," President Roosevelt reminds us in the message he has just delivered, "Japan's development was that of the middle ages. During that fifty years the progress of the country in every walk of life has been a marvel to mankind, and she now stands one of the greatest of civilised nations." Our educational system is still that of the middle ages—cannot we follow the example of Japan and reform it within the fifteen years that remain to make it fifty since Huxley made his appeal?

Some day perhaps people will awaken to the perception of the fact that most examinations serve the interests of those who prepare for and conduct them rather than the interests of the examined: they will grasp the simple fact that most examinations are highly remunerative, purely commercial undertakings, valued by teachers as giving bold advertisement. The present resistance to change is largely traceable to this circumstance. In those far off happy times, it will be possible, let us hope, to contract young people out of examinations by a money payment and so to save their souls. In the meantime, it is worth while to remember that Germany has achieved her

marvellous educational success entirely without adventitious
aids such as the University of London Matriculation, the
College of Preceptors, the Oxford and Cambridge Joint Board
and the University Local examinations afford us.

When Huxley's three central institutions are established in
London, we may hope that two of them will be attached to
hospitals : the technical flavour to be developed by mere
propinquity with the Clinic is considerable ; and it is of the
essence of success, in my opinion, that a technical flavour
be given to the training from the outset.   I know full well,
however, that opinions will differ widely on this point—
experiment alone can decide.   It is an unfortunate circumstance
that teachers of chemistry in the medical schools have generally
lacked the physiological instinct ; their work has had no
physiological trend.   *Hinc illæ lachrymæ*, in part, at all events ;
hence also the difficulty of instituting the necessary reforms.

To bring about reform in teaching the subjects of professional scientific study, some one must set to work to rough
out with considerable detail the elements of each course—no
light task, I am sure.   If appointed dictator in chemistry, I
should at once meet the demand of the medical faculty by fusing
inorganic with organic chemistry—there is no valid distinction
to be drawn between the two branches ; moreover, the study
of the former is rendered infinitely more difficult than it need
be by the neglect to introduce, at an early stage, considerations
derived from the latter—especially in discussing questions of
structure.   In teaching medical students, it will be desirable,
I believe, to introduce organic chemistry very early in the
course : at all events, for their purpose, the inorganic side may
be compressed into a very moderate compass and yet be made
to comprehend far more useful matter than is at present
included.   Thus, not a single instant need be wasted in
studying methods of preparing oxygen and similar trivialities—
there is more than sufficient oxygen in the air and in water
to satisfy human demands ; but the chemical properties of
oxygen, especially in water—the most wonderful, active and
useful of all oxygen compounds—will deserve most careful
consideration.   At present, water is just neglected : it is too
common a substance to consider.   Alcohol will also be worth
much attention—if only to defend it from undeserved strictures

and in order that it may be appreciated at its true value as the first cousin of water and a typical "alkaloid-ol."

Some of my scientific brethren will perhaps be aghast when I say that I would have the study of carbon compounds pursued mainly from the complex downwards—from the obvious to the unknown—that I would begin with the processes of digestion and fermentation and gradually develop a full understanding of the nature of every factor concerned in them.

Methods of preparing this, that and the other substance, all elaborate descriptions of physical properties, I would leave in the chemical cookery books—to be looked up when required. Function should be the main subject of study. The endeavour should be made to arrive at as clear an understanding as possible of vital processes—both of their nature and of the substances taking part in them.

The medical man is called upon to deal with the living human body and must gain the clearest insight into its workings: in the main, it is a chemical heat engine; and to acquire efficient control of its parts, such as the engineer has over the engine he attends, the medical man must be able to visualise the condition of each part and the processes which are operative throughout the body. It is to such ends that we must endeavour to teach chemistry—futile talk about the preparation of substances such as we now indulge in and mere constitutional formulæ must be abandoned and *discipline* substituted which will give rise to a feeling of true understanding and of sympathy with the subject-matter of study. It is possible, I fully believe, to engender such a state, if we will but give ourselves up to the work—provided always that students can be prepared at school to co-operate with their teachers at college. If the spirit of inquiry be developed in them and they can be induced to undertake the study of processes in a really serious manner, a true understanding of function—the indefinable sense termed feeling—will be gradually acquired.

It is difficult to say more: such things cannot well be expressed in words; the state can only be really known to those who enjoy it.

To summarise my conclusions: I am firmly of opinion that the medical curriculum, at least in so far as my own subject is concerned (if I be not mistaken in most others also), is in most urgent need of reform. The reform should take the direction of teaching the subject practically and with direct reference to its applications: as every branch of chemistry in turn must necessarily be laid under contribution, chemists need have no fear that their field of action will be thereby unduly limited. The course followed at present has little disciplinary value— there is much room for improvement in this direction at an early date. It will be unwise, however, to make radical changes until we are really clear what is desirable and what is possible—a scheme must be worked out thoroughly and carefully; and when put in action must be administered throughout a series of years without being subject to change by order. There must be no cast-iron rigidity in the scheme. As any system of control by external examinations would inevitably bar the path of progress, the central institutions should be free to administer their own affairs educational.

But no really effective change will be possible in the technical course of study until the schools have altered their ways: until proper habits are acquired at school the path of medical students at college must always be an uncertain one. It is needless to say more on so hackneyed a subject, except to point out that medical men may help their successors most materially if they will take every opportunity, each within his own sphere of influence, to affect public opinion.

Infinite possibilities are before us of ministering to the alleviation and cure of disease, as our knowledge of normal and pathological conditions increases. We are now wasting invaluable opportunities through our failure to institute systematic inquiry and in consequence of the lack of trained scientific observers among medical men. Every hospital affords abundant material for study: therefore every hospital should have its chemical laboratory devoted to scientific research work along physiological lines. The men in charge of such laboratories would, in course of time, become qualified to be teachers in the medical schools— for we need to recognise that teachers as well as students require training and full opportunity of gaining experience. It is the duty of the public, in protection of their own interests, to establish such laboratories.

THE CHEMICAL CO-ORDINATION OF THE ACTIVITIES OF THE BODY[1]

By ERNEST H. STARLING, M.D., F.R.S.

WE are accustomed to regard each act in the life of an animal as a link in a never-ending chain of adaptations to the environment, each act being a complex of a number of mutually adapted activities affecting very various parts of the body. This common action of many organs involves the existence of some nexus, some connecting or controlling mechanism. In many cases, and in every instance where the activity of one organ has to be swiftly adjusted to that of other parts of the body, the correlating mechanism is represented by the nervous system.

The *consensus partium* is, however, a characteristic, not merely of the higher animals, but of all organised existence, and is present throughout the whole of the plant and animal kingdoms, in many cases in the complete absence of a nervous system. In such cases the correlation between different parts of the organism must be effected by chemical means. The most marked reactions among the lowest organisms, such as bacteria, are those determined by chemical substances and spoken of generally as chemiotactic. Chemiotactic sensibility determines the aggregation of bacteria or other unicellular organisms around food, the collection of phagocytes in all classes of the animal kingdom round foreign bodies, and the union of the sexual cells both in plants and animals. In plants and the lower animals the transmission of an influence by chemical means from one part of the organism to another must be a relatively slow process. With the appearance of a vascular system and of a common circulating fluid bathing all the cells of the body, no chemical substance can be formed and discharged by any cell without being carried in a very short space of time

[1] The substance of this article formed the subject of an address given at the Naturforscherversammlung in Stuttgart, in September 1906.

to all the other cells of the body. It becomes possible, there-fore, for different parts of the body to carry out a common action, the regulating nexus being provided by chemical substances produced in the metabolism of one of the co-operating parties, and carried by the circulating fluid all over the body. The conception that among the constituents of the internal nutrient fluid of organisms are certain substances which function, not as foods in the ordinary sense of the term, but as excitatory substances (*Reizstoffe*), is one that has long been familiar to botanists, though even here we find an ambiguity of defining line between those bodies which are necessary, even in minimal quantities, for the building up of the framework of the cells, and those whose part it is to modify the functions of the already formed protoplasm. Foodstuffs are valuable in proportion as they furnish energy to the organism or material for its construction and growth. These " Reizstoffe " are, so far as we can tell, non-assimilable, and yield no appreciable amount of energy. It is their *dynamic* effects on the living cell which are of importance. In this respect they present a close analogy to the substances which form the ordinary drugs of our pharmacopœias. Since in the normal functioning of the body they have to be discharged at frequent intervals into the blood-stream, and carried onward by this to the organ on which they exercise their specific effect, they cannot belong to that class of complex bodies, which include the toxins, of animal or vegetable origin. These, which are supposed by Ehrlich to ape the part of a foodstuff and so be built up into the living framework of the cell itself, give rise, probably in consequence of this self-same property, when injected into the blood-stream, to the formation of antibodies. The formation of such bodies, in the case of the chemical agents of correlation, would annul their physiological effect. We must therefore conceive the latter as substances, produced often in the normal metabolism of certain cells, of definite chemical composition, and comparable in their chemical nature and mode of action to drugs of specific action, such as the alkaloids. This conclusion is borne out by the few in-vestigations which have been made as to the nature of the chemical messengers in the case of certain well-marked corre-lations of function in the higher animals. In consequence of the distinctive features of this class of bodies, and the important functions played by them in the higher organisms, I have proposed to give a special name to the class—viz. "*hormones*," from ὁρμάω = "I arouse or excite." A few examples of the most striking of these hormontic reactions may suffice to draw attention to the importance of this class of reactions, and to the promise, afforded by research in this direction, of further power in the hands of the physician.

The simplest example of a chemical correlation is seen in the mechanism by means of which the contracting skeletal muscle determines for itself an adequate supply of oxygen. Many years ago Miescher propounded the idea that the activity of the respiratory centre was determined by the tension of the CO_2 in the blood-plasma, and this in its turn would depend on the tension of the CO_2 in the alveoli of the lungs. This idea has been fully confirmed lately by Haldane and Priestley, and is borne out by the results obtained by Zuntz and his school. The effect of increased muscular activity, within physiological limits, is to augment the output of CO_2 by the muscles, and therefore to raise the tension of this gas in the blood. The immediate result is an increased activity of the respiratory centre. The respiratory movements are deepened and quickened, until the increased ventilation is just sufficient to reduce the CO_2 tension in the blood to its normal amount. If the muscular activity be carried to excess, so that the oxygen supply is insufficient to meet the total requirements of the muscles, there is a discharge into the blood of acid substances, such as lactic acid. These also will have the effect of raising the CO_2 tension in the blood and still more in the respiratory centre, so that the effect on the respiratory movements is even more marked than before. In this case the hormone is one of the commonest products of the metabolic activity of protoplasm. The chemical correlation, the fitting of the activity of the respiratory centre to the needs of the muscular system, is rendered possible by the development in the respiratory centre of a special sensibility to carbon dioxide. It is probable that the other hormones, whose action I propose to deal with, are also, in origin, products of the ordinary metabolic activity of some tissue, and that the evolution of the chemical correlation has been effected, not by the production of a special substance to act as a chemical messenger, but by the acquisition of a specific sensibility on the part of some other functionally related tissue.

It is in the alimentary tract that we meet with the most

typical examples of chemical adaptations. Take, for instance, the processes of digestion which occur in the duodenum. The researches of von Mering and others have taught us that, from half an hour to three hours after a meal, the pyloric sphincter yields at intervals to admit of the passage of the highly acid chyme, containing the first products of gastric digestion, into the duodenum. No sooner does this acid fluid enter the gut than there is an outpouring of the three juices which co-operate in intestinal digestion—viz. the pancreatic juice, the bile, and the succus entericus. The last named is the product of secretion of the glands lining the wall of the intestine itself. Their secretion might therefore be conceivably excited by the direct action of the acid chyme on the mucous membrane. No doubt a reflex contraction of the gall-bladder is an important factor in producing the inflow of bile. If, however, we establish a biliary fistula, we find that the entry of the chyme into the duodenum is followed in a minute or two by an actual increase in the amount of bile secreted by the liver itself. Here then are two glands, the pancreas and the liver, whose secreting portions are situated at some considerable distance from the primary seat of the stimulus—*i.e.* the duodenal mucous membrane. What is the nature of the connection between the mucous membrane and these two glands? A reflex secretion of pancreatic juice was observed by Claude Bernard on the introduction of ether into the small intestine, and was ascribed by him, as well as by all later writers, to the co-operation of the nervous system. The readiest means of inducing a flow of pancreatic juice, apart from the administration of a meal, was found by Pawlow to be the introduction of dilute hydrochloric acid into the duodenum, either directly or indirectly by way of the stomach.

In 1900 it was shown independently by Wertheimer and by Popielski that a secretion could be evoked by introduction of acid into the duodenum or upper part of the small intestine, even after section of both vagus and splanchnic nerves and destruction of the spinal cord. These authors concluded, therefore, that we had here an example of a reflex action carried out by the peripheral nervous system. It was to determine the conditions of this peripheral reflex that Bayliss and I took up the study of pancreatic secretion. We very soon found that the nervous system could play very little part in the so-called reflex.

We succeeded, for instance, in entirely severing all the nervous connections of a loop of intestine in the upper part of the jejunum, leaving it, however, still in vascular continuity with the rest of the body. The introduction of ·4 per cent. HCl into such a loop evoked a flow of pancreatic juice as profuse as that obtained early in the experiment, when the acid was injected into the loop while all its nervous connections were intact. We knew already from Wertheimer's experiments that the introduction of acid directly into the blood-stream was without effect on the pancreas. The only possible conclusion to be drawn from our experiments was that the acid acted on the epithelial cells covering the villi and separating the lumen of the gut from the blood-vessels, and that it was some substance, produced in these cells by the action of the acid, which was absorbed into the blood-stream, and carried to the gland to act as a specific excitant of the secretory cells. This conclusion was easily verified. On scraping off some of the mucous membrane, rubbing it up with acid, and injecting the hastily filtered extract into the jugular vein, we obtained within two minutes a flow of pancreatic juice greater than any we had observed as the result of the introduction of acid into the intestine. It was evident, therefore, that the nexus between the duodenal mucous membrane and the pancreas was not nervous but chemical. Under the influence of the acid, a new substance, which we may call pancreatic secretin, was produced in the epithelial cells to act as the special chemical messenger to call forth the activity of the pancreas. Although our observations have been fully confirmed by later workers on the subject, physiologists have not yet succeeded in isolating secretin. The fact, however, that it is not destroyed by boiling even in a strongly acid medium, that it is unaffected by gastric juice, that it is readily diffusible, and is not precipitated by the ordinary reagents for proteins and peptones, such as tannic acid or phosphotungstic acid, marks it out as a relatively stable body of definite composition, and probably of low molecular weight. It belongs, in fact, to the drug-class of physiological agents which we have designated as hormones.

Since the co-operation of the three juices—pancreatic juice, bile, and succus entericus—is necessary for the normal carrying out of the digestive processes of the duodenum, it would evidently be an economy of mechanism if the activity of the three sets of glands concerned in their production were aroused by one and the same means—*i.e.* if the secretin formed by the influence of acid on the mucous membrane of the duodenum were a secreto-motor agent, not only for the pancreas but also for the liver and the crypts of Lieberkühn. That this is the case for the liver has been shown by Bayliss and myself. In the case of the intestinal secretion, the evidence is not quite so clear. According to Delezenne, the intravenous injection of secretin causes a flow of succus entericus, at any rate in the duodenum and upper portions of the gut. On the other hand, Pawlow regards mechanical distension and the presence of pancreatic juice as the most effective agents in the production of succus entericus, while Frouin states that secretion of this juice can be excited by the injection of the juice itself, or an alkaline or neutral extract of the intestinal mucous membrane into the blood-stream. There is no doubt that the activity of the upper part of the gut differs markedly from that of the lower. In the one case, secretion predominates, in the other absorption, and it is possible that the varying results obtained by different observers really apply to different levels of the small intestine.

I may mention here one other chemical excitatory process in connection with the alimentary canal. Pawlow has taught us to recognise two phases in the secretion of gastric juice which follows the taking of a meal. The first phase of secretion is presided over entirely by the central nervous system, and is excited chiefly by appetite and by gustatory impressions acting through the brain and the vagus nerves. The second phase can be roused by the introduction of meat extracts or of the primary products of gastric digestion into the stomach, even after division of all its nerves. This second phase, which Pawlow regarded as probably determined by local reflex mechanisms, is really due, as shown by Edkins, to the absorption from the pyloric end of the stomach of some substance or other, a *gastric secretin*, produced by the action of the juice-arousing constituents of the food on the pyloric mucous membrane. From the cells of this membrane the gastric secretin is absorbed by the blood, and carried all over the body, exciting, as it passes again through the walls of the stomach, the activity of the glands lining the whole viscus.

In all these examples of chemical correlation the primary effect of the hormone is to excite increased activity of the responsive organ. Such a state of increased functional activity cannot be without significance for the *nutrition* of the tissues concerned. We know that the most effective means of exciting hypertrophy of any organ is to augment the calls upon this organ—*i.e.* to give it increased work to do. We should expect therefore that the indirect effect of these hormones or Reizstoffe would be an improved nutrition, and possibly increased growth, of the organs concerned. Another group of correlations exists in which increased activity is but an indirect effect, the primary result of the action of the hormone being diminution of activity, accompanied by increased assimilation and hypertrophy of the tissue.

The most striking instances in which *growth* is the primary effect of a chemical stimulus derived from some distinct organ are to be found in the correlations existing between the generative organs and the rest of the body. Although the manner in which this correlation is brought about has been the subject of speculation for many years, it is only quite recently that any attempt has been made to apply experimental methods to its explanation.

Especially interesting is the mechanism by which growth is aroused in the mammary glands. These organs are present in both sexes at birth in an immature condition. At puberty for the first time a difference appears between the mammary glands of the two sexes, a rapid growth taking place in the female simultaneously with the commencement of the ovarian functions. During the whole of sexual life the glands remain in the female at the same stage of development, unless pregnancy occurs. The onset of pregnancy acts as an impetus to a further great development of the gland substance, which continues with ever-increasing rapidity throughout the whole of pregnancy. At parturition the growth of the glands at once ceases, and one to three days later we find that the activity previously spent on growth is now applied to the secretion of milk—a secretion which, if the gland be emptied periodically, may last many months.

Since the whole cycle of changes can be prevented by removal of the ovaries, we must regard these organs as primarily responsible for the growth of the mammary glands, though whether they are the direct source of the impulses which determine the special growth during pregnancy, or whether these arise in the uterus, placenta, or fœtus, must be determined by experiment. That the impulses cannot be nervous in character is clearly demonstrated by the experiments of Eckhard and Ribbert, and especially by those of Goltz and Ewald on the effects of extirpation of the spinal cord. Since pregnancy occasions hypertrophy of the mammary gland, and parturition is followed by the onset of lactation, in the total absence of any possible nervous connection between the pelvic organs and these glands, it is evident that the correlated growth of the mammary glands must be determined by chemical substances arising in the pelvic organs and carried to the glands by the blood-stream. Knauer has shown that, whereas extirpation of both ovaries puts an end to the periodical changes in the uterus, which are responsible for the phenomena of "heat," both ovaries can be transplanted, thus dividing all their nervous connections, without abolishing the phenomena. In this case therefore the connecting link must be chemical rather than nervous.

In the case of the mammary glands we have to determine in the first place why the secretion of milk appears only at the end of pregnancy, and in the second place the origin of the stimulus which during pregnancy is responsible for the hypertrophy of the gland.

With regard to the first point, Hildebrand has suggested that during pregnancy some substance circulating in the blood exercises an inhibitory influence on the dissimilatory changes in the gland cells, which he regards as autolytic in character. Although it is extremely improbable that the chemical changes, which characterise activity, are identical with the autolytic changes occurring immediately after the death of the gland cells, the conception of a substance causing growth by acting as an inhibitory agent or, in Hering's nomenclature, having an assimilatory effect, is extremely valuable. According to this notion, so long as the inhibitory substance is circulating in the blood-stream, so long must growth of the mammary tissue proceed. With the removal of the source of the inhibitory hormone, which takes place at parturition, the gland tissue, built up to a high level of function, will undergo autonomous dissimilation, i.e. will enter into a state of prolonged activity. Miss Lane-Claypon and I have found that artificial cessation of pregnancy in the rabbit at any time during the first fourteen days—that is before any secretory alveoli have been formed—gives rise merely to retrogressive changes in the gland. If, however, pregnancy be brought to an end at any later period, activity is set up in the secretory alveoli, and a secretion of milk is the result. That this secretion is due to the removal of a stimulus and not to the production of a new stimulating substance in any of the involuting sexual organs is shown by the fact, well known to clinicians, that lactation ensues even after complete removal of the pregnant uterus with its appendages.

With regard to the second question, viz. the origin of the inhibitory hormone, the fact that double ovariotomy during pregnancy does not interfere with the hypertrophy of the mammary glands excludes the ovaries as the direct source of the stimulus. A careful study of the clinical evidence has led Halban to the belief that the source of the hormone is to be found in the chorionic villi and the placenta. His evidence is not, however, absolutely convincing, and we therefore sought to determine the question by the injection of extracts of fœtuses, of ovaries, of placentæ, and of uterine mucous membrane into virgin rabbits, in the hopes of producing, by one of these means, a hypertrophy of the mammary glands, similar to that which would be produced by the occurrence of pregnancy. It was evident to us, before we began our researches, that it would be difficult, if not impossible, to present any stimulus to the mammary glands which would be as effective as the normal one. For wherever the mammary hormone is manufactured, the manufacture must be assumed to proceed continuously. There is therefore a constant leakage of the active substance into the blood, and it is probable that the amount of this substance produced increases with the duration of pregnancy. At no time will the mammary gland be set free from the influence of this specific stimulus. On the other hand, however we might prepare our extracts of the tissues, we could not expect to get more than the amount residual in the tissue and caught, so to speak, in its progress through the placenta into the maternal blood-vessels. This amount we might inject into our rabbits, but it would probably be taken up and absorbed into the circulation long before we were ready for our next injection, so that, whereas under normal circumstances the mammary glands during pregnancy are being constantly stimulated to hyperplasia, we could not expect in our experiments to do more than give these glands a series of small shoves in the same direction.

In spite of the inherent difficulties of the research, we succeeded in six cases in producing in virgin rabbits a growth of the mammary glands, similar to that occurring during the early stages of pregnancy, and consisting in the proliferation of the epithelium lining the ducts, and the multiplication of these ducts by branching into the surrounding tissues. In one of these experiments, where our injections were carried out during five weeks, and the rabbit received the fluid extract of as many as 160 fœtuses, there was an actual formation of secreting acini towards the periphery of the gland. In all these cases the extract was derived from fœtuses. In a number of other experiments in which we injected extracts of uterus, placenta, or ovaries, we obtained no growth whatsoever. We are therefore justified in concluding that, under normal circumstances, the growth of the mammary gland during pregnancy is determined by a chemical substance, a hormone, which is produced mainly in the growing embryo, and is carried through the placenta by the blood-stream to the gland. The smallness of the effect obtained in our experiments, in comparison with the large amount of tissues employed, shows that the quantity of this hormone present at any given time in the tissues must be minimal, and that, in all probability, when injecting extracts of fœtus, we are simply injecting the small amount of material which is diffused through the juices on its way to the blood-vessels and into the maternal circulation.

Our experiments throw no light on the seat of production of the mammary hormone in the fœtus, nor do we yet know whether it may be split off by simple means from some precursor in the fœtal tissues, and so obtained in larger quantities, as is the case with pancreatic secretin. We have a certain amount of evidence that in one respect the mammary hormone resembles secretin or adrenalin, *i.e.* in the fact that it can be boiled without being deprived of its properties. The other questions as to the seat and nature of the specific substance, as well as the influence of various reagents in splitting it off from some possible precursor, must be left to future investigations.

These three examples may serve to show that it is possible by chemical means to influence either the functional or the nutritional condition of a tissue, in the direction either of increased or diminished activity, and that such means are normally employed by the animal body for co-ordinating the activities and growth of widely separated organs.

Other examples of chemical influences exerted by one organ on other parts of the body are known in which the final effect is not confined to one organ, but is manifested throughout the whole body. In some of these cases the diffuse character of the response is determined by the widespread distribution of some special reacting tissue or function. I need only mention in this connection the important control wielded by such organs as the suprarenals, the thyroid, the pancreas, and the pituitary body on the general metabolic processes of the organism. In the case of the first of these organs, we know that the medullary part secretes a drug-like body, adrenalin, into the blood-stream. This part of the suprarenals is derived in development from the sympathetic system, and is but one of a series of such organs. The function of its specific product is apparently limited to the sympathetic system. Adrenalin, as Langley and Elliott have shown, acts on every tissue of the body which receives a nerve supply from the sympathetic system, and in every case the effect of its injection is the same as that which would be produced on the tissue by electrical excitation of the sympathetic nerve. Thus it causes dilatation of the pupil, secretion of thick saliva, constriction of the blood-vessels, acceleration of the heart, relaxation of the muscular walls of the small and large intestine, contraction of the ileo-colic sphincter as well as of the uterus, and either contraction or relaxation of the bladder, according to the action of the sympathetic on this viscus, which varies from one species to another.

In the case of the thyroid gland, it is difficult to say whether its active principle, which is apparently the iodine-containing body, iodothyrin, first prepared by Baumann, is to be regarded as chiefly dissimilatory or assimilatory in its effects. It is certain that its presence in the circulating fluids is a necessary condition for the normal development of all the tissues of the body, especially of the bones, in the growing animal, but the effect of its administration to adults is to stimulate dissimilation. It increases the output of urea and may cause a rapid disappearance of fat from the body.

The essential influence exercised by the pancreas on the carbohydrate metabolism of the body was demonstrated nearly twenty years ago by von Mering and Minkowski, who described the fatal diabetes which ensued on total extirpation of this organ. Their experiments, with those of later observers, make it almost certain that the pancreas yields some internal secretion to the circulating fluids of the body, the presence of which is an indispensable condition for the assimilation of sugar either by the liver or the muscles. All attempts to imitate the action of the living pancreas by means of extracts of the organ have so far failed. If, however, this internal secretion is of the same nature as the other bodies, which I have included under the term "hormones," it should be possible to isolate, by some means or other, the active principle from the gland and, by the introduction of this substance into the blood-stream, to materially influence for good those cases of diabetes in man, which are due to pancreatic disease.

The important part played by internal secretions in the regulation of the activities of the entire body has long been realised by physiologists. The special point which I have endeavoured to emphasise in this article is that these internal secretions or hormones, as I have called them, are of a relatively simple chemical character, that they are susceptible of isolation and even, as in the case of adrenalin, of synthesis, and that their action is not that of a foodstuff, but of a drug, depending, as it does, on the chemico-physical configuration of the molecule, and not on the presence of haptophore groups, which would determine their assimilation into the living protoplasmic molecule. These hormones are widespread in their distribution and their effects, and we may hope that further investigations along these lines will place in our hands an armament of potent agents by which we may control many of the most important functions of the body.

ROCK-FOLDS

By PROF. ERNEST H. L. SCHWARZ, A.R.C.S., F.G.S.

Rhodes University College, Grahamstown, South Africa

A BED of sandstone or shale buried deep in the earth's crust becomes as adaptable to mountain-building forces as if it were so much putty, no matter how hard and intractable it may appear when exposed on the surface after the overlying material has been carried away by denudation. This adaptability is not inherent in the rock itself, in the same way as plasticity is a character of putty, elasticity that of india-rubber, and so forth, but is due to the solvent action of water, which at moderate temperatures and pressures is able to carry away portions that are under compression and to deposit the substance again in positions of tension. The action is well illustrated in the rounded limestone pebbles of the Nagelfluhe, on the foot-hills of the Jura, where the pebbles may be seen flattened; or, again, in the tiny ovoid grains of oolitic limestones which have been subjected to pressure, as in those of Ilfracombe. In the case of quartz and silicates which are not soluble in water at the surface of the earth, the case would appear to be more difficult of explanation, for we find that the originally water-worn pebbles of the older conglomerates are not only flattened, but are often pulled out into long fingers, as we may see in many of the hand-specimens of the Cango Conglomerate in Oudtshoorn, South Africa; in North American rocks the same feature has been beautifully illustrated by Mr. McCallie. Experimental evidence in favour of this solution of silicates by pure water has been afforded by Barus; on subjecting water contained in capillary tubes to considerable pressure and a temperature of 185° C., the volume of the water apparently decreased, but it was found that the loss was accounted for by the fact that some of the glass had been dissolved and redeposited in a crystalline state, the latter being a state in which, owing to the closer packing of the molecules, the material is denser and occupies less space than in the glassy condition. The rate of solution and redeposition was found to be half an hour for a given volume of water to dissolve and redeposit an equal bulk of glass, at the particular temperature of the experiment, namely, 185° C. If we now calculate supposing that the natural minerals in a rock are ten times as resistant to solution as the glass in the capillary tubes, and that the water held in the pores of the rock is one-tenth per cent. of the whole bulk, at 185° C. the rock could be wholly dissolved and redeposited in 5,000 hours, or considerably less than a year. As a matter of fact, however, most crystalline rocks, on analysis, yield more than ·1 per cent. of water, some schists and gneisses holding as much as 2 per cent. by weight or 5 per cent. by volume, which means that such rocks could be wholly dissolved and recrystallised every 100 hours or, say, once a week, at a depth at which a temperature of 185° C. exists, roughly some five miles beneath the surface.

In a homogeneous bed of sandstone at five miles depth within the earth's crust, a pressure acting upon it for a few days would fail to be transmitted, because every grain would allow itself to be transformed into disc, flattened normal to the pressure. Such an extreme case does not often happen, because the weight of the rocks above prevents the vertical flow which such a horizontal contraction would necessitate, but so much accommodation would be provided for in the bed of sandstone that a force striving to be transmitted through it would soon be lost. Supposing the rocks immediately above and below the sandstone were perfectly rigid and water-tight, then the force to be transmitted through the sandstone would act as if it were applied to a liquid; but in nature rocks are not water-tight, and they are so far from being rigid that, apart from their power of being dissolved and redeposited, they are always ready to give way to any pressure owing to the constant tremors which agitate the earth's crust. If then we recognise that silica and silicates become soluble under pressure force cannot be transmitted through a stratum of rock in the zone in which solution and redeposition can go on. Above this zone the rocks are fractured; cavities can and do exist, and water no longer permeates only the rock substance, but circulates in the fissures as well. Such material is not a fit one to transmit great and prolonged pressures; if we look at the folds in any considerable mountain range we shall find that if we are to imagine that this top fractured segment of rock is to account for the transmission of the lateral pressure which buckled up the mountains, it must have been thrust several miles distance over the underlying strata. The actual mileage does not concern us here, but the fact remains that rocks, fissured and structurally weak as we find them, must, on such a supposition, have been bodily transported, despite their weight, and have carried with them a battering power sufficient to crumple up rocks as resistant, at any rate, as themselves. It is evident, therefore, that neither in the zone of fracture nor in the zone of flow can lateral thrust be transmitted for any distance in the earth's crust.

The difficulty is beautifully illustrated by the behaviour of the sheet of lead in the bottom of Mr. Mellard Reade's famous sink. Lead is too pliant to transmit pressure, yet at one end of the sink, in response to a disturbing force represented by the alternate cooling and heating effect from the water let in by the hot and cold water-taps, there arose a well-defined anticlinal ridge at the other end of the sink. The explanation of this, I believe, lies in the expansion and contraction of the metal, which sets up compressional waves; these pass through the main body of the lead sheet, producing very little effect until they are retarded by the fixture of the sheet to the sides of the sink. The sum of these constantly repeated waves produces a shift in the particles of which the lead is composed, leading to permanent distortion; the return wave, which would have shifted the particle back to its original position, being damped out by the sides.

If we now fix our attention on a grain, say, in a sandstone which is undergoing elevation in a mountain chain, and if we suppose that there is a constant propagation of earthquake waves from north to south, and that these waves are retarded by some means, let us follow what would happen. The grain lies surrounded with water, which is ready to dissolve a portion of its substance directly some differential stress acts upon the particle. The wave comes along, drags at the obstruction, and produces a tendency to distortion. The sand grain under such circumstances would have a little of its substance dissolved from the area in which the pressure momentarily acts, and this would be shifted and redeposited in the positions of tension. For any one wave the substance affected might be very small indeed, but the aggregate in the centuries would allow the grain to be wholly transposed, and the effect, being cumulative, would not only allow the grain to travel onwards indefinitely, but, all grains in the rocks adjacent being similarly affected, the whole rock mass would move, and follow the direction which the wave surface would take in a perfectly elastic medium. This explanation of rock deformation accounts for the existence of perfectly unaltered rocks between the origin of the thrust which buckled the earth's crust, and the resultant

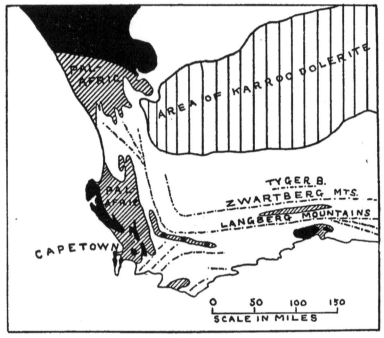

FIG. I.

mountain chain, and it provides an effect for the incessant tremors which sensitive seismographs, like the bi-filar pendulum, show are being propagated.

In the south-western portion of South Africa we find a very close imitation of Mr. Mellard Reade's sink, or rather, one of the far corners of it ; the newer Cape rocks—the Neo-afric beds, we may call them—rest on an older series, the Pal-afric, in such a way that these latter form a right angle, and they are further stayed by heavy intrusions of granite, which accentuate the angular outcrop. Keeping our illustration in view, this granite would represent the sides of the sink, and the Neo-afric beds the layer of lead; the Pal-afric beds also would represent the wooden bottom of the sink, for they pass under the Neo-afric series. The disturbing cause, the hot and cold water of our illustration, is represented by a very heavy intrusion of dolerite in the form of sheets in the central portion of the Neo-afric region. The disturbance is represented by several ranges of closely folded mountains which follow the lines of the granite external to them.

It was for a long while a matter of dispute as to whether the force which buckled up these mountains came from the interior or from the side of the ocean. But it has recently been found that on the southern coast, a little to the east of George, where the granite ends, there is a significant change in the character of the folds. Where the granite bars the way to the sea the folds in the mountains internal to it are closely huddled together, and follow the granite outcrop in trending east and west. Where the granite ends, the folds widen out, and curve seawards.

This feature on the south coast might be explained by the granite having been thrust northwards as an immense ram into the yielding strata, but then the same features exist on the west coast, where the granite runs north and south. It is unthinkable that two extended masses of granite should have travelled, the one northwards, and the other eastwards, and have clamped the later rock-formations between them. It has taken ten years' continuous field work to prove this point, but it has been accomplished. As regards the disturbing cause, the dolerite intrusions, it is hard to convey any sense of the magnitude of these. They cover an area of over 70,000 square miles, and the separate sheets are ranged one above another, sometimes forming laccolites, when the whole country for miles is one black mass of dolerite; more extensively they occur as sheets, one of which is exposed in the western Karroo, having a superficial area of 3,000 square miles, but the buried portion of this was probably four times as much again. The sheets, also, may be from 200 to 300 ft. thick. With so great an amount of molten material thrust into the Karroo rocks some effect must have been produced by the expansion, letting alone the space occupied by the intrusions themselves;

whether they melted out a passage for themselves, or simply occupied pre-existing fissures, is a problem with which we are not concerned here. The folded mountain ranges of the coast are a sufficient effect. Between the folded mountains and the escarpment, where the dolerite sills begin, there is a denuded area showing occasional flexures, but on the whole consisting of nearly horizontal shales and sandstones traversed by quartz veins. This last feature seems important, for it shows that, while the disturbance of the dolerite intrusion was being transmitted to the outer edge of the Neo-afric region, the intervening tract was so far from being in compression as to be traversed by open fissures, which subsequently became filled in with vein quartz. Curiously enough, this last feature

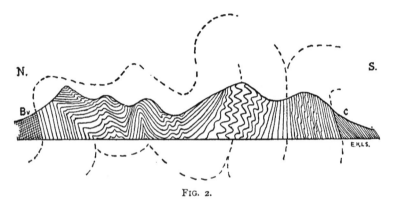

Fig. 2.

is faithfully reproduced in miniature in Mr. Mellard Reade's sink, for, in the centre of the sheet of lead, there are a number of transverse fissures, in this case gaping and empty.

Such, in brief outline, is the character of the folding in the south-western portion of South Africa ; the period during which it was in progress was approximately in late Jurassic times. The details are almost infinite in their variety, and of never-failing interest to the geologist, but I will confine attention here to the following types, which are of more general interest : (A) Folds in which the beds fail to adhere in the angles, (B) folds in which stronger beds have caused the weaker ones to flow at a greater rate than themselves, and (c) folds which have, as it were, tumbled on themselves, and have produced vertical pressure.

The area from which I will take my illustrations is that of

the Zwartberg Mountains; they form the innermost of the coast ranges of the south of Cape Colony, and extend from Ladysmith, in the Colony, past Oudtshoorn and Prince Albert, to Willowmore and Uniondale; that is to say, due east and west for a length of 170 miles. We can obtain some idea of the nature of the broad folding by laying an S on its back, thus: s. ∽ n., the syncline being on the south side and the anticline on the north. External to the anticline there are added, in places, several subsidiary folds which broaden out the range. On the south side there is a subsidiary anticline which, however, is only seen at the ends; in the middle, in the greater portion of the range, this anticline has been denuded away and the southern limb is found at a distance of several miles away. The rock is mostly coarse sandstone, the Table Mountain sandstone, bedded in various degrees of thickness; in one part, right in the heart of the mountains and almost inaccessible, there is a narrow valley filled in with the overlying Devonian shales. The ends plunge down beneath younger strata.

(a) *Folds in which the beds fail to adhere in the angle.*—Where well-bedded rocks are sharply folded there are frequently found spaces which are sometimes gaping, at other times filled in with veinstone and ore-bodies, as in the Bendigo goldfields and the Broken Hill mine, in Australia; but the feature I wish to illustrate here is one in which a series of beds have been bent, and have, as it were, acted as a nut-cracker on the rocks internal to them. This pre-supposes a certain tensile strength for the rock, which is a remarkable feature, considering the material is being folded at the same time that it is exerting this property. If we take a bar of iron and heat it red-hot in the centre so that we can bend it, we reduce its cohesion, but in these rock-folds the cohesion remains apparently the same during as well as before and after the movement.

A magnificent illustration of this is shown on the east side of the north end of Meiring's Poort, a tremendous gorge in the mountains. The fold is a gigantic "knee-bend," that is, a fold with the limbs straight, instead of being curved ; in this case one limb is horizontal and the other is turned under through an angle of 135°; that is to say, the whole forms an overfold. The face exhibiting the feature is practically vertical, and is, roughly, 2,000 ft. high. It will be seen that there has been a strong tendency for the beds to adhere to one another, and to retain their normal width, unlike many examples where there has been parting between the beds, and flow of the strata in the fold, so that the limbs become thin and the axis thick. It will be seen

FIG. 3.

that the beds in the angle of the bend have been crumpled backwards; to realise the tension on the outer beds which this has necessitated, take a pile of loose sheets of paper with each hand,

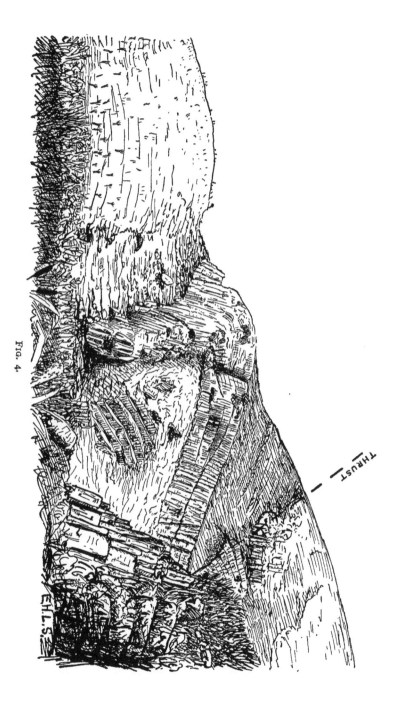

FIG. 4.

THRUST

holding the thumbs below and the four fingers above; the reverse ˙ fold on the interior can easily be produced, but the drag on the outer layers can be felt.

The second illustration I take from the Kouga Mountains, which lie to the east of the Zwartberg range. These form a wild and inaccessible block of country, traversed by wild kloofs and precipices, and were the last stronghold of the Bushmen in the south-west of the Colony. A narrow slip of Devonian shales lies folded in with the sandstone on the southern flanks, and the abrupt syncline shows many of the typically strange folds with which we are dealing. The particular fold I illustrate lies along the Schrikke river, and shows the topmost bed of the sandstone folded into a knee-bend, the southern limb dipping south at about 30° and the northern nearly vertical. The beds below the resistant bank have been caught in the nip and have been brecciated; only a small block of stratified sandstone appears through the rubble, representing, apparently, part of the rocks which were originally near the actual axis, and which have reached their present position by having been crushed backwards.

There are three ways of regarding this flexure: as having been produced by (1) the rock having been molten, (2) the rock having given way along slip or gliding planes, (3) the rock grains having been transposed by solution and redeposition. The first is clearly disproved by the conditions of the rocks; the second is an agent in the deformation of the rocks which is a potent factor in mass static conditions, but not, I think, under the conditions that prevail in rock-bending; while the third, which I assume here is the principal agency of rock deformation, renders the term "tensile strength" somewhat misleading. The actual tensile strength of rocks is measured by the modulus of elasticity, which in sandstones works out at anything between 30,000 and 400,000 lb. per square inch; that is, a force represented by these figures would be required to stretch a bar of the rock a square inch in section to twice its length. This is an actual property of the rock as we quarry it, but the tensile strength, used in the sense in which I have done, refers to the resistance to solution and redeposition, and can only be used comparatively, with regard to other rocks which simultaneously were subjected to the same temperature and pressure for the same time. It is, perhaps, more a co-efficient of the porosity, or amount of water held up within it, than of any property possessed by the component grains, since it is manifest that rock holding more water will have a greater amount of solvent to effect the transposition of material than one with less capacity. This feature comes out very clearly in the next example, where the nipping beds have broken along the axis and forced the nipped beds to flow between "jaws" of rock.

(B) *Folds in which stronger beds have caused the weaker ones to flow at a greater rate than themselves.*—The best example I know of illustrating this type of fold lies out on the Karroo to the north of the Zwartberg Mountains. Pressure has come upon

FIG. 5.

rocks divided into two groups, a lower one composed of close-grained sandstones, and an upper made up of hardened glacial till, the Dwyka Conglomerate. The latter has parted in the axis of the anticline, and the pressure has caused the lower sand-stones to squeeze through a narrow slit in the overlying forma-tion, and they now form an abrupt range of hills rising from the plains, called the Tygerberg. Doubts have been thrown on the anticlinal nature of this inlier, but the two ends plunge normally beneath the Dwyka Conglomerate and clearly demon-strate its nature. If we concede that the capacity for holding solvent water is a measure of a rock's readiness to flow under pressure, then this curious structure is easily explained, for a

sandstone composed of angular or rounded grains always contains more water in its interstices than a compact sandstone or
shale, and must, therefore, flow more readily.

Fig. 6.

A more convenient example for illustration is a small rock-
fold to be found at Angelier's Bosch, east of Prince Albert, along
the north of the Zwartberg Mountains. Here we have a small

S-fold, faulted between the syncline and anticline; the sandstone for the most part is thin-bedded, and would, from the multitude of bedding planes, contain a considerable amount of moisture. Interbedded with this is a massive bank of quartzite, about 4 ft. thick, and the rate of flow of this and of the thin-bedded sandstones caught in the nip of the bend was very different, as is apparent in the sketch. The actual tensile strength of a small bar cut from either the thick bed or from one of the thin beds would be about the same, and certainly the difference would not be sufficient to account for the flow in the axis of the anticline;

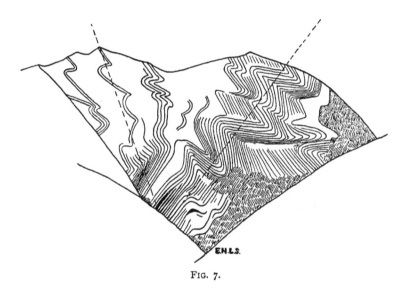

FIG. 7.

generally, therefore, taking the evidence from this small fold, together with that of the Tygerberg, this type affords abundant evidence in favour of the solution and redeposition theory.

(c) *Folds which have, as it were, tumbled on themselves, and have thereby produced vertical pressure.*—I take my illustration of this type from the great gap sawn through the Zwartberg Mountains near the eastern end, called Meiring's Poort, though similar examples may be seen in any of the western poorts. The mountains rise to nearly 7,000 ft. above sea-level, while the river which has sawn through them lies only between 1,000 and 2,000 ft. above. Towards the southern end the anticline has been raised above its supporting strength, that is, the weight of material piled vertically above the sandstone has been sufficient to crush it. The rock having been wet, the pressure has been accommodated by folding, not on a horizontal plane such as we are familiar with, for instance, in the Jura Mountains, but on a vertical plane, and as a result we have a miniature mountain range standing, as it were, on end. Seen from the point of view from which the sketch is taken, the folding appears to have been actually vertical, but on either side one may see similar folds whose planes lie inclined to the vertical and fan out from a centre near the level of the river. In the minor folds there are evidences of horizontal faults which look like thrust-planes due to tangential pressure, but I regard them as slip planes produced normal to the pressure, in the same way as slaty cleavage develops, that is, by compressional flow.

In offering the above illustrations I have been guided by the desire to select examples which have more than a local or purely geological interest. The study of rock-flow has a very important bearing on the true understanding of the strength of materials, and our South African folded mountains contain a vast number of instructive examples. The geological survey of the Cape Colony is carried out on too broad lines for details such as these to be adequately studied, but the questions of rock-flow have been so prominently brought to the notice of geologists by the writings of Van Hise, now collected together in that author's treatise on Metamorphism, that even while occupied in tracing boundary lines the examples in the Zwartberg Mountains could not be overlooked.

THE INTERNATIONAL FISHERY INVESTIGATIONS

By JAMES JOHNSTONE

University of Liverpool

WHEN, in July 1899, the Swedish Government invited the governments of the countries interested in the fisheries of the North Sea to join in a scheme for the joint investigation of that area, there appeared to be little hope of arranging concerted action of this kind. International regulation of the North Sea fishing area was, of course, a very familiar thing. Ever since 1842, when the Governments of Great Britain and France arranged a convention for the settlement of fishery disputes, some kind of international regulation had always been in force. But the necessity for this had always been apparent; disputes between fishermen of different nationalities working together on the high seas were then, as now, by no means infrequent, and the settlement of these by ordinary legal methods was always a difficult and protracted affair. So, although the arrangement of fishery legislation on an international basis was a somewhat laborious process, the absence of such was more inconvenient still. International conventions have therefore been arranged at different periods for the treatment of ordinary disputes between the fishermen of England and France, and later on of those of most of the other countries fishing in the North Sea, and these have, on the whole, worked very satisfactorily. It was, however, a very different affair to arrange for joint action in matters of international fishery research by scientific methods.

That this should be so is a rather surprising thing when one reflects that there must be many fishery restrictions which, if they are to be of any real use, must be based on an intimate knowledge of those facts of natural history which have reference to the life histories of the fishes and other marine animals

583

caught by the fishermen for the public markets. Fishery regulations have three main objects: (1) economic ones, such as the early "semi-commercial" legislation of the Scottish Fishery Board, which aimed at the building up of a national fishing industry by the application of a thorough-going system of bounties; or, in our own day, that of the Congested Districts Board of Ireland; (2) police regulations pure and simple, which have no other object than the prevention of disputes between the fishermen of the same or different countries working in close proximity to each other; and (3) restrictions on modes or times of fishing intended to conserve the public fish supply. It is always necessary when discussing fishery legislation to distinguish between these three main categories of regulations, for their aims are implied rather than explicitly stated in the statutes. The first two may conceivably be devised without any profound knowledge of marine natural history, but the latter must certainly be founded on a correct appreciation of the bearing of what we know of the habits, etc., of fishes in the sea. If our knowledge of these things is scanty, it must be extended before we can legislate with any degree of success.

Twenty-five years ago fishery restrictions were practically unknown. Those who had been concerned with the administration of the industry had been influenced by the economic doctrines current in the middle of the nineteenth century, and the earlier policy which taught that it was the duty of the State to foster by artificial means the national industries had fallen into disfavour. Previously to 1870 the fishing industry was the object of a great number of legislative restrictions, but under the influence of Huxley, Spencer Walpole, and others, a "liberalising and liberating" policy had been put into operation, and the fisheries were left to take care of themselves—no great hardship at the time. Bounties had before this time disappeared, and now fishery restrictions were with very few exceptions swept away. In the beginning of the 'eighties the fishing industry was about to experience a process of remarkable development which can be traced to three causes: the adoption of steam as a method of propulsion of fishing boats, the invention and general use of the otter trawl (this, however, came later on), and the application of methods of cold storage to the preservation of the catches made by fishing boats. The immense development of the fishing industry in the 'eighties and 'nineties

is to be attributed to these improvements and not to the removal
of restrictions on the modes or seasons of fishing.

But this expansion of the fishing industry soon led to a
reaction against the earlier policy of *laisser faire*. As fishing
increased the older fishing grounds began, one by one, to be
less productive than they had been. It was not an actual
exhaustion of these areas with which the fishing industry was
confronted, but rather a decrease in the density of the fish
inhabiting them; a decrease due, no doubt, to the greater
exploitation which they were undergoing. During the last
decade of the nineteenth century steam fishing boats were built
in great numbers, and it was found that whereas successful
"voyages" of fish could at one time be made in the home
waters—the North Sea, etc.—it was now necessary to go farther
afield. There is, of course, at the present time productive
fishing both in the North Sea and the other fishing grounds off
the British coasts, but nevertheless steam fishing vessels do
go now to the coasts of Iceland and to those of Spain and
Morocco to obtain the best catches. About 1895 this decreased
productiveness of the home grounds had begun to attract
attention, and many proposals were made for a renewal of
fishery restrictions on a large scale.

The feeling in favour of the reimposition of fishing restrictions had indeed produced legislation before this time. In 1885
the Fishery Board for Scotland obtained an Act which enabled
it to close portions of the territorial waters against trawling—a
method of fishing which was then regarded as prejudicial to the
fishing grounds if practised on a large scale. Economic reasons
were also responsible for this legislation, but with this we have
nothing to do in the meantime. Again in 1887 and in 1889 the
Board obtained further powers, with the result that, by the end
of the latter year, not only the territorial waters off the coasts
of Scotland, but also large areas lying without these limits were
made *maria clausa*. In England the same thing took place.
The decade 1880–90, which witnessed so great an increase in
the activities of local authorities, saw also the creation of local
fishery authorities and the multiplication of fishery officials
and restrictions. In 1888 power was given to the county and
borough councils to form fishery committees, and these were
soon taken advantage of, so that at the present time there are
over twenty of these authorities on the coasts of England and

Wales, administering some three hundred by-laws. Each authority has its own code of regulations, which may, or may not, have any relation to those in force in contiguous areas. Practically no form of fishing may be pursued anywhere round the coasts of this country unless regard is paid to the local by-laws. Sometimes this system of minute regulation is rather ridiculous. In one of our estuaries, for instance, a fisherman must not take mussels which are less than 2½ in. in length, but if he crosses an imaginary line he may take the same shell-fish if they are not less than 2 in. in length. On some parts of the coasts of Wales seaside visitors have had to submit the toy nets employed to catch prawns in the rock pools in order that the local fishery officer might satisfy himself that the latter were of the regulation size of mesh. No doubt some of these regulations are of utility, but it is still the case that most of them were founded, not on the results of precise investigation, scientific or otherwise, but only on the basis of local opinions and prejudices. There is, again, no doubt that a large number are not of any utility, and there is a growing disposition to inquire whether this system of minute regulation—a system which is both costly to the ratepayer and vexatious to the fishermen—has been justified in its results.

But while the growth of fishery restrictions during the last twenty years has been comparatively rapid, our knowledge of the natural conditions under which fishes live in the sea, and even of the economic conditions of the fishing industry, has accumulated very slowly. It is apparently the case that those who in this country have been responsible for the elaboration of fishery restrictions have not thought it necessary to base these on a statistical and scientific investigation of the local conditions of the industry for which they were legislating. The history of one of the largest of the English sea fishery committees is interesting in this respect. This authority began its career with an almost full complement of by-laws and a staff of fishery officers. With increasing experience it added to its code of regulations and its staff, but it was only after some time that it attempted scientific investigation, and then such work was regarded as a luxury, if not a fad. Then, under the stimulus of criticism, doubts began to arise as to the utility of the by-laws, and the stringency of these was gradually relaxed, and finally scientific investigation was begun. It seems incredible, but it is nevertheless the case, that in England the whole system of local fishery legislation was devised in almost entire ignorance of the conditions, scientific or otherwise, under which the local fisheries were carried on. The natural result is that now, after twelve or fifteen years of experience of local fishery supervision, we are still in ignorance as to the results of all the restrictions (save one or two at the most) which have been in operation during that time.

The scientific investigation of the sea fisheries began in Great Britain with the old "Board of British White Herring Fishery" which had charge of the herring fisheries of Scotland. When the Fishery Board for Scotland was reconstituted in 1882 definite provision for the prosecution of scientific research in connection with the industry was made, and under the various scientific members of the Board investigation was actively carried on. In England fishery research work was first instituted by the Marine Biological Association, which was founded in 1884. Much later, the local fishery committees were created, and one or two of these authorities began scientific investigations in a very tentative kind of way. It was not until the beginning of the present decade that a really adequate scheme of investigation was suggested. I have said that when the increased exploitation of the North Sea by modern fishing craft began, a distinct falling-off in the abundance of certain kinds of fish was experienced. This, of course, was much discussed by the trawling trade, and in casting about for a remedy attention was directed to the enormous quantities of small and comparatively valueless fish of certain kinds which were being landed from various fishing grounds in the North Sea. It was thought that the destruction of these small and immature fish was at least one of the causes of the growing impoverishment of many of the older fishing grounds, and attempts were made to procure legislation which would put a stop to this form of fishing. A number of bills were therefore introduced into the House of Commons, but none of these were successful in obtaining passage into law. That of 1900 probably went farther than any of its predecessors, for it resulted in the appointment of a Select Committee of the Lower House which made a recommendation which had important results.

The Select Committee of 1900 did not recommend the passage of the Immature Fish Bill of that year. They recognised that the evidence was not sufficient to justify such an interference with the conditions of the sea fishing industry, that there was not enough information as to the question, and that the contemplated legislation—a prohibition of the landing of fish of certain classes under certain specified sizes—was after all a measure which it was advisable should be treated internationally. They therefore recommended that steps should be taken to secure this international treatment, and the bill was dropped.

Almost at this time efforts were being made to obtain international investigation, at least, of the question of the connection of physical changes in the sea with biological changes, and with climatic phenomena. During the decade 1890—1900 various investigations into the cause of the movements of sea water in north European seas had been made by the Scandinavian and Scottish hydrographers with interesting results. It was, however, found that if decided results were to be obtained, it would be necessary to deal with the question on a much larger scale than had up to that time been possible. It was felt, too, that it was just these results which would in all probability supply the clue to much that had been very puzzling in connection with the migrations of fishes such as the herring. In 1899 a meeting was held at Copenhagen at which representatives from most of the States interested in the fisheries of the North Sea attended. The result of this conference was that a preliminary programme was drafted and submitted to the various governments. A further conference was held at Christiania in 1901, and some assurance of support was held out which resulted in a third meeting at Copenhagen in 1902, at which a definite scheme of research was agreed upon so as to include more of direct fishery investigation than was contemplated at the previous meetings.

In this year the Governments of Great Britain, Germany, Holland, Denmark, Norway, Sweden, Finland, and Russia (Belgium joined later on) consented to the institution of an international organisation for the study of the hydrography and biology of the sea from the point of view of the regulation of the sea fisheries, and the Conseil Permanent International pour l'Exploration de la Mer was finally created.

In its present form this body consists of representatives from each of the States participating in the investigations, with the inclusion of expert members. It delegates its powers to an

International Bureau and to a Central Laboratory at Christiania. The Bureau is under the direction of Dr. Hoek, a well-known Dutch zoologist, while the Central Laboratory is controlled by Nansen, the explorer and zoologist. The general direction and co-ordination of the investigations are carried out by the Bureau under the authority of the Council, and the principal reports are also published there. At the Central Laboratory apparatus of uniform pattern is devised and standardised, check analyses made, standard samples of sea water prepared and furnished to the various national staffs employed in the work, and investigations made with the object of securing uniformity of the analytical results.

The actual investigations are carried out by the various national staffs. A national organisation has been set up in each country which provides for the carrying on of the investigations by the establishment of laboratories, scientific staffs, and exploring vessels. In Great Britain the actual work was handed over to the Scottish Fishery Board and to the Marine Biological Association (the two best equipped bodies for the biological study of the sea in Britain). Each of these bodies has the control of one or more vessels for research at sea. Work ashore is carried out in Scotland at the Fishery Board Laboratory at Aberdeen, and at University College, Dundee; while in England the Marine Biological Laboratory at Plymouth and a similar institution at Lowestoft are the corresponding institutions. A similar arrangement exists in each of the countries participating in the investigations. The area over which the latter are carried on is a very extensive one, and comprises the whole of the North Sea, the Norwegian and White Seas, North Atlantic to the north and west of Scotland, the English Channel, and the sea off the west coast of Ireland. This extensive area is divided up between the various countries.

Three main lines of investigation were contemplated by those who drafted the Christiania programme. Recognising that a knowledge of physical changes in the sea—such as changes in temperature, salt contents, movements of great bodies of sea water and the like—was fundamental, particular attention was paid to the elaboration of a scheme of hydrographical research. Then came the biological work based on the collections made by the exploring vessels, and including such studies as those of the life histories of fishes and other marine animals of economic interest; their food, reproduction, and migrations; and the connection between the invertebrate population of the sea and the associated fish fauna. Lastly there were the statistical studies based on the fishing experiments made by the exploring vesels, or on the commercial statistics collected by the various fishery authorities in the North Sea countries. It is not to be wondered at that the value of each of these lines of research was differently appraised by the fishery people of the participating countries. The Scandinavian investigators laid much stress on the hydrographical researches not only from the point of view of the fisheries, but also from that of the assistance which such studies were likely to afford for the forecasting of weather changes. The Germans and Danes were more inclined to emphasise the value of the biological investigations into the life histories of fishes and other animals. Then in both England and Scotland particular attention was paid to statistical studies, as for instance the results of the fishing voyages of the exploring vessels and the commercial statistics of the fishing fleets. These have been worked up with the view of elucidating the movements of the fish shoals. There is, of course, much room for difference of opinion as to the relative merits of each of these lines of research, but one is safest in coming to the conclusion that each is essential and that none can be omitted if we wish really to understand the problems connected with the sea fisheries; and none can be neglected without serious detriment to the lasting value of the results hoped for in the end.

Hydrographical investigation did not begin with these international researches. Long ago, during the voyage of the *Challenger* in 1872-6 a great number of observations were made in the seas of the globe, and later on, in both Sweden and in Scotland, a considerable amount of work of this kind was effected. But it is safe to say that until the development of the methods of the last half-dozen years all such work was more or less imperfect and likely to lead to erroneous conclusions. It was essential that the hydrographical observations should be made simultaneously over wide areas, and by instruments and methods the accuracy of which was beyond reproach. What is actually done in these investigations is to determine the temperature of the sea at different levels from the surface down to the bottom—by no means a simple proceeding. Then the salinity—or salt contents—is also to be determined at corresponding situations. Temperature records are obtained by the use of delicate thermometers which have previously been standardised under the direction of the Central Laboratory. In order to ascertain the temperatures at the sea bottom and at intermediate levels it is necessary to procure a sample of water from the required depth in such a manner that its temperature does not appreciably rise while it is being obtained. The Nansen-Pettersen water-bottle is employed for this purpose, and the perfection of this instrument is one of the most creditable results of the international investigations. This apparatus is sent down to the required depth open, and when it has reached the level from which it is desired to obtain the sample it is closed by means of a "messenger," and the water, contained in an insulated vessel, is then brought on deck and the temperature is read on the contained thermometer. Salinity is determined from water samples collected in a similar way by ascertaining the amount of a standard solution of silver nitrate required to precipitate the contained salt. From tables prepared by the hydrographers of the Central Laboratory the total amount of solid matter contained in a unit quantity of the sea water is then calculated. Other properties of sea water are of course observed, but these two—the temperature and the salinity—are sufficient to define the hydrographical constitution of the water studied. Similar information is afforded by the identification of the plankton, or the microscopical fauna and flora of the water. By means of such observations it has been found that the water in different parts of the sea possesses slightly different properties.

In the North European area we find usually in the sea water which may be derived from three main sources: (1) Gulf Stream water, the salinity of which is always over 35 per 1000, and the temperature of which is relatively high; (2) colder and less dense water from the Arctic Sea ; and (3) fresher water from the great rivers and from the Baltic Sea. Over wide areas these various components may be traced, forming more or less well-defined drifts or streams, and often well-defined strata which lie on each other in the order of their specific gravities. They may remain distinct without mixing or they may diffuse into each other by the action of winds and tides, forming water bodies of composition intermediate to those of the three main categories I have mentioned above.

The key to the water circulation of the seas of Northern Europe is to be found in the study of the Gulf Stream. One may still read in the text-books of physiography that this great current reaches the shores of Britain and the Scandinavian countries. It is indeed true that sea water of sub-tropical origin does reach our shores, but the case is not so simple as it is usually stated. The Gulf Stream sweeping up from the Gulf of Mexico forms a closed eddy which never actually impinges on the shores of Great Britain, but turns round, forming a cyclonic circulation or eddy the northern boundary of which is in latitudes 40° to 50°. In the centre of this eddy is the well-known "Sargasso Sea." Now in some way or other a current does take origin in this eddy which reaches the shores of Northern Europe. This is the European Stream. Part of this, but a comparatively small part only, enters the English Channel and, passing through the Straits of Dover, flows into the North Sea. The greater body of water flows to the north, and passing through the Farőe-Shetland Channel enters the North Sea round the north of Scotland, and even the Baltic through the Skagerak and Cattegat. A considerable volume of this comparatively warm and dense water still flows to the north and east, and rounding North Cape enters the Barentz Sea, and soon cooling down, sinks beneath the surface and flows on as a deep current.

Add to this drift of North Atlantic water into the seas of North Europe a southerly current flowing down the east coast of Iceland into the Norwegian Sea, and we have the general scheme of the water circulation of the north Atlantic and the northern ocean. Now if these streams be attentively studied by means of temperature and salinity observations, it is seen that the flow of water is not uniform. Again, the key to the study of these seasonal variations is to be found in the variations in the extent of the Gulf Stream circulation, the boundaries of which undergo a periodic shifting. In March the northern boundary of the Gulf Stream touches the Azores, but does not impinge on the coasts of Africa or Europe. In November, however, the area of the stream has expanded greatly in a north-easterly direction and touches the coasts of Africa and Southern Europe, but not those of Britain. In the following March the stream again contracts to its former area.

It is these variations of the Gulf Stream current which are the causes of corresponding variations in the water circulation of the North Sea, the Norwegian Sea, and the Baltic. There is always a drift of relatively warm and dense water passing over the Wyville-Thomson ridge—the elevated ridge of sea-bottom which joins the Shetlands to the Faröe Isles. But the volume of Atlantic water which passes over this ridge is not always the same, but increases during the winter months and attains a maximum in the spring. Although the water in the "Norwegian branch of the European Stream" is heavier, owing to its greater salt contents, than the water normally present in the Norwegian and North Seas, it is, by reason of its higher temperature, lighter, and so it floats on the surface. Increasing in volume during the winter, it gradually covers the greater part of the central area of the North Sea, attaining its maximum expansion in March. Not only does it affect the hydrographic condition of this area, but just the same thing occurs in the outlying areas ; the Skagerak, the Baltic, the seas round Iceland, and the remote White Sea become annually invaded by a warm-water current, or heat wave, which at different seasons, according to the locality, attains a maximum and then subsides. It is the study of these seasonal variations, their periodic and unperiodic changes, which has now become so very important a department of marine research in relation to the changes in the condition of the sea fisheries. A hydrographic picture of the condition of the seas of North Europe would be as follows: an increasing flow of warm and dense water from the Atlantic during the winter months, covering the sea to a variable extent and attaining a maximum according to the topographical relations of the land and sea; then subsiding and becoming replaced in the North Sea by less dense and much colder water from the land and from the Baltic; and in the Norwegian and White Seas by cold and relatively light water from the Polar Sea.

Interesting as these changes are from the point of view of the oceanographer, they possess much greater interest for the student of meteorology and fisheries, for it is now beyond doubt that the hydrographic changes in the condition of the northern seas produce very important changes in the climate of the countries the coasts of which they affect. This is a very obvious thing when we consider that an enormous amount of heat is annually brought to northern countries by the vast body of warm water which washes their shores after drifting up from the sub-tropical Atlantic. Not only so, but it has clearly been proved that the barometric pressure of the atmosphere is also closely affected by the changing temperature of the sea, and obviously the boundaries of the icefields without the coasts of the ice-locked countries are to a very important extent also dependent on the temperature of the sea. But how can the abundance of fishes in the sea be affected by the temperature and salinity of the water? This a problem which until a few years ago had hardly come within the range of the methods of the naturalist. The Scandinavian zoologists did indeed attempt to investigate it with reference to the migrations of the herring off the coasts of Sweden, and *prima facie* evidence was furnished for the belief that this fish followed in its movements those of a body of sea water of well-defined physical constitution. When the international fishery investigations were initiated five years ago on a really adequate scale, it became possible to seek for the evidence that not only the herring but also a host of other fishes were affected by variations in the temperature and salinity of the water in which they lived.

The establishment of such a connection meant of course the investigation of the life histories of the fishes which form the material of the commercial fisheries. This enormous task had already been partially accomplished when the international investigations were commenced in 1902. Since 1880 many workers in Britain and on the Continent made out much of the habits and life histories of marine fishes. In Scotland, MacIntosh and the St. Andrews school of zoologists, in England the naturalists of the Marine Biological Association, in Germany the zoologists of the Kiel Kommission, and in Denmark the marine biologists of the fishery authority, all had done much to trace the migrations, spawning habits, development, and distribution of the commoner kinds of edible marine animals. The result was that at the beginning of the present century our knowledge of the life histories of fishes in the sea had already attained considerable dimensions. Still there were of course considerable gaps in this body of information, many of which at the present time are far from being filled up. Then much of this knowledge did not go very far towards the solution of the numerous problems which confronted the fishery legislators at the beginning of the 'nineties. Still less was a knowledge of the developmental histories of fishes likely to aid in the establishment of a connection between the physical changes in the sea which we have been considering and the migration habits of fishes and other animals. Quite another kind of information was required—that concerning the distribution and abundance of fishes at different times in the year and in different regions of the sea.

If we consider the life history of one of the better-known marine fishes such as the plaice, we will find it to be somewhat as follows : At a certain time in the year the fishes which have attained a certain age and size begin to spawn. Some degree of concentration of the fish on roughly defined spawning areas occurs and emission of the eggs then takes place. These eggs rise to the surface of the sea, drift about there while undergoing development—a process which requires about a fortnight or longer, according to the temperature of the sea. When the larva or young fish hatches out from the egg, it drifts about passively at, or near, the surface of the sea for a certain time until it undergoes the metamorphosis through which it passes into the juvenescent stage in its life history. It then seeks the shallow waters on the sandy coasts and there it remains for the next two years of its life, feeding and growing rapidly. Then begins the migration into the deeper parts of the sea, during which the fish is assuming the adolescent stage and the reproductive organs are arriving at their phase of maturity. In the fourth or fifth year of life the plaice has attained sexual maturity and begins to produce ripe eggs.

Such a life history, though applying precisely to the plaice, is also characteristic of many other marine fishes. Generally speaking, we can divide fishes into two main groups, one of which, the demersal or bottom-living fishes, have a life history which in its broad outlines resembles that indicated above. There is also a less numerous group of fishes with a life history which does not differ greatly from that of our typical fish, but which, in their adult stages, exhibit quite different habits —that is, they swim about at the surface and at intermediate depths of the sea more or less in shoals. The best examples of this class of pelagic fishes are the herring and the mackerel.

Again, we may divide fishes into two great categories according to the degree to which they perform migrations. One group, the best example of which is the plaice, may be regarded as semi-sedentary—that is, they do not perform lengthy migrations, but remain through the year on much the same fishing grounds with but little change of habitat. Soles and turbot also belong to this class. The other group contains fishes like the cod and hake, which are notorious wanderers, and visit with more or less regularity the same fishing grounds from year to year at approximately the same seasons.

It is just this general knowledge of the life histories of the fishes which is of importance to legislators. We require to know where and when the fishes spawn and where the eggs drift to, where the young stages are reared and nourished, what is the nature and extent of the annual migration (if the fish is a conspicuous migrant), how the fishes are distributed according to depth of water, what kind of food they prefer, and generally what are their habits and mode of life.

Many series of facts of this nature existed before the international research organisation came into existence ; but on reviewing the knowledge of these subjects which is indicated in the literature previous to 1900, it is surprising how many gaps existed. It would be foolish to say that these gaps have been filled up, but the contributions to the subject of the life histories of fishes during the last five years have been numerous and important. The defect of the fishery investigations carried out previous to the beginning of the present century was that the observations of the life history of fishes had mostly been such as were possible with but an imperfect equipment for extended work well out at sea, and, as we now know, with very imperfect fishing apparatus. The great merit of the newer investigations is that they have been prosecuted over wide areas, and simultaneously, by many workers and vessels : parts of the north European seas, which previously had never been adequately investigated, have during the last five years been subjected to a searching analysis of their fish populations. I may here refer to the investigations of the Russians in the Barentz Sea, and to the corresponding work prosecuted by the Norwegians in the seas round Iceland and in the Norwegian Sea itself. Before 1902 fishery investigations were practically restricted to such work as could be carried on in the immediate neighbourhood of marine laboratories in inshore waters, and by small vessels. Much indispensable knowledge was acquired by these means, but it is obviously essential that if the conditions of the sea fisheries are ever to be thoroughly understood, work over the whole of the fishing grounds must be undertaken.

Returning again to our typical fish, the plaice, we find that the investigations carried out by the fishing steamers of the international organisation during the last five years have enabled us to form a picture of the distribution of this fish which we could not previously have constructed. The fishing experiments and the study by the naturalists of the various countries of the statistics afforded by the commercial fisheries have shown us how the plaice is related, as regards its size and abundance, to the depth of water in which it is to be found. We have found that, with a few exceptional cases, the fish is smallest and most abundant in shallow waters, and largest and least numerous in the deeper parts of the North Sea. So constant is this relation that it is almost always possible to state what is the size of plaice on any fishing ground, given a knowledge only of the general depth of water. On the Dutch coast it has been possible to draw contour lines round the coast, and within each zone there are only to be found plaice of a certain size. Expressed in language borrowed from the mathematicians, we can say that the size of the plaice on any area of varying depth of water is a function of the depth.

A method of investigation first developed by the international naturalists on a large scale, that of marking and liberating living fishes, has enabled us to form some ideas as to the ordinary migrations made by the fish at different stages in its life history. The study of the deposition of limy matter in the bones and in the ear-stones has given us a convenient means of determining the age of the fish, so that by making fishing experiments on a large scale, and by the examination of the ear-stones or otoliths, and supplementing these observations by a method of statistically grouping the fishes caught according to their sizes, we are now able to map out the sea with reference to the prevailing size of the plaice to be found in the different zones and regions. The persistent fishing experiments of the exploring steamers have also enabled us to determine the limits of the areas where there is to be found a characteristic small plaice population : such areas are to be regarded as nurseries, from which the adjoining parts of the deep-sea area are recruited. But the further study of these small-fish areas has also shown that there is a definite relation between the prevailing size of plaice on a ground and the abundance of the fishes there, and, generally speaking, a small-fish area where the plaice are extraordinarily numerous is also an area where the fish are dwarfed and are smaller, and grow much more slowly, than in such regions where the density of the population is less.

A very great deal of the work of the international fishery staff has been of the nature of a survey of the fishing grounds of the seas of North Europe, made from different points of view. Thus the English and the Dutch have very largely concentrated their attention on the study of the distribution of the plaice, with results which I have just indicated above. The Norwegians have devoted some considerable attention to the cod fishery, which is *the* fishery in that country, just as the fishery for the herring is the predominant form of the industry in Scotland. The great cod fisheries round the Lofoten Islands on the coasts of Norway have been studied with particular care. Annually in the spring the cod shoals invade this area to deposit their eggs on the shallow sea bottom surrounding the islands and banks, and it is at this period that enormous numbers of the fishes are captured by the Norwegian fishing-boats. The fate of the eggs resulting from this spring spawning has been studied by utilising the Norwegian exploring steamer. For a limited period the eggs are to be found on the surface of the sea, just where the adult spawning fishes are most abundant at the bottom. Then the developing eggs drift slowly to the west and north under the influence of the prevailing current of water which sets in that direction. As the eggs drift to sea they undergo embryonic development, and after a certain time, according to the temperature of the sea, the little fishes hatch out and begin their pelagic career. As we pass outwards from the focus of distribution, the larvæ resulting from the hatching of the eggs become larger, and there are definite zones which have been mapped out by the Norwegian naturalists, each of which is characterised by the presence of cod-fry of different degrees of development. By-and-by these little cod disappear from the surface of the sea—it may be at a distance of about 120 nautical miles from the place where they were hatched—and settle down to their demersal stage of existence at the sea bottom. This is the outline of the life history of the cod shoals which inhabit the Lofoten banks at the spring of the year ; but corresponding investigations carried out in the North Sea, the Cattegat, and other regions have shown that the course of things is not always the same in other regions. The size of the spawning adult fishes is slightly different in these fishing grounds; the time of spawning is also different. It has been found, for instance, that the cod may spawn in some parts of the North Sea in the autumn; the distribution of the fry and the size of the latter, when it first seeks the bottom, may also vary. All these varying relationships of the adult fish to spawning-places, to depths of water, to currents, and the time of spawning, as well as the time taken for the young fish to assume the demersal habit, are essential for such a comprehension of the natural conditions of the fisheries as is necessary for the purposes of legislators. Most fisheries laws are, in fact, local ones, and must be founded on the conditions of the area to which it is intended to apply the legislation.

It has been the aim of these international investigations so to co-ordinate them that there is no unnecessary duplication of the researches. Just as the English and the Dutch have devoted particular attention to the study of the plaice, and the Norwegians to that of the cod, so the Danes, while by no means restricting their work, have been assiduous in the attempt to elucidate the life history of the fresh-water eel—a fish which has considerable importance for Danish fishermen. It has long been well known that the eel does not spawn in fresh water, but, after passing a certain time in the rivers, assumes a " bridal dress" of bright scales and descends to the sea. So much we have known for a considerable time, and we also know from the now classical researches of Grassi and Calandruccio in the Mediterranean that the eel goes down into abysmal depths before it deposits its eggs. It has been assumed rightly that the same process of spawning also takes place in our more northerly latitudes, that the eel also goes away somewhere into deep water to reproduce. But this is entirely an inference from the work of the Italian naturalists; and, indeed, these investigators did not actually obtain the larvæ of the eel from the depths of the Straits of Messina, but from the surface, where the little eels were brought up by reason of the exceedingly strong currents which prevail in that region. It remained for investigators on the Danish international exploring steamer to place the hypothesis of the Italians beyond the region of mere probability by discovering the larvæ of the European fresh-water eel in the open Atlantic Ocean south-west of the Faröe Islands, and at depths of over 1,000 metres. There is no doubt that the eels of North Europe must traverse the North Sea, or the English Channel, before they can reach the depth of water, and the other essential surroundings, for their reproduction. In a paper which is likely to be one of the classics of marine natural history, Schmidt has worked out the migrations of the eel for the purpose of the act of reproduction, and the converse migration of the young and immature fishes, or *Leptocephali*, from their birthplace in the deep Atlantic to the rivers of Northern Europe.

We have seen that an important part of the work of the International Fishery Organisation has been the study of the hydrographical changes in the waters of the seas of the north European area—that is, the determination of the yearly cycle of variation of temperature and salt-contents in the waters of selected representative areas. I leave aside the discussion of the value of such investigations for the student of the climate of North Europe, and will refer here very briefly to the connection which we now know to exist between these changes in the physical characters of the sea and the abundance and migrations of fishes. Some years ago, in the case of the migration of the herrings of the Swedish coast fishery, it was shown beyond doubt that this fish regularly inhabited a stratum of water of definite constitution—not Atlantic or strictly coast water, but a mixture of the two known as "bank water." The herring came and went with this stratum of bank water. Within the last few years there has been reason for the belief that the herrings which form the great summer fishery off the east coasts of Britain are also influenced in their annual migration and shoaling movement by the annual influx of Atlantic water into the North Sea from the European Stream. Not only so, but we have evidence that other well-known migratory fishes which appear to come to the British fishing grounds with a considerable degree of regularity are also under the influence of the same current of warm and dense water. The establishment of such connections between the migrations of the fishes of the type of the hake, cod, and herring with the hydrographical changes in the sea remains as the most difficult of fishery problems, and among the things which still await thorough investigation.

It is now quite certain that the metabolic processes of marine animals must be influenced by the changes in temperature and other physical properties of the water in which they live. The sea is to an animal inhabiting it exactly what the atmosphere is to a terrestrial animal such as a bird or a mammal. Indeed, a curiously close parallel can be drawn between a migratory bird living on the land and a migratory fish in the sea. In each case we have well-defined periods of reproduction, and periodic migrations, which are repeated from year to year with a certain regularity. Atmospheric changes, temperature, etc., affect the breeding seasons of the terrestrial animal as well as its migratory cycle. It is *a priori* probable that the changes · in the sea which we have already considered also affect the reproductions and migrations of the fish. We have, indeed, direct evidence of this in the influence that temperature exerts on the incubation period of the eggs of a marine fish. The period of development is so dependent on the temperature that it can almost be expressed mathematically. We have seen that in the case of the herring a body of sea water of a certain salinity has been associated with the migrations of this fish, and there is evidence that such a connection also exists in the case of the migrations of other fishes. We have here direct relationships, but it is quite certain that the influence of physical changes in the sea is usually an indirect one, and that fishes and other marine animals are affected by changing temperature and salinity, because these factors exert a more profound reaction on the animals and plants which form their food. On the land the great outburst of vegetable life in the spring has the most intimate effect on terrestrial animals. In the sea there is a similar development of vegetable life in the early months of the year. Among the commonest organisms in the sea are the diatoms or unicellular plants, and these are influenced in the closest manner by physical and chemical environment.

The increased light and temperature of the early spring leads to an increased assimilation by the diatoms of the inorganic materials which have been accumulating during the relatively cold and dark months of the winter. Under the lowered temperature of the winter bacterial activity has been reduced in the sea, and as a result the inorganic compounds of nitrogen on which the diatoms feed have been increasing. More intense sunlight leads to increased assimilation by the chlorophyll these organisms contain. A luxuriance of diatom life results in the sea during the months of February to April, and this has an indirect influence on the general fauna of the sea. The ultimate food matter in the sea is these inorganic compounds of nitrogen, but only the marine plants, and a few forms of animal life, can assimilate such materials. Upon these organisms, upon the diatoms practically, the whole of the animal life in the sea must ultimately depend. Small animals, and whole groups of others, feed chiefly on these lowly organised plants, and are themselves preyed upon by other forms of life. Thus the sedentary mollusca living at the sea bottom live almost entirely on diatoms, and microscopic animals like foraminifera and radiolaria, and in their turn these mollusca form the food of fishes like the plaice. It is evident that the abundance of fishes like the plaice may be conditioned by changes in the sea which lead to a more or less luxuriant growth of diatom life.

Thoroughly to demonstrate the connection between the physical phenomena taking place in the sea and the abundance of useful animals therein requires the study of hydrographic events on the one hand, and of fishery statistics on the other. It is only by the consideration of the catches made by the fishing fleets that we can form really useful estimates of the varying abundance of fishes on the various grounds ; and it is also by observations of the movements of the fishing vessels from ground to ground in pursuit of the fish shoals that we can properly follow the migrations of the latter. Both these lines of investigation are necessary if we wish to establish thoroughly the causes of the migrations of the fishes, or determine what are the factors which operate in producing local abundance or scarcity in particular fisheries. These are preeminently the kinds of investigations for which an international organisation is necessary. It is quite impossible that the hydrographical researches can be carried on otherwise than by a number of vessels working simultaneously over the whole of the north European area. Fishery statistics are at the present time in a most unsatisfactory state, both in Great Britain and in the other North Sea fishing countries. Much information that is quite essential in the inquiry into fundamental fishery questions is not afforded by the fishery statistics of this or of any other country; the statistics that do exist are collected and tabulated differently in the various fishing countries. It is certainly most desirable that in the reconstruction of the organisation for the joint prosecution of fishery investigations, which will be entered upon in the course of next summer, the necessity for the retention of some machinery on an international scale, both for the collection and collation of fishery statistics, and for the continuance of the present hydrographical investigations, should be considered.

BIBLIOGRAPHY

The more important publications dealing with the International investigations are:

1. Published by the International Council—
   Rapports et Procès-verbaux, i.-v., particularly iii. and v.
   Bulletin des Résultats.
   (Both the above in English and German.)
   Publications de Circonstance (English, French, and German).
2. The Danish publications—
   Meddelelser fra Kommission for Havundersøgelser. Series *Fiskeri, Hydrografi,* and *Plankton* (in English).
3. The German publications—
   Beteiligung Deutschlands an den Internationalen Meeresforschung.
   Wissenschaftliche Meeresuntersuchungen, Kiel Kommission.
4. The English Blue-Books—
   Fishery and Hydrographical Investigations in the North Sea and adjacent regions. 1. Cd. 2612, 1905; 2. Cd. 2670, 1905.
5. *Journal of the Marine Biological Association.*

Most of the above are periodical publications, appearing annually or more frequently.

THE RELATIONSHIP OF MINING TO SCIENCE

By W. E. LISHMAN, M.A., M.Inst.M.E.

The meetings of the British Association for the Advancement of Science always leave behind them food for reflection, and last year's meeting has been no exception.

The ordinary observer will, in all probability, have been struck by the fact that these meetings, the embodiment in the highest degree of our national, and, we may say, of our Imperial investigations in the realm of science, are year by year becoming more and more concerned with our industrial and economical aspirations. Last year especially have we had (in addition to purely speculative and technical subjects) excursions into matters ranging from Education, the Unemployed, Railway Management, the Steam Turbine, down to questions of dietary and coffee-making, and a host of other allied and unallied subjects,—all contributed to by persons competent to handle their subject in a more or less masterly fashion. Radium nowadays always commands a respectful hearing. There is something comforting in the thought that this globe of ours, and the sun which shines upon it, may yet be spared to us for another few million years. But other matters of a purely scientific nature have also engrossed attention ; and, although we may be tempted, at times, to question whether the hitherto high level of scientific thought is being fully maintained, there is, I think, no evidence to the contrary.

Nevertheless, the question does arise, and it is worth the asking, how far it is advisable that science and industry should meet on the same plane, and that plane the platform of the British Association ; whether, that is, pure science should step down from its lofty pedestal and indulge in overtures with the more mundane affairs of social and industrial welfare. Is it in the highest interests of science that it should be so ?

This involves the further question, which we cannot here go

into, as to whether science has any other justification for its existence than the furtherance of industrial and economic welfare. But, whatever diversity of opinion there may be as to the propriety of a body such as the British Association concerning itself with these matters, there can be none as to the simple fact that science and industry should go hand in hand and mutually subserve each other. It is in the interest of industry that science should be brought to bear upon it; and it is equally in the interest of science that industry should prosper. It was the purpose (Dean Church tells us) of Francis Bacon in his book entitled *Advancement of Learning*, "to make knowledge really and intelligently the interest, not of the school or the study or the laboratory only, but of society at large." And this, says Prof. Ray Lankester, in his Presidential Address, "is what our founders also intended by their use of the word 'advancement.'" One word, not in any sense of determent, but of warning is, however, necessary: Science must ever be on her guard against losing her identity and becoming merged beyond recognition in a merely scientifically inclined industrialism. Such a state of things would be little better than were she to stand aloof altogether. Science and industry have each their own function, and the best results may be expected by exercising such function independently, and comparing notes as often as possible.

Assuming, however, that it is within the province of the leading organisation of British science to minister to the welfare of industrial enterprise (and few, I think, will doubt it), we are further led to inquire how far it really does so, and how far, in turn, there is any response on the part of national industry as a whole, and sectionally of its component parts.

That there has been a response, and a marked one, is evident from a glance at the various subjects dealt with at the meetings. But a closer examination will reveal the fact that such response has by no means been general. Many departments of our industrial life are keenly alert to what they owe to science, and are ever ready to return the compliment by contributing the results of practical experience, thus either confirming the *a priori* reasoning of scientists, or, it may be, disproving it,—in either case forming a more substantial basis for future operation, and advancing our knowledge step by step. In the Engineering section, for example, at the recent meeting the subject of the

39

Steam Turbine was ably handled by Mr. Stoney, and its " pros "
and "cons" subsequently discussed under the searchlight of
science. So too the merits, demerits, and future possibilities of
gas-engines were taken up, laid down, and no doubt advanced
a step. In other branches also of engineering there has been
evinced a disposition to acknowledge, and to profit by, the
methods of science.

While, however, there is evinced this confraternity of interest
between engineering and science, there is one important branch
of our industrial life which is practically unrepresented. I refer
to that of mining. There is not, throughout the whole list of
industrial enterprises for which we are world-famed, and upon
which our national greatness so largely depends, an undertaking
of greater importance than that of turning the raw material of
the earth's crust into the wealth which ministers to the wants
of man. It requires no laboured argument to show—it cannot
indeed be gainsaid—that it is upon our coal that the superstructure of our industrial life depends. Add to this the exploitation
also of other minerals, and the importance of the operation of
mining needs no further comment. And yet it is the fact that it
is only within recent years that this importance has begun to
have due weight attached to it, and is becoming a matter of more
general interest. It is with a view to furthering this general
interest, and emphasising the importance attaching to it, that
these pages are contributed.

Mining is one of those industries (if industry we may call it)
which, rooted and grounded in practical experience, has all
along been loath to part with the past and adapt itself to an
ever-changing environment. It has not readily assimilated the
advantages placed before it. It is essentially conservative, and
its evolution has consequently been slow. This probably is
natural enough when we consider its essentially practical character, and the reasonable apprehension, on the part of those
who provide the necessarily large amount of capital involved,
lest any radical change should prove disastrous. It is this
wariness, this circumspection (natural up to a certain point)
combined with a certain instinctive attraction for the old and
indifference to the new, that has caused mining to stand, as it
were, so much in its own light. But the spirit of change, and a
new environment, have at last proved equal to the occasion ;
and modern thought and modern ideas have now gained a

footing in the mining industry, and bid fair to retain it. The change, too, is not simply one of improved machinery and mechanical application generally, but also one of *personnel.* The time was, and that very recently (and in some quarters still is), when the only men thought fit to manage a colliery or a mine were those who were matured and brought up in the closest connection with it, who had "risen from the ranks," and had gone through every detail of the work; the demand, in short, was for the "practical man" pure and simple. No one is going to quarrel with the importance attached to the practical man— the man who has "been through the mill." It is essential that one who is responsible for the safety of men's lives, and for the right use of a large amount of capital, should have a practical and intimate knowledge of his work. But this is not the only requirement, and merely "going through the mill" will not meet it. Modern mining is a very different thing from what it was even a few years back. The old rule-of-thumb methods no longer apply. New conditions have arisen, and these call for new methods of meeting them. One need only compare a modern-equipped colliery with its prototype of, say, fifty years ago, in order to realise the strides that have been made. Winding, hauling, ventilation, the winning of coal, and the rendering it fit for the market, were all in a primitive condition, while the introduction of electricity as a motive power was not even dreamt of till a much later date. Nor does the comparison end here. Not only has the controlling head of a modern mine to have an intimate acquaintance with all the latest phases of mechanical appliances, but he must also have regard to the altered conditions with which he is confronted underground. These are vastly different from what they were in the early days of mining, when the coal or the mineral lay within easy reach of the miner, and required, perhaps, only the turning over of a few sods in order to obtain them. Now, however, 3,000 or 4,000 ft. of intervening earth and rock have often to be penetrated before access can be gained to the hidden wealth. And not only so, but, having reached this depth, the question of raising the material to the surface is one which requires fresh treatment.

The problem of deep mining is one which is largely engaging the attention of mining men at the present time. There is no known limit to the depth at which minerals occur. As far as the earth has been penetrated they have been found; but their extraction is attended with a new set of difficulties which come in the train of increased depth. The problem of raising from depths hitherto unattempted is mainly a mechanical one, while the correlative problem of dealing with increased temperature due to such depth is largely a scientific one. It therefore becomes apparent that, as the more easily worked seams and veins gradually approach exhaustion, the need for scientific and ingenious methods of reaching those less accessible becomes the more pressing, and will demand all the resourcefulness we are capable of. This means that the directing head must be something more than a mere practical man: he must be a man of technical training and scientific ideas. Here then we have at least one point of contact between mining and science.

Great as has been the advance on what we may call the engineering side of mining in recent years—an advance it has shared in common with other industries—it should not be lost sight of that there has been an advance also in another direction, and of equally great importance. It is one of the objects of science to discover and interpret the laws of nature. The operations of mining, far from being exempt from these laws, stand in their own peculiar relationship to them. The conditions are such as do not obtain at the surface, and independent investigation becomes therefore more or less necessary. Lengthened experience of underground conditions has led to a much better understanding of the *laws* which govern the various operations of mining than was the case fifty or sixty years ago. And the effect of this has been mainly in the direction of safety and economy—two primary objects to be aimed at in all industrial concerns. Our increased knowledge in this respect has made more salutary conditions possible, and has, even within the last few years, been the means of greatly reducing those dire calamities which with awful suddenness too often dealt death to hundreds of underground workers. Investigation into coal-dust explosions, and the behaviour of gases under conditions approximating to those obtaining underground ; a better knowledge of the laws of ventilation, and the means of applying them, together with the introduction of safer means of illumination, have largely contributed to this result. Indeed, had it not been for this enlightened view of matters, much of the purely *mechanical* improvements would have been inapplicable. Again we see an adjustment, a relationship, between mining and science; and again, the need not only of a practical, but of a theoretical training.

What is being done to meet this new situation, to meet the demand for a more highly trained class of men ? Technical education has of late years no doubt made great strides, and a brief retrospect bearing upon the subject will assist in showing our present position. The relationship between science and industrial enterprise[1] to-day is very different from what it was a few years ago. Both have undergone the natural process of evolution, and that process, like all natural processes, has been for the benefit of mankind. Looking back over a period of years, it is not a difficult matter to trace out the services that science has rendered to humanity. But what is noticeable is that they were services bestowed, as it were, by the scientific few upon the unscientific many. Many of the world's greatest achievements and industrial revolutions have sprung from the genius of a few men. Such men as Kepler, Newton, Watt, Stevenson, Davy, each express for us some clearly defined acquisition of knowledge. Mankind during these periods were for the most part mere onlookers, and reaped the fruits of those gifted few. And what we wish to observe here is that, in spite of the scientific genius of these men, the age which produced them can scarcely be called scientific.

It is three hundred years since Bacon published *The Advancement of Learning.* At that date (1605) science, as we have come to regard it, did not exist. Bacon's ideas were far ahead of his time, and the atmosphere in which he lived was not conducive to the progress of knowledge. Nevertheless, the publication of his work did much to lay the foundation of modern natural science, and in 1662 this took practical shape in the birth of the Royal Society of London. Then followed Newton with his theory of gravitation, as set forth in his *Principia,* published in 1687, and the seventeenth century closed with a considerable measure of scientific activity—a partial realisation of Bacon's anticipations.

Since then other two centuries have passed, and again we mark the change. The age, we are glad to think, is becoming a

---

[1] Much of what follows is an amplification of what formed the basis of an address ("Mining : is it a Science ?") given before the Nova Scotian Institute of Science on December 11th, 1905, and I am indebted to that Institute for permission to make use of it.

scientific one. We still have—and let us hope will continue to have—our Keplers, Newtons, Watts, Kelvins, etc., in the forefront of science, but the scientific spirit, instead of being confined to the few, is now diffused amongst the many, with the result that hardly any undertaking is considered complete unless it is stamped with the hall-mark of science. Science has gone steadily forward, until to-day one discovery comes after another with such rapidity, and with such far-reaching effects, that we are apt to lose the true sense of perspective in the very frequency with which they are brought before us.

Moreover, I take it to be a fact that discoveries in modern times have a more direct, and a more instantaneous, effect upon the nation at large than like discoveries would have had in times gone by. And the reason is not far to seek. Our use of the word "technical" shows it. For what does it imply? Simply that an intimate connection exists, and is acknowledged to exist, between science in theory and science in practice. It is the word we employ to denote our sense of the intimate and necessary relationship between knowledge gained and the application of it ; a truth the acknowledgment of which has been somewhat tardy, but one which now marks a new phase in the history of scientific development. It marks the *democratising*, so to speak, of science.

I take, then, the influence of science at the present day to be twofold : (1) There is the direct effect of scientific discoveries themselves; the *material* gain. (2) There is the influence due to the diffusion of scientific knowledge, or, as we may say, to the equality of opportunity of acquiring that knowledge and training ; in other words, the influence due to technical education.

Such being the effect of science generally, let us look for a moment at the sciences in particular. To trace the evolution of the various sciences is an interesting matter in itself, but it is beyond the scope of this article, except to notice it in the briefest manner. In days gone by, when the sum-total of knowledge was small as compared with that of the present time, few divisions sufficed to express it, and there were consequently few sciences. But as time went on, and man discovered more and more of nature's hidden things, it became necessary to specialise and form branches of knowledge ; and hence, by degrees, arose the various sciences—astronomy, physics, chemistry, geology, etc. This evolution of the sciences has gone on in common with the evolution of things in general. It is the usual development from the general to the special, the low to the high, the simple to the complex.

But here we must observe that this marking off of our knowledge into different departments has no existence in actual fact. Nature does not work under the limitations which man, to suit his convenience, would sometimes ascribe to her. She is of one piece woven throughout. The different sciences are not so many distinct and isolated compartments, having no inter-relationship; but, on the contrary, are intimately connected and dove-tailed into one another, so that in reality there is no clear line of demarcation between one science and another. The nomenclature of the sciences is nothing more than a convenient mode of labelling and expressing our observations of the phenomena taking place around us. But it is instructive as showing the increased range of our knowledge; for as soon as any department becomes too unwieldy to be dealt with as one, an allied department is created, and the birth of a new science is the result.

The birth-rate in the scientific world shows no evidence of decreasing, nor is there any sign of diminishing vitality. On the contrary, fresh territory is daily being won, and the barrier of ignorance thrust further and further back. In this march of science what place does mining occupy, and what has been its contribution towards it?

It is not my purpose to claim for mining any more than its just merits; but to place it before the public in perhaps a somewhat different light from that in which it is usually presented, and to show that the profession (for so I think it should be regarded) is one not simply of commercial, but also of national concern.

No one will deny the claim of mining to rank as an art; the drawback, indeed, is that it is liable to be too exclusively so regarded. Originally, as Herbert Spencer has pointed out,[1] science and art were one; and it was only with the advent of complicated processes, arising out of experience in handicraft, that they became differentiated. Since then science has been supplying art with truer generalisations and well-defined principles, while art on its part has been supplying science with improved materials and more perfect instruments. Further, the sciences may even act as arts to each other; in surveying, for example,

[1] *Essays*, vol. i.: The Genesis of Science.

mathematics, geometry, etc., are all called into requisition ; so that it is often difficult to say where art ends and science begins. We are not indeed far wrong in saying that art is simply an applied science—science in action, so to speak ; and this marks the passing of theory into practice.

Mining is one of the oldest of the arts, as astrology was one of the oldest of the sciences. It was for long of a very crude character. But, with accumulated experience, came also a knowledge of general principles, and these also accumulating, in the course of time furnish the abstract principles upon which scientific mining is based, and with which the success of any mining enterprise is indissolubly bound up. It needs but to ask ourselves what would be our position to-day were we to turn to the old rule-of-thumb methods instead of those based upon science, in order to realise how much of our success is due to the latter, and at the same time how much of our failure is due to the want of it.

We may briefly define mining as *the art of transforming the potential wealth of the earth's crust into real wealth.* As part of that real wealth we may even include science itself. For it is by investigation into the phenomena of nature that science is advanced ; and this investigation could never have taken place to the extent it has done, had it not been for that exploration into the body of the earth which is the special province of mining. Radium, for instance—that latest of scientific wonders— it is questionable indeed whether it would ever have been heard of, had not mining given us the metal uranium. And if we turn from a special instance of this kind to science itself ; if, for instance, we take geology, that complementary science, so to speak, of mining, and ask ourselves what it would have been without the miner's experience in the body of the earth, it is akin to asking what physiology would have been without anatomy. We may say the same of the other sciences in varying degrees ; for, besides the direct benefit derived from the extraction of coal, iron-stone, precious metals, and minerals of all kinds, there is the even more important indirect benefit derived from the investigation into the properties of these various substances.

The intrinsic value of the substance brought to the surface may be great, or it may be small, but its value as a factor in physical research may be immeasurable. The net result to mankind is not to be measured by commercial results alone, but also by what is of greater importance—those more lasting effects which tend to broaden the basis of science.

But there is the other side of the picture. Whatever contributions mining has made to science, they have been returned tenfold. And this only serves to emphasise the intimate and reciprocal connection that exists between the two. Referring again to geology, consider how it has abundantly repaid the debt due to mining. Many a mining enterprise, but for the light afforded by geology, would never have been embarked upon. The Kent Coal-field (although in its infancy) is a case in point. Here mining, on each side of the Channel, contributed its quota to geology. Geology generalises upon the information thus acquired, and offers a theory which mining accepts and carries into practice.

So also with gold mining—as for instance in the case of "saddle reef" formations such as those of Bendigo and Nova Scotia,—geology is able to lay down general principles which point to the occurrence of rich quartz veins in the anticlinal domes of the folds rather than in the "legs" on either side.

Then, in regard to chemistry, the same mutual dependence is evident. Ores which were once regarded as valueless have, by its aid, been rendered valuable. New and improved processes of extracting and recovering metals from their ores have vastly extended the field of mining operations, and have brought within its scope minerals which before were untouched. To mention nothing else, the cyanide process of dealing with refractory ores of gold has completely revolutionised that branch of mining; and it is rather curious to observe that, whereas *alluvial* and *placer* mining were at one time the chief sources of gold, and *reef* mining quite a secondary source, the reverse is now the case, and the most important gold-fields of the world are those where reef mining is carried on. Here, then, is a significant example of the application of science to industry; as soon as the demand of the gold-mining industry came with sufficient force the resources of chemistry were ready at hand to meet the requirements.

Even medical science has its own peculiar relationship with mining. Special industries have special diseases attaching to them, and require special treatment and investigation. Thus we have miner's phthisis, nystagmus, and that disease ankylostomiasis (miner's anæmia), which a short time ago worked such havoc in Westphalia, besides many cases of poisoning peculiar to the working of certain ores ; while ambulance work further brings this honoured profession into close association with the mining industry.

We might carry this inter-relationship between mining and the various sciences further ; enough, however, has been said to show that the relationship exists. But what is the advantage of it all ? Does it matter to the community generally, and to the mining profession in particular, whether or not the industry is placed upon a scientific basis ? For in our matter-of-fact way we usually judge of the value of a proposal rather by the results immediately calculable from it than by those more remote, but none the less discernible, if we take the pains to look for them. Unless my contention thus far can be refuted, or the facts explained away, the gain derivable from inculcating a scientific spirit into industrial affairs is obvious. By raising the status of the mining profession it would attract a class of men who have hitherto held aloof from it, and whose services will be the more required as the exigencies of mining become greater. This will render mining more efficient, and the result will be reflected in the nation at large. That the economic aspect, for example, of our coal supply is a question of national importance a glance at the recent report of the Royal Commission on the subject will make clear. But the economical getting of coal depends upon scientific mining ; and the economical *use* of it depends upon scientific treatment, and this in a large measure comes within the province of the mining engineer. Our coal is a national asset of great value, and it is the nation's concern to see that a right use is made of it. Chemical engineering, by various improved methods of dealing with hitherto discarded material, is doing much towards annihilating waste. Mining engineering must help to make these methods effectual.

I do not lose sight of the fact, already intimated, that technical education is doing a great deal towards inculcating the scientific spirit we are insisting upon. And it is also a source of satisfaction that the State has encouraged this spirit by enacting that two years of college training may stand for two years' practical work in regard to the requirements for the examination for a colliery manager's certificate. This has given a considerable impetus to university work, and has led to the institution of special degrees in mining. Nor do I agree with the conclusion to which Prof. Ray Lankester appears to have come, when he says in his Presidential Address : " It is a fact which many of us who have observed it regret very keenly, that there is to-day a less widespread interest than formerly in natural history and general science, outside the strictly professional area of the school and university."

Nevertheless, although we are no doubt doing a great deal towards popularising science, and driving home the importance of it, there is still "a less widespread interest " in it than there should be; and no opportunity should be lost of making the interest more ·widespread wherever possible.

Important and indispensable as is the bearing of science upon mining, some years' acquaintance with the profession as one of its members has convinced me that there is not that proper appreciation of the value of scientific method which ought to accompany it. I make this as a general statement, knowing full well that there are many laudable exceptions, as a glance at some of the Presidential Addresses and Papers delivered before our Mining Institutes will testify. But there the matter usually ends : the appreciation is not " widespread." It is not that there is any actual disregard of science ; on the contrary, it is gratifying to observe an increasing readiness to adopt scientific methods, and it is a healthy sign ; but there is lacking that *positive*, aggressive attitude of taking the initiative, and going out to meet science on its own ground. The attitude is one rather of passivity than anything else, and it is not an attitude calculated to further any matter in hand. It is reflected in the absence (referred to in the early part of this article) of any specially mining contribution of a scientific character at the British Association meeting. (I shall only be too glad to be corrected on the point, but so far as I am aware no subject of this nature was dealt with[1]). And surely, for a country which is pre-eminent in mining, this is not as it should be. There ought to be a spontaneous desire on the part of the mining community to be heard ; and let us hope that ere another meeting comes round this will be the case.

With the object in view of encouraging the " Advancement

[1] Mention however should be made of Mr. J. Backhouse's Paper in the Economic Section on " Lead Mining in Yorkshire."

of Science" amongst the mining fraternity, and of forming a link between it and the representative institution of British Science, the suggestion is thrown out that a standing committee composed, in part, of those practically engaged in mining (agents, mine managers, and others), and, in part, of those on the professional or teaching staff of universities and colleges, be formed, which would be representative of the mining community throughout the kingdom. Colonial representatives should also be added. The business of the committee should be to keep in touch with science on the one hand, and with mining on the other, and to see that suitable papers and other matter are brought before the meetings of the British Association. Some such policy as this would, I am inclined to believe, raise the scientific tone of the mining profession, and would, at the same time, further the object for which the British Association exists.

It is gratifying to observe that the opportunity of emphasising the importance of this inter-relationship between science and the community was not let slip at the recent meeting of the Association. Principal Griffiths, indeed, devoted the concluding portion of his Presidential Address in the Mathematical and Physical Section to the matter. He points out that it is "a national duty to seek out and, when found, utilise the latent scientific ability of the rising generation for the purpose of adding to our stores of natural knowledge," and lays stress upon "the necessity of more free communication between the laboratory and the market-place."[1]

It is worthy of note also that on the same day (August 1st) that Prof. Ray Lankester was lamenting the lack of interest in science in England, Prof. G. Lippmann, as President of the French Association, was calling the attention of his brother scientists to precisely the same thing. He urges "the need at every factory for a scientific staff provided with research laboratories."[2] It is worth more than a passing thought whether some such laboratories could not with advantage be instituted at certain of our mining centres, where research work specially appertaining to the particular mining district could be carried out. If such work were then co-ordinated it would prove of considerable value both to science and to mining. A

[1] *Nature*, August 9, 1906.
[2] *Ibid.* August 16, 1906.

laboratory of this kind, if instituted even for the sole purpose of research into such a matter as that of coal-dust explosions, would justify its existence. Accidents such as the recent lamentable explosions both at home and abroad cannot fail to suggest lines upon which investigation might be pursued for some time to come. Does not the question of the spontaneous combustion of coal dust, for example, open up a field of enquiry?

A word in conclusion as to the commercial aspect of this matter. It is frequently laid to our charge that we in England are slow to adopt the improvements of scientific research. Questions of an abstract nature do not arrest our serious attention until their practical value is clearly demonstrated, and we thus lay ourselves open to an unfavourable comparison with some of our Continental rivals. The Germans, for example, display a remarkable aptitude for seizing upon new theories and inventions, and turning them into practice in a remarkably short time. This is in part, no doubt, due to their different conditions, which compel them to adopt means which, in our own case, are neither so necessary nor so urgent. Pressure of circumstances has thus, for example, made them pioneers in modern coke-making; and in the application of electricity they also take the lead. In this latter connection it may be of interest to quote a remark made by Mr. Cunyngham some time ago, when referring to the evidence given before the Departmental Committee on Electricity, of which he was chairman, to the effect that three-phase systems were being employed in Germany, which were, he believed, "partly invented by an Englishman, developed by a Swiss engineer, and worked out and practically applied by the Germans."

The moral of this is that no greater mistake can be made than to suppose that because the connection between theory and practice is not immediately self-evident it therefore does not exist. And were modern industry established upon a more scientific basis, it is probable that much of this hesitancy over the application of theory to practice would disappear. The result would be greater efficiency, greater economy in production, and greater ability to sell in the open market in competition with others : all conducive to the maintenance of our position as the leading industrial nation. To quote Principal Griffiths once more: "If we can convince the men of business of this country that there are few more profitable investments than the encouragement of research, our difficulties in this matter will be at an end." Many, we are glad to think, are already convinced of this. It ought not to be a difficult matter to bring home to the rest that one of " the greatest of national assets is scientific discovery."

RECENT EXPERIMENTAL WORK ON OSMOTIC PRESSURE

By JAMES C. PHILIP, M.A., PH.D., D.Sc.

Lecturer on Physical Chemistry, Royal College of Science, London

THE number of scientific workers who have made an experimental study of the phenomena of osmotic pressure is not great. The phenomena themselves are of fascinating interest and of fundamental importance, especially on their quantitative side, and the paucity of available data must be attributed to the experimental difficulties which arise when an accurate measurement of osmotic pressure is attempted. Among these difficulties must be reckoned the unreliable character of the rigid media which have been employed for the support of the all-important semi-permeable membrane. Pots or tubes of unglazed porcelain are commonly used for this purpose, but out of a given consignment of pots or tubes, presumably more or less uniform, only one or two probably will permit the deposition of an absolutely satisfactory membrane. The reasons for the possibility or impossibility of depositing a satisfactory membrane on a given pot or tube seem to be imperfectly understood, but of the facts there is little doubt. Examples of this difficulty may be found even in the most recent work on osmotic pressure. Thus Messrs. Morse and Frazer[1] found that of a lot of five hundred pots made for them, not one was suitable for the measurement of osmotic pressure, while in the case of another lot of one hundred pots some twenty-five or thirty were fairly satisfactory. Again, the Earl of Berkeley and Mr. Hartley,[2] who use porcelain tubes in their experiments on the osmotic pressure of concentrated solutions, state that of one hundred tubes of various makes only three permitted the deposition of a really efficient membrane. Morse and Frazer, it is true, indicate that they have some evidence as to the conditions essential for a useful porous wall,

[1] *American Chemical Journal*, 1905, **34**, 1.
[2] *Philosophical Transactions*, 1906, **A**, **206**, 481.

and record the production in the laboratory of several cells of first-class excellence. It may therefore be possible in the near future, when these essential conditions are better understood, to obtain trustworthy porous vessels for the demonstration and measurement of osmotic pressure.

Preparation of the Semi-permeable Membrane

A feature of recent work on osmotic pressure is the use of the electric current in the deposition, or in the "remaking," of the copper ferrocyanide membrane—the membrane which is almost exclusively employed in the absolute measurement of the osmotic pressure of aqueous solutions. When the porous walls have been completely freed from air, a result that may be effected by electrical endosmose, the cell is filled with a solution of potassium ferrocyanide and placed in a solution of copper sulphate. The electrodes of platinum and copper, the former dipping in the ferrocyanide solution and acting as cathode, are connected with the terminals of a 100-volt circuit, and electrolysis is continued until a satisfactory membrane of copper ferrocyanide has been deposited. Provision is made for the renewal of the potassium ferrocyanide solution during the electrolysis, since the alkali produced at the cathode has naturally a deleterious action on the membrane.

Unless hard-baked cells are used, such a method of procedure as that just described yields a membrane embedded in the interior of the porous walls, and to this there are serious objections. It is preferable, if not essential, to have the membrane deposited close to that surface of the porous wall which is to be in contact with the solution. Pfeffer, in his original investigations of osmotic pressure, laid stress on this point, and deposited his copper ferrocyanide membranes close to the inner surface of the porous pots employed.

In the experiments made recently by Lord Berkeley and Mr. Hartley, the porous tubes used were first immersed in a copper sulphate solution in a desiccator, which was subsequently evacuated. The result of this treatment was that air was completely removed from the porous walls, and the pores of the porcelain were filled with the solution of copper sulphate. After being superficially dried, the tubes were plugged at both ends and rotated rapidly in a solution of potassium ferrocyanide,

In this way a regular deposit of copper ferrocyanide, *very close to the outer surface* of the porcelain tubes, was obtained. The tubes were allowed to soak in the ferrocyanide solution, and were then set up for electrolysis, the electrolysis being carried out on the lines described previously. This electrolytic treatment of the membrane seems to contribute materially to its efficiency, and it is even found advisable to "remake" a tube electrolytically after it has been used in an actual measurement of osmotic pressure.

Another factor that appears to contribute to the efficiency and permanence of a membrane is the presence, in small quantity, of the membrane-forming salts. In the complete absence of these, the membrane tends to dissolve in a colloidal form, and so deteriorates.

The degree of efficiency of a copper ferrocyanide membrane may be adjudged by several criteria. An actual test made with a strong sugar solution will show whether the pressure developed in the cell is of the expected order of magnitude, and a subsequent examination of the solvent outside the cell will show whether the membrane is strictly semi-permeable, or whether there has been a leak of the sugar. Again, a determination of the electrical resistance of a membrane will, in a general way, indicate its probable efficiency. This is, however, not an infallible criterion, for, although as a rule the resistance of a good membrane deposited in a porous cell is of the order of 100,000 ohms, efficient membranes have been produced the resistance of which was considerably lower.

The Measurement of the Pressure

As regards the actual measurement of the osmotic pressure, recent work has brought some new developments. The usual procedure is to put the solution in the porous pot which carries the semi-permeable membrane, to connect a closed manometer with the pot, to immerse the latter in the pure solvent, and then to observe the pressures indicated in the manometer. This sounds simple enough, but in reality the greatest difficulty is experienced in securing an efficient junction between the pot and the manometer, for the pressures developed in the cell, even in the case of fairly dilute solutions, are such as to put all joints to a very severe test. Morse and Frazer, who have adopted this usual method in their recent work, refer to the great

40

manipulative difficulties involved in effecting a satisfactory connection between the cell and the manometer, and suggest that the largest single experimental error in their work is that due to the lack of rigidity in the rubber stopper employed to close the cell.

In the experiments carried out by Lord Berkeley and Mr. Hartley, a porous tube with the semi-permeable membrane deposited close to its outer surface passed tight through the centre of a metal chamber containing the solution. This metal chamber consisted of two halves screwed together, and the joints were rendered tight with the aid of dermatine rings. The metal vessel was connected with an apparatus by means of which a known hydrostatic pressure, sufficient to balance the pressure of the inflowing water, could be applied to the solution. The porous tube, occupying a horizontal position and filled with water, was connected at each end with a vertical glass tube. Observation of the level of the water in one of these tubes showed whether water was entering the solution through the membrane, or whether the applied hydrostatic pressure was squeezing water out of the solution. In the latter case the applied hydrostatic pressure must be greater than the osmotic pressure, in the former case less. By adjusting the apparatus, therefore, so that water is neither entering nor leaving the solution, an equilibrium pressure is found which is equal to the osmotic pressure.

Magnitude of the Pressures observed with Copper Ferrocyanide Membranes

The number of substances for which there is no passage through a membrane of copper ferrocyanide is very limited. Recent work on the osmotic pressures observed with such membranes deals almost exclusively with solutions of cane sugar, although the behaviour of dextrose and one or two other sugars has also been the subject of investigation to a limited extent. The statement that a copper ferrocyanide membrane is absolutely non-permeable for cane sugar should perhaps be qualified; but there seems to be no doubt that when the conditions are favourable and due care has been taken in the setting up of the osmotic cell, the amount of sugar which passes through the membrane, even from a highly concentrated solution, is practically nil.

The time necessary for a solution to develop its maximum osmotic pressure appears to vary very considerably with the membrane employed. This is especially the case when the osmotic pressure is measured by the manometer method, as employed by Messrs. Morse and Frazer. These workers speak of one cell in which the maximum pressure was usually developed within seven hours, and of two other cells for which over forty hours were similarly required. It is noteworthy that these times appear to be independent of the concentration of the solution in the cell. In the method of measuring osmotic pressure adopted by Lord Berkeley and Mr. Hartley, the time required for the actual measurement is shorter and more uniformly independent of the particular membrane employed. In only a few cases was the membrane kept under pressure for more than seven hours, while in many experiments a much shorter time was found sufficient.

The results obtained for the osmotic pressures of cane sugar solutions by Messrs. Morse and Frazer on the one hand, and by Lord Berkeley and Mr. Hartley on the other, are not directly comparable, for two reasons. In the first place, the measurements of Messrs. Morse and Frazer were made at temperatures of 20° to 24° C., whilst those of Lord Berkeley and Mr. Hartley were carried out at 0° C. In the second place, the ranges of concentration differ in the two cases, for the solutions examined by Messrs. Morse and Frazer were between 0·1 normal and 0·82 normal, while the range of concentration of those examined by Lord Berkeley and Mr. Hartley was 0·527–2·2 normal, and only one of the latter solutions falls within the range of concentration covered by Messrs. Morse and Frazer's experiments. An idea of the magnitude of the osmotic pressures of the various solutions may be best gained by a glance at the following tables. Table I. gives a summary of Messrs. Morse and Frazer's results, Table II. a summary of Lord Berkeley and Mr. Hartley's results:

TABLE I		TABLE II	
Gram molecules of cane sugar per litre of solution.	Osmotic pressure in atmospheres.	Gram molecules of cane sugar per litre of solution.	Osmotic pressure in atmospheres.
0·098	2·53	0·527	13·95
0·192	4·77	0·877	26·77
0·369	9·66	1·23	43·97
0·533	14·74	1·58	67·51
0·684	19·34	1·93	100·78
0·825	24·23	2·19	133·74

The first thing that strikes one on looking at the foregoing tables is perhaps the fact that copper ferrocyanide membranes are capable of withstanding pressures up to 100 atmospheres and over. The results obtained in both the investigations described are in fact a striking commentary on the efficiency of copper ferrocyanide as a semi-permeable membrane, and somewhat discount the recent criticism of an American chemist who speaks of " so-called semi-permeable membranes."

The pressures observed by Messrs. Morse and Frazer, by Lord Berkeley and Mr. Hartley, far exceed any recorded in the earlier investigations of osmotic pressure. Pfeffer, in his classical researches, records a pressure of 436·8 cm. of mercury for a 3·3 per cent. solution of potassium nitrate, and one of 307·5 cm. for a 6 per cent. solution of cane sugar ; but even these maximum pressures are greater than any observed during the thirty years that have elapsed since Pfeffer's work was published. The fact that very high pressures are now attainable is doubtless due in the main to the care that has been bestowed on the deposition and electrolytic consolidation of the membrane. It is possible also that the highest pressures recorded in Table II. have been attainable only because of the low temperature at which the experiments were carried out. Indications are not wanting that the membranes become less efficient if kept or used at higher temperatures.

The question naturally arises whether the determinations of osmotic pressure obtained by the two methods, and recorded in Tables I. and II. respectively, are concordant. As already indicated, it is scarcely possible to answer this question satisfactorily, since the ranges of concentration in the two cases overlap to a small extent only. At the same time it may be interesting to compare the 0·533 normal solution in Table I. with the 0·527 solution in Table II. The value 14·74 atmospheres is the osmotic pressure of a 0·533 normal solution at 24°, and if it is supposed that osmotic pressure is proportional to the absolute temperature, the osmotic pressure of the solution at 0° would be about 13·6 atmospheres, a value smaller than the pressure recorded in Table II. for a less concentrated solution. Interpolation also for other points on the common range of concentration indicates that the pressures observed by Lord Berkeley and Mr. Hartley are in general rather higher than those observed by the American workers. Too much weight, however, must not be laid on this conclusion, since the scope of the comparison is limited.

INDIRECT DETERMINATION OF THE OSMOTIC PRESSURES OF CANE SUGAR SOLUTIONS.

As is well known, there are certain properties of solutions which are intimately related to their osmotic pressure. Thus it is possible, from the difference in vapour pressure of the solvent and the solution at the same temperature, and from the difference in boiling point or freezing point of solvent and solution, to calculate the osmotic pressure. The difficulty is that the methods usually available for the determination of the vapour pressure, the boiling point, and the freezing point of solutions are not very accurate, for dilute solutions at least, and consequently most of the data obtained by these methods are of little use in checking the values of the osmotic pressure determined by direct observation. In the case of a cane sugar solution, for example, containing $\frac{1}{100}$th of a gram molecule per litre, the osmotic pressure at $0°$ would be 17–18 cm. of mercury, a very considerable and easily measurable quantity, whereas the freezing point of the solution would be only $0·02°$ lower than the freezing point of water; the boiling point of the solution would be only $0·005°$ higher than the boiling point of water, and the vapour pressure of the solution would, even at $100°$, be only $0·15$ mm. lower than the vapour pressure of water.

By a dynamical method, however, it is possible to find the ratio of the vapour pressures of solvent and solution, by determining differences of weight instead of differences of pressure. Such a method was employed some years ago by Walker, who adopted the device of simply bubbling air for some time through a series of four absorption tubes or bulbs. If, in such an arrangement, the first and second tubes contain the aqueous solution for which the vapour pressure is to be determined, the third tube contains pure water, and the fourth tube contains strong sulphuric acid, then, on the assumption that in the first and second tubes the air becomes charged with water vapour up to the vapour pressure of the solution, the loss of weight in the third tube, observed after a current of air has passed through the apparatus for some time, is a measure of the difference between the vapour pressures of solvent and solution, while the gain in weight of the fourth tube is a measure in equivalent terms of the vapour pressure of water at the temperature of experiment.

Recent investigation by the Earl of Berkeley and Mr. Hartley[1] has shown that the accuracy of the method just described may be considerably improved by passing the current of air *over* instead of *through* the liquids in the absorption vessels. In the bubbling method there is a gradual fall of hydrostatic pressure throughout the train of vessels, and experiment shows that if a current of air is bubbled successively through two absorption tubes containing water, there is invariably a loss of weight in the second tube, even although the air is saturated on leaving the first tube; the bubbles increase in volume and take up an additional quantity of water in the second tube. It may also be shown that in the bubbling method particles of fine spray are given off and cause loss of weight. Both these sources of error are avoided by passing the air *over* the liquids, and providing that the exposed surface is large and constantly renewed.

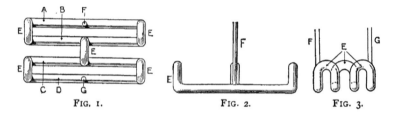

FIG. 1.   FIG. 2.   FIG. 3.

Each of the special absorption vessels devised by Lord Berkeley and Mr. Hartley to fulfil these conditions consists of four parallel and horizontal glass tubes, A, B, C, and D, sealed together by five inverted U-tubes, E. Of these five U-tubes, two make connection between the adjoining ends of A and B, two make a similar connection for C and D, while the fifth connects the middle of B to the middle of C. The air enters and leaves the vessel by two vertical glass tubes, F and G, sealed to the middle of A and the middle of D respectively. Fig. 1 is a plan of such an absorption vessel, figs. 2 and 3 are side and end elevations respectively. Each vessel is filled to the extent of about one-third with the required liquid, and is so fitted on a platform in a constant temperature bath that it can be slowly oscillated, and its two ends raised and lowered alternately. The con-

[1] *Proceedings of Royal Society*, 1906, **A, 77,** 156.

sequent flow of liquid from end to end secures thorough mixing, and periodically wets the branches through which the current of air passes. Since the air must be passed slowly through the train of vessels, the experimental work is rendered very tedious, and the time consumed in determining the vapour pressure of one solution runs into days. The results are accurate to within 5 per cent., and from the lowering of the vapour pressure which is thus ascertained the osmotic pressure P of the solution may be calculated by the formula $P = \dfrac{As}{\sigma} . \log_e \dfrac{p}{p_1}$, where p and p_1 are the vapour pressures of the water and the solution respectively, s is the density of water at the temperature of the experiment, and σ is the vapour density of water under the standard atmosphere A.

It is very interesting to compare the values of the osmotic pressure of concentrated cane sugar solutions reached in this indirect way with those obtained by direct measurement, and in Table III. the data necessary for the comparison are recorded; the figures are those obtained at 0° C. by the Earl of Berkeley and Mr. Hartley:

TABLE III

Gram molecules of cane sugar per litre of solution.	Osmotic pressure in atmospheres.	
	Directly observed.	Calculated from the Vapour pressure.
1·58	67·51	69·4
1·93	100·78	101·9
2·19	133·74	136·0

The amount of material so far available for a comparison of the two sets of values is not very large; but the agreement is sufficiently striking, and shows that the direct method of determining osmotic pressure yields values which may be regarded as very nearly correct. It is noteworthy that the osmotic pressure calculated from the vapour pressure is, in all the three cases just cited, higher than the osmotic pressure observed directly. This is only what might be expected, for all the experimental errors associated with the direct determination of osmotic pressure would tend to give too low a value. In so far also as the sugar solutions are compressible, the observed value of the osmotic pressure will be less than the real value.

The direct determination of the osmotic pressure of a solution is conditional on the discovery of a satisfactory semi-permeable membrane. Since most solutes will pass, more or less rapidly, through a copper ferrocyanide membrane, and since no superior membrane has yet been found, it follows that the scope for direct determination of the osmotic pressure of aqueous solutions is somewhat limited. But when once it has been shown that the values of the osmotic pressure deduced from the vapour pressure agree closely with those found by direct observation, it will be possible to apply the vapour-pressure method with confidence over a wide field, for this method is subject to no such limiting condition as the existence of a semi-permeable membrane.

Theoretical Value of the Osmotic Pressure

One of the most notable and interesting investigations associated with the name of van't Hoff is that in which he deals with the analogy between solutions and gases. The main conclusions he reached were that osmotic pressure obeys the laws of gases, and that the osmotic pressure exerted by a dissolved substance in dilute solution is equal to the pressure it would exert if it were in the gaseous state at the same temperature and occupied the same volume as the solution. When van't Hoff published his memoir in 1887, the experimental material available for the verification of his conclusions was somewhat scanty, for the only direct measurements of osmotic pressure recorded up to that date were those made by Pfeffer. There is, therefore, some justification for the criticisms passed in certain quarters on the slender experimental basis on which the gas theory of solutions originally rested.

In an examination of van't Hoff's conclusions in the light of modern work on osmotic pressure, it must be remembered that, according to van't Hoff himself, they are valid in their simple form only for dilute solutions, in which the volume of the dissolved molecules is negligible compared with the total volume of the solution. Where this condition is not fulfilled, deviations from the simple rule may be expected, just as gases subjected to high pressures exhibit considerable deviations from the behaviour required by the simple gas laws.

The cane sugar solutions examined by Lord Berkeley and Mr. Hartley were certainly concentrated, and therefore it is not surprising that these investigators found very marked differences between the directly observed osmotic pressures and the theoretical values calculated on the basis of the validity of the simple equation $PV = RT$. The following Table IV. shows that the observed osmotic pressure is found in all cases to be greater than the theoretical pressure, and that the deviation between the two sets of values increases with increasing concentration of the cane sugar solutions:

TABLE IV

Gram molecules of cane sugar per litre.	Osmotic pressure in atmospheres.	
	Observed.	Calculated.
0·527	13·95	11·8
0·877	26·77	19·6
1·23	43·97	27·5
1·58	67·51	35·4
1·93	100·78	43·2
2·19	133·74	49·1

The solutions used by Messrs. Morse and Frazer, although the most of them can hardly be called dilute, furnish more suitable material for an examination into the validity of van't Hoff's propositions, and it is interesting therefore to learn that the results obtained by these investigators are in harmony with the view that osmotic pressure obeys the laws of gases. This statement, however, is subject to the reservation that the standard osmotic volume is not 1 litre of solution, but the volume occupied by 1,000 grams of water. It is further found that cane sugar, dissolved in water, exerts an osmotic pressure equal to that which it would exert if it were gasified at the same temperature, and the volume of the gas were reduced to that of the solvent in the pure state. It must be remembered that this is not the first attempt to allow for deviations from normal behaviour which are due to the volume occupied by the solute in a solution. Cohen[1] has proved that the increase in the velocity coefficient for sugar inversion with rising concentration of the sugar is conditioned by the gradually increasing fraction of the total solution volume which is occupied by the dissolved sugar molecules.

How far the important result reached by Morse and Frazer is borne out by their experimental work will be clear from a study of Table V. In this table the concentrations of the solutions are recorded in two ways, firstly as gram molecules

[1] *Zeitschrift für physikalische Chemie*, **23**, 442.

per litre of solution, and secondly as gram molecules per 1,000 grams of water. Solutions made by dissolving a gram molecule of solute, or some fraction of the same, in 1,000 grams of water are called *weight-normal,* to distinguish them from *volume-normal* solutions, prepared by dissolving the given weight of solute in water and then diluting to 1 litre. The third column gives the osmotic pressures recorded by Messrs. Morse and Frazer for the various solutions, while the fourth and fifth columns contain the calculated values (I. and II.) of the osmotic pressure. The values under I. are those calculated on the basis of the simple van't Hoff proposition, the values under II. are calculated on the basis of the modified proposition enunciated by Morse and Frazer.

TABLE V

Gram molecules of sugar.		Observed osmotic pressure in atmospheres.	Calculated osmotic pressure in atmospheres.	
Per litre of solution.	Per 1,000 grams of water.		I.	II.
0·098	0·1	2·53	2·39	2·42
0·192	0·2	4·77	4·64	4·80
0·369	0·4	9·66	8·91	9·62
0·533	0·6	14·74	13·0	14·55
0·684	0·8	19·34	16·7	19·35
0·825	1·0	24·23	20·1	24·23

There is no doubt which set of calculated values is the more in harmony with the observed osmotic pressures, and the agreement between the figures in the third and last columns leaves little to be desired. At the same time it must be said that the standard osmotic volume should almost certainly be, not the volume occupied by 1,000 grams of water *in the pure state,* but the volume occupied by 1,000 grams of water *in the solution.*

It would be interesting to know whether the osmotic pressures of still more concentrated solutions, such as those examined by Lord Berkeley and Mr. Hartley, could be calculated on the same principle, but the data necessary for the calculation are not available at present.

The general result, then, of Messrs. Morse and Frazer's work is to show that when weight-normal solutions are considered, the osmotic pressure is proportional (1) to the concentration, and (2) to the absolute temperature—in other words, the gas laws are applicable to solutions of cane sugar. This result is all the more interesting in view of the recent criticism of Prof. Kahlenberg, who, on the strength of some experiments made with pyridine solutions and a semi-permeable membrane of vulcanised caoutchouc, finds that the gas laws do not hold. The osmotic pressures observed by him are far short of the values required by the gas laws, and the changes of pressure that accompany changes of temperature are very much greater than they would be if osmotic pressure were proportional to the absolute temperature. If the experimental work brought forward by Prof. Kahlenberg were adequate and sound, these results would be of the utmost importance, but some of the experiments are open to serious criticism, and the work is not weighty enough to overthrow the results reached by other workers, and already described in this paper. Future investigation will probably bring some explanation of the conflicting evidence.

The Osmotic Pressure of Colloidal Solutions

Among the investigators who have contributed to the comprehension of osmotic phenomena are to be found many physiologists and botanists. Pfeffer himself, whose classical researches form the foundation of our knowledge of the subject, was Professor of Botany at Basle, and many others besides him have attacked the problem from the biological side. Quite recently another interesting contribution to our knowledge has been made from a similar quarter in the form of an investigation of the osmotic pressure of colloidal solutions by Prof. Benjamin Moore and Dr. Roaf.[1]

Colloidal solutions have been the subject of extensive study in recent years, and the current view of a colloidal solution as a suspension of very finely divided particles has led frequently to the impression that it would possess no osmotic properties such as are shown by solutions of true crystalloids. While it is undoubtedly true that in some cases the colloid "solution aggregate" is so large that measurable readings of osmotic pressure cannot be obtained, the work just referred to has shown that many colloids in solution give a distinct and definite osmotic pressure.

The membrane used by Prof. Moore and Dr. Roaf consisted of parchment paper, which, although it allows free passage to water and such dissolved substances as uric acid and even

[1] *Biochemical Journal*, 1906, **2**, 34.

the sugars, is non-permeable to gelatine, starch, and various other colloids. A definite osmotic pressure was found for gelatine, gum acacia, and the serum proteids, while starch and gum tragacanth give rise to no measurable osmotic pressure. A 10 per cent. solution of gelatine gives an osmotic pressure of about 70 mm. of mercury at 30° C., rising to about 140 mm. at 80°. This increase is greater than it would be if the osmotic pressure were proportional to the absolute temperature, and it is exceedingly probable that the large "solution aggregates" present at lower temperatures are subject to dissociation as the temperature rises. Re-cooling does not seem to lead at once to the re-formation of the large aggregates, for the osmotic pressure at 30° of a gelatine solution which has been heated to 80° for a short time is higher than before heating, and returns only gradually to its former value. Prolonged heating at 80°, on the other hand, leads to a permanently higher value of the osmotic pressure, a phenomenon that is accompanied also by changes in the physical properties of the gelatine solution.

The objection might be raised that the osmotic pressure exhibited by gelatine solutions is due to the inorganic matter invariably associated with the colloid; but it must be remembered that parchment paper is permeable for crystalloids, while Moore and Roaf have shown that the maximum osmotic pressures observed for gelatine solutions at a given temperature remain steady for days together.

The Nature of Osmotic Pressure

It is not proposed to discuss here at any length the various views that have at one time or another been advanced as to the cause and nature of osmotic effects. No general agreement has been reached on the matter, and it is noteworthy that some of those who contribute most to our experimental knowledge of osmotic pressure are the least ready to dogmatise on the theoretical side of the question.

This much may be said, however, that the older way of regarding osmotic pressure as due to the impact of the dissolved molecules on the semi-permeable membrane has been generally abandoned. This point of view was the result of pressing too far the undoubted analogy between gases and solutions, and makes too little of the part which must be played by the solvent. Many, however, in their anxiety to avoid the view which practically neglected the presence of the solvent, have gone to the opposite extreme, and have suggested that the solvent alone is responsible for the osmotic effects. It is, of course, true that the primary osmotic effect consists in a passage of the solvent into the solution through the semi-permeable membrane; but inasmuch as the occurrence of the osmotic effect is conditional on the presence of the solute, and the magnitude of the effect is determined by the amount of solute in the solution, it is obvious that the part played by the solute must be put in the foreground. From the point of view of thermodynamics it is all the same whether the solute molecules are linked to the solvent molecules or not. Larmor, for example (see *Nature*, 1897, **55**, 545), supposes that each molecule of solute forms a loosely connected complex with the surrounding solvent molecules, and concludes that "provided the solution is so dilute that each such complex is, for very much the greater part of the time, out of range of the influence of the other complexes, then the principles of thermodynamics necessitate the osmotic laws." He adds: "It does not matter whether the nucleus of the complex is a single molecule, or a group of molecules, or the entity that is called an 'ion'; the pressure phenomena are determined merely by the number of complexes per unit volume."

A view that finds acceptance in some quarters at the present time regards osmotic pressure as due solely to the difference in the surface tensions of solvent and solution. The advocates of this conception reach the result that solutions with equal surface tensions are in osmotic equilibrium, even though they may not be of equimolecular concentration—a conclusion which it should be possible to bring to the test of experiment.

Apart from the problem of the nature or cause of osmotic pressure, there is the question as to the function of the semi-permeable membrane. The substances for which copper ferro-cyanide is non-permeable are substances generally of high molecular weight, and it has accordingly been maintained that the membrane acts as a sieve, allowing small molecules to pass through, but preventing the passage of large molecules. It was with the aid of this conception that Pickering in 1897 interpreted some osmotic experiments in which he had found that a porous vessel was permeable to propyl alcohol or water separately, but not to a mixture of the two substances. The experiments were supposed to furnish conclusive evidence of the existence of some molecules (hydrates) in solution, which were larger than those of either substance, and were consequently unable to pass so readily through the meshes of the porous diaphragm. Recent investigation, however, by Barlow, and by Findlay and Short, has shown that the experimental basis of this conclusion is unsound.

Evidence is accumulating in favour of the view that a semi-permeable membrane dissolves or imbibes only the substance to which it is permeable; such a membrane, therefore, is efficient in proportion as it selectively absorbs one of the two substances in contact with it. But the fact must be emphasised that the result of applying thermodynamics to solutions is independent of the way in which the semi-permeable membrane acts.

THE REFORM OF THE MEDICAL CURRICULUM :

A REJOINDER

BY JOHN WADE, D.Sc. LOND.

Lecturer on Chemistry at Guy's Hospital, University of London

THE attack which Prof. Armstrong has made on the medical curriculum of the University of London, in the preceding number of this journal, might well lead the reader who is unacquainted with the history of the subject to believe that our whole system is radically wrong ; that medical education has remained unaffected by the rapid march of science during the past fifty years ; and that unless the present curriculum is entirely remodelled the issue will be fraught with disaster. The deep interest which Dr. Armstrong has taken in the complex problems of elementary education and the wholesome reforms in school teaching which have been effected of late years, in part owing to his teaching, justly lend weight to his views on any educational matter ; and it is with some diffidence that the present writer ventures to challenge his conclusions. Perhaps, however, association with medical students and medical men for upwards of a quarter of a century, a daily experience of sixteen years in teaching such students, and a position as one of the senior teachers of chemistry in the medical schools of the metropolis, will be allowed to constitute sufficient justification.

What is the question at issue ? It is agreed on all sides that the advance of medical science has for many years past been on chemical lines, and for many years to come must continue to be so. Prof. Armstrong, after quoting several prominent physiologists in support, himself proceeds : " The physiologists see quite " clearly that lack of chemical technique is at present the greatest " hindrance to progress from which they suffer. The problems " they have to solve are the most difficult problems in chemistry. " . . . Medical men require a far deeper, more intimate knowledge "of chemistry than that they possess at present." Quotations such as these might be multiplied indefinitely; the very text-books of physiology and therapeutics abound in chemical formulæ and the discussion of chemical methods. There is no disagreement on the fundamental point that chemistry is a necessary and integral part of the medical course.

The real question is therefore one of detail: With what aims is chemistry to be taught to medical students? How and to what extent is it to be taught? To whom is its teaching to be entrusted? It is on some of these points that the writer feels compelled to join issue; and in doing so he proposes to confine himself mainly to the methods of the University of which Dr. Armstrong and he are recognised teachers.

On the first of the points indicated surely no two opinions are possible, for all must agree with Prof. Armstrong that teaching after the school stage should have a distinct and carefully directed bias towards the lifework of the student—that in the case of the medical student illustrations should wherever practicable be selected from medical and biological sources. There are a few indeed who, proceeding to the other extreme, openly avow their conviction that the teaching of the pure sciences should be limited to such fragments as may happen for the time being to be of direct utility in medical practice. At a recent conference of medical teachers of this University a surgeon expressed the opinion that botany for medical students should be limited to the study of moulds, on the ground that these might be useful as an introduction to bacteriology. He might also have proposed to limit the study of Latin at school to such words as might be useful in writing prescriptions. Such a policy would indeed be retrograde, and if carried into effect could only result in degrading the professional man into a mere mechanic, incapable of independent action, and bound for life to empirical methods learnt unthinkingly.

Rank empiricism of this kind is very different, however, from an intelligent shaping of the course to the special needs of the student, and would scarcely be worthy of notice, but for the danger that Dr. Armstrong's remarks might be adduced as those of a scientific man in its favour. Our newspapers and magazines have been flooded of late years with protests against the lack of breadth and of scientific method amongst our manufacturers, and their obstinate adherence to antiquated, rule-of-thumb methods. Dr. Armstrong has himself frequently contributed to such discussions, and his views are so well known that it is inconceivable that he should advocate in medicine the very principles against the application of which in other technical occupations he has so often protested. His strictures as to the alleged irrelevancy of the chemical teaching in medical schools relate, it is clear, to a bygone period, too remote to serve as the basis of an authoritative expression of opinion; their only effect could be to lend weight to antiquated views of the most retrograde description. In direct opposition to what he would have us believe, chemistry in the London medical schools is now taught with direct reference to its scientific application in physiology and in medicine; the curriculum laid down for medical students has been specially constructed, and, except in the purely elementary parts, is essentially different from that laid down for chemical students.

And when we come to consider how chemistry is to be taught, and to what extent it is to be taught, we find that Prof. Armstrong's allegations as to the unsuitability of our methods are again vitiated by what the writer cannot help feeling, with all respect, is an inability to divest himself of old memories, and an unwillingness to pay attention to the views of those whose ideas as to progress may not altogether coincide with his own. The teaching of chemistry has improved enormously within the writer's own recollection. Twenty years ago, as Dr. Armstrong must surely remember, laboratory teaching was confined to a summer course of test-tube "analysis," and training in chemical method was non-existent. The section of chemistry most intimately connected with physiology and medicine—namely, that dealing with the carbon compounds— was taught by means of blackboard demonstrations, without experimental illustration, and the student was afforded no opportunity of verifying his teacher's statements. Such teaching was absolutely destructive of the scientific spirit, and teachers who, like Dr. Armstrong, worked under these depressing conditions must naturally find it difficult to realise that others under happier circumstances have obtained correspondingly happier results. Prof. Armstrong's contentions on this point cannot be maintained; at the present day they are devoid of foundation. Much progress has been made both in method and in material since he was actively engaged in teaching medical students, and with the steady advance in scientific method such progress must inevitably continue.

How, then, should chemistry be taught to medical students in order to ensure the best possible knowledge for future application? In every science there are certain fundamental principles which underlie all its branches and must be mastered before any of its applications can be appreciated. The ordinary laws of leverage are the same for the engineering student who afterwards applies them in constructing cranes and bridges as for the surgeon who employs them for studying the localities of greatest strain in bones, and the conditions under which they are most liable to be broken. The medical student and the embryo analyst or chemical manufacturer must alike be familiar with such fundamental principles as the neutralisation of acids by bases, oxidation and reduction, and the laws of chemical combination; and with those simple typical carbon compounds which by their coalescence or modification lead to such diverse products as the complex aniline dyes on the one hand and the constituents of foodstuffs on the other. This is the legitimate domain of the chemist, and no one who is not engaged in the furtherance of chemistry can possibly keep before him with that vividness which is essential to effective teaching the conditions which define and limit the existence and transformations of such bodies.

But after this common elementary part has been mastered the paths of the various classes of students diverge widely. The teachers of chemistry in medical schools have been accused of desiring to turn their pupils into chemists. The mere utterance of this accusation is sufficient proof of the inability of those responsible for it to comprehend the problem; the thing is impossible, the idea ludicrous. Chemistry is too vast a subject to allow more than the fringe of it to be mastered by a student who is not making it his lifework. The aim of the medical teacher is, and can only be, once the common elementary stage is passed, to limit the student in the very manner which Prof. Armstrong proposes as a novelty—namely, to those parts of the subject which have a practical bearing on his subsequent work.

But even supposing that Dr. Armstrong's contentions were correct at the present day, and the methods of teaching chemistry were in the deplorable state he hints at, what are the remedies he proposes? The efficacy of concentrating the teaching of the scientific subjects in three central institutions is, as he admits, a matter of controversy. Two of these institutions would preferably be attached to hospitals, he states, in order that a technical flavour might be given to the training at the outset; but many will fail to see in what respect a class of eighty or ninety students, such as is frequently assembled in some of the larger of the metropolitan schools, would be benefited by removal from a laboratory attached to a hospital to a similar laboratory attached to another hospital or in a separate institution. In the latter case, indeed, the technical flavour would be completely eliminated, and what is more, clinical teachers working in the research laboratories which Prof. Armstrong is evidently unaware have long existed in many of the medical schools would no longer be able to confer with their scientific colleagues, as the writer can testify they constantly do at present.

The suggested elimination of examinations is a course the expediency of which is again open to grave doubt; while the student of exceptional ability might possibly be left in ignorance of his own deficiencies, the average student would be liable to lose the concentration of purpose and the orderly habits which examinations undoubtedly inculcate; such a student would learn little about his subject and still less about its methods. There is much to be said on both sides, but most teachers will agree that on the whole the abolition of examinations would lead to far greater harm than is caused by their imposition; in schools where the teaching is good there is rarely much complaint on this account. But that examinations can be improved is obvious; the writer well remembers, indeed, how Dr. Armstrong himself some fifteen years ago, when examiner in this University (a position which it is difficult to believe he sought as a "bold advertisement"), expressed emphatically the opinion that examinations if conducted in a scientific manner were by no means the artificial tests they were currently reputed to be. And progress has steadily continued: the elaboration of practical examinations; the introduction of questions involving a knowledge of laboratory work, and therefore incapable of being crammed; the allowing of a wider latitude to candidates by means of a choice of questions—all these have made examinations in science very different from what they were when Dr. Armstrong taught medical students.

Nor is it possible to take his proposals as to the rearrangement of the chemical course very seriously, for in several instances they are mutually destructive. He says in one place that "futile talk" about constitutional formulæ must be abandoned, whilst in another place he considers that organic chemistry should be utilised in discussing in inorganic chemistry this very question of structure or constitution. He would pursue the study of carbon compounds mainly from the complex downwards, beginning with the processes of digestion and fermentation, the elucidation of which he elsewhere and rightly describes as one of the most difficult problems in chemistry. What does our knowledge of these processes really amount to? Little beyond long words, it is to be feared. It is this very treatment of complex problems without adequate preparation in the underlying sciences that has kept back physiological chemistry for so many years, and, except for the admirable work of two or three notable pioneers, is still keeping it back in England. But as he admits that no really effective change will be possible until the ways of primary schools are altered, his somewhat unique remedies need not be discussed further.

As to his proposal to fuse organic with inorganic chemistry, the scheme has been tried, and with deplorable results. What is feasible with a student who contemplates a career in pure chemistry is not always practicable with a medical student: the one gives the main part of his time to the subject, the other at most a third of it. Dr. Armstrong would probably agree that the most intelligent student of pure chemistry could hardly be expected to start on the carbon compounds within six months of commencing his chemical studies, and experience has shown that with the limited time available a still longer interval must be allowed to the University medical student. Of course, one may say in a certain sense that there is no distinction between the two branches, any more than there is any real distinction between arithmetic and algebra; but the teacher who began a course of algebra before his students had mastered the ordinary rules of arithmetic would probably find that his labour was wasted.

Prof. Armstrong's assertion that chemistry in the medical schools at the present day is taught in an unsympathetic manner is thus unjustifiable, and his remedies both uncalled for and impracticable. The system to which he would revert is one which was in vogue some thirty years ago, and has long been superseded. Once the foundation is securely laid, the teacher now loses no opportunity of working along physiological and medical lines, and of tracing the lines of development in a direction which will tend to elucidate the later work of the student. So far from studying such irrelevant matter as the aniline dyes, as has been fantastically suggested, the attention of the student is directed to the chemical nature of such materials as the proximate constituents of food—sugars, fats, and the amino-acids from proteins; of the degradation products of the body—urea, the purine derivatives, lactic acid; and of materials such as will enable him to form an intelligent idea of the chemical relations of the natural and artificial drugs with which the pharmacopœias - abound. Surely it cannot be seriously maintained that such matters are devoid of professional interest, or that teaching which devotes a large amount of time to them is unsympathetic.

As to the extent to which chemistry should be taught to medical students, the writer has long held very definite opinions, which, through the courtesy and co-operation of his colleagues, he has to a large extent succeeded in putting into practice. Chemistry for medical students bears precisely the same relation to physiology and pathology as anatomy does, the chemist dealing with the nature and relation of the materials, and the anatomist with the nature and relation of the structures, the function of which in both cases it is the part of the physiologist to investigate. All treatment of the nature and relation of the carbon compounds met with in the animal economy is thus the business of the chemist, whilst the physiologist and pathologist apply the knowledge thus obtained to explain the working of the organism in health and disease; in the writer's laboratory, for example, students dissect chemically, and, where possible, re-construct artificially, compounds such as urea, which they obtain in the physiological laboratory as natural products. Without a knowledge of the chemical nature of the substances they deal with, the physiologist and pathologist are helpless, and this knowledge it is clearly the place of the chemist to impart. Those who attempt to work at too many branches of science run a serious risk at the present day of being declared masters of none.

We turn to the relative proportion of time which should be

allotted to the various parts of the curriculum. One need hardly discuss seriously the views of those extremists who advocate the relegation of the preliminary sciences to the ordinary school curriculum, and who hold that a medical student should have completed his study of chemistry, physics, and biology before entering upon his medical course. It is questionable, indeed, whether the average schoolboy will ever be capable of specialising to this extent; in the present state of the teaching of science in the ordinary schools, in which it is too often classed as a mere subsidiary subject, this is certainly impossible, whatever it may be in the future. Whether this is but a temporary phase in the development of the subject is an open question; but it is significant that many teachers find students who have "learned science at school" far more difficult to deal with than those who have confined themselves to the ordinary school curriculum. School methods, in spite of the efforts of reformers, are still essentially didactic, and it is to be feared that years must elapse before the average schoolboy can be taught to disregard authority and think for himself, without at the same time upsetting the whole system of obedience and respect for authority on which school discipline essentially depends. It must be remembered, moreover, that the medical student usually conforms to the biological or visualising rather than the mathematical or idealising type of mind, and that his powers of abstraction are therefore more or less latent at the time he commences his medical studies; these have to be developed before progress can be made.

Although the organisation of science teaching in the public schools is steadily improving, the number of really efficient pupils who enter directly at the London medical schools is as yet extremely small. For the present, as Dr. Armstrong admits, the medical curriculum must be framed on the assumption that the student on commencing his course is for practical purposes ignorant of the preliminary sciences. The Conjoint Examining Board of the Royal Colleges has for some years past offered every encouragement to students to work at these before entering at a medical school, but although the standard is far lower than that required by the Universities, the number of those who take advantage of this opportunity is surprisingly small. The fact is that the average parent does not decide as to the future of his boy until he is about to leave school, and that boys who show early a marked aptitude for science rarely enter the medical profession, but prefer a vocation in which the work in pure science is less limited.

Dr. Armstrong, in opposition to many of the clinical teachers, maintains that more time should be devoted to chemistry even than is allotted at present, but it is doubtful whether this is really necessary. The time cannot be reduced: that is certain; nor will the other subjects allow of its material extension. The only adjustment desirable would be that afforded by the gradual transference of the purely chemical parts of the physiological course, as these become defined by the advance of chemical science, to the hands of those best qualified to deal with them. Far from extension being possible, an outcry has arisen of late, as Prof. Armstrong remarks, that the proportion of time allotted to the clinical period is inadequate, and ought to be increased at the expense of the scientific period, preferably by limiting the study of chemistry. It is probably true that in some cases there is a certain amount of congestion, but this is largely due to want of co-ordination and as a whole has been greatly exaggerated. Fifty years ago anatomy was the only preliminary subject which was taught practically, and the early part of the medical course was at that time divided between the dissecting-room and the medical and surgical wards. Physiology as a separate subject was nonexistent, and chemistry was taught only by means of lecture experiments. The great benefit which accrued to the student by filling his days with practical work led to the elaboration of anatomy to an extent which is scarcely conceivable by those who have not come in contact with it, and eventually imposed such a task on the memory as to unfit all but the intellectually most robust for work demanding connected thought. Hence the almost instinctive opposition to scientific method which is still discernible in a small and now rapidly diminishing section of the older school of medical practitioners. The early 'seventies witnessed the introduction of laboratory teaching in chemistry, followed at a somewhat later period by laboratory work in physiology. The time required for this work naturally had to be taken from that which had formerly been spent in the dissecting-room, although by no means entirely so, for it is impossible even for the most willing to work day after day at one subject without incurring that form of mental fatigue known as staleness. The clinical work was thus more or less tacitly postponed, and eventually parted from the scientific work and assigned to a separate period.

Although human anatomy is now assigned to the scientific period, it must be remembered that it is not recognised by the Universities as a separate scientific subject, and is only accepted as part of the scientific curriculum when associated with comparative morphology. As Dr. A. P. Beddard, one of the writer's clinical colleagues, remarks: ".I have always supposed that "it is the intention of the University of London to educate its "students so as to understand not only the medicine of to-day, "but also the steps by which the medicine of the time when "the student is in actual practice will have been reached. Few "would deny that at the present time medicine, pharmacology, "bacteriology, and physiology are advancing along lines which "are broadly chemical. The relative importance of chemistry "with regard to anatomy has, in fact, been increasing for many "years, and is now greater than ever it was. Yet those who "oppose the adequate teaching of chemistry and physiology on "the ground of their alleged uselessness to the practitioner are "among the first to insist on teaching the ordinary student "anatomy as if he were to be an operating surgeon. Speaking "from personal experience as a student, the amount of organic "chemistry deemed necessary by the University of Cambridge "for those men who had not previously taken chemistry at the "Tripos is ridiculously inadequate, and speaking as a lecturer "in pharmacology the same is true of the chemistry required "by the Conjoint Board; the curricula of the Universities of "London and Oxford alone approximate to what is needed in "this respect."

It is very significant that University men who come to London for the clinical period after completing their scientific studies at Oxford or Cambridge find, with the above exceptions, no difficulty in completing the necessary clinical work within the same time as is at present allotted to the undergraduates of our own University. The fact is that the thorough training in scientific method afforded by the early part of the University medical curriculum has an enormous influence on the rate at which the student masters its practical applications; the extra time which is spent on the scientific period, as compared with that required from the candidates for the ordinary licences, is far more than counterbalanced by the gain in time and in thoroughness ensured in the clinical period. The University man, moreover, has formed a habit of reading and inquiry which enables him to keep in touch with the advance of his profession for the rest of his career, even if he has not time to share in it ; but the average licentiate remains bound to what has been taught him during his pupilage.

The picture which Prof. Armstrong would draw of care-worn students harassed by the imposition of an impossible task has in fact no counterpart in reality. University students are expected to work hard, and do work hard ; their degrees would not otherwise be worth the paper on which they are written. But they work with a will, and, what is more, they work with trained intelligence and keen interest if their course is properly arranged ; yet they manage withal to find abundant time for physical and mental recreation. Not only can chemistry be studied along with anatomy and physiology, but it reacts to their mutual advantage. Mr. F. J. Steward, M.S., the writer's colleague in Anatomy, is "convinced that the work in chemistry can be done without in any way interfering with the study of anatomy"; "indeed, it is an agreeable change," he continues, "and so helps to prevent staleness." Dr. M. S. Pembrey, his colleague in Physiology, is "convinced from experience, both as a student and teacher, that a knowledge of organic chemistry is of the utmost importance for the chemical aspects of physio-logy"; he thinks "that it is actually an advantage that men should study organic chemistry at the same time that they work at chemical physiology," and "does not find them delayed thereby." When teachers are willing to co-operate, no difficulty arises, and there is no complaint of congestion.

We come finally to the question of *personnel*—to what class of teachers the chemical course .is to be entrusted. Apart from the few who object to the teaching of chemistry in medical schools in any form whatsoever, there is a large section who think that the chemistry which they agree is indispensable should be taught by those who are engaged in applying it rather than by those who spend their leisure in endeavouring to extend its boundaries—by the physiologist and pathologist rather than by the chemist. This argument, although plausible, is unsound, for by the same reasoning physiology ought to be taught by the physician who is engaged in applying it rather than by the professional physiologist who is engaged in extending it—which physiologists are hardly likely to admit. The system has, in fact, been tried, and in both cases discarded as impracticable; just as the clinical teacher could no longer keep sufficiently in touch with physiology to render him an efficient teacher of this subject, so neither he nor the professional physiologist who took his place could make an efficient teacher of chemistry; in both cases the necessary vividness was lost through lack of circumstantial detail.

There is no doubt that the majority of the clinical teachers have less objection to the presence of physiologists in the medical schools than of that of their colleagues the chemists. The distinction probably arises from the fact that the majority of professional physiologists—though there are some notable exceptions—have been through the medical course, and are qualified to practise medicine, whereas the majority of the teachers of chemistry have been trained in pure science; one heard of no objection to chemistry in the days when it was taught by medical men. It is comparatively few, however, who after the seven years' training required to make an efficient chemist are able to devote another four years or more to obtaining a medical qualification, and those who do so naturally find an outlet in applying their chemical knowledge to physiological and pathological problems, rather than to those of pure chemistry. But it is noteworthy that those who have done this are emphatic in their condemnation of the proposal to relegate the teaching of chemistry to medical men. Dr. F. G. Hopkins, F.R.S., Reader in Chemical Physiology in the University of Cambridge, in a letter to the writer expresses himself as follows:

" I feel strongly that a thorough training in the general "principles of organic chemistry is a necessary antecedent to "the study of physiological and pathological chemistry. It is "certain, too, that efficient instruction in inorganic chemistry "(of which subject the biological chemist is as a rule profoundly "ignorant) should precede the organic, and that meanwhile "(such is the urgent need of biology) as much physical chemistry "should be imparted as possible. All this teaching should be "in the hands of the pure chemist. As regards the amount of "it which is necessary to prepare the medical student for a "proper grasp of the facts of physiology, I am quite aware that

"the ground covered must be limited, and that it should be
"carefully chosen with regard to the special needs of the student.
"It seems to me that the principles you have laid down are
"indisputable, and indicate clearly as a minimum some such
"curriculum as is at present in force in the University of
"London; remembering my own case, and judging from my
"own experience of students in London and Cambridge, I cannot
"well conceive of less being efficient.

"I understand it is being urged that the physiological
"or pathological chemist, from his superior knowledge of its
"points of application, is a fit and proper person to assume
"responsibility for chemical teaching at an earlier stage than
"has hitherto been usual. My own feelings and experience lead
"me to doubt the correctness of this view. The current literature
"of biological chemistry is enormous, and as it is necessary for
"the physiological chemist to keep himself also well informed in
"the progress of general physiology, his time is well filled. It
"is well-nigh impossible that he should keep his knowledge of
"general chemistry in a condition which will equip him as an
"efficient teacher of the subject. As a chemist he must remain
"highly specialised, and therefore should not be allowed to
"intrude too early on the student. My very definite opinion
"is that the specialised physiological chemist is seldom able to
"teach the groundwork of chemistry efficiently, and can only
"hope to teach his own subject well when his pupils come to
"him already well versed in fundamental principles."

Stronger support than this is scarcely possible. Despite
the attitude of a certain section of the clinical teachers, chemistry
is an integral part of the medical curriculum, which for
University men, at all events, must be arranged with the view
of securing that the student shall not merely learn empirically
the methods obtaining at the time of his pupilage, with which
the ordinary licentiate perforce remains content, but shall also
understand the underlying scientific principles. For this
purpose a training in scientific method is essential, and can
only be given by men whose daily work is in the domain of
the science they teach. In framing a University curriculum
the voices of all the teachers must be heard, the teacher of
science equally with the clinical teacher who applies his
methods; no difference in status can be recognised. It would
be an evil day for University education if the statutory

inspectors were to report, as they did not so long ago with certain of the licensing examinations in the curricula of which the teachers of science have little or no voice, that the standard in chemistry was below the minimum they could recognise.

To conclude: Prof. Armstrong's contention that chemistry is not taught in the medical schools with a proper sense of its applications is not warranted by the facts, and could apply only to his experience of twenty years ago; his experience of the teaching of the present day is far too limited to give authority to his sweeping generalisations. For the present, and probably for many years to come, the curriculum must be framed on the assumption that the student on commencing his medical course has no knowledge of science; and whether this science is taught in London in institutions connected with hospitals, or in separate institutions as at Oxford and Cambridge, the same course must be followed. Prof. Armstrong's proposals as to the elimination of examinations and the re-arrangement of the chemical curriculum are certainly not practicable at present, and it is doubtful whether, in the nature of things, they ever can be. The chemical curriculum cannot be lengthened as Prof. Armstrong proposes, nor can it be shortened without injuriously affecting the remainder of the course. The necessary training in chemistry must be given by expert teachers, for the day is long past when clinical teachers or others not engaged in working at the advancement of pure chemistry could be entrusted with it. Despite Prof. Armstrong's statements, our progress has been steady and continuous, and the teaching of the medical sciences at the present day is as far removed from that of twenty years ago as this, in turn, was from that of the purely anatomical era.

The future of medical teaching depends on co-operation between the professional teachers of science and the teachers of clinical work, such as for many years it has been the good fortune of the present writer to enjoy. Without long and friendly discussion with his colleagues this would indeed have been impossible, and their broad and generous spirit is gratefully acknowledged. Would that it were universal! By quietly discussing their points of difference, medical men and men of science would then arrive at a common standpoint, and controversies such as these would no longer be possible: it is to be hoped that this spirit of contention will fade away, and that it will be recognised that, as science progresses, even further subdivision may be necessary ; the day is for ever past when a single mind could traverse the whole domain of know-ledge, and the future progress of the world must depend on the co-operation of specialists, who, being aware of their own limitations, will be the more eager to welcome help from their brother-workers.

THE RELATIONSHIP BETWEEN THE COLOUR AND CONSTITUTION OF ORGANIC COMPOUNDS

By S. J. MANSON AULD, Ph.D.

Scientific and Technical Department, Imperial Institute

SIMULTANEOUS with the growth of Organic Chemistry, and the consequent wide extension of synthetically formed bodies with definite constitution, there has grown up the desire to draw a direct connection between their physical properties and chemical constitution. In no case has this been more exemplified than in the property of "colour," which from its very nature is, as a rule, the first to become apparent to the senses. Its investigation has been all the more attractive and has received greater attention because of the necessity of building up a certain "chemistry of colour" to aid in the systematic synthesis of the organic dye-stuffs and colouring-matters. The properties of colour and dyeing-power must, however, be kept distinctly apart, and in discussing the former there is no need to inquire whether the substance possessing it has also the property of a dye.

The investigations of Graebe, Liebermann, Kehrmann, Kostanecki, Thiele and others have all tended to show that, in general, the presence of double bonds in one form or another is necessary for the production of colour.

Of the simplest group of organic compounds, the *hydrocarbons*, the greater number are colourless; this is invariably the case with the derivatives of methane, *i.e.* the saturated hydrocarbons. The presence of one or two pairs of doubly-linked carbon atoms is also insufficient to produce substances absorbing light in the visible spectrum, but a certain class of hydrocarbons containing three such ethylene linkages has been discovered, all of which are highly coloured.[1] These hydro-

[1] Thiele, *Ber. d. d. Chem. Ges.* **33**, 666 (1900).

carbons are known as *Fulvenes*, and the mother-substance *Fulvene*, which is isomeric with benzene, has the formula—

$$\begin{matrix} CH = CH \\ | \qquad\quad \\ CH = CH \end{matrix}\!\!\Big\rangle C = CH_2.$$

Especially highly coloured are the methyl phenyl and dimethyl derivatives, substituted in the CH_2 group. The presence of this five-membered ring and the third double bond seem to be essential to the production of colour, for when the latter is transformed into simple bonds by reduction or addition (*e.g.* of halogen), colourless substances ensue. Accepting the Kekulé formula for the benzene ring also in the polycyclic compounds, *e.g.* in the gold-yellow acenaphthylene,[1]

$$CH = CH$$

we find that all the coloured hydrocarbons contain the group

$$\begin{matrix} = C \\ = C \end{matrix}\!\!\Big\rangle C =.$$

That the reverse of this is not true is witnessed by the colourless diphenyl.

CHROMOPHORE-AUXOCHROME THEORY

Much earlier than the discovery of these coloured hydrocarbons, similar appearances with other groups than C=C led Witt[2] to the formulation of his Chromophore-Auxochrome Theory.

Certain groups exist which, when bound up with hydrocarbon radicles in sufficient number and in a suitable manner, tend to produce coloured compounds. These groups are termed *Chromophores*. As a rule the Chromophores by themselves do not cause colour, but require the presence of other groups, called by Witt *Auxochromes*, before their action becomes noticeable. The auxochromes alone have no colour-producing properties.

[1] *Ber. d. d. Chem. Ges.* **26**, 2354 (1893).
[2] *Ibid.* **9**, 522 (1876).

Chromophores

The principal chromophoric groups are $-C = O$, carbonyl, $-C = S$, $-C = N$, nitrile, $-N = N$, azo, N_2O, azoxy, $N = O$, nitroso, and NO_2, nitro, but other chromophores, some containing sulphur and tellurium, are also known. The most important of the groups just mentioned are the carbonyl, azo, and nitro groups; and it may be as well to describe them more closely.

The *Carbonyl* group by itself does not produce colour, but a second $-C = O$ group in juxtaposition strengthens the action to such an extent that coloured compounds almost invariably result. In the β and γ positions[1] this influence is not felt. Thus we have—

$$CH_3 \cdot CO \cdot CH_3 \quad CH_3 \cdot CO \cdot CH_2 \cdot CO \cdot CH_3 \quad CH_3 \cdot CO \cdot (CH_2)_2 \cdot CO \cdot CH_3$$

Diacetyl.	Acetyl acetone.	Acetonyl acetone.
Yellow.	Colourless.	

In the aromatic series many cases are known where a single carbonyl group apparently acts as a chromogenic centre, but it must be remembered that, accepting Kekulé's formula, many double bonds are present, and these might well act as additional " chromophores." Thus in fluorenone, besides the CO group we have a distribution of double bonds similar to that in fulvene.

Coloured.	Colourless.

It is noteworthy that on replacing the carbonyl oxygen with two chlorine atoms the chromophore is destroyed. Fluorenone dichloride is colourless.[2]

In both the aliphatic and aromatic series it has invariably been found that compounds containing the *azo group*, $-N = N-$ are highly coloured. The simplest members of the series,

[1] Reckoned from the first carbonyl, *e.g.*—

$$CH_3 \cdot CO \cdot \overset{\alpha}{CH_2} \cdot \overset{\beta}{CH_2} \cdot \overset{\gamma}{CO} \cdot CH_3.$$

[2] Smedley, *Journ. Chem. Soc.* 1905, 1249.

diazomethane $CH_2\overset{N}{\underset{N}{\big\langle\|}}$ and diazoethane, are yellow-coloured, but it is in the aromatic compounds that the colour is most noticeable. Azobenzene $C_6H_5 - N = N - C_6H_5$, the simplest aromatic derivative, forms orange-red crystals. In the light of this theory the azo group is consequently regarded as one of the strongest chromophores, although when bound up in a closed chain, and even in presence of the above-mentioned chromophoric triple ethylene linkage, it is practically inactive. Tolazon,[1] for example, is only slightly coloured.

Probably no other group is more intimately bound up with the production of colour, or more important in its bearing on the relationship to constitution, than the *nitro group*, NO_2. Regarded for the present merely as a chromophore, it does not, for instance, exert so strong an action as the azo group, but it is most important owing to the ease with which it can be introduced, and the generality of its action.

Quinones

Before proceeding to give an account of the auxochromes, it will be as well to devote a few lines here to a short discussion of that class of compounds known as quinones.

The quinones may be regarded as derivatives of benzene, in which two hydrogen atoms have been replaced by two atoms of oxygen. The replacement may be either in the *ortho-* or *para-* position, when we distinguish between ortho-quinones and para-quinones. Meta-quinones are wholly unknown. All quinones and quinone derivatives so far known are coloured to a greater or less extent. The simplest quinone is *para*-benzoquinone or ordinary quinone, which is formulated thus :

or simply

---

[1] L. Meyer, *Ber. d. d. Chem. Ges.* **26**, 2230 (1893).

The ring of this gold-yellow substance is therefore composed of four chromophoric groups, viz. two ethylene linkages and two carbonyl groups. The presence of the $-CO$ groups alone is insufficient to produce colour, as may be seen from the similarly constituted but colourless diketo hexamethylene.

$$
\begin{array}{c}
CO \\
\diagup \quad \diagdown \\
H_2C \qquad CH_2 \\
| \qquad\quad | \\
H_2C \qquad CH_2 \\
\diagdown \quad \diagup \\
CO
\end{array}
$$

It will be observed that although *ortho*-quinone can also be readily formulated, it is an impossibility to construct a similar formula for *meta*-quinone owing to the distribution of the double bonds. All compounds derived from quinone, whether by substitution of hydrogen by monovalent radicles or of oxygen by divalent radicles, are also coloured and are termed *quinonoids* or said to possess a quinonoid structure. This applies only as long as they contain the skeleton nucleus.

or

Among the more important oxygen substituents may be mentioned derivatives of $=CH_2$, *i.e.* compounds like the well-known dye aurin which are derived from the hypothetical quino-methane $O=C_6H_6=CH_2$. Either or both of the oxygen atoms may also be replaced by nitrogen as in quinone diphenylimide and allied substances. The former is brown-red in colour.

$$C_6H_5 \cdot N = \left\langle \overline{} \right\rangle = N - C_6H_5.$$

A number of polycyclic colouring-matters and dye-stuffs contain the same nucleus and are now formulated in a similar manner as *ortho*-quinones. To this class belong the derivatives of acridine and apo-safranin :

Acridine.

Apo-safranin.

Auxochromes

The present accepted theory of auxochromic action assumes that on the addition of an auxochrome to a chromogen[1] (*i.e.* a substance which contains a chromophore group and gives a deeper-coloured compound on the addition of an auxochrome), a strain is induced on the benzene ring, varying in degree according to the auxochrome employed. In this way, according to Kauffmann,[2] the benzene ring is "no dead unalterable structure, but a most delicate picture which on the slightest touch at once alters its properties to a greater or less extent in reply."

In the strict optical sense all benzene derivatives are coloured in that they all show more or less well-defined absorption spectra. In many of the chromogens of course the absorption bands lie far out in the ultra-violet, but each substituent alters the positions of the lines, some in such a manner that they are moved into the visible portion of the spectrum. To these latter belong the auxochromes. The scope of this article does not admit of the consideration of these theoretically coloured bodies. The relationship between absorption-spectra and constitution is a wider subject, and the term "coloured" is here confined to bodies giving an absorption in the visible spectrum.

The most important auxochromes are the hydroxyl ($-$ OH) and amino ($-NH_2$) groups, and in virtue of their respective acid and basic properties, many of the coloured substances produced by their action are dyes. In the light of the auxochrome theory the $-NH_2$ group is much more powerful than $-$OH. *Para*-nitraniline is deep yellow, whereas *para*-nitrophenol is nearly colourless. Substitution of the hydrogen of the amino group by alkyl or aryl residues produces new auxochromes which are more powerful in action than the parent group. The commonest of these new auxochromes are $-N(CH_3)_2$, $-N(C_2H_5)_2$, and $-NHC_6H_5$, their action increasing with the molecular weight. Similar substitution of the hydroxyl group has no settled effect. Liebermann[3] has shown that alkylation of alizarin and isoanthraflavic acid causes the absorption bands to move towards the violet, whilst similar treatment of anthraflavic acid and quinizarin moves them

---

[1] *Ber. d. d. Chem. Ges.* **9**, 522 (1876).

[2] *Ibid.* **39**, 1959 (1906).

[3] *Ibid.* **23**, 1566 (1890).

towards the red. Auxochromes are generally supposed to support each other, and the number of groups present is consequently of importance. In general also it has been found that the auxochromes exert their strongest action when in the position *ortho* to the chromophor. As pointed out by Kauffmann[1] this is plainly seen in the case of the fuchsin dyes.

Here we have the quinonoid ring acting as chromophore and the animo groups as auxochromes. When in position 3 (see formula above) the NH_2 produces a violet dye-stuff, when in position 2 a bluish green, and in position 1 a blue-green.

As a measure of the condition of the benzene ring and consequently of the effect of the auxochromic substituents, Kauffmann[2] claims the relative magneto-optical rotations to be of great importance. He draws the conclusion that the greater the rotation the greater the auxochromic action, *i.e.* the greater the strain on the benzene ring caused by an auxochrome, the higher will be its colour-producing value. As a consequence Kauffmann places the auxochromes in the following order, based on the magneto-optical constants—

$-O\cdot COCH_3$	OCH_3	$NHCOCH_3$	NH_2	$N(CH_3)_2$	$N(C_2H_5)_2$
-0.260	$+1.459$	1.949	3.821	8.587	$8.816.$

Salt Formation

The auxochromes, being as a rule groups strongly acid or basic in character, produce coloured bodies readily capable of forming salts. Chromophors also in some cases cause salt formation, but generally only with strong acids or bases. The effects of salt formation on the colour are far reaching, and vary considerably with the conditions.

In the case of bases giving salts with colourless acids it has

[1] *Zusammenhang z. Farbe u. Konstitution; Sammlung chem. u. chem.-tech. Vorträge,* **IX.** 29.

[2] *Ber. d. d. Chem. Ges.* **39**, 1959 (1906).

been shown that, in general, salt formation in the auxochrome lessens the colour considerably, whilst a similar combination in the chromophore deepens it.[1] As examples of the former class may be mentioned the nitranilines, aminocoumaric acids, amino-anthraquinone, etc. *Ortho*-aminobenzophenone is deep yellow, its sulphate and hydrochloride, on the other hand, colourless. The deepening of the colour on chromophoric combination with acids is evidenced by the acridines, the quinoxalines,[2] and other compounds, which though themselves colourless give coloured salts. Phenazine again gives red salts, although the base itself is only yellow-coloured. Most of these substances will be mentioned again later.

When salt formation is due to combination with a " colourless " metal, the effects produced are reversed. In the auxochrome a most marked deepening of the colour is noticed; in some cases indeed the colour is first called forth by salt formation. The nitrophenols, the phthaleins, the oxyazobenzenes and many triphenylmethane derivatives are all examples of this class of compounds. Very few examples are known of metallic salts produced directly from chromophoric groups, but those examined all show that the colour tends to disappear. Thus oxytoluquinoxaline carboxylic acid is yellow, but gives colourless ammonium, potassium, and sodium salts :

$$C_7H_6 \left\langle \begin{array}{c} N \diagdown C \cdot COOH \\ | \\ N \diagup C \cdot OH \end{array} \right.$$

The exponents of Witt's chromophore-auxochromic theory have endeavoured to embody these results in several general statements, and regard the influence of salt formation as being merely passive in character, simply changing the state of strain in the compound, and therefore also the colour either positively or negatively, according to the position of the salt-producing groups.

Any chemical explanation of the chromophore-auxochrome theory, which as generally accepted is really less a theory than a well-chosen statement of facts, has been left to quite recent years. It has been the endeavour of Kauffmann[3] to show that the benzene ring alters its constitution in its different

[1] Kauffmann and Beisswenger, *Ber. d. d. Chem. Ges.* **36**, 561 (1903).
[2] Hinsberg, *Ann. d. Chem.* **237**, 327.
[3] *Ber. d. d. Chem. Ges.* **33**, 1725 (1900) ; **34**, 682 (1901) ; **35**, 3668 (1902).

derivatives and corresponds in the chief cases to the following three formulæ:

Diagonal.	Kekulé.	Dewar.
I.	II.	III.

The action of the colour-producing groups consists in altering the contour of the benzene ring to produce those forms (II. and III.) which contain the double bonds. These, as previously shown, have a strong chromophoric action under certain conditions.

In many cases the auxochrome-chromophor theory is very unsatisfactory and leads to a number of contradictions. In the azo compounds, for example, although in general the colour intensity of the azo chromophore is continually increased by the auxochromes combined with it, from the nearly colourless azoisobutyricester to the red azobenzene and azodicarboxylic acid derivatives, yet it is frequently found that the strong auxochromes $-OH$ and $-NH_2$ have no action, or indeed sometimes a negative action, whilst other lesser known groups produce strong colour intensification.

The Quinone Theory

The first attempt to find an explanation of colour apart from Witt's chromophore-auxochrome theory was occasioned by Armstrong's original discussion on the origin of colour about fifteen years ago.[1] Armstrong pointed out that in the case of those coloured compounds which had at that time been well studied, by far the greater number could be given a *quinonoid* structure, *i.e.* be formulated as derivatives of quinone. This it was maintained was the cause of their colour, and indeed of coloured organic compounds in general, the converse of this, that quinone derivatives are always coloured, being known to be true. Although these views were put forward without any definite practical confirmation, they at once found a very wide acceptance, and the formulæ of many coloured bodies which did not accord with them have been reconsidered and reconstructed.

---

[1] *Proc. Chem. Soc.* **1888**, 27, **1892**, 101, 103, 143, 189, 194, **1893**, 206, **1896**, 42.

In this way was built up the "quinone" theory of colour, which commends itself particularly because of its extreme simplicity compared with the intricacies of the auxochrome-chromophore theory, and above all because it gives a chemical explanation of the phenomenon of colour. It is needless to give detailed examples of the method of quinonoid formulation. Some have already been mentioned, and by making assumptions in some cases it has so far been possible to formulate nearly all the coloured compounds in this way. Instance might be made of the interesting series of compounds known as *fulgides*. Stobbe[1] explains the colour of the fulgenic acids and similar compounds by the number and position of the chromophoric phenyl and carboxyl groups, and also by the influence of the conjugated double linkages. The colour can, however, be brought into harmony with the quinone theory, as the substances may be looked on as *ortho-* and *para-* "quinonoid" derivatives of tetrahydrofurfuran—

$$
\begin{array}{cc}
\begin{array}{l}
CH_2 - CH_2 \\
| \qquad\quad | \\
CH_2 - CH_2
\end{array}\!\!\Big\rangle O
&
\begin{array}{l}
R \cdot CH = C - C=O \\
| \qquad\qquad | \\
R \cdot CH = C - C = O
\end{array}\!\!\Big\rangle O
\end{array}
$$

It is interesting to note that Thiele[2] has succeeded in preparing hydrocarbons of undoubtedly "quinonoid" structure. The best known of these compounds is tetraphenyl-*para*-xylylene—

$$
(C_6H_5)_2 \cdot C = C\!\!\Big\langle\begin{array}{l} CH = CH \\ CH = CH \end{array}\!\!\Big\rangle C = C(C_6H_5)_2,
$$

which forms orange-yellow needles.

In some cases the generally attributed quinonoid structure has been disputed on chemical grounds. This has been the case with the phthaleins, a class of compounds produced by the condensation of phthalic acid with phenols and which give deeply coloured alkali salts. In the free state these bodies are supposed to contain a lactone[3] ring which by the action of dilute alkali undergoes a desmotropic change into a quinone containing a carboxylic group.

[1] *Ann. d. Chem.* **349**, 333 (1906).
[2] *Ber. d. d. Chem. Ges.* **37**, 1463 (1904).
[3] The name *lactone* is given to the type of compounds produced from one molecule of hydroxy acids by loss of water.

These formulæ have been disputed by von Baeyer [1] as being not completely in accord with their chemical behaviour, and quite recently Silberrad [2] has brought forward evidence against the quinonoid character of the salts. Similar to the phthaleins, Silberrad has produced melliteins and pyromelliteins, highly coloured bodies produced from mellitic and pyromellitic acids. Some derivatives of these substances which by analogy should possess a quinonoid structure due to rupture of a lactone ring, cannot, according to Silberrad, be formulated as quinones, and admit of no tautomerism. For example we have—

K salt of octabrom 3 : 3′ : 6 : 6′ : 9 : 9′ hexahydroxydixanthylbenzene tetracarboxylic acid.

para-xanthyl derivative.   meta-xanthyl derivative.

Arguing from these halogenated derivatives to the free hydroxyl compounds (which, by the way, can be formulated as quinones) and thence to the phthaleins in general, Silberrad regards the present formulæ of the latter as unsatisfactory and, favouring the theory of Baeyer, ascribes latent chromophoric properties to the 9-carbon atom of the xanthyl derivatives. It must be remembered, however, that these substances are of

[1] Ber. d. d. Chem. Ges. **38**, 569 (1905).
[2] Journ. Chem. Soc. **89**, 1787 (1906).

very high molecular weight and above all non-crystalline, and this makes them unsatisfactory substances to employ in the present discussion. Green[1] has also pointed out that assuming quadrivalent oxygen it is possible to formulate these substances as *ortho*-quinones as follows :

One or two other cases might be mentioned where the quinone theory offers no real explanation of the presence of colour so far as the substances have been investigated. Particularly interesting is the case of diaminoterephthalicester, where the " quinone" and the Witt theories come in opposition. This red substance which gives colourless salts might, on the quinone theory, be formulated as—

when it would be assumed that salt formation produces a true benzene—*i.e.* a colourless derivative. Kauffmann[2] argues that this is improbable, as the free amine should certainly also be colourless. He explains the colour as being the strong intensifying action of the NH_2 groups on the weakly chromophoric $CO_2C_2H_5$ —.

Although, as stated above, the quinone theory of colour was originated by Armstrong, yet it has been left to others,

[1] *Proc. Chem. Soc.* **319**, 12 (1907).
[2] *Loc. cit.* p. 33.

and principally to Hantzsch, to give an experimental proof of
the quinonoid structure of the chief dyes and colouring matters.
As an example of this may be cited the derivatives of triphenylmethane which Hantzsch [1] proved to be derived from triphenylcarbinol in the colourless leuco-bases, and from a quinonoid
body in the coloured salts. The leuco-base of crystal violet,
for example, has the formula—

$$(CH_3)_2N \cdot C_6H_4 \diagdown_{C} \diagup^{C_6H_4N(CH_3)_2,} \diagdown_{OH}$$
$$(CH_3)_2N \cdot C_6H_4 \diagup$$

the hydrochloride, however, being—

$$(CH_3)_2N \cdot C_6H_4 \diagdown_{C = C_6H_4N(CH_3)_2 \cdot Cl.} \diagup$$
$$(CH_3)_2N \cdot C_6H_4$$

Similar changes in the acridine series were explained by
Hantzsch [2] in the same way.

Nitro Compounds

Bound up inseparably with the quinone theory and indeed
serving, to a certain extent, as a test case for its validity we have
the whole class of nitro compounds and especially the nitrophenols. The nitrophenols themselves are in the solid state
either colourless or only slightly coloured; their salts, on the
other hand, are invariably highly coloured, strongly absorbing
light in the visible portion of the spectrum. The **normal** alkyl
and aryl derivatives again are quite colourless. On the Witt
theory no explanation of this can be given beyond the statement
that the auxochromic nature of the hydroxyl group is only
brought strongly into evidence by salt formation. It was first
proposed by Armstrong [3] to ascribe to the salts a quinonoid
structure produced by an intra-molecular change, and in this
way to explain their colour.

ortho-nitrophenol.   Sodium ortho-nitrophenolate.

This view was advanced on very slight grounds, and very
little experimental evidence was offered in its favour. On the

[1] Ber. d. d. Chem. Ges. 33, 278 (1900).
[2] Hantzsch and Kalb, Ber. d. d. Chem. Ges. 32, 3120 (1899).
[3] Loc. cit.

other hand, Hantzsch has succeeded in bringing forward a mass of evidence to prove that the nitrophenols are pseudo-acids and that the salts are derived from *aci*-nitrophenols—*i.e.* true acids. In this way they fall into line with the coloured salts of the nitrolic acids. These latter compounds were discovered by V. Meyer,[1] who gave them the formulæ of nitroximes, or nitro-isnitroso bodies—

$R . C \diagup_{NO_2}^{N . OH}$. Dissolved in alkalies they undergo a change into an isodynamic form with production of red salts, which on standing are gradually converted into more stable colourless salts corresponding to the free acids.[2] Hantzsch expresses the various changes as follows:

True nitrolic acid derivates. Nitro-isnitroso compounds.		Enythronitrolic acid salts.		Leuconitrolic acid salts.
$CH_2 . C \diagup_{NO_2}^{N . OH (R')}$	$\xrightarrow[\xleftarrow{HCl}]{KOH}$	$CH_3 . C \diagup_{N}^{N} \diagdown_{O\ OK}^{O}$	$\rightarrow$	$CH_3 . C \diagup_{NO}^{NO_2K}$
Colourless, acid stable.		Coloured, labile.		Colourless, alkali stable.

Hantzsch has also shown that for the nitrophenols there exists a series of isomeric labile red quinonoid *aci*-ethers, corresponding to the colourless stable true nitrophenol ethers.[3] This, taken in conjunction with his proof that all constitutionally unalterable nitrobenzene and nitrophenol derivatives are colourless,[4] leads him to the belief that the weak coloured nitrophenols are solid solutions, and the aqueous solutions equilibria of nitrophenols and *aci*-nitrophenols (or their ions).[5]

The intensely coloured salts of the nitrophenols are practically completely *aci*-salts, as their colour and absorption-spectra are almost identical with those of the ethers. Thus the *rôle* of the "auxochrome" is transformed from a passive to an active one, and leads us again to a chemical theory of colour. For this Hantzsch wishes to discard the term "quinone" theory, owing to the number of non-benzene derivatives which must be included, and to replace it with *umlagerungs* theory, implying

[1] *Ann. d. Chem.* **175**, 88 ; **180**, 170.
[2] Graul and Hantzsch, *Ber. d. d. Chem. Ges.* **31**, 2854 (1898).
[3] *Ber. d. d. Chem. Ges.*, **39** 1084 (1906).
[4] Cf., however, Kauffmann, *Ber. d. d. Chem. Ges.* **39**, 4237 (1906).
[5] *Ber. d. d. Chem. Ges.* **39**, 3072 (1906).

that to produce a colour-change a rearrangement of the molecule is necessary. This explains the fact that the alkyl and acyl derivatives of the nitrophenols, $X \cdot OCH_3$ and $X \cdot O \cdot COCH_3$ are colourless, on the ground that they have an equal action in rendering the hydrogen atom immovable—*i.e.* preventing an isomerisation from $X \cdot OH$ to $X <^O_H$. Similar explanations are given to the colour changes of the nitranilines, violuric acid and similar oximido ketones, the dinitro and trinitro paraffins, etc. .

As required by the dynamic theory, Hantzsch contends that the " auxochromic " value of the $-OH$ group is exactly equal to that of the methoxyl, and he points out that Baly's[1] work on absorption-spectra confirms this view. The absorption-spectra of *para*-nitrophenol and *para*-nitroanisol fall together in the remote ultra-violet.

An indirect proof of the quinonoid nature of the nitrophenols has been given by the discovery of the coloured mercury-*aci*-nitrophenols[2] which contain mercury in a closed chain. From their behaviour they must have the formula—

In a paper entitled " Colour and Constitution of Acids, Salts and Esters,"[3] Hantzsch extends his views on the dynamic theory of colour to include the classes of compounds named. For most of these substances the second tautomeric forms have not yet been isolated, but great differences have been observed in the colour and behaviour of closely related derivatives, chiefly salts. In the case of *salicyl aldehyde*, the true phenol is colourless $C_6H_4 <^{OH(R')}_{CHO}$. The ions and the alkali salts are, however, coloured, showing a transformation into $C_6H_4 <^O_{CH \cdot OM''}$, whilst the ammonium salt again is colourless, *i.e.* $C_6H_4 <^{ONH_4}_{CHO}$.

In the case of substances of acidic nature, Hantzsch

[1] *Journ. Chem. Soc.* **89**, 518 (1906).
[2] *Ber. d. d. Chem. Ges.* **39**, 1105 (1906).
[3] *Ibid.* **39**, 3080 (1906).

differentiates between the following classes: (1) *Acids constitutionally unchangeable.* These give colourless or equally coloured hydrogen compounds, salts, alkyl and acyl derivatives. (2) *Pseudo-acids,* which together with their alkyl and acyl derivatives are colourless but which give coloured salts and ions. (3) *Coloured acids,* giving colourless alkyl compounds, but coloured ions and salts. Here the acid is a *mero*-compound ("*mero*" denotes the above-mentioned solid solution of two dynamically isomeric compounds, one colourless, the other coloured); the salts are derived from the real acid and the alkyl derivatives from the pseudo-acid. (4) *Acids of the type of anthraquinone,* which together with its non-ionisable derivatives (acetate and methyl ether) are coloured, but which yield deeper coloured isomerisable hydroxyl bodies (alizarin) and still deeper coloured salts. This may be represented as follows—

$$C_6H_4 \!\!\!\begin{array}{c}\diagup CO \diagdown \\ \diagdown CO \diagup \end{array}\!\!\! C_6 {}_2 \!\!\!\begin{array}{c}\diagup OR \\ H \diagdown OR \end{array} \qquad C_6H_4 \!\!\!\begin{array}{c}\diagup C(OH) \diagdown \\ \diagdown CO - \diagup \end{array}\!\!\! C_6H_2 \!\!\!\begin{array}{c}\diagup O \\ \diagdown OH \end{array}$$

Yellow.   Red.

$$C_6H_4 \!\!\!\begin{array}{c}\diagup C(OH) \diagdown \\ \diagdown C(OH) \diagup \end{array}\!\!\! C_6H_2 \!\!\!\begin{array}{c}\diagup O \\ \diagdown O \end{array}$$

Violet.

Very interesting from the point of view of Hantzsch's theory are the derivatives of succino-succinicdiethylester, which gives intensely coloured salts. In an analogous manner to salicyl aldehyde Hantzsch formulates the metallic derivatives as—

$$
\begin{array}{c}
C_2H_5O \cdot C \cdot OM \\
\| \\
C \\
HC \diagup\diagdown C = O \\
O = C \diagdown\diagup CH \\
C \\
\| \\
MO \cdot C \cdot OC_2H_5
\end{array}
$$

abandoning the formulæ of Baeyer[1] and Hermann.

HALOCHROMISM

Closely related to colour-change on salt formation is the phenomenon known as halochromism, by which is meant the ability of colourless or weakly coloured substances to combine

[1] *Ber. d. d. Chem. Ges.* **19**, 428 (1886).

with acids to form highly coloured salts without the necessary formation of a chromophoric group—*e.g.* a quinonoid. Halochromism has been chiefly studied by Baeyer and Villiger,[1] Kehrmann,[2] Kauffmann,[3] and others, and has generally been connected with certain oxygen bodies. The oxygen in these compounds possesses weakly basic properties, and in salt formation is supposed to pass from the divalent to the tetravalent condition. The class of compounds showing halochromism is very varied and extensive, and includes in particular many ketones and phenols and their ethers. In the case of triphenylcarbinol, which dissolves in concentrated sulphuric acid with a deep yellow colour, Baeyer and Villiger ascribe metallic properties to the $(C_6H_5)_3C$ radicle, so that consequently sulphuric acid produces a carbonium-salt.

It seems highly probable that halochromism is also due to intra-molecular rearrangement, but the absence of sufficient material leaves this a matter of hypothesis.

FLUORESCENCE

Although the subject of fluorescence is, strictly speaking, outside the scope of this article, yet it possesses a certain interest when viewed simply as a variation of the property of colour. This is particularly the case regarding the theory of Hewitt.[4] It has been shown by Hewitt in a large number of cases that fluorescent substances possess the ability to tautomerise, and from this he infers that fluorescence is produced by the absorption of light of a certain wave-length by one form, and the emission of the light in a different wave-length by the other form. As an example may be quoted the members of the fluorescein group—

[1] Baeyer and Villiger, *Ber.* 35, 1189, 1754, 3013 (1902) ; 36, 2774 (1903).
[2] Kehrmann and Wentzel, *Ber.* 34, 3815 (1901).
[3] *Ber.* 35, 1321 (1902) ; 36, 561 (1903).
[4] *Zeit. f. physik. Chem.* 34, 1 (1900).

It will be observed that in this system of "symmetrical double tautomerism" the two forms to which the activity is ascribed are *para*-quinonoid in structure.

Conclusion

Although a direct and fully proved relationship between colour and constitution does not at present exist, yet the "chemical theory" has apparently gone a considerable way towards the discovery of such a connection. From the accumulated evidence to hand it certainly appears that colour, or at any rate change of colour, is intimately bound up with a corresponding change in the structure of the molecule. Baly, in his work on the relationship between constitution and absorption spectra,[1] ascribes the origin of colour to "isorropesis," by which is meant an oscillation between the residual affinities of adjacent atoms.

In many cases the auxochrome-chromophor theory is still necessary, owing to the colour phenomena being inexplicable in any other way, or at least by means of definitely set out formulæ. In such cases a retention of the Witt theory is unavoidable. Hantzsch is of the opinion, however, that the chemical theory will gradually absorb the other as more facts come to light. At present they exist side by side, and in many cases mutually support each other.

To a limited extent the presence of colour may be taken as a criterion of constitution, but conclusions drawn solely from colour phenomena and in direct opposition to theories built up on chemical experiment must of necessity (as yet) be received with reserve. This is the case with views advanced by Armstrong,[2] arising out of certain colour appearances among the azo-compounds and the oximes. As the result of his arguments Armstrong arrives at the conclusion that the Hantzsch-Werner hypothesis is untenable, and the stereochemistry of nitrogen unnecessary. One is almost tempted, however, to use this as a *reductio ad absurdum* proof of the falsity of his premises.

---

[1] *Journ. Chem. Soc.* **89**, 489, 502, 514, 618, 982, 966 (1906).
[2] Armstrong and Robertson, *Journ. Chem. Soc.* **87**, 1272 (1905).

THE RECENTLY DISCOVERED TERTIARY VERTEBRATA OF EGYPT

By C. W. ANDREWS, D.Sc., F.R.S.

British Museum (Natural History)

As in most branches of science, the growth of our knowledge of the fossil vertebrates of the world takes place, as a rule, by the slow accumulation of isolated facts ; but occasionally some fortunate discovery not only leads to the bridging over of long recognised gaps, but also throws much light on points the significance of which was previously obscure. The discovery that the remains of vertebrates are comparatively common at several horizons in the Tertiary formations of Egypt was such a happy chance, and has resulted in the solution of several long-outstanding problems.

Until within the last few years the palæontological history of Africa, so far at least as the mammalia were concerned, was an almost complete blank. It is true that so long ago as 1875 Owen described the occurrence of a primitive Sirenian in the Middle Eocene of the Mokattam Hills, near Cairo, and a few years later Schweinfurth discovered bones of Zeuglodonts in the Middle Eocene deposits of the Fayûm ; but in both instances the animals in question are of aquatic habits, and therefore threw no light on the mammalian fauna of the Ethiopian land-mass that must have existed throughout Tertiary, and probably also Secondary, times.

The highest horizon in the Egyptian Tertiary beds at which vertebrate remains are found is the Middle Pliocene, beds of this age occurring in the Wadi Natrun, a depression in the Libyan desert some sixty miles from Cairo. From this locality collections have been made by Captain Lyons and Mr. Beadnell, and also by Drs. Stromer and Blanckenhorn. They have been described by Dr. Stromer and the present writer. The chief mammals recorded are *Hipparion, Hippopotamus, Libytherium* or *Samotherium*, and *Mastodon*, as well as carnivora, including a sabre-toothed tiger and members of the Canidæ, Lutrinæ, and Phocidæ. The next bone-bearing horizon is the Lower Miocene, fluvio-marine beds of this age at Mogara, about a hundred and fifty miles west of Cairo, and the Wadi Faregh, nearer the Nile valley, having yielded a number of interesting forms. Of these, *Brachyodus africanus*, an animal closely allied to *Hyopotamus*, bones of which are common in the Oligocene beds of the Isle of Wight, was discovered in 1898 by Dr. Blanckenhorn, and seems to be the first Tertiary land-mammal recorded from Egypt. Later, Mogara was visited by Mr. Beadnell and the late Mr. Barron, who was accompanied by the present writer. Many specimens were collected, including remains of a Rhinoceros, and also of a Proboscidean closely allied to, if not identical with, *Tetrabelodon angustidens*, from beds of similar age in Europe.

Although the mammals and other vertebrates found in the beds above referred to are of considerable interest, they are only such as might have been found in any European deposits of similar age, and afford no clue to the real autochthonous mammalian fauna of the Ethiopian region ; in fact, it is only in the Middle and Upper Eocene beds of the Fayûm that we find remains of animals that can be regarded as representing that fauna. Considering the importance of these fossils, it is proposed to give a brief account of their discovery, of the locality in which they are found, and, finally, of the more important forms represented in the collections which have been made up to the present.

The first remains of land-mammals from this locality were collected in 1900. In this year the present writer had the privilege of visiting the district with Mr. Beadnell, of the Egyptian Survey, who was engaged in mapping this area. On this occasion remains of marine animals, including a Sirenian (*Eosiren*) and large snakes (*Gigantophis* and *Pterosphenus*), were collected, accompanied by traces of an ungulate, to which the name *Mœritherium* was afterwards given. These seemed to be of such interest that a further visit was made, resulting in the discovery of many new forms, including *Barytherium* from the Middle Eocene, and *Palæomastodon* from the Upper Eocene. Towards the end of the same year Mr. Beadnell discovered remains of an extraordinary ungulate, to which he gave the name *Arsinoitherium*, and he also obtained portions of the skeleton of several other new forms. Since then the locality

43

has been visited on several occasions by Mr. Beadnell on behalf of the Egyptian Geological Survey, and by the writer for the British Museum. The large collections made on these occasions have been described and figured in the *Catalogue of the Tertiary Vertebrata of the Fayûm*, published last year.

The Fayûm is a province of Egypt lying about sixty miles south of Cairo, to the west of the Nile valley, from which it is separated by a strip of desert traversed by a canal, through which practically the whole water supply of the district passes. It consists mainly of a depression in the desert, the lowest portion being occupied by a large lake of brackish water—the Birket-el-Qurun—which is, in fact, the remnant of the much larger body of water described by Herodotus under the name Lake Mœris. From early historic times, for various reasons, this lake has been decreasing in size, and there are to-day numerous evidences of its former extent, such as traces of the old shore lines marked by stumps of tamarisk bushes, which then, as now, fringed its margin; but still more eloquent witnesses of its former size are the ruined towns and temples, now lying in the desert far from any water supply. To the north of the lake the land rises in a succession of escarpments separated by plains of varying width, to a height of about 340 metres above the sea; the surface of the lake itself being about forty-four metres below the level of the Mediterranean. The lower escarpments are carved in beds of Middle Eocene age, the higher in the Upper Eocene, the actual summit of the escarpments being formed by the outcrop of a sheet of inter-bedded basalt, above which are the gravelly fluvio-marine Oligocene beds which form the undulating surface of the high desert stretching away towards the north.

The vertebrate remains are found some distance to the north and west of the lake, and they occur at several horizons, the lowest being near the bottom of the Middle Eocene. At this horizon the beds are almost exclusively marine, and the only vertebrates found are aquatic types, the most interesting being a primitive toothed whale, *Prozeuglodon*. The next bone-bearing beds are at the top of the Middle Eocene, and consist of a series of marine and estuarine deposits, which contain the remains of both marine and terrestrial mammals, the most important of the latter being *Mœritherium*, the earliest known Proboscidean, and *Barytherium*, a remarkable ungulate of which the affinities are uncertain.   It is, however, from the Upper Eocene fluviatile beds that by far the greater number of forms have been obtained.   These beds are obviously the deposits of a great river, probably flowing from the south-west, and carrying down in its floods the carcasses of drowned animals inhabiting its banks, together with vast numbers of tree-trunks which to-day, in a silicified state, are strewn over the plains formed by the dip slopes of these beds.   This series of fluviatile beds seems to have continued with some interruptions throughout the Oligocene and Miocene periods, continuing, probably, till well on in the Pliocene; and it is in such deposits at Mogara and the Wadi Natrun that the Miocene and Pliocene faunas above referred to are derived.   In fact, the conditions seem so favourable to the preservation of vertebrate remains, that it is almost certain that only further exploration of the region to the north of the Fayûm depression is necessary to lead to the discovery of faunas at other horizons.   If this should prove to be the case, then it seems certain that in Northern Africa we shall have a succession of mammalian types second in interest only to the wonderful series found in North America.

A brief account of some of the more important of the fossil Vertebrata, more especially the mammals, at present known from the Fayûm, may now be given.   In the first place it should be noted that, in addition to early forms of groups already known, several entirely peculiar types of mammalian life have been found.   Amongst these the most important are *Arsinoitherium*, which has been regarded as representing a new order of mammalia, most nearly allied to the Hyracoidea, and *Barytherium*, which not improbably may also represent a new sub-ordinal group, but of which the affinities are at present quite uncertain.

Arsinoitherium is one of those extremely peculiar types which, as in so many other instances, shows by its extreme specialisation in certain directions that loss of adaptability to new conditions of life which almost inevitably leads to extinction.   Many similar instances might be quoted, one of the most notable being the Titanotheriidæ of North America.   In its general appearance *Arsinoitherium* must have much resembled a large rhinoceros, but instead of having one or two horns in the median line, it not only possessed a pair of small horns situated over the orbits, but also a pair of enormous nasal horns, both pairs, unlike the horns of *Rhinoceros*, being formed by bony outgrowths of the skull, that were probably covered with a horny sheath during life. The posterior surface of the skull slopes forward, and is deeply hollowed for the attachment of the powerful muscles necessary to support the heavy head. The front of the snout is narrow and pointed, a circumstance which, coupled with the character of the incisor teeth, makes it at least probable that the animal did not graze, but browsed on bushes and low herbage, most likely with the assistance of a mobile upper lip, like the Black Rhinoceros of to-day. The teeth were of very peculiar structure : the dentition is complete, and forms on either side of the jaw a closed series, the crowns of all the teeth wearing to a common level, with the exception of the anterior upper incisors, which form slight hook-like projections, and no doubt helped in seizing the food. All the teeth are high-crowned, the molars especially so, and it is further remarkable that the upper molars differ entirely from the premolars in form. The type of molar structure here found is quite unknown elsewhere, but it may have been derived from the deepening of the crowns of molars like those of *Hyrax*, though some writers are inclined to regard it as a specialisation of the type found in *Coryphodon* and other primitive Amblypoda.

The limbs were short and massive, and the feet were much like those of the Elephant, all five toes being retained. At the same time this resemblance with the Elephant, in the hind feet at least, is only superficial, the actual arrangement of the tarsal bones being widely different. As remarked above, the affinities of this remarkable creature are uncertain, and it was considered necessary to establish a new subdivision of the Ungulata for its reception, though at the same time relationships with the Hyracoidea were pointed out. Winge, on the other hand, in a recently published memoir on the Ungulata, boldly refers it to the Hyracoidea. Probably its real position will remain doubtful till some earlier and less specialised members of the same stock have been discovered.

Barytherium, from the Middle Eocene beds, is another large and heavily built ungulate, of which unfortunately very little is yet known. Only the upper and lower jaws, with the cheek teeth, and a few limb bones, have yet been found. All are characterised by their immensely massive construction. The teeth have comparatively low crowns, with two transverse ridges. The humerus has all its ridges and processes for the attachment of muscle greatly developed, indicating a fore limb of great strength, and—judging from its form—possibly employed in digging. The relationships of this creature are unknown; it is by some regarded as belonging to the Proboscidea, and it has even been suggested that there may be some relationship with the South American Pyrotheria.

Although *Arisinoitherium* and *Barytherium* are interesting for the peculiarities they present, their very isolation detracts considerably from their importance, for they throw no light on the earlier history of any of the previously known groups of mammals. From this point of view the remains of primitive Proboscideans from these Egyptian deposits are of vastly greater interest, for they at once settle the point of origin of the groups, and carry back the line to a generalised type of ungulate showing only the beginning of the extraordinary specialisations characteristic of the later forms. Previous to the discovery of these Egyptian forms, the earliest Proboscideans known were species of *Tetrabelodon* and *Dinotherium* from the lowest Miocene beds of Europe, where they appear suddenly at this horizon, no trace of any related form being found in the earlier Tertiary deposits of that continent. The sudden appearance in the European fauna of these and members of some other groups led to a number of speculations as to where these animals had originated. Osborn, Stephen, and Tullberg for various reasons all came to the conclusion that the evidence pointed to the existence of an Ethiopian land-area in early Tertiary times, and they considered that not only the Proboscidea, but several other groups—notably the Sirenia, Hyracoidea, certain Edentates, the Antelopes and Giraffes, the Hippopotami, several divisions of the Rodentia, and lastly the Anthropoidea—originated in that region. Of many of these, the early forms have still to be found; but the predictions of the above writers have already been fulfilled in the case of the Proboscidea, the Sirenia, and the Hyracoidea, so that there is good reason to hope that ancestral forms of some of the other groups will yet be discovered in Northern Africa.

The earliest Proboscidean yet known is *Mœritherium*, remains of which are found in the Middle and Upper Eocene. This animal was about the size of a Tapir, which, moreover, it must have much resembled in general appearance. The skull presents none of the striking peculiarities of the later Proboscidean skull, though traces of the beginnings of some of these characters can be seen. Thus, the nares are already a little removed from the front of the snout, and the nasal bones are small; again, the bones of the occipital region are somewhat swollen by the development of cellular tissue in their interior, a development that reaches enormous dimensions in the modern Elephants. In the upper jaw all the teeth of the full Eutherian dentition are present, with the exception of the front premolars. The second incisors are much larger than the others, and form downwardly directed tusks, the beginning of the great tusks of the later types. The premolars are all simpler than the molars, the low crowns of which bear two transverse ridges, each ridge being formed by the fusion of two tubercles; so that in fact the teeth may almost be said to be quadrituberculate—a very primitive condition. The anterior portion of the mandible is spout-like, and bears two pairs of incisors, which project forwards. Of these the inner pair are small, while the outer are enlarged, and become the lower tusks of later forms. The canine is lost. The description of the upper-cheek teeth given above applies equally well to the lower, except that, as usual, the last lower molar has a third lobe or heel. The skeleton is imperfectly known, but it is certain that the neck was relatively long, so that the animal could reach the ground with its mouth in the usual way. The limb bones, so far as known, are practically those of a diminutive elephant. In this animal, therefore, we have a comparatively generalised type, but at the same time some of the characters which developed to such an extraordinary extent in later forms are already recognisable. Such are the transverse ridging of the teeth, the enlargement of one pair of incisors to form tusks, the beginning of the shifting back of the narial opening, owing to the development of a short proboscis and the commencement of the inflation of the bones at the back of the skull.

Although remains of *Mœritherium* are first found in the Middle Eocene beds, it persisted till the Upper Eocene period; but there it is accompanied by an animal, *Palæomastodon*, which shows a considerable advance towards the later proboscidean type. Probably *Mœritherium* still continued to inhabit the swamps, while *Palæomastodon* represents a form becoming adapted to existence on dry ground. Although referred either to *Mœritherium* or *Palæomastodon*, several forms, intermediate both in size and in some other respects between these two genera, are known to have existed, but the remains by which they are represented are at present scanty.

Palæomastodon is represented by several species, the commonest being *P. wintoni*, which must have been rather larger than a large cart-horse. In this animal the skull approximates in many respects to that of the Elephants proper. Thus the nostrils have shifted back till they are only a little in advance of the orbits, and the nasal bones are very short. At the same time the bones at the back of the skull are much more enlarged than in *Mœritherium*, owing to the increased development of spongy tissue within them. The upper incisors are now reduced to a single pair, the second, and form moderately large downwardly directed tusks, with a band of enamel on their outer side. The canines have disappeared. There are three upper premolars, the last having a pair of transverse ridges, while the molars have three transverse crests. The mandible is in many respects peculiar; the anterior spout-like portion is greatly prolonged, so that it projected considerably beyond the skull, and its extension is increased by the large procumbent incisors, corresponding to the second pair of *Mœritherium*. The other incisors, the canine, and the first two premolars have disappeared, and there is a long edentulous interval between the tusks and the third premolars. The fourth premolar is two-ridged, the first and second molars three-ridged, while in the third molar there may be as many as four transverse crests. The neck was a little longer than in the Elephants, and the animal could doubtless reach the ground with its lower incisors, which (with the portion of the mandible projecting beyond the skull) were covered by the fleshy upper lip and nose, the terminal portion of which may have been more or less free and prehensile. The limb-bones are essentially similar to those of *Elephas*, particularly in the largest species, *P. beadnelli*. The animal must have much resembled in its general appearance a gigantic pig, with a short neck and elongated snout.

Mœritherium and *Palæomastodon* are the only genera of Proboscideans known from the Eocene beds, and at present no member of the group has been found on any Oligocene strata;

but when the Lower Miocene beds are reached, Proboscidean remains are abundant, and we find them not only in African but also in European and probably Asiatic and American localities, the group having become widely spread since the Upper Eocene. In the Lower Miocene deposits of Europe two genera, *Tetrabelodon* and *Dinotherium*, are found, of which only the first is at present known in Egypt, where remains have been found at Mogara and in the Wadi Faregh. In this animal, which is as large as an elephant, the skull is practically the same as that of the later Elephants; the tusks are now very large, though they are still directed somewhat downwards, and have a band of enamel on their outer side. The milk molars, as in the earlier forms, are still replaced by premolars; but these are soon pushed forward and shed through the great increase in size of the permanent molars. Of these the first and second, though large, still have crowns with only three transverse ridges; the third molar, on the other hand, is still more enlarged, and its crown may be made up of five or six transverse crests; it is in fact so large, that when it is fully cut, not only the premolars but also the first molars are displaced, there being no room for them in the jaw. The anterior part of the mandible, with the procumbent incisors, has now attained an extraordinary length, projecting still farther beyond the skull than in *Palæomastodon*; in fact in this genus we have the culmination of the specialisation in this direction, and the long straight snout must have presented a remarkable appearance, the animal having resembled an elephant in which the lower jaw was so elongated that it could reach the ground, and was covered with the fleshy snout, the end of which alone was free. So far as the Egyptian deposits are concerned, this is the last of the Proboscideans found; but it may be permitted to give a short summary of the subsequent changes which ended in the evolution of the modern genus *Elephas*. During the Miocene the long mandibular symphysis—probably because it had attained an unwieldy length—became rapidly shortened up, leaving the upper lip and snout free, as the movable proboscis so characteristic of the group. *Tetrabelodon longirostris*, of the late Miocene, represents a stage in this process. In this animal the symphysis is comparatively short, and although the two lower tusks attain a considerable size, they certainly could not reach the ground. At the same time the number of ridges in the molar teeth is increased to four in the two anterior ones.   In the Pliocene the mandibular symphysis becomes still more shortened up; but in some species of *Mastodon* the lower incisors still persist, though of small size and usually soon shed.   The number of transverse ridges in the molars increase and become deeper, till in *Stegodon* (from the Pliocene of the Siwalik Hills) the anterior molars may have six or seven ridges, the last one eight or nine.   The valleys in these teeth are deepened, and may be more or less filled with cement.   At the same time the milk molars are displaced by the development of the molars behind them, before they can be replaced from below by the premolars.   In *Elephas* proper the elongated mandibular symphysis of the early forms is represented by a small process forming the chin of the mandible, and never bearing any trace of lower incisors.   The molars acquire a much greater number of transverse ridges, and become higher in the crown.   In *Elephas primigenius* there may be as many as twenty-seven ridges in the last molar.   All this long series of changes is illustrated by specimens shown in the Palæontological Galleries of the Natural History Museum, and representing perhaps the most complete history of any Mammalian group yet known.

In the Upper Eocene beds the Hyracoidea are represented by two genera, *Megalohyrax* and *Saghatherium*, including several species.   None of these throw any light on the relationships of the order; but some of them are of large size, and indicate that formerly the group was of far greater importance than it is to-day, when its only representatives are a few comparatively small species, all of which (according to some authorities) should be placed in a single genus, *Procavia* (*Hyrax*).

The occurrence of remains of Sirenians in the Middle Eocene beds of Egypt has long been known, Owen having described —under the name *Eotherium*—a brain cast of one of these animals from the Mokattam Hills near Cairo, so long ago as 1875, and further remains from the same locality were noticed by Filhol in 1878.   Within the last four or five years not only have skulls and other portions of *Eotherium* been found, but remains of other genera have come to light, both from the Mokattam Hills and from the somewhat later deposits of the Fayûm.   These early forms have been described by Dr. O. Abel and the present writer.   Their chief points of interest are those in which they show approximation to the land-mammals from which the group arose. Thus in *Eotherium* the pelvis has a complete obturator foramen enclosed by the pubis and ischium, and judging from the acetabulum there must have been a fairly well developed hind limb. In the later forms, even at the top of the ·Middle Eocene, the pelvis has undergone considerable further reduction, the pubis and ischium not enclosing a foramen, and the acetabulum being so small and indefinite that the hind limb must have been rudimentary. In these early Sirenians also the dentition approaches the primitive Eutherian type, there being three incisors, a canine, four premolars, and three molars on each side of the upper jaw. In the later types there is at most one pair of incisors, often much enlarged, while the canines and some of the premolars also are lost. This more normal structure of the pelvis and the character of the teeth show that a Sirenian such as *Eotherium* is not very remote from the terrestrial ancestor from which the group must have sprung; and it is very interesting to note that in the form of the pelvis and of the teeth this ancestral form must have much resembled *Mœritherium*, a fact strongly supporting Blainville's suggestion that the Sirenia and Proboscidea are closely related. Many other points of similarity might be pointed out, such as the form of the brain in *Mœritherium* and in *Eosiren* or *Eotherium*, and modern representatives of the two groups also agree in a number of points. Thus in both there are (1) pectoral mammæ; (2) abdominal testes; (3) bilophodont molars, with a tendency to the formation of additional lobes behind; (4) the same arrangement of the intestines, and (5) to some extent the same character in the placenta. Altogether there seems to be very good reasons for regarding the Proboscidea and the Sirenia as offshoots of a common stock, the one being adapted for a terrestrial, the other for an aquatic mode of life.

All the carnivora collected belong to the primitive group, the Creodonta, and all the species can be referred to genera already known from Europe or North America. A few of the limb-bones found seem to indicate that some of these animals had adopted a semi-aquatic mode of life, and it is possible that the Seals originated from some such type.

Far more interesting than the Creodonts themselves is a group that is now definitely known to have originated from them, namely, the Zeuglodonts, usually—and probably rightly—

regarded as primitive-toothed whales. Remains of the later members of this group are found widely spread over the world in the Eocene beds, occurring in North America, New Zealand, and Europe. It is only quite lately that any light has been thrown on the origin of these animals. Prof. E. Fraas described, from the lower Middle Eocene limestones of the Mokattam Hills, a skull which in all essential respects is that of a Zeuglodon, but at the same time the dentition is that of a Creodont carnivore, none of the peculiar characters of the Zeuglodont teeth being present. This specimen, to which the name *Protocetus* was given, proves fairly conclusively that the Zeuglodonts originated from some Creodont ancestor which acquired aquatic habits, and probably this happened on the northern coasts of the Ethiopian continent. From beds of a little later age in the Fayûm an animal, almost precisely intermediate between *Protocetus* and *Zeuglodon*, has been described under the name *Prozeuglodon*. In this creature the skull approximates still more closely to that of the true Zeuglodonts, and the teeth have also acquired the serration characteristic of the group, though at the same time some of the premolars and molars have a third inner root, which is lost in the later forms. In the upper beds of the Middle Eocene of the Fayûm typical Zeuglodonts, *e.g. Z. osiris*, occur ; so that in this region we have a complete passage from *Protocetus*, in which the teeth are those of a Creodont, to *Zeuglodon osiris*, in which they are typically Zeuglodont—that is, the molars are two-rooted, and have serrated cutting edges. At the same time the narial opening shifts back, and although it is still well in front of the orbits and the nasal bones are long, the change is in the direction of the type of skull found in the early Odontoceti ; and although the relationship of these animals to the Zeuglodonts has frequently been doubted, there seems much to be said in its favour. This question has lately been discussed by Fraas and by Abel.

Remains of birds are very rare, and with the exception of fragments of the skeleton of a heron-like wader, the only specimen of importance is the distal end of a tibio-tarsus, which is interesting, because it shows that probably a true Ratite (*Eremopezus*) existed in the Eocene in this region, and may be the ancestral type from which the Struthiones and Æpyornithes sprung—numerous common characters between the two groups having been pointed out by Burckhardt. A relationship with the South American Rheas is also possible. On the other hand, this bird may be merely another instance of the results of retrogressive change, leading to loss of flight and increase of size in some group of Carinate birds, such as has happened in the case of the Stereornithes and Gastornithes.

Among the Reptilia no very important new forms have been discovered. In the Middle Eocene remains of large and probably marine snakes are found, one of these (*Gigantophis*) having probably attained a length of 30 to 40 ft. Another (*Pterosphenus*) is of interest, because a closely allied species is found in North America also associated with Zeuglodonts. From the same horizon there have been collected remains of numerous Pleurodiran tortoises, a group formerly widely spread, but at the present day found only in Madagascar and South America. The most remarkable of the Egyptian Pleurodires is *Stereogenys*, in which the palate and mandible are modified to form broad crushing surfaces, probably for breaking the shells of the animals which formed its food. In the Upper Eocene beds Pleurodiran tortoises are likewise found, the modern genus *Podocnemis* being represented by several species; but at this horizon the most notable chelonian is a gigantic land-tortoise (*Testudo ammon*), shells of which are comparatively numerous. This species approximates most nearly to the Aldabara and Madagascar giant tortoises among living forms. Numerous Crocodiles are found both in the Middle and Upper Eocene, and include both long- and short-snouted forms. One (*Tomistoma gavialoides*) seems, to some extent, to bridge the gap between the true *Tomistoma* and the Gharial.

Remains of fishes are found in several horizons, but none are of special interest. From the Middle Eocene are several peculiar Saw-fishes, and also several large Siluroids, which are curiously like species now existing in the Nile.

From the above account it will be gathered that a considerable number of Tertiary vertebrates are already known from Egypt, and include forms of great interest. At the same time, these must constitute a mere fraction of the faunas that have inhabited this region, and therefore, since the conditions seem to be very favourable to the preservation of vertebrate remains, it is to be hoped and expected that many new types of vertebrate life will be discovered, especially when the desert between the Fayûm and the Wadi Natrun can be thoroughly explored—a matter of no very great expense or difficulty.

The question of the relations of the Ethiopian land-mass to other regions during the Tertiary period is one of great interest. That it is almost certain that Africa and South America were united during the Secondary period has been pointed out by many writers, and there is considerable probability that this union may have persisted till early Tertiary times, though there is considerable difference of opinion as to the position of the connection. It has even been suggested that a belt of shallow water and a chain of islands may have existed between Africa and Brazil so late as the Miocene. This connection between Africa and South America would account for a number of curious facts of distribution, as, for instance, the presence of the Hystricomorphine rodents and the Pelomedusid chelonians on both continents. The occurrence in the Santa Cruz beds of Patagonia of *Necrolestes*, a close ally of the Cape Golden Moles (*Chrysochloridæ*), has also been pointed to as evidence of this former union, but this has been considerably discounted by the discovery in the Miocene of North America of *Xenotherium*,[1] an animal which is almost certainly closely allied to the Chrysochloridæ, though it was described by its discoverer as probably a Monotreme.

Some South American palæontologists have asserted that certain groups of Ungulates found in the Tertiary beds of Patagonia are closely allied to, if not the actual ancestors of, some of the African subdivisions of that order—*e.g.* the Hyracoidea. There seems, however, to be no real ground for this belief, and it is far more probable that the two continents were separated before the main divisions of the Ungulata had become differentiated, and that such resemblances as do exist are merely the result of parallelism in the course of evolution of the group in the two areas.

The late Oligocene or early Miocene union between Africa and the Palæarctic continent has already been referred to in connection with the migration of the Proboscidea; but it is certain that other unions, probably of a temporary nature, must have occurred in earlier Tertiary periods. The presence in both the European and African Eocene of the same genera of

[1] Douglass, " The Tertiary of Montana," *Mem. Carnegie Museum*, vol. ii. (1905) p. 204.

Creodonts (*Hyænodon*, *Pterodon*, etc.) and of an Anthracotheroid approximating to *Brachyodus*, is evidence of this earlier junction.

The relations of Africa with Madagascar are also interesting. The mammalian fauna of Madagascar is a comparatively poor one, and is completely wanting in many of the typically African groups of mammals. Tullberg has accounted for this by supposing that the eastern part of Africa, with Madagascar, was separated from the main southern and western African continent by a belt of sea, and that it was only after the isolation of Madagascar that these two parts of the Ethiopian continent united, and the richer fauna of the southern and western portions spread over the whole. This probably occurred in the Oligocene, at which time the union with South-western Asia and Europe took place, followed by the dispersal into the northern continent of the Proboscidea and other groups.

The importance of Africa in the history of the Mammalia is further increased by the fact that, as Stromer has pointed out, some part of the region has probably been above the sea since Permo-triassic times, during which a great variety of land reptiles existed, some of which, the Theriodonts, approximate very closely to the mammalian type, and, in fact, are probably the stock from which the Mammalia sprung. This being so, it is by no means improbable that somewhere in this continent beds of Jurassic and Cretaceous age will be found, containing remains of animals which will completely bridge the gap between the two great and now widely distinct groups—the Mammalia and Reptilia.

THE OPPORTUNITY OF THE AGRICUL-TURIST ·

By HENRY E. ARMSTRONG

[An address delivered at the opening of the session of the S.E. Agricultural College, Wye, Kent]

SOME among you who have learnt to read—not merely to play with books but to know them as real friends, to consult them and consider them—will remember the lines in Tennyson's *In Memoriam*:

They say,
The solid earth whereon we tread

In tracts of fluent heat began
    And grew to seeming random forms,
    The seeming prey of cyclic storms.

The day was, we are led to suppose, when our globe was a fiery fused mass. Its atmosphere was a strange one in its early stages —full of steam and acid fume. Gradually, the mass cooled down and the crust became solid; there is some reason to think that contraction took place along lines which determined the formation of the great ocean basins very much in their present positions on the earth's surface. As the acid steam condensed, falling as rain on the land, it dissolved out certain materials, washing them into the oceans and at the same time carving out deep valleys in every direction. The salt we know to-day may well be the actual salt made in those far-off times—while all else is changed. To quote again from Tennyson's poem:

There rolls the deep where grew the tree,
    O earth what changes hast thou seen!
    There where the long street roars, hath been
The stillness of the central sea.

The hills are shadows, and they flow
    From form to form and nothing stands;
    They melt like mist, the solid lands,
Like clouds they shape themselves and go.

I ask those who can read and who have read to ponder over these magnificent lines; I would urge all to begin who cannot, in order that they may understand their extraordinary significance. The wonderful art the poet exercises of compressing a whole history into a few short sentences, the marvellous power of words to excite long trains of thought, is nowhere better displayed; but the power of the words in themselves is *nil*—they acquire meaning only when the will exists to give meaning to them.

My object, however, in quoting Tennyson's lines is to ask you to think of them as descriptive of the genesis of agricultural soils. The primary soils were undoubtedly sands and clays formed by the decay and disintegration of the erstwhile molten crust—of rock like that known to us now as granite, made up of at least three different crystalline materials: felspar, mica and quartz. Quartz is indestructible except mechanically; mica is not easily affected; felspar, however, under ordinary atmospheric influences, is sooner or later broken down into clay and soluble matters: granite therefore gradually disintegrates on exposure, owing to the decay of its felspar.

In the earliest times, as now, the clays thus formed, being in a very fine state of division, would have been carried far out to sea and only slowly deposited; but the streams which conveyed the clay out to sea would also grind down the mica and the quartz until eventually the one was converted more or less into clay and the other into quartz pebbles and sand-grains, which also were conveyed greater or less distances out to sea and deposited as sediment on the ocean floors: in this way, gravels, sands and clays were laid down contemporaneously but in different regions. The thickness of rock which underwent disintegration must have been enormous, considering the enormous thickness of the sedimentary rocks.

But sand and clay are not the only rocks known to us: those of you who are at home in this district are aware of the existence in it of a vast mass of chalk. In what relation does this stand to sand and clay? How will you find out? No detective attempts to unravel the mystery of a crime from his armchair—he goes out into the world and seeks for signs: and so must you if you desire to solve the many problems which limestone presents.

No doubt there are chalk pits to be found and the coast is open to inspection. Suppose you visit the pits, hammer in hand : using your eyes, you will sooner or later light upon shell remains and Tennyson's lines will come home to you :

> There where the long street roars, hath been
> The stillness of the central sea.

Under the microscope, chalk bears most distinctly the impress of a manufactured article ; far from being formless like clay and sand, it consists for the most part of particles which are obviously remains of organisms (*Foraminifera*). Within recent years the origin of chalk has been disclosed by the discovery over vast regions of the ocean floor, in the Atlantic and Pacific areas, of an ooze or deposit precisely similar in appearance to chalk ; the formation of this ooze has been traced to the warm surface, where the organisms themselves flourish : as they die, their outer shelly casings sink slowly through the deeps and are deposited, forming what is now called *Globigerina* ooze. Chalk, therefore, is also a sedimentary deposit but of organic origin, laid down in clear water in regions beyond those to which clay is carried—in central seas, in fact. As chalk yields lime when burnt, it is a form of limestone. But limestones occur in all geological formations—of whatever age ; the very earliest contain animal remains which tell that they were formed in the sea but their minute structure has suffered defacement at the hand of time and we cannot always trace them to their origin.

But what is limestone ; of what is it made ? Bred out of sea-water by minute organisms, surely it is a material deserving our careful study ; and it invites attention from so many points of view. The farmer, guided by that shrewd power of observation which he so often displays, has noted its agricultural value. Walk, say, in the Isle of Wight along the chalk downs, and note how frequently quarries occur abutting on the clay or sandy lands bordering the ridge : to what end has so much chalk been dug out—in the past, we must suppose, as the signs of recent removal are few ? On inquiry we learn that it was once the custom to spread chalk on such lands with great advantage to the crops ; but of late years the practice is deemed too expensive by most farmers.

Here and there in all limestone districts kilns are to be found in which the stone is burnt into lime—and lime confronts us everywhere in admixture with sand, as mortar. Whitening, which is simply very finely ground soft chalk washed free from

44

all gritty matter and dirt, is used all the world over in making whitewash, whilst when mixed with oil it forms putty. We have some reason to be grateful, therefore, to the *Globigerina* for producing so universally useful a material as chalk. But it is a substance which sometimes makes itself awkward. There are few districts in which the water does not contain more or less dissolved limestone, being thereby rendered "hard," as it is termed: such hard water, we know, is less suitable than rain water for washing purposes and is often productive of considerable difficulty when used in steam boilers.

How is such a substance to be dealt with to discover its nature? Is no clue to be found by considering what is commonly known about it?

A man's character is not be found out by merely looking at him: his actions must be studied. To determine the character of a substance, its capabilities must be inquired into. Every one knows that limestone gives lime when burnt—the name implies as much. Some profound change must be involved—the lime is so very different from the limestone. On the Kentish coast but a few miles away, twice during every twenty-four hours, the chalk cliffs are wetted by the waves—yet nothing happens. But every bricklayer's labourer who is called upon to make mortar has learnt the trick of reducing lime to the finest of powders without any effort to himself—by merely pitching a pailful or two of water upon it: the man who made the observation originally and applied his discovery in practice must have been a transcendent genius; he scarcely wore trousers, but had he done so he would have realised that he could put his hands quietly into his pockets and look on at the water doing his work for him and doing it far more effectively than he could himself. Yet the beauty of the process is rarely appreciated. Carry it out for yourselves and see what happens. The water is greedily absorbed: then, after a time, the wetted lime grows hot, clouds of vapour are given off—and gradually the lumps fall into an impalpable dry powder, riven in every direction by the expansive force of the steam.

The lime is said to be "slack" or slaked—its thirst is satisfied. But where does the water go? How is so marvellous a change to be explained? The time should be near at hand when even the bricklayer's labourer will ask such questions. I trust many of you take real interest in the spectacle and desire to understand what has happened. If I ask you to suggest some means of finding out, what will you answer? I fancy some one says: "Make sure that the water really is all expelled, weigh the lime, then pour a known quantity of water upon it, let it slake and weigh the dry powder. Don't work in a bricklayer's empirical fashion but be scientific; control all your actions by weighing and measuring." Good! Such advice is worth following. With very little trouble we find that although some of the water is driven off as steam, much of it remains with the lime— the weight increasing by about one-third. It is thus discovered that lime has a great affection for water; but the union is marked by considerable display of temper, by heat. Was the lime in the limestone? If so, how was its affection for water masked? What happened in the burning to confer so powerful a thirst upon it? Let us again be scientific and watch the limestone burning with the aid of the balance—the man in charge of the limekiln tells us that the lime produced always weighs less than the limestone which is burnt. Taking whitening, as being the cleanest form of limestone procurable, on burning it in a crucible in a gas muffle furnace until it no longer changes in weight, it loses slightly less than 44 per cent., leaving about 56 per cent. of lime; however often the experiment be repeated with the whitening, the result is always practically the same, though different limestones give different results according as they contain more or less clay or grit. Assuming that the lime is precontained in the limestone, it would seem that, unless it be in some way destroyed, about 43–44 per cent. of the whitening is driven off into thin air—since there is no visible smoke such as would be noticeable if a condensible substance were expelled.

I have chosen these two simple examples as illustrations of the way in which with a little care and consideration you can make use of ordinary commonplace but suggestive occurrences as arguments in devising experiments, with the object of solving problems relating to common objects and common phenomena; our scientific method of working is nothing but common sense writ large systematically applied to practice.

But this by way of parenthesis. We are only beginning to understand limestone. Let us look further into its doings; it seems to be full of interest.

In limestone districts, where hard water is used, people know quite well that there is often a deposit from such water when it is kept in water-bottles in hot weather and that invariably a crust is formed in the kettles and boilers in which it is heated. It is said vaguely that the water contains lime and such deposits are often referred to as "lime out of the water." In most households it is known that although the deposit from the water sticks hard to a bottle and cannot be simply washed out, it is easily removed by acid, even by vinegar. Is it right to speak of a natural water as containing lime? How comes it that acid removes the deposit referred to? Surely such questions may be answered by a few simple experiments—the solubility of lime is easily contrasted with that of limestone, their comparative behaviour with acids easily examined. It turns out that lime is much the more soluble in water and that limestone is scarcely soluble at all, to a far smaller extent than will account for the amount of solid dissolved in hard waters; the solution of lime is quite unlike any ordinary water, peculiar in taste and in many other ways. Acids are found to dissolve both lime and limestone with extraordinary ease but to affect them very differently, the former disappearing quietly although the liquid becomes very hot, whilst the latter disappears with much spluttering and fizzing—something being forced out from the seething mass. It is easily ascertained that a gas is given off. A prudent man desires to know how much money he spends—to know merely that his cash balance is diminishing is not enough: the scientific inquirer will not be satisfied with the observation that something escapes on dissolving the limestone in acid but will question how much. It is as easy as not to make the experiment in such a manner that it affords rough quantitative results. The loss in repeated trials with whitening amounts to from 42–43 per cent. on the amount of the limestone dissolved.

The habit of putting two and two together is the one before all others to be acquired. Thinking back, we remember that the limestone lost about 44 per cent. in weight when burnt; now we find that it apparently loses nearly as much when dissolved in acid: is this a mere coincidence or has the resemblance a deeper meaning? We have no proof as yet that lime is precontained in limestone—none that the gas comes from the limestone alone; but it is at least conceivable that the loss which it suffers when burnt is due to the gas which is given off when the stone is dissolved in acid and which may be collected easily. The conclusion is a legitimate one from the facts—it is at least a sound provisional working hypothesis. How shall we attempt to demonstrate its truth? Obviously, if by bringing together lime and the gas which is given off when limestone is dissolved in acid we can produce a material having the properties of limestone, we shall have verified the hypothesis and solved the problem as to the nature of limestone-stuff. This turns out to be easy of accomplishment. On merely exposing lime in a vessel full of the gas, it gains in weight just up to the point required on the assumption that it is converted into limestone-stuff; at the same time it acquires all the properties of this material while losing those of lime. Sherlock Holmes could do no more! Or, arguing from the fact that lime is so much more soluble than limestone, we may pass the gas into a solution of lime in water: soon a solid falls out—a precipitate, we call it—as we should expect would be the case if the gas were to unite with the lime in solution. Having carried out the experiment on a scale such that 2 or 3 grams of the product are available, it is possible to ascertain what weight of lime it yields when burnt and what amount of gas is given off when it is dissolved in acid—and so to characterise it as chalk-stuff,[1] for the mere appearance of the precipitate proves nothing.

And proceeding in this way, discovery follows discovery; even carelessness sometimes has its place. Thus, when through an oversight or forgetfulness, perhaps, the gas is passed into the lime-water during some considerable time, the precipitate seems to lessen instead of increasing in amount, as if it were soluble in presence of the gas. Does not such an observation serve to recall the fact that natural water sometimes contains an inex-

---

[1] Objection has more than once been taken to the use of expressions such as "chalk-stuff" or "limestone-stuff" and "chalk-" or "limestone-stuff" gas: nevertheless I continue to use them and do so advisedly. Chalk and limestone are not definite substances but species of natural material which, if destroyed, cannot be reproduced: the names have a definite geological connotation but no chemical meaning. The varieties of limestone are composed of a definite substance in greater or less proportion: to call this substance from the outset by its recognised scientific name, when the game of an inquiry into its nature is being played, is simply to give away the game entirely from the outset and make it a worthless one to play from the point of view from which alone the game is worth playing—as a means of gaining knowledge of method. Let me urge that provisional names must be used throughout every inquiry. "Stuff" is an honest old English word of which far too little use is now made. Probably if I had called the gas *Gypsogen*, little objection would have been raised: but it is difficult to be rational in these modern times.

plicably large amount of what appears to be limestone-stuff? Now, we can take water such as is available in this district and ascertain whether the solid which is deposited on boiling it is in reality limestone-stuff; but will it not be well first to boil the solution prepared by passing the gas into the lime-water to see how it behaves? It soon appears that it behaves like a natural hard water. On the other hand, the deposit from a hard water —the fur from a kettle or boiler—behaves exactly like limestone when burnt and towards acids. May we not, therefore, also expect that the hard water will contain the limestone-stuff gas? As soon as the idea that such may be the case is evoked in our minds, it occurs to us—since the limestone-stuff is deposited on boiling the water—to boil hard water and pass the steam into lime-water : a precipitate soon appears and, therefore, we can write Q.E.D.—as at the conclusion of an exercise in geometry. The gas is obviously soluble in water : we take boiled water which is not rendered turbid by lime-water and pass in the gas —lo and behold! on adding lime-water the precipitate appears. But the deposit forms in water-bottles without the water being heated : may we not, therefore, infer that the solution of the gas in water is an unstable one—may we not expect to find the gas in the air? Does not this question recall to our recollection the fact that when we have carelessly left clear lime-water exposed to the air a deposit has collected on the surface? Does not this serve to suggest the experiment of exposing a considerable quantity of lime-water to air so as to obtain sufficient solid for examination? When the experiment is made, sufficient solid having been obtained and examined, we soon satisfy ourselves that it is chalk-stuff: consequently that air does contain the gas, although in relatively small amount.

But the magic circle is not yet complete. We speak of lime as made by burning limestone : is this correct—do we burn it as we burn wood and coals and other things? You reply, "Perhaps not; we can kindle wood and coal and when once on fire they continue to burn—limestone will not take fire ; it is baked rather than burnt." We have learnt that it is simply resolved into two other materials by this baking—what becomes of wood and coal when burnt? Seemingly they are destroyed —but has not our experience with limestone taught us to be careful : that disappearance does not necessarily mean destruction? What do we know of the burning of coal—what does every cook know? Why do we poke the fire—why is a chimney provided to the fireplace? We know that air must have access to the kindled coal—that one function of the chimney is to cause a draught; every one knows the effect of so holding a newspaper as to close the opening above the grate and force the air to pass rapidly over the burning coal. Clearly the air is in some way necessary and promotes the burning. Such an argument as this would lead to an experiment being made to ascertain what happens to the air when things burn in it and ultimately to the discovery that air consists to the extent of about one-fifth of a gas which has an extraordinary affection for all combustible substances, just as water and the gas set free from limestone by acids have for lime. When this constituent of air and a combustible substance meet under appropriate conditions, they contract a firm union with great display of warmth. The products of such unions vary much in character, some being solid, some liquid, some gaseous; some few are neutral like water, some are alkaline like lime and soda, whilst others are acid. The gas is named *Sour-stuff*, because when the part which it plays in combustion first came to be understood, the substances with which the experiments were made happened to be such as gave rise to acids. The Germans, who are direct, simple folk, call the gas to the present day *Sour-stuff*—the stuff which makes sour things; so do we, without knowing it, as—following the lead of the French—we use the Greek equivalent oxygen. But having so long enjoyed the advantages of an exclusively classical education, we no longer attempt to apply our knowledge and so use the word without thought and without being reminded of its significance.

When coal or wood, indeed anything of vegetable or animal origin, is burnt, it is seemingly destroyed—that it takes wings to itself and dissolves into thin air is the discovery of but little more than a century. The winged product we know is the gas imprisoned in limestone, on this account called by Black—who was the first to recognise its individuality—*Fixed air*, now known in our technical jargon as Carbon Dioxide, the compound which gives rise when joined to water to the very weak and unstable acid, carbonic acid. To the instructed ear, these two names carry oceans of meaning. If you know how the expression *dioxide* is arrived at you are acquainted with the theoretical basis—the molecular-atomic theory—on which chemistry rests.

But when acids are mentioned, always think of sour-stuff—oxygen is *par excellence* the acid-giving element; the text-books may tell you of hydrogen as characteristic of acids and even mislead you into a temporary belief in non-existent nonsensities called hydrogen ions: trust them not but pin your faith to oxygen; learn to think of oxygen as the most wonderful element known to us and water as containing it in perhaps its most active form; at the same time, bear in mind that it has not only an acid aspect but is also the progenitor of earths. Iron, you know, is converted only too easily into the earthy substance iron-rust by combination with atmospheric oxygen under the conjoint influence of liquid water and carbonic acid—hence the need of protecting it by paint or of keeping it in a dry, warm place. At the smithy forge, iron and oxygen burn together into a brittle black scale—again an earthy substance. Those who have played with magnesium know that it not only burns brilliantly but that it is converted into a white earth. Zinc is systematically burnt to form zinc oxide, which is largely used as the basis of white paint. Both magnesium and zinc oxide look like lime and the former tastes like lime: whence Brer Rabbit—who argued, you will remember, that what looked like Sparrer Grass and tasted like Sparrer Grass was Sparrer Grass —would say they are limes or earths; and *vice versa* that an earth like lime must belong to the oxides. This kind of argument has in fact led chemists—led Humphrey Davy—to discover that lime is the oxide of a metal much like magnesium.

Just think what is got out of limestone by considered argument and measured study! Itself a neutral substance and a typical salt, it is resolved into two typical oxides—an acidic oxide and an earthy, alkylic oxide. In deciphering its story, you may learn almost all that is necessary for an agriculturist to learn of chemical method. I say this because I feel that it in no way receives the attention it deserves—because it is one of the substances which should always live in your thoughts.

Ordinarily, it is neglected, scarce seen; the attempt is rarely made really to decipher its unobtrusive mysteries:

> The world is too much with us; late and soon,
>   Getting and spending, we lay waste our powers:
> Little we see in Nature that is ours;
>
>   .   .   .   .   .
>
>   .   .   .   for everything we are out of tune;
>   It moves us not.

Colleges such as this are intended to remedy the state of things complained of by the poet; to attune their students to a real appreciation of Nature. Your gain will be in proportion to the extent to which you avail yourselves of the multitude of opportunities the College affords for the study of methods and the cultivation of the faculty of insight.

Let me now touch on some of the problems which should come under your notice on the vital side. The amount of carbon dioxide in our atmosphere is but three parts in ten thousand—yet this is the whole trading capital of the agriculturist the wide world over. That so much can be made of so little is very wonderful; and when we learn something of the mechanism by which Nature effects her purpose, it borders on the miraculous. Watching on a sunny day in spring the quivering leaves in a woodland district or the waving blades in a wheat field, we may well fancy that we are contemplating an ideal state of lazy enjoyment: the forces at work, the ceaseless activity of the leaf-surfaces, are in no way apparent. The plant is thought of merely as growing. Oliver Wendell Holmes's beautiful lines:

. . . God has made
This world a strife of atoms and of spheres ;
With every breath I sigh myself away
And take my tribute from the wandering wind,
To fan the flame of life's consuming fire—

are not only descriptive of the process of animal respiration regarded as a destructive process but also of plant respiration regarded as a constructive process—tribute being taken by the plant from the wandering wind of its $\frac{3}{10000}$ of carbon-containing gas as well as of a certain modicum of oxygen.

It is taught in the schools that under the influence of sunlight and with the assistance of the green colouring-matter, the carbon dioxide which the plant inhales is resolved into carbon which is retained and oxygen which escapes. The warmth displayed and dissipated when carbon and oxygen unite must in some way be recovered and restored to the system: the sun affords the necessary energy; hence it is that we regard vegetable matter as bottled-up sunshine. At present the nature of the initial process is entirely unknown to us; we can only form very general ideas as to the changes which supervene. Personally, I incline to the view that it will eventually be recognised that carbon dioxide is only indirectly deprived of its oxygen: it seems to me probable that water undergoes electrolysis under the influence of light and chlorophyll and that whilst oxygen is evolved, hydrogen becomes attached to the chlorophyll, subsequently acting on carbonic acid as a reducing agent and converting it into formic acid and then into formaldehyde. The evidence that this aldehyde is the primary material with which the plant works is almost overwhelming.

From this point onwards, change undoubtedly proceeds to some extent as in the laboratory but mainly along lines which require us to assume that directive forces are at work: the chief substances elaborated, the carbohydrates (sugars, starch, etc.) and the albuminoids or proteins, also the essential oils and alkaloids, are all optically active substances.

The significance of this circumstance can be appreciated only when certain geometrical considerations are understood. The carbon atom has the power of combining with four other atoms or groups of atoms; a compound in which four different systems are associated with a carbon atom may be symbolised by a four-sided prism or tetrahedron. Two such models may be constructed—you can make them easily from cardboard—in which the systems are arranged in a different order, so that the one is to the other what an object is to its reflected image or a right-hand to a left-hand glove—such models will not fit over one another and are said to be without a plane of symmetry, as they cannot be divided into halves. A carbon atom in such a condition is termed *asymmetric*.

All asymmetric compounds have the remarkable property of deflecting polarised light; hence they are said to be optically active. Whenever substances containing asymmetric carbon are produced artificially from inactive material, the two forms of opposite but equal activity are always produced in equal proportions — the one form turning polarised rays to the right, the other turning them to the left, to equal extents although in opposite directions: the product is therefore optically inert, as the equal opposite activities of the two forms neutralise one another. If change took place in the plant in the same way as in the laboratory, it is to be expected that mixtures in equal proportions of the two optically active forms of each particular compound would be produced. As a matter of fact, the carbohydrates are of one series only—the dextro- series; at one time it was thought that perhaps these alone were preserved and that the plant for its own purposes made use of and destroyed those of the other series; but of late years this conception has given place to the idea that only members of the one series are formed: in other words, that the course of change is so directed that only *d*-glucose and compounds derived therefrom are built up. The nature of the directive force is even surmised.

In plants, as well as in animals, the digestion of the more complex compounds such as starch, the biose sugars and the albuminoids—that is to say, their simplification and assimilation—takes place under the influence of enzymes. These enzymes are kittle folk to deal with; at present, we are quite unaware of their precise nature, as they cannot be separated in a state of purity and are very easily destroyed. One of the most typical is the *diastase* of malt, which is present in all cereal grains and in green leaves: this enzyme acts on starch alone, converting it into soluble sugar (maltose). Invertase, which is present in most yeasts and in growing leaves, has the specific property of converting cane sugar into dextrose and levulose or fructose. In all cases these enzymes act by promoting the hydrolysis of the compound affected; they, as it were, enforce the absorption of the elements of water and so determine the resolution of the compound into simpler substances. Their activity is altogether wonderful, a minute quantity sufficing gradually to transform a relatively large amount of the hydrolyte—*i.e.* the compound hydrolysed. Although the effect they produce is similar to that produced by acids, they are distinguished from these by the fact that, whereas the acids act indiscriminately and only with varying degrees of readiness, the enzymes act selectively—a given enzyme acting only on a particular compound or related series of compounds. The only rational explanation yet given of this specialised behaviour is that the enzyme and hydrolyte are compatible substances, that one is in some way built to fit the other; the closest approach to a picture of the manner in which an enzyme acts is afforded by a Geometer caterpillar clasping a twig with its many legs while it bites away the adjacent leaf-edge with its mouth: we may think of an enzyme as clinging to its compatible hydrolyte and presenting water at the appropriate centre at which the resolution of the complex molecule is to be effected.

Supposing the enzyme or something very near to it were to preside at the birth of the compound, acting after the manner of the last upon which the shoemaker makes the shoe, we have a picture which probably is not far removed from that which would be revealed to us could we see the process at work in plant and animal—the directive agency—the outcome of which is the formation of one particular type of asymmetric compound.

Such in brief appears to be the philosophy both of progressive and of retrogressive change in plant and animal.

Questions of extraordinary interest and importance centre around the problem of the origin of the nitrogenous constituents of plants, especially of the albuminoids or proteins. First comes the question of the supply of nitrogen to plants. As you know, in practice this takes the form either of ammonia salts or of nitrates or of vegetable or animal matter containing assimilated nitrogen; in whatever form it be supplied, however, there is reason to suppose that it comes into action in the form of nitrate. The ammonia of ammonia salts and the ammonia derived from nitrogenous organic matters by their decay— apparently under the influence of organisms in the soil—under- goes oxidation within the soil at the instance of nitrifying bacteria; it does not seem to be available directly. Possibly— nay, probably—the nitrate undergoes reduction within the plant into hydroxylamine. But in the case of leguminous plants, a marvellous symbiotic mechanism is come into existence, whereby the plant is enabled to utilise nitrogen derived from the atmosphere; you probably know that nodular growths form upon the roots and that in these bacteria abide which, we believe, are the active assimilating agents. From a chemical point of view, the whole process is most mysterious and marvellous; but there is no problem in agriculture which is so important a one to solve in the interest of future generations. Our stores of ammonia and of nitrates are rapidly being depleted; it is scarcely probable that the methods of making these substances artificially will satisfy the demand that must arise in the future. A great extension of the natural process of accumulating nitrogen from the atmosphere is therefore to be desired.

As the albuminoids, like the carbohydrates, are all optically active and apparently members of one class, it is not improbable that their genesis is in some way bound up with that of the carbohydrates; the process is certainly a directed process and doubtless they also are constructed on an enzymic model or last.

Hitherto, little attention has been paid to the production of quality, in the chemical sense, either in plants or animals; quantity and certain characteristics have mainly been in demand. Now we are beginning to realise that quality must be attended to—that we know little or nothing at present, except empirically, of our own requirements and still less of those of agricultural stock. Man has just felt his way by picking and choosing and reason has played but a small part in determining his choice of food. The time is at hand when in feeding ourselves and our stock we must be guided more by scientific principles.

Food is of value to animals from two points of view—as the source of energy and as constructive material. It is only of value in so far as it is digestible—that is to say, in so far as the enzymes are at our disposal which are capable of taking its complex constituents to pieces and rendering them soluble and transmissible to various parts of the body. The carbohydrates are of value mainly as fuel; they are required in relatively small quantity as constructive materials.

The function of the albuminoids or proteins as constructive materials is clear—that they serve also as fuel cannot be doubted; but we have little idea at present to what extent they are necessary, if at all, as the source of energy. The animal mechanism is of extraordinary complexity—its parts differ much in composition, being constructed of very different nitrogenous materials; and there are also great differences between the nitrogenous materials of different origin, whether animal or vegetable.

We may compare the digestion of food with the pulling down of a building by housebreakers. Not long ago it was supposed that digestion—at least, in its earlier stages—involved the breaking down of the complex albuminoid structures into a few simpler but, at the same time, still complex sub-groups; and that these were then assimilated and re-united to form the animal tissue. This view has given place of late to the idea that the simplification of the albuminoids is carried very far—in fact, down to the ultimate units, such as leucine, tyrosine, etc.; it is supposed that the tissues are reconstructed from these very simple materials.

The albuminoids generally may be contrasted with buildings

which not only differ in construction but which—when taken
to pieces—turn out to be made of bricks of very different sizes
and shapes, perhaps twenty or more; and not only so, for the
different buildings are found to contain the different-sized bricks
in different proportions. This point is a very important one.
To give an illustration. Wheat gluten on digestion yields about
a third of its weight of a particular amino-acid—glutaminic
acid—which is quite a minor constituent of most animal
albuminoids. What becomes of this particular kind of brick—
of this particular product of animal digestion—when flour is
consumed by us? It is certain that it is only in part assimilated;
but before the unassimilated portion can escape from the body, it
must be completely burnt up. The value of flour as food must
depend on whether it is burnt up to a considerable extent
usefully or merely in order that it may be got rid of. If the latter
be the case, work is thrown unnecessarily upon the organs which
effect the change. In any case, much of the nitrogen is wasted,
being useless as fuel. *Per contra*, certain constituents of our
food, present in very minute quantity, are nevertheless probably
of utmost consequence as furnishing the necessary means of
constructing some particular all-essential part of the animal
mechanism. To supply the required amount of such material,
it may be necessary to consume considerable quantities of the
ordinary foods; the use of a more concentrated form of the material
would presumably be of great advantage. Until questions such
as these are solved, stock-raising and the maintenance of health
must be subject to purely empirical rules. Some of you will
know that there has been considerable discussion of late on the
nutritive value of various kinds of bread and on the selection of
seed calculated to give a good wheat. Probably the discussion
is entirely premature—we simply don't know what constitutes
a good wheat; there is no accepted standard of goodness.
What is called strength is a purely fictitious expression with
regard to the food-value of flour; it has reference merely to
the power of the flour to carry water and to its stickiness—the
" strongest " flour being one from which the baker can make
the greatest number of loaves and the biggest loaf; it is therefore the weakest as a food material.

While the office of nitrogenous manures as contributing
a necessary material to the formation of albuminoids and that of
phosphates as contributing to the formation of nuclear elements

can be understood, that of the third mineral substance necessary for plant-growth—potash—is at present veiled in mystery. It is only clear that in some way both potash and phosphate play an all-important part in the production of sugars and starch. Most plants are pronounced gypsophilists and lime is taken up in considerable amount by some plants. The importance of lime-stone as a regular constituent of soils is scarcely appreciated as yet by agriculturists; the results attending the exhaustion of the soil of this constituent as a consequence of the repeated application of ammonia salts are now only beginning to attract attention: I trust that what I have said will lead you to see that it is a substance worthy of your most serious attention at all times.

In terminating this address, let me point out the objects I have had in view. I want you to see how wonderful a field lies before you—if you will but learn to see into the heart of things. No career has greater opportunities than yours; none is lived under more favourable, more interesting conditions. The problems to be solved are innumerable and it is of infinite importance to the wellbeing of mankind that many of them should be solved without delay. Nothing could be farther from the truth than to suppose that farming is played out—the opportunity of the agriculturist is but beginning to come. Pursuing as he does one of the most difficult of businesses, his being the one profession which is indispensable to mankind, he should be among the most intelligent of workers.

But if you are to take advantage of your opportunities, you must be strenuous workers while at college here. Do not be too anxious to accumulate mere knowledge of facts but do every-thing you can to learn to think for yourselves and to help yourselves; cultivate your powers of observation; devote yourselves as far as possible to the study of method. In my earlier remarks I have sought to indicate the way in which a simple inquiry may be conducted; perhaps when you think over my remarks they will be helpful to you and lead you to see the need of a motive for every act and the way in which motives arise almost spontaneously—when they are looked for.

The danger you have most to guard against will be that of allowing yourselves to be taught dogmatically; text-books are pernicious company unless you learn to read them critically, to consider and question their every statement, to inquire into the methods and understand the train of reasoning which have led to the conclusions you are asked to accept. On no account confine your attention to text-books—do all you can to acquire the art of consulting larger books and especially original memoirs : it is not easily gained. Two books I would specially recommend to your notice in connection with the earlier part of my address—(1) Kingsley's *Town Geology*, a shilling edition of which has recently been published by Messrs. Macmillan, (2) Black's *Experiments on Magnesia Alba, Quick Lime, etc.* (1755)—one of the Alembic Club reprints (1s. 6d.), published by Simpkin, Marshall & Co.

The former not only contains interesting matter but is valuable on account of its argument ; the latter is one of the few classics we have in Chemistry—no book deserves more careful attention at the hands of the student of scientific method.

In my later remarks I have dealt very briefly with some of the more important problems which should claim your attention. They are subjects to which you should give special consideration. The field covered by the sciences which bear upon agriculture is so wide that some limitation must be made if you are to study effectively ; but let me urge you not to take too narrow a view of your requirements.

The men who go from this College should be the leaders of agriculture throughout our empire and must be trained broadly and thoroughly ; only those will be worthy to serve as guides who become imbued with the spirit of research while here. If you will but work with the earnestness I have advocated, you will learn to appreciate the difficulties as well as the opportunities which must attend agricultural inquiry— the conditions, as a rule, are so complex that the necessary insight is only to be obtained by the most patient and prolonged investigation.

In an empire such as ours the problems awaiting attack are extraordinary in their variety and such as to tax in a high degree the powers of all who attempt their solution. When in the United States I have been struck, as every one is, by the wonderful care and precision with which inquiry is organised and prosecuted in every possible direction by the Agricultural Bureau in Washington. The far-reaching influence which the Bureau is exercising on the agricultural community is altogether remarkable. I am lost in amazement at the supineness of our authorities and at the failure of our public to appreciate the immensity of the interests at stake. At present we do little more than drift along; no organised attack is being made even on problems of urgency.

Efforts such as are being made in this College are therefore all the more praiseworthy. But students need to realise how great a burden is cast upon them if they are to live up to their opportunities and in order that they may in the end achieve for agriculture the recognition and respect it merits as the one necessary industry of the world during all time: the sole connecting link between us and the far-distant luminary which is the mainspring of all vital activity upon the earth:

> . . . the great sun,
> Girt with his mantle of tempestuous flame,
> Glares in mid-heaven ; but to his noontide blaze
> The slender violet lifts its lidless eye
> And from his splendour steals its fairest hue,
> Its sweetest perfume from his scorching fire.

45

MODERN PLANT-BREEDING METHODS:

WITH ESPECIAL REFERENCE TO THE IMPROVEMENT OF WHEAT AND BARLEY

By R. H. BIFFEN, M.A.

Agricultural Department, Cambridge University

At the opening of the twentieth century the subject of breeding had reached a critical stage in its development, and the few years that have since passed have seen it placed upon a satisfactory basis. In place of fantastic hypotheses the breeder now possesses well-grounded facts on which it is possible to base comprehensive plans for the betterment of the plants which are the objects of his experiments. The main line of attack, until recently, has been that known as "selection." For many years the breeders have recognised the fact that the offspring of a given individual differ amongst themselves to a certain extent, and they have considered that if those individuals showing— from their point of view—the most favourable variations were selected, further generations raised from them, and this process of selection repeated year by year, it was possible to effect vast improvements. Many examples might be quoted of attempts made by breeders to sum up these small differences. So simple was the process that the breeder hardly appeared to recognise its possible limitations, and so readily understandable was it that no seedsman could afford to be without his "pedigree cultures." If this process of selection was indeed capable of one-half that its advocates imagine, then it is incredible that so little progress has been made with plant improvement. By now we might fairly have expected the demands in many directions to have been met : a hardy marrowfat pea, for instance, should not still be wanted, or yet a potato which a late May frost would not spoil. There are many such desiderata which selection has brought no nearer to us.

At this time the diverse phenomena grouped together under the term "variation" were beginning to be analysed in detail, more particularly by De Vries, Johannsen, and Nilsson, with results which throw a great deal of light upon this process of selection. It is impossible to discuss these researches in detail in a brief manner, but the outcome of them may be shortly stated. Each cultural variety of any crop, though pure to the casual observer, now appears to be composed of a number of definite types distinct from one another in minor characters, and the most that can be effected by selection is to pick out the best of these types.

Limits are set at once to the results which may be obtained by selection, and as the production of radically fresh types by mutation is not practicable at present, a halt will have to be called in the process of plant improvement once the best types have been isolated.

A second method employed by breeders is hybridisation, and the first year of the new century saw the rediscovery of Mendel's researches, and with that the proper understanding of this complex subject. The phenomena of hybridisation had been investigated with great patience by a number of scientific workers, such as Kolreuter, Gärtner, Naudin, Godron, and Rimpau, but it appeared impossible to draw any conclusions of general application from their researches. To a great extent this was due to the complexity of the cases which they examined and to the fact that they were concerned with such problems as the conversion of one species into another by repeated artificial fertilisation, or with the part played by sex in determining the form of the hybrids.

The economic breeder with nothing to guide his work trusted to chance. He recognised the fact that the offspring of his hybrids showed great " variability " and amongst these " variations " he hoped to find some showing improvements on the parents. Once found, the new types had to be fixed. This meant as a rule years of selection and possibly failure at the end. So uncertain were the results that comparatively few seedsmen have devoted much attention to the subject, and much of the improvement we see in our garden plants we owe to the patience of the amateur who has made some class of plants an object of special study. Further, this difficulty of fixing the new types has led largely to their vegetative propagation, and it is now amongst plants which can be readily multipled in this way that we find most hybrids.

Half a century ago Mendel faced these difficulties and won from a simple series of experiments results which we are now beginning to realise are not merely invaluable to the breeder, but which affect every subject in which heredity plays a part. Mendel saw the weak point in all previous experiments, and in the introduction to his paper on *Experiments in Plant Hybridisation* he stated that " not one [experiment] has been carried out to such an extent and in such a way as to make it possible to determine the number of different forms under which the offspring of hybrids appear, or to arrange these forms with certainty according to their separate generations, or to definitely ascertain their statistic relations." In this sentence Mendel outlined the whole plan of the series of experiments which have placed the subject of inheritance on a definite basis. Thanks to these, we now know that the phenomena which appeared almost too complex for any analysis are in reality simple, and that, given the knowledge of Mendel's laws, they may even be predicted. Seven years ago the results of crossing together two varieties differing in several characteristics would have been summed up in the statement that "the type was broken," now in many cases one can state so exactly what the progeny of the hybrid will be that it is in the power of the breeder to make his crosses with the certain knowledge of obtaining the results he requires.

From the first Mendel saw the necessity of keeping the problems as simple as possible, and in his earlier experiments he was satisfied to follow out the inheritance of single pairs of differentiating characteristics, such as the round or wrinkled shape of peas or the inflated or normal form of their pods. Time after time he found these characteristics appearing among the progeny of the hybrids in definite proportions. Thus he found on the average that three round peas were produced to each wrinkled pea or three individuals with normal pods to one with inflated pods, and so on. Searching for a reason for these statistic relationships, he was led to adopt the hypothesis, which, later, he proved beyond all manner of doubt, that the germ cells of the hybrids, *i.e.* the pollen grains and egg cells, carried such characters in a pure and not a blended condition, each gamete bearing one of the pair only. As will be shown later, this simple fact, for such we may now term it, accounts for the varied phenomena of hybridisation in an extraordinarily complete way. Writing soon after the rediscovery of these laws, Bateson ventured to declare that Mendel's experiments were worthy to rank with those which laid the foundations of the atomic laws of chemistry, and now that many workers have shown that they apply not solely to the plants with which Mendel himself worked, but to all the plants and animals with which experiments have now been carried out, few who realise the importance of an exact knowledge of heredity will be found to dispute the statement.

With the knowledge that law and order underlie the phenomena of hybridisation came the realisation that breeding must soon depart from the haphazard methods of old, and in their place careful studies of the inheritance of all the characteristics in which the breeder was interested must be made in order to obtain definite instead of chance results. The experiments to be described now were planned with the object of testing Mendel's laws and determining to what extent they would afford help in solving the complex problems which the breeder has to face.

They are concerned with most of the crops of the farm, but at present the investigations on cereals only will be described. These plants are particularly suitable for the study of such problems on account of the fact that they are so invariably self-pollinated that complications due to vicinism are not likely to occur. The one drawback to their employment is the fact that each crossing gives one grain only, and where the gametic output of any plant has to be investigated the mere making of sufficient crosses becomes a matter of some difficulty.

The results have proved unusually simple, and certain of them may be quoted here in order to give an explanation of the main points in the Mendelian story and to show how its principles may be employed for the improvement of our cultivated plants. For this purpose the details of a cross between two varieties of barley, *Hordeum vulgare* and *H. Steudelii* happen to be particularly suitable.

In *Hordeum Steudelii* the medium floret of the group of three which is characteristic of all barleys is hermaphrodite, whilst the two lateral florets are rudimentary and entirely devoid of sexual organs. The ear is thus flat or two-rowed. Its awns and paleæ are a dead-black colour. In *H. vulgare* all three florets of each group are hermaphrodite and fertile. The barley is thus six-rowed: the paleæ and awns are white in colour.

There are then two pairs of differentiating characters to be considered—the presence and absence of fertile lateral florets and the black or white colour.

The cross was made in both directions, *H. Steudelii* being used as both male and female parent. The resulting hybrids resembled one another and were very similar to *H. Steudelii* inasmuch as they were black in colour and the lateral florets were grainless. A closer examination showed, however, that these florets were slightly larger than in that parent, and on dissection they were found to contain stamens. In this respect, then, the hybrid differs from either parent. On sowing the grains produced as the result of self-fertilisation they produced plants in which the lateral florets were either hermaphrodite, staminate, or completely sexless. Both black and white forms of each of these types occurred. For the present the colour characters can be neglected and attention confined solely to the development of the lateral florets. The statistics showed that the three classes were present in the ratio of 1 : 2 : 1. Numbers of plants of each type were harvested separately and their grain again sown, with the result that all those with hermaphrodite and all those with sexless lateral florets were found to breed true to these particular characters. Those with staminate lateral florets, on the contrary, in all cases gave a mixed offspring in which the three groups with hermaphrodite, staminate, and sexless lateral florets again appeared in the proportions of 1 : 2 : 1.

This pair of characters provides a simple example of Mendelian segregation, and they may be taken as a starting-point for explaining the details of Mendel's discoveries from. Calling the hermaphrodite character A and the sexless lateral a, then the hybrid with staminate laterals produced when A and a meet will be Aa. The heterozygote Aa produces gametes which are either A's or a's, not a blend of the two. Where gametes carrying the A character meet, the embryo which is produced can give rise solely to plants of this character—that is, with hermaphrodite lateral florets : similarly, where a and a meet, the plant which results from this combination can only bear sexless lateral florets. Where A and a meet, as in the original crossing, the form with staminate lateral florets is again the result. No other combinations are possible besides AA, Aa, aA, and aa, and as Aa and aA are known, from the reciprocal crosses, to be identical, we have a complete explanation of the fact that the progeny of the hybrid consists of individuals with hermaphrodite, staminate, and rudimentary lateral florets in the ratio of 1 : 2 : 1, and further that the individuals with staminate lateral florets break again in the next generation, whilst those with the forms characteristic of the parents breed true from the first.

Turning now to the colour of the individuals composing the generation raised from the hybrid, they were found to be either black or white with no signs of any intermediate colour. On ascertaining the frequency of these colours there were found to be three black individuals present to each white one, the colouring being distributed impartially over the three groups just described.

This, then, is a case similar to those investigated by Mendel in peas in which one character appeared in its full intensity in the heterozygote whilst its fellow was apparently lost until the succeeding generation. Mendel summed up the behaviour of such characters by terming the one which appeared in the hybrid " dominant," the other, which did not appear until the next generation, " recessive." Black, then, is dominant over white, and it is not practicable to distinguish between the black heterozygote and the black parent by colour alone.

Instead of being able to pick out one pure black to two heterozygotes to one white, we find three blacks to one white. On growing cultures of individual black plants on the average one in three was found to breed true to the black colour, whilst the other two threw off whites in the same proportions as in the previous generation. All the whites, on the contrary, bred true. Two of three blacks were therefore heterozygotes formed by the union of the black-carrying and white-carrying gametes produced by the F. 1. Black and black meeting give a homozygote, black and white or white and black heterozygotes showing the dominant black colour.

Turning now to the consideration of both pairs of characters together, the importance of these phenomena to the breeder will become evident. Black and white were distributed impartially over the forms with fully fertile, staminate, or rudimentary lateral florets. In other words, the characters of the parents appear in fresh combinations, and in consequence types differing from the parents have arisen. The all-important question to the breeder is the possibility of obtaining these in a stable condition. Such a form as that with staminate lateral florets is, as already demonstrated, unfixable, and consequently the black and white types of this may be neglected as valueless. Forms with fully fertile or with rudimentary lateral florets breed true from the outset, the recessive white character also breeds true, so that the combination of white and either of these forms of laterals should be fixed. Numerous trials have shown that this is the case. The corresponding black forms might or might not breed true, the complication being introduced by the fact that the black may be either homozygous or heterozygous. To isolate the homozygote it was, however, only necessary to raise a further generation of these black types, sowing the produce of each plant separately, when one in three was found to breed true to the combination of blackness and either the hermaphrodite or sesters lateral florets.

It is thus clear that a plant breeder knowing these facts could pick out in the generation raised from the hybrids the white forms with either hermaphrodite or rudimentary lateral florets knowing that these would breed true, and in the following generation he could distinguish with certainty the corresponding black forms. Until recently it would have been a difficult matter, following the ordinary practice, to isolate these latter types in a pure condition. A number of, say, the type with rudimentary lateral florets would have been gathered at hazard and the seed sown. The heterozygous blacks would throw off the recessive whites, and though these would have been rogued out, a fresh generation of heterozygotes would repeat the phenomena season by season. In the same way no roguing would ever fix the form with staminate lateral florets, for the possession of such is, in this case, the mark of the heterozygote. On the contrary, the recessive white, in combination with either of the forms of the lateral florets which are characteristic of the parents, would come true from the first—an illustration of the fact so puzzling to breeders that some hybrid forms would breed true whilst others of the same descent were unfixable.

The fresh combinations of characters occurring in the parents which are produced on hybridising are perfectly pure. Naturally, owing to the briefness of the period for which such experiments have been carried out, the matter has not been tested as thoroughly as the hypercritical might wish for, and there is still among practical people much suspicion as to the fixity of hybrids —a view not to be wondered at when the seed lists contain numbers of plants which have to be described as "inclined to sport a little," a fact not unconnected in most cases with the occurrence of heterozygotes.

Numbers of cases, however, have now been observed where thousands of individuals have been raised of fixtures isolated in the first and second generations from the actual hybrids without showing any signs of reversion to the parent types. Such facts do away at once with the conception that the fixity of a given type is dependent on the number of years for which it has been cultivated. The inheritance of the following pairs of characters has been shown to be on Mendelian lines:

Wheat.

Dominant.	Recessive.
Lack of awns (beardless).	Presence of awns (bearded).
Red colour in paleæ and glumes.	White colour in paleæ and glumes.
Grey colour in paleæ and glumes.	White colour in paleæ and glumes.
Grey colour in paleæ and glumes.	Red colour in paleæ and glumes.
Presence of hairs on glumes, or rough chaff.	Lack of hairs.
Keeled glumes.	Rounded base to glumes.
Red colour of grain.	White colour of grain.

In the case of the following characters the heterozygote is intermediate between the two parents:

Long and short glumes.
Long and short grains.
Lax and dense ears.
Brittle and tough rachis.

The various characters in the barleys have been investigated with the following results:

Dominant.	Recessive.
Black colour in the paleæ.	White paleæ.
Purple colour in the paleæ.	White paleæ.
Colour in the grain.	White grain.
Trifurcate paleæ.	Paleæ ending in awns.
Lack of awns.	Paleæ ending in awns.
Brittleness of the rachis.	Toughness of the rachis.
Narrow glumes.	Broad glumes.

Whilst in the following the heterozygote is intermediate in character :

Lax and dense ears.

Varieties with hermaphrodite lateral florets crossed with those with rudimentary lateral florets give a heterozygote with staminate lateral florets.

Those with hermaphrodite lateral florets when crossed with varieties with staminate lateral florets give a form in which the laterals are hermaphrodite but less developed than in the parent with hermaphrodite lateral florets.

Barleys with staminate laterals crossed with varieties with sexless laterals give a heterozygote with small but sexless lateral florets. The F. 2 generation in these cases consists of three types varying in the degree of development of the lateral florets, and as a general rule readily distinguishable from each other.

It is also certain that the inheritance of many less readily appreciable characters, such as those dealing with the surface of the leaf, the structure of the straw, the general habit of the plant, and so on, can be shown to follow the same Mendelian laws. As a matter of fact these morphological characters have relatively little economic importance, though in a few cases among the cereals the substitution of one character for another, such as smooth for rough chaff, might have a certain value. The breeder of the future will have to direct his attention more to such features as time of ripening, the quality of the grain for special purposes, flavour, hardiness, etc., than has been done in the immediate past. How far a knowledge of Mendelian phenomena will aid him in such work still has to be determined, and each case will have to be investigated on its own merits.

The need for such investigations has become increasingly obvious of late years, for in spite of the continuous introduction of new varieties of our staple crops, the actual improvements are slight or too often non-existent. The majority of these new introductions fail in the competition with the older types, and in the course of time disappear ; but the weeding-out is a costly matter to the cultivator. This in itself has led to the extensive variety tests which have become too much a feature of the work of our agricultural stations. One need only point to the results of the recent potato boom to emphasise this fact. In the course of a few years hundreds of so-called improved varieties have been foisted upon the growers, and of all these it is questionable whether half a dozen at the outside will survive the competition of the varieties already in existence. Further than this, the improvements of recent years often fail to stand critical investigation. To take one instance, we are told that the varieties of mangels grown in this country are the finest in the world, and undoubtedly they are, if their shape is to be the criterion of excellence. A recent examination of the crop, however, brings out the startling fact that in the essential respect of feeding value these varieties are now on much the same level as when the crop first came into general cultivation. As in so many cases, the introducers have been concerned far more with the appearance of their exhibits at the agricultural shows than with the really essential features of the roots. In some cases one even has to record that quality is inferior to what it was twenty years ago.

If, working along the lines laid down for us by Mendel, some knowledge of the inheritance of the features of economic importance could be obtained, it would go a long way to remedy this state of affairs. Mendel's own work provides us with an excellent example showing that such investigations are possible. Amongst other characters, he investigated the inheritance of the time of flowering and the nature of the reserve stores in the seeds of peas. His experiments with wrinkled and round peas are, from the consumer's point of view, with peas of good and indifferent quality, wrinkled seeds in gardening phraseology being marrowfat peas. These experiments show at once that the laws of inheritance do not apply solely to the morphological characters which most observers have been content with tracing, but to physiological characters as well. The researches of late years have made it certain that these are not exceptional cases, and that the breeder can reasonably hope to take such characters into consideration in his efforts to improve our crops.

With the object of demonstrating the possibility of this, a series of investigations of a highly technical nature have been commenced at the experimental farm of the Cambridge University Department of Agriculture.

Experiments of this kind demand a certain amount of co-operation, for no plant breeder can be expected to possess that special knowledge of grain which the miller or maltster, for instance, acquires only by years of experience, neither is the breeder always in a position to appreciate the good and bad qualities of any variety from the point of view of the farmer. In the case of the wheat-breeding experiments to be described, the hearty co-operation of the National Association of British and Irish Millers has been of inestimable value. In the first case, a broad survey of the crop was made to determine what features were most in need of improvement. From the farmer's point of view, the varieties in general cultivation may be described as satisfactory on the whole. They give unusually large crops per acre, both of grain and straw, whilst the crops stand well under adverse conditions of weather. Such wheats as, for instance, Square Heads Master, Rivet, Browick, etc., are more or less ideal for most of the conditions under which their cultivation is carried out in this country. The miller, on the other hand, does not appreciate these wheats, and he is not prepared to give the same prices for them as for certain varieties imported from abroad, more particularly from some parts of the United States and Canada. A sufficient explanation of this fact is afforded by the statement that saleable bread cannot be produced from these English varieties unless they are blended with considerable proportions of foreign grain. The miller and baker sum the matter up by saying that the wheats cultivated here are lacking in "strength" or the capacity to yield a light, well-piled loaf. Obviously, then, one of the features which the breeder's attention has to be directed to is that of strength. At the outset the difficulty arises that we have no definite knowledge of what factors determine whether a wheat is strong or not; consequently there is no simple test at the breeder's disposal to aid him in discriminating between the good and bad, and at present the final test is only to be given in the bakehouse. There the flour of different varieties of wheat can be converted into bread, and the loaves compared with those made from flours whose behaviour is well known to the baker. Using such a flour as a standard, the experienced baker can express his opinion of the variety under examination by means of a scale of marks. Such a method has its drawbacks, inasmuch as the personal equation is a large one, and the judgments have to be made, as far as possible, by one man. Nevertheless, repeated duplicate tests have given such closely agreeing results that the method is one in which confidence can be placed. On the scale actually employed in the baking trials average English wheat is marked at about 60, London Households (*i.e.* a good bread-making flour) at 80, whilst the strong wheats imported from Canada may mark from 90 to 100. Higher marking than this is exceptional, though it occurs at times. If the quality of our wheat crop could be raised to that indicated by a mark of 80, English-grown wheat would be suitable for most of the purposes of the baker.

The obvious suggestion is that strong varieties should be imported and grown in this country in place of Square Heads and such varieties. There are, however, so many points which determine whether a variety can be profitably cultivated under any given set of conditions that at first sight it appeared improbable that such suitable varieties would be discovered. A minor point may be chosen to indicate this difficulty. If, for instance, the introduced variety differed in its time of ripening from ordinary English wheats by, say, ten days either side of the usual limits, its cultivation might be rendered impracticable, from the agriculturist's point of view, by the attacks of sparrows. The most serious difficulties to be apprehended, though, were from deficiencies in yielding capacity, for our wheats as a whole are characterised by producing unusually large crops. The attempt has, however, been made with some peculiarly interesting results.

A large number of strong varieties have been introduced and tested from season to season, and their cropping capacity and strength determined. In the majority of cases the strength has been found to deteriorate rapidly, and after a couple of years the grain when tested has proved no stronger than ordinary English wheat. Some noteworthy exceptions have been found, which show without question that strength is not so completely determined by climatic conditions as these trials would appear to suggest. Galician wheat, for instance, retains its strength perfectly under our conditions, which are indeed somewhat similar to those of parts of Western Europe in which it is grown. This variety is the wheat known as Fife in Canada and parts of the United States, and the value of such wheats as those described in commerce as Manitoban depends chiefly on the amount of Fife wheat they contain. Tests extending over several seasons and over a considerable range of soil conditions have shown that English-grown Fife or Galician wheat is fully as strong as that grown in Canada. No deterioration is to be anticipated in this respect, for it has been shown lately that this same variety has actually been cultivated in this country for some fourteen years, chiefly as a spring-sown crop, for which its rapid maturation makes it peculiarly suitable.

The cultivation of this wheat will prove sufficiently profitable in some localities, but it will be found unsuitable in many. The difficulty of finding foreign wheats adapted to our conditions is well shown in the remarks of growers who have tested Fife for the Home-grown Wheat Committee of the Millers' Association. All agree as to the value of the grain, but some find that the sparrows take too heavy a toll owing to its ripening too early for the district; others that the straw, though bright and of good colour, is brittle; others that it lays in bad weather; and many that the yield is insufficient to remunerate the grower even when the enhanced prices are taken into consideration. If only these defects could be avoided by building up a variety with the quality of Fife in its grain and the good features of Square Heads Master in its general habit, an ideal wheat for both the farmer and the miller would be obtained. It rests with the plant breeder to determine whether such a combination is indeed possible. The first step to take is to determine whether strength and its opposite are Mendelian characters. For this purpose a number of wheats known to be strong have been crossed with ordinary English wheats, and the strength of the grain of the hybrids and then of their descendants for successive generations determined. Most of the best English varieties, such as Square Heads Master, Browick, Stand-up White, Rivet, etc., have been used for this purpose.

The strength of the hybrids has been determined for the most part by eye, supplementing this on occasion by such rough-and-ready tests as chewing the grain to determine to a certain extent the physical characters of its gluten. It is difficult to put into words the distinguishing features of strong and weak grain, although on comparing the two one can distinguish them apart with a fair degree of certainty. Strong grain is, as a rule, more or less translucent, and its surface has a dull bloom. Under pressure it tends to break into angular, glassy fragments. A weak wheat, on the other hand, may have an opaque, starchy appearance, and under pressure it becomes powdered. Under some circumstances, which need not be discussed here, weak wheats are frequently as translucent as the strong ; but in such cases the surface of the grain has an oily appearance which is fairly characteristic. Further, it sometimes happens that really strong wheats have a few starchy-looking grains mixed with the translucent ones. It is evident, then, that difficulties must arise when these characteristics have to be distinguished by eye alone. To a certain extent, but not entirely, by growing the grains of parents and hybrids under the same conditions they may be avoided.

Taking the evidence as a whole, it is safe to say that the grain of the hybrid plants was as strong as, or nearly as strong as, that of the strong parents. Certainly no hybrids have yet been raised in which the grain resembled the weak parent in this respect. Further, the grain of the hybrid plants was always of the same type, and no cases have been met with where strong and weak grain occurred in the same ears. In the generation raised from the hybrids segregation into strong and weak forms occurred. In one case so far the proportion in which these two types occurred was three strong to one weak, but in the other cases it has proved impossible to grade the types into two definite groups, owing to the occurrence of individuals in which judgment by eye pointed to the existence of an intermediate stage between strength and weakness. Under such circumstances it is impossible to be precise, and the suggestion that they occur in the proportion of one strong to two intermediate to one weak individual must be taken with a certain amount of reserve. This is most probably correct, for in the majority of cases the strong wheats picked out from this generation have retained this characteristic in the next. Where the segregation into the two types was simple, then the test of the subsequent generation showed that one in three, on the average, bred true to the strong character. Comparison with the strong parent showed that these types were fully as strong as itself. In order to test this once for all, those which were fixed in other respects as well as strength were grown in quantity in the open field, and the grain from six varieties was milled and baked.

To make the results as conclusive as possible the operators were given no information with regard to the origin of the grain, and they were simply requested to report on it. The man in charge of the milling operations gave it as his opinion that the varieties milled like Manitoban wheat, this in itself being an indication of considerable value. The flour was then tested in the bakehouse under the usual experimental conditions. Throughout the various operations it was evident that strong wheats were being dealt with, and the marks assigned to the loaves left no doubt about this. They were as follows : 88, 84, 84, 84, 70, and 45, or four of the six tested were fully equal to the blended flours known as London Households. The strong parent was estimated to have a strength represented by 80–90, whilst the English parent would mark at 55 or 60. The variety earning 45 marks offers a certain amount of difficulty. Two seasons previously it had been selected more for cropping capacity and other features than its strength. There was then some doubt as to its actual baking value : in the following season after some hesitation it was selected for further trials, and produced a crop of grain which appeared to be satisfactory. The test of the bakehouse negatived this view, however, and 45 must be taken to represent the strength of the English variety under the somewhat unfavourable conditions in which it was grown and harvested. That marked at 70 is being further tested; it is in all probability a mixture of dominant strong and recessive weak wheats.

A considerable number of these strong hybrids have been grown, and they include types showing all the possible combinations of the characters present in the parents. Attention has to be directed to one of these, namely, cropping capacity. The two parents differ markedly in this respect, for the strong one only produces about one-half the crop of the English parent. At present little is known as to the inheritance of cropping capacity, and it is proving a difficult character to investigate. Under these circumstances it is satisfactory to find that some of the hybrids grown under the same conditions as a control plot of the strong parent produced double the quantity of grain to the same area. In fact, many of the varieties would have passed muster as the English parents had the grain not been examined. This negatives the view so often brought forward by those who prefer speculations to actual experiment, that strength is associated with diminished yielding capacity. The evidence, then, shows that the inheritance of this intangible feature strength can be traced with sufficient certainty for the plant breeder's purposes, though it cannot yet be claimed that the whole of the details are known.

If a knowledge of Mendel's laws of inheritance will enable one to handle such an indefinite characteristic as the quality of the grain of wheat, in spite of our ignorance of the factors which determine this quality, there can be little doubt that the plant breeder now has his subject more under control than even the most optimistic could have hoped for some few years ago. The mere fact that one can with certainty solve problems of this kind where the majority of previous attempts have been failures is not without its moral value, and it tempts the breeder to apply his methods to equally abstruse characters. Most of the improvements which will be effected as time goes on will be concerned with features of this kind, which offer problems the haphazard methods of the immediate past gave one little hope of attacking with any prospect of success.

Disease Resistance

Such a feature is that of disease resistance. To breed disease-resisting plants has always been one of the problems which has appealed to the small band of workers who have concerned themselves with plant-improvement. Knowing that crossing "broke the type," they have raised numerous hybrids in the hope of finding amongst the progeny some free from the ills the parents were subject to. Looking back at the existing records, the Mendelist often sees clearly where their experiments were bound to fail, for this character was not necessarily present in the plants they used as parents.

Experiments to test the possibility of raising disease-resisting wheats and barleys have been in progress for some five seasons, with results which indicate that the problem is far from being hopeless. For the most part they have been concerned with the yellow rust, *Puccinia glumarum*, this being the commonest and the most serious cereal disease in this country. In other parts of the world the rusts, again, are, on the whole, the most serious of the diseases of these crops, and in some districts an epidemic is such a certainty that it is useless to attempt their cultivation. The rusts sometimes belong to different species from the one with which the experiments to be recorded were carried out, but from trials already made it would appear that the problem is the same for all.

A variety immune to the yellow rust under the ordinary

conditions of cultivation was secured and crossed with slightly, moderately, and extremely susceptible varieties. Further crosses were made between slightly and extremely susceptible varieties, and between extremely susceptible varieties. It is difficult to give any accurate description of the differences in the degree of susceptibility. In practice they have been distinguished with sufficient accuracy by grouping the varieties into five classes, ranging from no disease, through varying degrees of susceptibility, up to excessive liability.

The degree of liability to the attacks of the rust appears to be a definite and constant character. Some varieties, for instance, never show more than a few scattered pustules of uredospores, whilst others are so suitable a pasturage for the parasite that the whole plant becomes orange-yellow with the spores, and the ground below it is even coloured in the same way. During the course of the experiments, the intensity of the epidemic has been somewhat variable. For three seasons it has proved over-average, one average, and one under-average, but the complication produced by external conditions of which, at present, we know little, has not interfered to any extent with the results. One case only need be described in any detail. A wheat belonging to the subspecies *Triticum compactum* was found to be entirely free from yellow rust, even in rust years when it was grown in the midst of other varieties which were so badly attacked that the whole of the foliage, and even the chaff, was coloured yellow. This was crossed with a variety known as Michigan Bronze, which is so excessively susceptible to the attacks of this rust that it rarely produces any grain under our conditions, though in some parts of the continent it can be cultivated successfully. The hybrid between these varieties was particularly vigorous. It was grown alongside small plots of the parents, partly to ensure a plentiful supply of spores reaching it from the Michigan Bronze, and partly to be certain that the conditions of cultivation for parents and hybrids were the same. Late in the spring the hybrids began to show the first yellow flecks on the foliage which herald the formation of rust pustules, and in the course of a few weeks there was no difference between Michigan Bronze and the hybrid as far as the rustiness of the plants was concerned. All parts of the plant—the foliage, the stems, and even the awns on the chaff—were smothered with the yellow pustules.

In spite of the severity of the attack, a fair quantity of grain was obtained at harvest, though very few grains of Michigan Bronze were produced.   It is possible that the exceptional vigour of the hybrids enabled them to withstand the attacks of the parasite better than the parent, but on the other hand, it may be that the actual severity of the attack was not so great as in the case of Michigan Bronze, though, if amount of rust on the foliage is the reliable indication we take it to be, this was not the case.   Whether we are dealing with the ordinary phenomena of dominance, or whether the heterozygote is more or less intermediate between the parents in this respect, is a matter of no immediate importance; the essential fact is that the hybrid was excessively susceptible to the disease which attacks one of its parents.

All the available grain was sown in the following season. The rust was late in appearing, and the epidemic was comparatively slight, but the plots taken as a whole were badly attacked. About midsummer, when the epidemic was considered to be at its height, the plots were examined plant by plant, with the result that amongst this mass of rust-laden individuals numbers were found to be free from any sign of disease.   That this freedom from disease was in any way due to lack of opportunity for infection was disproved by the fact that rust-coated leaves were continually rubbing against the plants from their diseased neighbours, with the result that loose spores were found in abundance on the foliage.   These disease-resisting plants were kept under observation until the time of harvest.   They retained their immunity till the end, their clean vigorous foliage and stems standing out in vivid contrast to the diseased and shrivelled plants around them.   The proportion of susceptible to immune plants was approximately as three to one, the expected Mendelian ratio if immunity is recessive to susceptibility.   Experiments with slightly susceptible and extremely susceptible parents showed in a similar manner that the excessively susceptible habit was again dominant, and crosses between very susceptible varieties gave either all diseased offspring in the generation raised from the hybrids or failed outright owing to the severity of the disease.   Wherever it was possible to obtain statistics this has been done, and the nett result is that, if figures are any guide, immunity and susceptibility to the attacks of yellow rust form as sharply a differentiated pair of characters as Mendel himself would have wished for.

The two parents differed from one another in the shape of the ears as well as in their behaviour with regard to the attacks of yellow rust, the immune parent being dense-eared and the susceptible lax-eared. In other respects they were very similar to one another. Amongst the descendants of the original hybrids there were three morphologically distinct types with dense, lax, or intermediate ears. Both the power of resisting disease and susceptibility to it were distributed uniformly over these three groups. Thus dense-eared types similar to the parent, but liable to the attacks of yellow rust, have been raised together with immune forms similar in external appearance to the susceptible Michigan bronze.

The property of disease resistance, being recessive, will breed true in succeeding generations. This has not been tested in the case of the experiment described, but there is abundant evidence for it obtained from other crosses. In one of these cases four separate generations from the hybrids have been cultivated, some of the forms in field plots in the open, and they still show the same degrees of susceptibility to disease as the two parents.

Experimental evidence has also been obtained which indicates that the immunity to other diseases to which cereals are liable can be transferred in this same manner, so that it is not an extravagant hope that one of the functions of the breeder will be to mitigate the effects of these scourges by providing them with unsuitable hosts.

Up to the present most of the crosses have given unusually simple results, and few of those complications due to the presence of invisible factors such as have been investigated in sweet peas and stocks, and in some cases in the animal world have been met with, though it is by no means improbable that a number exist. One interesting example of a compound character has been seen, namely, grey colouring in the chaff, together with the presence of hairs. The colouring is due to the presence of pigment in the glumes. It is not due to the hairs themselves, for many varieties of wheat have a rough or felted chaff and no such colouring. On segregation the felting and the hairy character always go together, all the rough chaffed forms being grey and the grey chaffed rough. All attempts to separate these characters have so far failed, and at present it appears to be impossible to breed a smooth, grey chaffed wheat. In the case of a cross between the rough grey Rivet wheat and the smooth white Polish wheat, however, rough chaffed white wheats have resulted. The cross was made in both directions, giving in each case a hybrid with rough chaff which was slightly tinged with grey. Several thousand plants were raised from the hybrids, but not one of these showed the dark grey colour of the parent Rivet. The plots appeared to contain white chaffed wheats only. An examination, individual by individual, showed that here and there the merest traces of grey could be detected in some few cases, and one plant was found with awns of a brownish colour approaching that of Rivet wheat. The individuals with the most pronounced colouring were selected for raising a further generation from, but they again produced white chaffed wheats, with either no traces of colour or the merest tinge. The grey colouring then appears to have disappeared. This is probably the nearest approach to the phenomenon of monolepsis or false hybridism which has been seen since Millardet's much-discussed experiments with strawberry hybrids. In pre-Mendelian times a fair summary of the result would be to say that the hybrids bred true directly to the white colour. It is probable that further crosses will throw some light on this suppression of the grey colouring.

A statistical examination of the crop showed that there were three rough chaffed individuals to one smooth chaffed— the result expected if smoothness is recessive. The smoothness of the chaff was, however, confined to those individuals which were characterised by the possession of long glumes like those of the Polish wheat. Consequently all the plants with short glumes similar to those of Rivet wheat, and the heterozygotes with glumes of an intermediate length, had rough chaff. In this case, then, there is a coupling of the rough chaff character with the glume length which will entail further investigation.

From the account of the experiments given above, it is clear that Mendel's laws of inheritance are of wide application, and their value to those who would attempt to improve our cultivated plants is obvious. The breeder now has to recognise the fact that his plants are built up of a series of units for the most part capable of separate segregation when crosses are made. As a direct consequence of this, fresh combinations of these units can be obtained. The methods of the breeder of the future now become clear. In the first case, he will have to become familiar with all possible cultural varieties of the plant he is concerned with. From these he will have to choose a variety here, another there, showing the features he desires to build together into one ideal type. Once these characteristics have been decided on, it only remains to make the necessary crosses and ultimately isolate the type. The whole matter appears simple, and, as far as one can foresee, the main difficulty will be to secure new types in which all of the many features are satisfactory. This is particularly the case with the crops of the farm, where a slight failure in any one respect means, from the agriculturist's standpoint, complete failure. Still, given sufficient workers, the overcoming of such difficulties is merely a matter of time, and it rests with those who desire to see their crops improved to provide those who are willing to acquire the technical knowledge necessary for such work with sufficient opportunities to carry it on. Fortunately the signs are not wanting that this will soon be the case ; and in countries where the cultivation of crops is still a matter of national importance, experimental stations devoted to this purpose are being founded.

THE DANGER OF FLIES

By ARTHUR E. SHIPLEY, M.A., D.Sc. (PRINCETON), F.R.S.

Fellow and Tutor of Christ's College, Cambridge

It is one of those facts which not unfrequently occur in science that we know less about the life-history and habits of the commonest insects than we know about scarce and remote species. For instance, the life-history of the common house-fly, one of the most widely distributed insects in the world, is as yet very incompletely known.

It was Linnæus who first described this insect and named it *Musca domestica*, and De Geer who, in the middle of the eighteenth century, first described its transformation. In 1834 Bouché described the larva of the insect as living in the dung of horses and fowls. In 1873 the well-known American entomologist A. S. Packard[1] reinvestigated the question, and L. O. Howard[2] has recently written on the subject. In our own country C. Gordon Hewitt has published a short preliminary account[3] of what will fill a long-felt want when published—his monograph on the House-fly. Packard noted that in the August of 1873 the house-fly was particularly abundant, especially in the neighbourhood of stables. He was able to observe the insects laying their ova in clumps containing some 120 eggs in the crevices of stable manure, " working their way down mostly out of sight." The eggs hatched in about twenty-four hours, but he noticed that those hatched in confinement required from five to ten hours longer, and that these larvæ when hatched were smaller than those hatched out in the open. The eggs are oval and cylindrical, one twenty-fifth to one-twentieth of an inch long and about one-hundredth of an inch wide, and of a dull, chalky white colour.

The little larva has not been seen emerging from the egg-

---

[1] *Proceedings of the Boston Society of Natural History*, xvi. 1874.

[2] U.S.A. Department of Agriculture, Division of Entomology, *Bulletin* 4, *New Series*.

[3] *Manchester Memoirs*, li. 1906, p. 1.

case, but probably, as in the case of the meat or blow-fly, *Musca vomitoria*, the egg-shell splits longitudinally and the maggot pushes its way out. The length of the newly hatched larva in its first stage (or instar) is seven-hundredths of an inch, and it remains in this stage about twenty-four hours, when it casts its skin and appears as a larger maggot three-twentieths of an inch long. In this condition it remains from twenty-four to thirty-six hours. After a second moult the maggot attains the length of one-quarter of an inch, and in this stage it remains five or six days. During its life the larva moves actively about amongst its surroundings, eating up the decaying matter, but avoiding bits of straw and hay. There is some evidence to believe that if pressed for food larvæ may devour one another. After living altogether some five to seven days the larva somewhat suddenly turns into a dark brown pupa or chrysalis. The transition takes place very rapidly—in the course of a ˏew minutes—and the pupa remains enclosed in the last larval skin. After another period of five to seven days in normal circumstances the insect hatches out, at first running around with soft and baggy wings which, however, soon stretch out, harden, and dry. It is worthy of note that whereas Howard found the complete metamorphosis to take ten days, and Packard from ten to fourteen days, in the cooler climate of Manchester Hewitt finds it takes from twenty to thirty days. The last named gives some interesting particulars as to the effect of the weather upon the rate of development. It is believed that many flies pass the winter in the pupa state; the adult fly also survives the cold weather hidden away in cracks and crevices, from which it may from time to time emerge when the sun shines warmly.

When the larvæ are reared in too dry manure, they attain only one-half their usual size. Too direct warmth and the absence of moisture and available semi-liquid food also tend to dwarf them.

A word may be said about the distribution of the insect. It is practically cosmopolitan. As Mr. Austen records, "the British Museum collection, though very far from complete, includes specimens from the following localities : Cyprus; North-West Provinces, India ; Wellesley Province, Straits Settlements; Hong-Kong; Japan; Old Calabar; Southern Nigeria; Suez; Somaliland; British East Africa; Nyassaland;

Lake Tanganyika; Transvaal; Natal; Sokotra; Madagascar; St. Helena; Madeira; Nova Scotia; Colorado; Mexico; St. Lucia; the West Indies; Pará, Brazil; Monte Video, Uruguay; Argentine Republic; Valparaiso, Chili; Queensland; New Zealand." It is carried all over the world in ships and trains, and seems to be equally at home in the low latitudes of Finmark or in the humid heat of equatorial Brazil.

The diseases which flies convey from man to man—which rendered them by no means the least formidable of the plagues of Egypt, and fully justified Beelzebub's title of the "Lord of Flies"—are for the most part conveyed mechanically. The proboscis acts as an inoculatory needle. No part of the life-history of the disease-causing organism must necessarily be carried on in the body of the fly; it is conveyed mechanically and without change from an infected to a healthy subject. The mouth-parts can pick up the anthrax bacillus, and if the fly then alight upon a wounded surface it will set up "wool-sorters' disease." It, together with the flea, is accused of transmitting the plague-bacillus, not only from man to man, but from rat to man. Flies are active agents in disseminating cholera; and any one who has watched them clustering around the inflamed eyes of the children in Egypt, or in Florida, will not readily acquit them of being the active agents in the spread of inflammatory ophthalmia or of "sore eye."

It is worthy of note that after exhaustive experiments on the Tsetse-fly (*Glossina palpalis*), which conveys that most fatal of diseases, sleeping sickness, Prof. Minchin and his colleagues, Mr. Gray and Mr. Tulloch, have come to the conclusion that the Protozoon (*Trypanosoma gambiense*), which causes the disease, does not—as might be expected—pass through certain stages of its life-history in the fly, but is mechanically conveyed upon the biting mouth-parts of the insect. The deadly parasite is, indeed, so easily cleaned off these appendages that a single bite is sufficient to wipe them off. A tsetse-fly which has bitten an infected person will set up the disease in the next person (or monkey) it bites; but the insertion of the proboscis, quick and instantaneous as it is, serves to clean it—to wipe off adhering trypanosomes, and if it now bite a second person (or monkey), it fails to infect them. This is a most important discovery, and contrary to what we should have expected; but our knowledge of the history of the genus *Trypanosoma* is still too small to justify generalisation, difficult as it is to avoid it.   The diseases which in our country are disseminated by flies are all bacterial and all mechanically conveyed.

In passing it is worth recording that, contrary to the usual statement that tsetse-flies are confined to the continent of Africa, Captain R. M. Carter[1] has recently brought some back from the Tabau River and from other localities in South Arabia. Mr. Newstead has recognised the specimens as belonging to the species *Glossina tachinoïdes*.   It evidently does not live on big game, since, except the gazelle, game is absent.   The Bedouins say that it bites donkeys, horses, dogs, and man, but not camels or sheep.   It is at times so troublesome as to force the natives to shift their camps.

The common house-fly has been known for some time to be an active agent in the dissemination of bacterial diseases.   In intestinal disorders, such as cholera and enteric fevers, which are caused by micro-organisms, the flies convey the bacteria from the dejecta of the sick to the food of the healthy.   In the recent war in South Africa they are described in the standing camps as dividing their activities "between the latrines and the men's mess-tins and jam rations."[2]   In the Spanish-American War in Cuba, and in the South-African War, and in several recent outbreaks of enteric fever in the British Army in India, flies have been proved to be the carriers of the *Bacillus typhosus*. Dr. Veeder[3] writes : "In a very few minutes they may load themselves with dejections from a typhoid or dysenteric patient, not yet sick enough to be in hospital or under observation, and carry the poison so taken up into the very midst of the food and water ready for use at the next meal. There is no long roundabout process involved.   It is very plain and direct; yet when thousands of lives are at stake in this way the danger passes unnoticed."   Similar records come from the Boer camp at Diyatalawa in Ceylon.   The bacilli are conveyed direct just as they might be by an inoculating needle. They do not pass into the body of the fly, neither do they undergo any part of their life-history in its tissue.

Dr. Sandilands[4] has recently investigated outbreaks of

[1] *Brit. Med. Journ.*, No. 2394, November 17, 1906, p. 1393.
[2] Austen, *Journal of the Royal Army Medical Corps*, ii. 1904, pp. 651-67.
[3] *Medical Record*, vol. liv. 1898, pp. 429-30.
[4] *Journal of Hygiene*, vi. 1906, pp. 77-92.

epidemic diarrhœa. He points out that the prevalence of diarrhœa follows the earth's temperature, and does not follow the temperature of the atmosphere. It is a well-known fact that this illness is more prevalent in the houses of the poor than in the mansions of the rich. As Dr. Newsholm, Medical Officer of Health for Brighton, said: "The sugar used in sweetening milk is often black with flies which have come from neighbouring dust-bins or manure heaps; often from the liquid stools of diarrhœa patients in the neighbouring houses. Flies have to be picked out of the half-emptied can of condensed milk before it can be used for the next meal. When we remember the personal uncleanliness of some mothers and that they often prepare their infants' food with unwashed hands, the inoculation of this food with virulent colon bacilli of human origin ceases to be a matter of surprise."

Compared with cow's milk, which nourishes a very numerous progeny of bacteria, the bacterial content of Nestlé's milk is very low, according to Dr. Sandilands. In certain seasons the cow's milk is exposed to temperatures which favour an enormous multiplication of bacteria, and yet it is not then a frequent source of diarrhœa, in fact mere numbers have little or no influence on the incidence of the illness. The greater number of cases are due to infection conveyed from some patient in the near neighbourhood and conveyed mechanically by flies.

The great attraction of the sweetened condensed milk for flies to some extent explains the greater prevalence of infantile diarrhœa among children fed on this preparation.

As was stated above, one of the most remarkable features in the prevalence of infantile diarrhœa is that it follows the rise and fall of the earth's temperature, and not that of the air. In the same way the number of house-flies does not reach its maximum with the first burst of hot weather. The prevalence of these insects follows rather than coincides with periods of great heat. The flies, in fact, lag behind the air temperature and persist for a time after the hot weather has ceased. In other words, the meteorological conditions associated with an increase or a diminution of the prevalence of diarrhœa exercise a similar influence on the prevalence of flies.

The transference of the *Filaria bancrofti*, whose presence in the human body in the adult stage is associated with various diseases of the lymphatics, the most pronounced of which is the terrible elephantiasis, is due to more than one species of gnat or mosquito. It is true that no one has ever seen the actual transference of the *Filaria* from the biting organs of the *Culex*, *Anopheles*, *Panoplites*, or *Stegomyia* into the human body, but the circumstantial evidence is so strong that on it any jury would convict. Noè and Grassi have demonstrated a similar mode of infection for the *Filaria immitis*, which exists, in the adult stage, in such incredible numbers in the cavity of the right side of the heart of dogs, especially in tropical and in sub-tropical countries, that it is difficult to see how the circulation can be maintained at all. It is therefore interesting to note that the proboscis of our common house-fly frequently harbours a larval nematode which has been described by Carter[1] under the name of *Habronema muscæ*; and again (if it be the same species) by Generali[2] under the name *Nematodum sp.* (?); and again by Piana,[3] who is inclined to think it is the larval form of *Dispharagus nasutus* (Rud.). What the further history of this parasite is we do not conclusively know; but judging by analogy—and in the case of the grosser parasites it is not always wise to do that—the nematode probably develops in some higher animal which eats the fly. Piana brings forward a good deal of evidence that this is the domestic fowl.

Another parasite which attacks flies is the fungus or mould *Empusa muscæ*, whose growth is fatal to the insect. The hyphæ penetrate into the body, and, as they grow, weaken the fly until it is unable to lift a leg, but remains glued by its viscid feet to the object upon which it rests. The fungus spreads and radiates out in all directions, covering the fly as with a velvety pile, and giving off countless minute spores, which are blown away to alight, if they are lucky, on a further victim.

I think enough has been said to prove that flies are a very real danger to our community. I have refrained from giving the appalling statistics of our infant mortality, partly because of the difficulty of discriminating between the claims of the flies and those of other agencies which affect the lives of our babies— *e.g.* the insurance companies which do a large trade in insuring infants. Legislation has not attempted to control the

---

[1] *Ann. Nat. Hist.*, ser. 3, vii. p. 29.
[2] *Atti Soc. Modena*, ser. 3, ii. *Radiconte*, p. 88.
[3] *Atti Mus. Milano*, xxxvi. 1896, p. 239.

latter. Sanitation might do much to destroy the former. In well-administered towns slaughterhouses no longer "fill our butchers' shops with large blue flies"; they have been replaced by abattoirs under proper inspection. Stables should also be segregated or controlled. The practice of backing the mansions of Berkeley Square by stable yards should either be given up, or the manure heaps in which the flies breed should be under cover so close as to prevent the access of the fly. A layer of lime spread over the manure effectively prevents the fly laying. Creolin, in its cheap commercial form, is also recommended, sprayed over the manure heaps every two or three days. It not only deters flies from ovipositing, but should they succeed in doing so it kills the resulting larvæ.[1]

Ross has shown us how to clear Ismailia of malaria; the Americans have rid Havana for the first time in a century of yellow fever; the same could be done with flies if only the people liked to have it so. The motor car, with all its destruction of nervous tissue, its prevention of sleep, its danger to life and to limb, has one great merit—it affords no nidus for flies.

[1] Theobald, *Second Report on Economic Entomology*, British Museum (Natural History), London, 1904, p. 125.

ON THE USE OF SOLUBLE PRUSSIAN BLUE IN INVESTIGATING THE REDUCING POWER OF LIVING ANIMAL TISSUES

By DAVID FRASER HARRIS, M.D., B.Sc. (LOND.)

Lecturer on Physiology and Histology in the University of St. Andrews

THE reducing power which living tissues possess is a property of the utmost importance from the biochemical standpoint. All animal bioplasm has in greater or less degree an oxygen avidity[1] in virtue of which it abstracts oxygen from some substance holding that element in a dissociable union. In the higher animals it is oxyhæmoglobin that is thus partially reduced: this is a case of direct deoxidation: on its efficiency depends the continued existence of the bioplasm.

Living tissue can, however, act as a chemical reducer under conditions which do not admit of deoxidation. To this aspect of the vital metabolism I directed my attention in 1896,[2] and found that the reducing power of bioplasm can be rather strikingly shown by the use of the soluble Prussian blue and gelatine "mass" well known to histologists for injecting blood-vessels. Organs injected with abundance of this blue material frequently appear "failures" from the histological point of view, in that their smaller vessels seem so pale or so devoid of colour as not to be capable of being demonstrated. This change of colour from blue to pale green or even to a "white" (leuco) condition, I believe is due to the *intra vitam* reduction of the potassio-ferric-ferrocyanide,

$$\mathrm{Fe''} \underset{(CN)_3}{\overset{(CN)_3 - K}{\diamondsuit}} \mathrm{Fe'''}$$

[1] Paul Ehrlich, *Das Sauerstoff-Bedürfniss des Organismus*, Berlin, 1885.
[2] D. F. Harris, *Proc. Roy. Soc. Edin.*, Session 1896-7, vol. xxi.

to the di-potassio-ferrous-ferrocyanide,

$$\mathrm{Fe''} \Big\langle \begin{array}{l} (\mathrm{CN})_3 \Big\langle \begin{array}{l} \mathrm{K} \\ \mathrm{K} \end{array} \\ \\ (\mathrm{CN})_3 = \mathrm{Fe''} \end{array}$$

In some cases the vessels of the animal or organ were, and in other cases were not, washed out with warm normal saline solution ('75 per cent. NaCl). In some of the experiments, the mixture was, used in full strength, *i.e.* saturated solution of soluble Prussian blue with 5 per cent. of gelatine added; in others it was diluted with an equal volume of '75 per cent. NaCl.

The apparatus I use consists of a very large glass bottle in which water from the tap can accumulate and so raise the pressure of the air above it, this pressure being recorded by a manometer connected by a T-piece through the indiarubber cork.

The compressed air is driven over on to the surface of the liquid blue gelatine in a Woulff's bottle: this forces the material out by a tube passing through the second aperture in the bottle which leads to the cannula. When the whole animal is injected, the cannula is inserted in the aortic arch as soon as the heart ceases beating under deep chloroform anæsthesia.

It is in the liver that the most energetic reduction of the ferric to the ferrous salt seems to take place: on concluding the injection under a pressure of 230 mm. of mercury in the portal vein (the hepatic vein being tied), the lobes of the congested liver present no green or blue appearance whatever, but are of that grey-brown colour which they assume on having had all their blood washed out. At the thin margins of some of the lobes, the access of air here and there restores the colour.

I have perfused the living kidney (pig, sheep, cat, rabbit) both *in situ* and isolated from the body, and have not only found the smaller vessels and capillaries filled with the leuco compound which irrigation with H_2O_2 immediately restores to the blue condition, but have been able to induce an artificial excretion of leuco gelatinous urine to drop from the ureter. This material, on treatment with H_2O_2, became blue: on cutting open the chilled, injected kidney after twenty-four hours, the pelvis and calices were found filled with masses of absolutely colourless gelatine which on irrigation with H_2O_2 at once became blue.

Here the material excreted is produced trans-epithelially : it must have traversed at least (1) the endothelium of the capillary wall, and (2) the epithelium of some part of the uriniferous tubule : these two tissues, or one of them, evidently being sufficiently energetic in their reducing power to transform the deep blue ferric to the colourless ferrous salt.

In several experiments on the kidney, the gland was practically *perfused* rather than injected with the gelatine and Prussian blue (diluted one-half with ·75 per cent. NaCl) under pressures varying from 100 to 300 mm. mercury ; and these demonstrated in a very interesting way the toxicity of soluble Prussian blue (used, as it was in these cases, in the absence of oxygen), as well as the consequent inability of devitalised protoplasm to effect the reduction of the same material which a few moments before it had completely reduced.

One experiment may be taken as typical on this point— kidney of sheep : blood washed out of vessels by warm ·75 NaCl; blue gelatine injection through renal artery, begun at 3.46 p.m. (outflow from renal vein unobstructed) : at 3.55 leuco gelatine began to drip from the ureter (pressure in renal artery 250 mm. Hg) and continued to do so for fifteen minutes thereafter, when the flow began to show a green hue and finally a green-blue : epithelium was now devitalised and was reducing no longer.

In the liver and kidney the fully reduced or leuco compound obtained (di-potassio-ferrous-ferrocyanide) is the result of *intra vitam* reduction by the protoplasm of the hepatic and renal epithelium respectively. The tissues are alkaline reducers; the alkaline salts of lymph and tissues furnish the alkaline medium in which alone the reduction can take place.

The change to green or white is, in the first place, not due to putrefaction, for the tissues are not yet dead. [We had a very instructive demonstration of putrefaction-reduction in a tube of decomposing gelatine and Prussian blue in which the lowest stratum was "leuco," the middle green, while that next the air in the tube was blue.] The change of colour to green or white is, in the next place, not due to the action of the salts of the lymph or tissues—is not a mere "fading due to alkalinity of tissues." No doubt strong solutions of the hydrates of the alkalies cause a rapid " fading " of the Prussian blue and gelatine, but K·OH, NaOH, and NH₄·OH do not exist in the fluids of tissues.

The following salts which are found in blood or lymph were investigated with a view to observe their action on the blue gelatine both individually and when more than one was present, viz. NaCl; KCl; Na₂CO₃; Ca₃(PO₄)₂; Na₃PO₄. The strength of solutions used was less than 1 per cent. None of them added alone to the blue gelatine caused the appearance of the green or leuco stage : the subsequent addition of a little $\frac{N}{1}$ Na·OH caused slow fading of the blue, whereas if $\frac{N}{1}$ K·OH was used instead, there was a rapid fading to the colourless condition.

The addition of hydroquinone or pyrogallol to a mixture of Berlin blue gelatine and any one of these salts produced a change of colour first to blue-green and then to green which was ultimately discharged, the white of the leuco stage being rapidly masked by the brown colour assumed by these reducers.

By subsequent treatment with H₂O₂, the leuco or the green material can be brought back to the blue condition, *i.e.* the potassio-ferric-ferrocyanide reconstituted. A mixture of all these salts (in which there was some haziness probably due to CaCo₃ being precipitated) caused a slow fading of the blue colour not through a green stage ; this latter was, however, at once produced by the addition of pyrogallol to such a mixture.

Ringer's solution,[1] added warm, produced no change beyond that due to a corresponding dilution : the subsequent addition of pyrogallol caused the appearance of a blue-green which was restored to the original blue by H₂O₂. Ringer's solution with Na₂CO₃ added produced no fading, but, as before, the addition of pyrogallol at once caused the mixture to become green.

I therefore conclude that the green or leuco salt (the di-potassio-ferrous-ferrocyanide) is produced by the reducing activity of the living protoplasm acting, like the pyrogallol or hydroquinone in the above experiments, as a reducer in an alkaline medium.

Thus I do look upon the " failure " of histological preparations injected with Berlin blue where the leuco ferrous

[1] ·75 per cent. NaCl saturated with calcic phosphate and with 2 c.c. KCl added for each 100 cc.

salt is formed in the capillaries as not due merely to "fading in contact with alkaline tissues."[1]

The change, if it must still be called "fading," is really *reduction* by living tissues acting on the blue ferric salt in an alkaline medium, the green or white ferrous salt being produced.

My observations show that we can be more explicit as to the fading, and assign it to its cause—*intra vitam* reduction. No doubt the salts of the tissues constitute the requisite alkalinity of the medium in which the reduction takes place, and hence it is that in the (acetic) acid medium recommended by Rawitz[2] the reduction is very much less complete: this is, of course, what for histological purposes is desired.

Reducing power is not absent from the blood, although its energy is far below that possessed by tissues. It is demonstrated in those vessels from which blood has not been washed out by salt solution prior to injection, the mixture of the blood and blue gelatine being greenish and not purple. If into defibrinated blood *in vitro* a little warm, liquid blue gelatine be poured, a green liquid is observed to be the result; and if into excess of liquid gelatine a few drops of blood be let fall or a few flecks of blood-clot dropped, the blood in both conditions will be seen to have become green. Either the living leucocytes or the "reducing substances" (Pflüger) in blood, or both conjointly, are responsible for this reduction of the Prussian blue; but this change of colour to green cannot be attributed to the action of the inorganic salts of the blood.

The red of blood and the blue of the soluble Prussian blue, if physical factors alone co-operated, would produce a purple, reddish or bluish according as the colour of the blood or the gelatine prevailed; but not a green, which indicates that a chemical factor has been at work.

In blood-vessels, other than capillaries from which all the blood has been previously washed out, the blue colour of the gelatine is unaltered, whereas in the capillaries the substance is of the palest green or is colourless. This is precisely what we should have expected, since blood normally and blue gelatine artificially is being brought within the sphere of the biochemical (reducing) activity of the living cells only in the capillaries. The thick-walled vessels, in that they merely convey blood to

[1] Mann, *Physiological Histology*, Oxford, 1902, p. 160.
[2] *Ibid.* p. 160.

and from these regions of active metabolism, are not the seats of interchanges between their contents and the tissues.

As might be expected, very little reducing power can be *demonstrated* by the method of immersing even the most active tissues in warm Prussian blue and gelatine, for only the cells at the surface of the solid piece of tissue (liver, kidney, gland) exert any reducing power, and thus produce a greenish appearance only in their immediate neighbourhood.

This method is vastly inferior to that of perfusing the substance to be reduced: by perfusion through the capillaries every cell-district equally throughout the cell-mass has the reducible material brought within the sphere of biochemical activity.

This point needs no further elaboration; but it must be borne in mind by subsequent workers with this Prussian blue method. Living tissue cannot be expected to act like so much pyrogallol, for instance, which, dropped into a solution of Prussian blue, reduces it with extreme rapidity; a piece of living liver not being distributed throughout the liquid medium cannot be expected to act like the liquid-reducer.

In one or two experiments on perfusion of kidney (lamb) I used *in*soluble Prussian blue (Williamson's blue) in tap-water suspension (or " colloidal solution "); no flow through the ureter was obtained, and a very much less perfect degree of reduction in the capillaries. I regard it as quite unsuitable for reduction experiments owing to its insolubility preventing its passage through vascular endothelium in order to come within the sphere of the reducing power of the living epithelial cells.

In investigating the reducing power of an organ or in attempting to assign to a given organ its relative position in the scale of energy of reduction, one must avoid the fallacy of overloading a poisoned organ with injection-mass. For instance, in the case of the lungs, it would be incorrect to say that because the lungs are found filled with a perfectly blue mass after having been perfused for half an hour or so, they have no power of reduction.

As living protoplasm, pulmonary tissue has certainly some reducing power; but this is exerted in the earliest stages of the perfusion and, the protoplasm being shortly thereafter poisoned, no subsequent Prussian blue is reduced, and so we end with the lungs poisoned and overloaded with the unaltered blue ferric salt.

To judge of the possession of reducing power one must observe the particular organ from the earliest moment of perfusion, so that the organ, if *in situ*, must not be allowed to be covered over by tissues which in its normal position hide it from view.

I am not prepared to give a table of the degrees of energy of reducing power possessed by the various organs, but liver and kidney stand high in the scale of possession of reducing power ; the skeletal muscles and cortex cerebri would come next in order, and then the glands. Hitherto I have studied liver and kidney the most carefully by the Prussian blue method, but I hope to investigate by this method the reducing power of muscle, pulmonary tissue, spleen, and heart.

The living tissues and organs must be regarded as reducing agents in virtue of the same property by the exercise of which they are deoxidising agents during their normal life *in situ*. By the continued deoxidation of oxyhæmoglobin they obtain the oxygen necessary for their existence, but in the absence of oxygen in such a loose chemical union as in oxyhæmoglobin, they can either deoxidise compounds in which the element is much more firmly bound than in HbO_2—*e.g.* alizarine blue, silver nitrate, osmium tetroxide ; or they can reduce such a compound as methylene blue (which has no oxygen) to its pale green chromogen, or, as in the case before us, a salt of iron in trivalent form to one of iron in divalent. Viewed from the physical standpoint, this last is a degradation of energy in that the electrical charge on the ferri-ion is reduced to that on the ferro-ion through the agency of the bioplasm.

The continued existence of living bioplasm is ensured by that avidity for oxygen which the living substance possesses ; the taking of the gas into chemical union with itself constitutes the inspiratory phase of tissue or internal respiration ; but if oxygen, free or in loose union, be not available, the bioplasm can, in virtue of this same oxygen-craving property, reduce from " higher" to "lower" conditions certain substances which contain no oxygen.

It is this power of reduction as distinguished from the power of deoxidation possessed by protoplasm which my experiments demonstrate in regard to a soluble ferric salt ; undoubtedly both phenomena are expressions of the same chemical activity on the part of animal bioplasm.

INDEX TO VOL. I

PAGE

Agriculturist, The Opportunity of the (ARMSTRONG) 683
Ammonites, Decadence of (OSWALD) 400
ANDREWS, C. W. The Recently Discovered Tertiary Vertebrata of Egypt . 668
ARBER, E. A. NEWELL. The Origin of Gymnosperms 222
ARMSTRONG, H. E. (a) The Reform of the Medical Curriculum—a Problem
in Technical Education 544
(b) The Opportunity of the Agriculturist . . . 683
ASHLEY, W. J. A Science of Commerce and some Prolegomena . . . 3
AULD, S. J. MANSON. The Relationship between the Colour and Constitu-
tion of Organic Compounds 650
Australian Mining-fields—Geological Plans (GREGORY) 117

BAYLISS, W. M. The Nature of Enzyme Action 281
BIFFEN, R. H. Modern Plant-breeding Methods : with especial reference to
the Improvement of Wheat and Barley . . . 702
Blood-platelets (BUCKMASTER) 73
BUCKMASTER, G. A. The Blood-platelets 73

Ceylon : Botanical and Agricultural Science (WILLIS) 308
Ceylon : Rubber Exhibition and Cultivation (WILLIS) 538
Chemical Co-ordination of the Body (STARLING) 557
Chloroform a Poison (COLLINGWOOD) 12
Commerce, A Science of (ASHLEY) 3
COLE, G. A. J. On a Hillside in Donegal : a Glimpse into the great Earth-
caldrons 343
COLLINGWOOD, B. J. Chloroform a Poison 12
Colour and Constitution of Organic Compounds (AULD) 650
Corn Smuts and their Propagation (JOHNSON) 137
Crystallography, Chemical and Structural (TUTTON) 91

DAVIS, W. A. The Rusting of Iron 408
Double Fertilisation in Plants (THOMAS) 420

Education, Economics of University (RÜCKER) 365
Education : Reform of the Medical Curriculum (ARMSTRONG) . . . 544
Education : Reform of the Medical Curriculum—a Rejoinder (WADE) . . 635
Electrolytic Dissociation Theory (SENTER) 381
Enzyme Action (BAYLISS) 281
EVANS J. W. The Quantitative Classification of Igneous Rocks . . 259

737

2 1 0 ˉ 8 9

PAGE

Fishery, International Investigations (JOHNSTONE) 583
Flies, The Danger of (SHIPLEY) 723
Fungi, Study of (SMITH) 530

Geography (Physical) as an Educational Subject (MARR) 27
GREEN, J. REYNOLDS. Recent Work on Protein-hydrolysis . . . 427
GREGORY, J. W. The Geological Plans of some Australian Mining-fields . 117
Grew, N., and Plant Anatomy (ROBERTSON) 150
Gymnosperms, Origin of (ARBER) 222

HALL, A. D. The Solvent Action of Roots upon the Soil Particles . . 51
HARDY, W. B. The Physical Basis of Life 177
HARRIS, D. FRASER. On the Use of Soluble Prussian Blue in investigating
             the Reducing Power of Living Animal Tissues . . 730
HENRY, T. A. On the Occurrence of Prussic Acid and its Derivatives in
             Plants 39
HILL, T. G. Stelar Theories 325
Hillside in Donegal (COLE) 343
HOPKINS, F. G. The Utilisation of Proteids in the Animal . . . 159

Igneous Rocks, Quantitative Classification (EVANS) 259
Indiarubber, Chemistry of (PICKLES) 496
INMAN, A. C. Science in Medicine 238
Insects, Distribution of Injurious (THEOBALD) 58
International Fishery Investigations (JOHNSTONE) 583

JOHNSON, T. The Corn Smuts and their Propagation 137
JOHNSON, T. The Principles of Seed-testing 483
JOHNSTONE, J. The International Fishery Investigations . . . 583

KERSHAW, J. B. C. The Artificial Production of Nitrate of Lime . . 361

Life, Physical Basis of (HARDY) 177
LISHMAN, W. E. The Relationship of Mining to Science . . . 604
LOCKYER, W. J. S. Some World's Weather Problems . . . 206
LYDEKKER, R. A Year's Progress in Vertebrate Palæontology . . . 448

MARR, J. E. Physical Geography as an Educational Subject . . . 27
Medicine, Science in (INMAN) 238
Medical Curriculum Reform (ARMSTRONG) 544
Medical Curriculum Reform (WADE) 635
Mining and Science (LISHMAN) 604

Nencki, Marcel (SCHRYVER) 512
Nitrate of Lime, Artificial Production (KERSHAW) 361

Osmotic Pressure (PHILIP) 619
OSWALD, F. The Decadence of Ammonites 400

PAGE

Over-strained Materials (RANKINE) 465

PHILIP, J. C. Recent Experimental Work on Osmotic Pressure . . 619
PICKLES, S. S. The Chemistry of Indiarubber 496
Plant-breeding Methods (BIFFEN) 702
Preface (ALCOCK and FREEMAN) 1
Proteids, Utilisation in the Animal (HOPKINS) 159
Protein-hydrolysis (GREEN) 427
Prussian Blue, Use of Soluble, in investigating the Reducing Power of
            Living Animal Tissues (HARRIS) 730
Prussic Acid and Derivatives in Plants (HENRY) 39

RANKINE, A. O. The Behaviour of Over-strained Materials . . . 465
ROBERTSON, A. Nehemiah Grew and the Study of Plant Anatomy . . 150
Rock-folds (SCHWARZ) 569
Roots : Solvent Action on Soil Particles (HALL) 51
Rubber, Chemistry of (PICKLES) 496
Rubber : Exhibition and Cultivation in Ceylon (WILLIS) . . . 538
RÜCKER, SIR ARTHUR. The Economics of University Education . . 365
Rusting of Iron (DAVIS) 408

SCHRYVER, S. B. Marcel Nencki, 1847—1901 512
SCHWARZ, E. H. L. Rock-folds 569
Seed-testing (JOHNSON) 483
SENTER, G. Some Recent Developments of the Electrolytic Dissociation
            Theory 381
SHIPLEY, A. E. The Danger of Flies 723
SMITH, A. L. Recent Advance in the Study of Fungi 530
STARLING, E. H. The Chemical Co-ordination of the Activities of the Body 557
Stelar Theories (HILL) 325

Tertiary Vertebrata of Egypt (ANDREWS) 668
THEOBALD, F. V. Some Notable Instances of the Distribution of Injurious
            Insects by Artificial Means 58
THOMAS, E. N. Some Aspects of "Double Fertilisation" in Plants . . 420
TUTTON, A. E. H. Some Recent Progress in Chemical and Structural
            Crystallography 91

Vertebrata, Tertiary, of Egypt (ANDREWS) 668
Vertebrate Palæontology—A Year's Progress (LYDEKKER) 448

WADE, J. The Reform of the Medical Curriculum—a Rejoinder . . 635
WILLIS, J. C. The Ceylon Rubber Exhibition, and Rubber Cultivation in
            the East 538
WILLIS, J. C. The Progress of Botanical and Agricultural Science in Ceylon 308
World's Weather Problems (LOCKYER) 206

PRINTED BY
HAZELL, WATSON AND VINEY, LD.,
LONDON AND AYLESBURY.